AF589522

QUANTUM STATISTICAL MECHANICS

Many-body theory stands at the foundation of modern quantum statistical mechanics. It is introduced here to graduate students in physics, chemistry, engineering and biology. The book provides a contemporary understanding of irreversibility, particularly in quantum systems. It explains entropy production in quantum kinetic theory and in the master equation formulation of non-equilibrium statistical mechanics.

The first half of the book focuses on the foundations of non-equilibrium statistical mechanics with emphasis on quantum mechanics. The second half of the book contains alternative views of quantum statistical mechanics, and topics of current interest for advanced graduate level study and research.

Uniquely among textbooks on modern quantum statistical mechanics, this work contains a discussion of the fundamental Gleason theorem, presents quantum entanglements in application to quantum computation and the difficulties arising from decoherence, and derives the relativistic generalization of the Boltzmann equation. Applications of statistical mechanics to reservoir ballistic transport are developed.

WILLIAM C. SCHIEVE is Professor Emeritus in the Physics Department and Center for Complex Quantum Systems at the University of Texas, Austin. His research interests lie in non-equilibrium statistical mechanics and its applications to areas such as quantum optics, relativistic statistical mechanics, dynamical models in biophysics, and chaos theory.

LAWRENCE P. HORWITZ is Professor of Physics Emeritus in the School of Physics at Tel Aviv University. He is also Professor of Physics at Bar Ilan University, and Research Director in Theoretical Physics at Ariel University. His research interests lie in the theory of unstable systems, foundations of quantum theory, quantum field theory, particle physics, relativistic mechanics and general relativity, quantum and classical dynamical systems, and chaos theory.

QUANTUM STATISTICAL MECHANICS

Professor Emeritus WILLIAM C. SCHIEVE
Physics Department, University of Texas at Austin, Texas
Center for Complex Quantum Systems

Professor Emeritus LAWRENCE P. HORWITZ
School of Physics Tel Aviv University Ramat Aviv, Israel
Department of Physics Bar Ilan University Ramat Gan, Israel

CAMBRIDGE
UNIVERSITY PRESS

University Printing House, Cambridge CB2 8BS, United Kingdom

One Liberty Plaza, 20th Floor, New York, NY 10006, USA

477 Williamstown Road, Port Melbourne, VIC 3207, Australia

314-321, 3rd Floor, Plot 3, Splendor Forum, Jasola District Centre, New Delhi - 110025, India

79 Anson Road, #06-04/06, Singapore 079906

Cambridge University Press is part of the University of Cambridge.

It furthers the University's mission by disseminating knowledge in the pursuit of education, learning and research at the highest international levels of excellence.

www.cambridge.org
Information on this title: www.cambridge.org/9780521841467

First published 2009

A catalogue record for this publication is available from the British Library

Library of Congress Cataloging in Publication data
Schieve, W. C.
Quantum statistical mechanics : perspectives /William C. Schieve, Lawrence P. Horwitz.
p. cm.
Includes bibliographical references and index.
ISBN 978-0-521-84146-7
1. Quantum statistics. I. Horwitz, L. P. (Lawrence Paul), 1930– II. Title.
QC174.4 S35 2009
530.13′3—dc22
2008047238

ISBN 978-0-521-84146-7 Hardback

Contents

Preface

This book had its origin in a graduate course in statistical mechanics given by Professor W. C. Schieve in the Ilya Prigogine Center for Statistical Mechanics at the University of Texas in Austin.

The emphasis is *quantum* non-equilibrium statistical mechanics, which makes the content rather unique and advanced in comparison to other texts. This was motivated by work taking place at the Austin Center, particularly the interaction with Radu Balescu of the Free University of Brussels (where Professor Schieve spent a good deal of time on various occasions). Two Ph.D. candidate theses at Austin, those of Kenneth Hawker and John Middleton, are basic to Chapters 3 and 4, where the master equations and quantum kinetic equations are discussed. The theme there is the dominant and fundamental one of quantum irreversibility. The particular emphasis throughout this book is that of open systems, i.e. quantum systems in interaction with reservoirs and not isolated. A particularly influential work is the book of Professor A. McLennan of Lehigh University, under whose influence Professor Schieve first learned non-equilibrium statistical mechanics.

An account of relatively recent developments, based on the addition in the Schrödinger equation of stochastic fluctuations of the wave function, is given in Chapter 13. These methods have been developed to account for the collapse of the wave function in the process of measurement, but they are deeply connected as well with models for irreversible evolution.

The first six chapters of the present work set forth the theme of our book, particularly extending the entropy principle that was first introduced by Boltzmann, classically. These, with equilibrium quantum applications (Chapters 7, 8, 9 and possibly also Chapters 14 and 15), represent a one-semester advanced course on the subject.

As frequently pointed out in the text, quantum mechanics introduces special problems to statistical mechanics. Even in Chapter 1, written by the coauthor of this work, Professor Lawrence P. Horwitz of Tel Aviv, the idea of a *density operator* is required which is *not* a probability distribution, as in the classical case. The idea of the density operator lies at the very foundations of the quantum theory, providing a description of a quantum state in the most general way. Statistical mechanics requires this full generality. We give a proof of the Gleason theorem, stating that in a Hilbert space of three or more real dimensions, a general quantum state has a representation as a density operator, based on an elegant construction of C. Piron. This structure gives the quantum $\mathfrak{H}$ theorem, a content which is essentially different from the classical one. This makes the subject surely interesting and important, but difficult.

Quantum entanglements are quite like magic, so to speak. It is necessary and important to see these modern developments; they are described in Chapter 15. This is one chapter that might be used in the extension of the course to a second semester. One- and two-time Green's functions, introduced by Kadanoff and Baym, might be included in the extended treatment, since they are popular but difficult. This is included in Chapter 16 with an application in Chapter 19.

An extension to special relativity is described in Chapter 10. This is a new derivation of a many-body *covariant* kinetic theory. The Boltzmann-like kinetic equation outlined here was derived in collaboration by the authors. The covariant picture is an event dynamics controlled by an abstract time variable first introduced by both Feynman and Stueckelberg and obtains a covariant scalar many-body wave function parameterized by the new time variable. The results of this event picture are outlined in Chapter 10.

Another arena of activity utilizing quantum kinetic equations for open systems is the extensive development in quantum optics. This has been a personal interest of one of the authors (WCS). This interest was a result of a Humboldt Foundation grant to the Max Planck Institute in Munich and later to Ulm, under the direction of Professors Herbert Walther, Marlon Scully and Wolfgang Schleich. The particular area of interest is described in the results outlined in Chapter 11. This material can be included as an introduction to quantum optics in an extended two-semester course.

The idea of spontaneous decay in a quantum system goes back to Gamov in quantum mechanics. This irreversible process seems intrinsic, introducing the notion of the Gel'fand triplet and rigged Hilbert spaces states. The coauthor (LPH) has made personal contributions to this fundamental change in the wave function picture. It is very appropriate to include an extensive discussion of this, which is the content of Chapter 17, describing, among other things, the Wigner–Weisskopf method and the Lax–Phillips approach to enlarging the scope of quantum wave functions. All of this requires a more advanced mathematical approach than the earlier discussions in this book. However, it is necessary that a well-grounded student of quantum mechanics know these things, as well as acquire the mathematical tools, and therefore it is very appropriate here in a discussion of quantum statistical mechanics.

Chapter 18 is in many ways an extension of Chapter 17. It is an outline of what has been called extended statistical mechanics. Ilya Prigogine and his colleagues in Brussels and Austin, in the past few years, have attempted to formulate many-body dynamics which is intrinsically irreversible. In the classical case this may be termed the complex Liouville eigenvalue method. As an example, the Pauli equation is derived again by these nonperturbative methods. This is not an open-system dynamics but rather, like the previous Chapter 17 discussion, one of closed isolated dynamics. This effort is not finished, and the interested student may look upon this as an introductory challenge.

The final chapter of this book is in many ways a diversion, a topic for personal pleasure. The remarkable objects of our universe known as black holes apparently exist in abundance. These super macroscopic objects obey a simple equilibrium thermodynamics, as first pointed out by Bekenstein and Hawking. Remarkably, the area of a black hole has a similarity to thermodynamic entropy. More remarkable, the S-matrix quantum field theoretic calculation of Hawking showed that the baryon emission of a black hole follows a Planck formula. Hawking introduced a superscattering operator which is analogous to the extended dynamical theory of Chapter 18.

To complete these comments, we would like to thank Florence Schieve for support and encouragement over these last years of effort on this work. She not only gave passive help but also typed into the computer several drafts of the book as well as communicating with the coauthor and the editorial staff of the publisher. The second coauthor wishes also to thank his wife Ruth for her patience, understanding, and support during the writing of some difficult chapters.

We also acknowledge the help of Annie Harding of the Center here in Austin. Three colleagues at the University of Texas—Tomio Petrosky, George Sudarshan and Arno Bohm—also made valuable technical comments. WCS also thanks the graduate students who, over many years of graduate classes, made enlightened comments on early manuscripts.

We recognize the singular role of Ilya Prigogine in creating an environment in Brussels and Austin in which the study of non-equilibrium statistical mechanics was our primary goal and enthusiasm.

Finally, WCS thanks the Alexander von Humboldt Foundation for making possible extended visits to the Max Planck Institute of Quantum Optics in Garching and later in Ulm. LPH thanks the Center for Statistical Mechanics and Complex

Systems at the University of Texas at Austin for making possible many visits over the years that formed the basis for his collaboration with Professor Schieve, and the Institute for Advanced Study at Princeton, particularly Professor Stephen L. Adler, for hospitality during a series of visits in which, among other things, he learned of the theory of stochastic evolution, and which brought him into proximity with the University of Texas at Austin.

1

Foundations of quantum statistical mechanics

1.1 The density operator and probability

Statistical mechanics is concerned with the construction of methods for computing the expected value of observables important for characterizing the properties of physical systems, generally containing many degrees of freedom. Starting with a formally complete detailed description for these many degrees of freedom, probability theory is used to obtain effective procedures. Quantum statistical mechanics makes use of *two* types of probability theory. One of these is the set of natural probabilities associated with the quantum theory which emerges from its structure as a Hilbert space. For example, the Born probability is associated with the square of a wave function. The second is the essentially classical probability associated with an ensemble of separate systems, each with an *a priori* probability assigned by the frequency of occurrence in the ensemble. The quantity which describes both types of probability in an efficient, convenient way is the density operator.

As an example which illustrates many of the basic ideas, consider a beam of particles with spin $\frac{1}{2}$. We shall repeat the resulting definitions later in complete generality.

The spin states of these particles are represented by two-dimensional spinors which we denote by the Dirac kets $|\sigma_z\rangle$ for $\sigma_z = \pm 1$, corresponding to the z component of the spin $\boldsymbol{\sigma}$ of the particle. If we perform a filtering measurement to select a particle of spin $\boldsymbol{\sigma}'$ with spin $\sigma'_z = \pm 1$ in the z direction, the outcome of the measurement on a beam of particles with spin σ_z is

$$\left|\langle \sigma'_z \mid \sigma_z \rangle\right|^2 = \delta_{\sigma'_z, \sigma_z} .$$

This result can be written as

$$\left|\langle \sigma'_z \mid \sigma_z \rangle\right|^2 = \operatorname{Tr} P_z\left(\sigma'\right) P_z\left(\sigma\right),$$

where the projection operator $P_z(\sigma) = |\sigma_z\rangle\langle\sigma_z|$ represents the state of the beam of particles with spin σ of definite value σ_z, and the projection operator $P_z(\sigma')$ represents the experimental question of which value, ± 1, this set of particles has.

If we measure instead a different component of spin and, for example, ask for the fraction of particles in the ensemble with spin in the $\pm x$ direction, the measurement is represented by a projection operator $P_x(\sigma) = |\sigma_x\rangle\langle\sigma_x|$, with $\sigma_x = \pm 1$. In terms of the eigenvectors of σ_z,

$$|\sigma_x = \pm 1\rangle = \frac{1}{\sqrt{2}}\left(|+1\rangle \pm |-1\rangle\right).$$

It is true (for any of the values of σ_x and σ_z) that

$$|\langle\sigma_x \mid \sigma_z\rangle|^2 = \frac{1}{2}.$$

We can write this result as

$$|\langle\sigma_x \mid \sigma_z\rangle|^2 = \mathrm{Tr}\left(P_x(\sigma)\, P_z(\sigma)\right).$$

Let us now consider a beam of spin $\frac{1}{2}$ particles with a fraction γ_+ with spin up and γ_- with spin down in the z direction $\left(\gamma_+ + \gamma_- = 1\right)$. The probability to find spin up as the outcome of the experiment is

$$\begin{aligned} P_+ &= \left|\langle\sigma'_z = +1 \mid \sigma_z = +1\rangle\right|^2 \gamma_+ + \left|\langle\sigma'_z = +1 \mid \sigma_z = -1\rangle\right|^2 \gamma_- \\ &= \gamma_+, \end{aligned}$$

since the second term vanishes. If $\gamma_+ = \frac{1}{2}$, the result is indistinguishable from the probability to find a spin $\pm\frac{1}{2}$ in the x direction in a beam of particles with definite spin in the z direction.

We can write the result of the second example as

$$\begin{aligned} P_+ &= \gamma_+ \mathrm{Tr}\left(P\left(\sigma'_z = +1\right) P\left(\sigma_z = +1\right)\right) + \gamma_- \mathrm{Tr}\left(P\left(\sigma'_z = +1\right) P\left(\sigma_z = -1\right)\right) \\ &= \mathrm{Tr}\left(\rho P\left(\sigma'_z = +1\right)\right) \end{aligned}$$

for

$$\rho \equiv \gamma_+ P\left(\sigma_z = +1\right) + \gamma_- P\left(\sigma_z = -1\right).$$

The operator ρ is called the *density operator*, representing a state consisting of a *mixture* of components with spin up and spin down in the ensemble of possibilities. We see that, with a slight generalization of the procedure used above with $\rho_z \to \rho_0$, no matter what direction 0 we test in the experiment, the outcome P_0 (a linear combination of γ_+, γ_- with coefficients less than unity) can never reach unity if γ_+ or γ_- is not unity. In the first example, where we have a beam with definite

σ_z, the state is represented by a *vector*, and the measurement of the spin in the z-direction can yield probability one. For a general choice of $\gamma_\pm$, there is *no* vector that can represent the state. In the first case the state is called *pure*, and it can be represented by a projection into a one-dimensional subspace (in the previous example, $P_{\sigma_z} = |\sigma_z\rangle\langle\sigma_z|$). This is equivalent to specifying the vector, up to a phase, corresponding to the one-dimensional subspace. In the second case, it is called *mixed* and does not correspond to a vector in the Hilbert space.

It is clear from the discussion of these examples that the *a priori* probabilities $\gamma_\pm$ are essentially classical, reflecting the composition of the beam that was prepared in the macroscopic laboratory.

Although a density operator ρ of the type that we have defined in this example appears to be a somewhat artificial construction, it is actually a fundamental structure in quantum statistical mechanics (Dirac, 1958). It enables one to study a complex system in the framework of an ensemble and in fact occurs on the most fundamental level of the axioms of the quantum theory.

It was shown by Birkhoff and von Neumann (1936) that both quantum mechanics and classical mechanics can be formulated as the description of a set of questions for which the answer, as a result of experiment, is "yes" or "no." Such a set, which includes the empty set ϕ (questions that are absurd, e.g. the statement that the system does not exist) and the trivial set I (the set of all sets, e.g. the statement that the system exists), and is closed with respect to intersections and unions, is called a lattice. A lattice that satisfies the distributive law

$$a \cap (b \cup c) = (a \cap b) \cup (a \cap c)\,,$$

where $\cup$ represents the union and $\cap$ the intersection, is called Boolean. These operations have the physical meaning of "or" (the symbol $\cup$), in which one or the other of the propositions is true, and "and" (the symbol $\cap$), for which both must be true for the answer of the compound measurement to be "yes." An example of such a lattice may be constructed in terms of two-dimensional closed regions on a piece of paper. This is discussed again in the appendix to this chapter.

Both classical and quantum theories may be associated with lattices in terms, respectively, of the occupancy of cells in phase space or states in the subspaces of the Hilbert space. The questions a correspond, in the first case, to the phase space cells (with answer corresponding to occupancy) and in the second to the projection operators P_α associated with a subspace M_α, with the answer corresponding to the values ± 1 which a projection operator can have. These values correspond to evaluating the projection operator on vectors which lie within or outside the subspace.

Birkhoff and von Neumann asserted that the fundamental difference between classical and quantum mechanics is that the lattices corresponding to classical mechanics are Boolean, and those corresponding to quantum mechanics *are not.* The non-Boolean structure of the quantum lattice is associated with the lack of commutativity of the projection operators associated with different subspaces:

$$a \cap (b \cup c) \neq (a \cap b) \cup (a \cap c)\,. \tag{1.1}$$

This is a fundamental difference between classical and quantum statistics.

Let us illustrate this point by a simple example, again using the spin $\frac{1}{2}$ system. Each of the Pauli spin matrices has eigenvalues ± 1 and is therefore associated with a set of projection operators of the form

$$P_i = \frac{1}{2}\left(1 \pm \sigma_i\right)$$

for $i = x,\ y,\ z$. Let us consider three closed linear subspaces associated with the projections into the subspaces with the σ_i positive, i.e. with the P_i defined as above with positive signs. We call these subspaces $M_x,\ M_y,\ M_z$; they correspond to propositions which are not compatible, i.e. the corresponding projection operators do not commute. We shall show explicitly, for this simple example, that

$$M_z \cap \left(M_x \cup M_y\right) \neq \left(M_z \cap M_x\right) \cup \left(M_z \cap M_y\right),$$

that is, this set of propositions is not Boolean. The construction is interesting in that it illustrates the special structure of the topology of Hilbert spaces as well as the notion of the non-Boolean lattice.

We start by constructing the union of the manifolds M_x and M_y by their joint linear span. Taking the standard definition of the Pauli matrices,

$$\sigma_x = \begin{pmatrix} 0 & 1 \\ 1 & 0 \end{pmatrix}, \qquad \sigma_y = \begin{pmatrix} 0 & -i \\ i & 0 \end{pmatrix}, \qquad \sigma_z = \begin{pmatrix} 1 & 0 \\ 0 & -1 \end{pmatrix},$$

the projection operators into the subspaces with positive eigenvalues are

$$P_x = \frac{1}{2}\left(1 + \sigma_x\right) = \frac{1}{2}\begin{pmatrix} 1 & 1 \\ 1 & 1 \end{pmatrix}$$
$$P_y = \frac{1}{2}\left(1 + \sigma_y\right) = \frac{1}{2}\begin{pmatrix} 1 & -i \\ i & 1 \end{pmatrix}$$
$$P_z = \frac{1}{2}\left(1 + \sigma_z\right) = \frac{1}{2}\begin{pmatrix} 1 & 0 \\ 0 & 0 \end{pmatrix}.$$

The corresponding eigenvectors are given by projecting a generic vector v into the respective subspaces. For

$$v = \begin{pmatrix} v_1 \\ v_2 \end{pmatrix},$$

using the result just given,

$$P_x v = \frac{1}{2}(v_1 + v_2)\begin{pmatrix}1\\1\end{pmatrix},$$

so that M_x is represented by the linear span of the normalized eigenvector:

$$v_x = \frac{1}{\sqrt{2}}\begin{pmatrix}1\\1\end{pmatrix}.$$

Similarly,

$$P_y v = \frac{1}{2}(v_1 - iv_2)\begin{pmatrix}1\\i\end{pmatrix},$$

so that the corresponding (normalized) eigenvector is

$$v_y = \frac{1}{\sqrt{2}}\begin{pmatrix}1\\i\end{pmatrix}.$$

Finally,

$$P_z v = v_1\begin{pmatrix}1\\0\end{pmatrix},$$

so the corresponding eigenvector is

$$v_z = \begin{pmatrix}1\\0\end{pmatrix}.$$

The union of the subspaces M_x and M_y is the closed linear span of vectors in both subspaces. By taking the combination $v_x + iv_y$, it is easy to see that the vector v_z (and hence the subspace M_z) is contained in $M_x \cup M_y$. To construct the distributed operation

$$(M_z \cap M_x) \cup (M_z \cap M_y),$$

we must use the construction for which the projection operator corresponding to the intersection of two noncompatible subspaces is generated by an alternating succession of projections into the two subspaces (Jauch, 1968). The products $P_z P_x$ and $P_z P_y$ are, it so happens, idempotents up to coefficients less than one, i.e.

$$P_z P_x = \frac{1}{2}\begin{pmatrix}1 & 1\\0 & 0\end{pmatrix}$$

$$(P_z P_x)^2 = \frac{1}{4}\begin{pmatrix}1 & 1\\0 & 0\end{pmatrix}$$

and

$$
\begin{aligned}
P_z P_y &= \frac{1}{2}\begin{pmatrix} 1 & -i \\ 0 & 0 \end{pmatrix} \\
\left(P_z P_y\right)^2 &= \frac{1}{4}\begin{pmatrix} 1 & -i \\ 0 & 0 \end{pmatrix},
\end{aligned}
$$

which implies that both $(P_z P_x)^n$ and $\left(P_z P_y\right)^n$ go to zero as $n \to \infty$. Therefore,

$$M_z \cap M_x = M_z \cap M_y = 0.$$

Clearly,

$$M_z \cap \left(M_x \cup M_y\right) \neq (M_z \cap M_x) \cup \left(M_z \cap M_y\right).$$

Although $P_z P_x$ and $P_z P_y$ are not zero (the two corresponding vectors are not orthogonal), the closed subspace that is common is empty. One can think of this geometrically in terms of two lines that have some projection on the other, but the intersection of the two lines is just a point of zero measure. Physically, this implies that we cannot have a definite statement of the joint values of σ_z and σ_x or σ_y. The noncommutativity of the associated projections is essential; if they were commutative, the product of projections would be a projection, and the products would not converge to zero. It is clear from this example that compatible subspaces would satisfy Boolean distributivity.

We shall later discuss the Wigner function, which appears to provide joint distributions over noncommutative variables such as q and p; however, these functions are not probabilities, since, although they are the coefficients of what might be called the Weyl basis for the operator algebra of the quantum theory which appear in expectation values, they are not positive (Wigner, 1936).

1.2 The Gleason theorem and consequences

The axioms of quantum mechanics are implicitly developed in the fundamental work of Dirac (1958). Let us focus here on probability. Given P_i $(i = 1, \ldots)$, a sequence of projections $P_i P_k = 0$ for $i \neq k$, then the probability measure w

$$w : P \to [0, 1]$$

satisfies

$$\text{(a)} \quad \cup_i w(P_i) = w\left(\sum_i P_i\right) \tag{1.2}$$

$$\text{(b)} \quad w(\phi) = 0, \quad w(I) = 1$$

$$(\phi \text{ is the zero projection})$$

$$\text{(c)} \quad w(P) = w(F) = 1 \rightarrow w(P \cap F) = 1$$

Piron (1976) added another axiom, namely that partially ordered (by inclusion) sets of the non-Boolean lattice of the quantum theory form Boolean sublattices, and with this he was able to show a converse result, i.e. that such partially ordered lattices can be embedded in a Hilbert space (or a family of Hilbert spaces if there are superselection rules), thus inducing the full structure of the quantum theory.

Along with the sets of "yes-no" questions that form the basic elements a of the quantum lattice, one may assume a function $w(a)$ with values between zero and unity, with the interpretation of a probability measure, which has the so-called sigma additivity property

$$w(a \cup b) = w(a) + w(b) \tag{1.3}$$

when a and b have no intersection, i.e. $a \cap b = \phi$. This idea is consistent with the notion of probability for the "yes" answer for a and b. Gleason (1957) showed that for any Hilbert space of three or more real dimensions, there is a density operator, self-adjoint and positive, ρ, such that

$$w(a) = \mathrm{Tr}\rho P_a, \tag{1.4}$$

where P_a is the projection operator into a subspace corresponding to the question a. This existence theorem is one of the most powerful and important theorems in the foundations of the statistical quantum theory. The function $w(a)$ is called a *state,* a notion completely consistent with Dirac's definition of a state in the quantum theory, i.e. for any a, this function provides the probability of its truth and therefore corresponds to maximum knowledge.

The original proof of Gleason is rather long and involved, but Piron has given a simple and elegant proof, which is given here in an appendix to this chapter for the mature student.

The density operator (often called "density matrix") has the properties

$$\mathrm{Tr}\rho = 1 \tag{1.5}$$
$$\mathrm{Tr}\rho^2 \leq 1.$$

The first follows from the fact that the sum over all disjoint a of $w(a)$ is the total probability measure on the set of all questions (and the sum over all disjoint P_a is the unit operator). The second follows from the first; all eigenvalues of ρ are real and positive with values less than or equal to unity. With these properties, one can prove that the spectrum of ρ must be completely discrete.

Mackey (1963) has given a converse theorem. If the function $w(a)$ can reach the value unity on a one-dimensional subspace of the Hilbert space, the corresponding density operator is just a projection into this one-dimensional subspace and can be put into correspondence (up to a phase) with the vector of the Hilbert space generating this one-dimensional subspace. Such a state is called *pure.* A state which cannot reach the value of unity on any one-dimensional subspace is called *mixed.*

The proof is very simple. Let P_0 be the projection onto a one-dimensional subspace generated by the vector ϕ_0, and let us use the representation, taking into account the discrete spectrum of ρ,

$$\rho = \sum_i \gamma_i \left|\psi_i\right\rangle\left\langle\psi_i\right| . \tag{1.6}$$

Here we use the Dirac ket $\left|\psi_i\right\rangle$ to signify an element of the Hilbert space. Then if $\mathrm{Tr}\rho P_0 = 1$, it follows that

$$\mathrm{Tr}\rho\,(1 - P_0) = 0,$$

or

$$\mathrm{Tr}\sum_i \gamma_i \left\langle\psi_i\right|(1 - P_0)\left|\psi_i\right\rangle = \mathrm{Tr}\sum_i \gamma_i \left\|(1 - P_0)\left|\psi_i\right\rangle\right\|^2 = 0,$$

where $\|\chi\rangle\,\|^2$is defined as $\langle\chi \mid \chi\rangle$, the norm of the vector $|\chi\rangle$. Since the γ_i are positive, this implies that

$$(1 - P_0)\left|\psi_i\right\rangle = 0$$

for all of the $\left|\psi_i\right\rangle$, i.e.,

$$\left|\psi_i\right\rangle = \lambda_i \left|\phi_0\right\rangle$$

for all i. Substituting into Eq. (1.6), we see that in this case we must have

$$\rho = \sum_i \gamma_i \left|\lambda_i\right|^2 \left|\phi_0\right\rangle\left\langle\phi_0\right| .$$

Furthermore, if the $\left|\psi_i\right\rangle$ and $\left|\phi_0\right\rangle$ are normalized, $|\lambda_i|^2 = 1$. Then, by Eq. (1.5) and Eq. (1.6) (for the $\left|\psi_i\right\rangle$ orthogonal), one sees that the sum of the γ_i is unity; hence

$$\rho = \left|\phi_0\right\rangle\left\langle\phi_0\right| ,$$

which is the projection operator into the subspace generated by $\left|\phi_0\right\rangle$. This theorem therefore identifies the pure states with vectors of the Hilbert space, and it is for this reason that one often calls the vectors of the Hilbert space "states." Every vector in the Hilbert space corresponds to a pure state.

If w_1 and w_2 are two different states, then

$$w = \lambda_1 w_1 + \lambda_2 w_2$$

with $\lambda_1 + \lambda_2 = 1$ and with λ_1, λ_2 positive also is a state; the set of states form a convex set (Jauch, 1968). Such a state is called a *mixture.* A state which cannot be represented in terms of two others is called *pure;* the pure states are the extremal subset of a convex set. These definitions are, of course, consistent with Mackey's result.

1.3 Calculation of averages of observables

Let us now consider an observable represented by a self-adjoint operator A on the Hilbert space with a spectrum of discrete eigenvalues a_k. Such an operator can be represented as a sum over projections into its eigenstates, i.e.

$$A = \sum_k a_k P_k, \tag{1.7}$$

where, if $P_k = |\phi_k\rangle\langle\phi_k|$ and the $|\phi_k\rangle$ form a normalized orthogonal set, we clearly have

$$A\,|\phi_k\rangle = a_k\,|\phi_k\rangle.$$

The expectation of this operator in some pure state represented by $|\psi_i\rangle$ is then

$$\begin{aligned}\langle\psi_i|\,A\,|\psi_i\rangle &= \sum_k a_k\,\langle\psi_i|\,P_k\,|\psi_i\rangle \\ &= \sum_k a_k\,|\langle\psi_i \mid \phi_k\rangle|^2,\end{aligned} \tag{1.8}$$

with the usual quantum interpretation that $|\langle\psi_i \mid \phi_k\rangle|^2$ is the quantum mechanical probability that a system in the state described by $|\phi_k\rangle$ is found in the state $|\psi_i\rangle$. The weighting of the eigenvalues of A by this probability then gives the expected value of this observable in the state described by $|\psi_i\rangle$. Suppose now that we prepare a system which contains subsystems in the states $|\psi_i\rangle$ according to the *a priori* probability distribution γ_i. This can be arranged by preparing a system with the number of subsystems in each state $|\psi_i\rangle$ proportional to the γ_i. This is an ensemble. We emphasize here that this step, as in our previous example, is entirely *classical.* We build an ensemble of subsystems with *a priori* probabilities based on their frequency of occurrence, a completely classical notion of probability, i.e. the frequency interpretation.

The overall expectation of the value of the observable A is then given by the sum over all of the expected values in each of the quantum states, with coefficients equal to the classical probabilities of the occurrence of each quantum state in the ensemble, i.e.

$$\langle A \rangle = \sum_i \gamma_i \langle \psi_i | A | \psi_i \rangle .$$

This result is obtained directly by computing

$$\langle A \rangle = \mathrm{Tr} \rho A, \tag{1.9}$$

where

$$\rho = \sum_i \gamma_i | \psi_i \rangle \langle \psi_i | . \tag{1.10}$$

Viewing this in a slightly different way, we see that

$$\langle A \rangle = \sum_k a_k \mathrm{Tr} \left(\rho P_k \right), \tag{1.11}$$

where

$$\begin{aligned} \mathrm{Tr} \left(\rho P_k \right) &= \sum_i \gamma_i \langle \psi_i | P_k | \psi_i \rangle \\ &= \sum_i \gamma_i \left| \langle \psi_i \mid \phi_k \rangle \right|^2 \end{aligned} \tag{1.12}$$

is the probability of finding the system in the subspace associated with P_k. This probability is composed of two types of expectation: the quantum probability to find the P_k in each state ψ_i, and the classical probability for the occurrence of the state ψ_i (determined by the relative number of subsystems in that state).

The results that we have given can easily be extended to the most general case of an observable with both discrete and continuous spectra without change in the formal structure, although as we shall see later, there are special technical aspects that arise in the continuous case (for example, in scattering theory). To see this, we use the spectral representation theory of von Neumann. It was shown by von Neumann (1955) that every self-adjoint operator A, corresponding to a physical observable, has a spectral representation of the form

$$A = \int a \, dE \left(a \right), \tag{1.13}$$

where a takes on a continuous set of values (the real line), and the self-adjoint set of operators $E\left(a\right)$ is called a "spectral family." It satisfies the property

$$E\left(a\right) E\left(b\right) = E\left(\min\left(a, b\right)\right), \tag{1.14}$$

with $E(-\infty) = 0$ and $E(\infty) = I$. It easily follows from these properties that

$$dE(a)\,dE(b) = \begin{cases} dE(a), & \text{if } a = b; \\ 0, & \text{otherwise} \end{cases} \tag{1.15}$$

where a and b now refer to names given to infinitesimal intervals along the line (i.e. for Δa small, $dE(a) = E(a + \Delta a) - E(a)$). The integral Eq. (1.13) is considered to be of Stieltjes–Lebesgue type, in the sense that if the weight function $\langle \psi |dE(a)| \psi\rangle = d\,\|E(a)|\psi\rangle\|^2$ has a jump discontinuity at some point a_0, the integral is evaluated as the difference between the values of $\|E(a)|\psi\rangle\|^2$ above and below the point a_0. If, in particular, $d\,\|E(a)|\psi\rangle\|^2$ is zero in the neighborhood of the point a_0 (except at the point itself), so that the jump is isolated, one obtains a contribution to any expectation value of A just from the point $a = a_0$ (in this neighborhood). The coefficient, since $E(a)^2 = E(a)$, is $\langle\psi|\, E(a_0 + \varepsilon) - E(a_0 - \varepsilon)\,|\psi\rangle$, where ε is infinitesimal. The operator $E(a_0 + \varepsilon) - E(a_0 - \varepsilon)$ may then be identified with one of the discrete projection operators appearing in Eq. (1.7). Hence, the representation Eq. (1.11) includes both discrete and continuous spectra. In Eq. (1.8) one then uses

$$\langle\psi_i|\, A\,|\psi_i\rangle = \int a d\,\|E(a)\,|\psi_i\rangle\|^2,$$

and Eq. (1.9) remains valid quite generally.

We now turn to time evolution, which is the central issue of this book. The quantum states ψ_i from which the density operator is constructed evolve under Schrödinger evolution as

$$i\hbar\frac{\partial}{\partial t}|\psi_i\rangle = H\,|\psi_i\rangle. \tag{1.16}$$

It follows simply that for ρ of the form of Eq. (1.10), acting with the time derivative on both factors $|\psi_i\rangle$ and $\langle\psi_i|$, using Eq. (1.16) and its conjugate, we see that

$$\frac{d\rho}{dt} = i\hbar\,(\rho H - H\rho) = i\hbar\,[\rho, H], \tag{1.17}$$

a time evolution similar to the evolution of a Heisenberg operator but with opposite sign.

Eq. (1.17) forms the basis for the description of the dynamical evolution of a system in statistical mechanics, the analog of the classical Liouville equation (Tolman, 1938). Since the Schrödinger equation is reversible in time, this evolution is reversible (Farquahar, 1964). Under such an evolution, a pure state remains pure, and a mixed state does not change its character (this follows from the fact that the change in time of $\mathrm{Tr}\rho^2$, given by $2i\hbar\mathrm{Tr}(\rho\,[\rho, H])$, vanishes). We shall discuss in later chapters evolution given by, for instance, master equations, the Pauli equation and the Lindblad equation, describing *irreversible* processes. Such equations can describe the evolution of a density matrix for a pure state into a density matrix corresponding to a mixed state. (For this more general evolution, $\mathrm{Tr}(\rho\dot{\rho})$ does not vanish.)

Although, as we have previously emphasized, the density operator might appear to be a somewhat artificial construction, combining both classical and quantum probability notions to achieve an overall expectation value, it actually arises on the most fundamental level of the quantum theory. Methods for the construction and study of this operator and its time evolution are the essential goal of the techniques of statistical mechanics; the theory is constructed on this basic foundation.

Good general references to the topics of this chapter are the books of Tolman (1938), Dirac (1958), Farquahar (1964), Landau and Lifshitz (1970), Balescu (1975), Dvurecenskij (1993), and Huang (1987). Extensive pertinent references are given at the ends of later chapters.

Appendix 1A: Gleason theorem

The Gleason theorem (Gleason, 1957) is concerned with the calculation of the probability w of obtaining the answer "yes" as a result of carrying out an experiment which is an ideal measurement of the first kind on a system in some given state. In working out the proof of this theorem, we shall follow closely the presentation given by C. Piron (1976).

To study and prove the result, we shall need some definitions already implicit in previous sections.

The logical propositions of the quantum theory correspond to equivalence classes of questions $\{\beta\}$ which are realized in terms of measurements. A question β is called a measurement of the first kind if, every time the answer is "yes," the proposition b, corresponding to the equivalence class defined by $\{\beta\}$, is true immediately after the measurement. (Measurement will be taken up again in Chapter 13.)

A question β is said to be ideal if every proposition b defined by such a β, which is true beforehand, is again true afterwards when the response of the system is "yes."

We shall assume that the probability w is the same for every question β defining the proposition b, for β (or $\beta^{\sim}$, its complement) is an ideal measurement of the first kind. We may then denote this probability by $w\,(p, b)$, where p is the initial state in which the experiment is carried out, and b is the proposition defined by the equivalence class $\{\beta\}$.

The Gleason theorem applies to the construction of the function w in the framework of a Hilbert space, on which the operators of the quantum theory are represented. The closed subspaces of a Hilbert space, with their associated projection operators, form a set subject to the operations of intersection and union, and contain the empty set and the set of all subsets, i.e. a structure called a lattice, isomorphic to the lattice of propositions (Birkhoff and von Neumann, 1936; Birkhoff, 1961; Piron, 1976), as mentioned earlier. For an irreducible proposition system, in which there is only one minimal proposition (no superselection rules), every self-adjoint operator corresponds to an observable. Let $P(H)$ be such a Hilbert realization.

We now state the Gleason theorem (Gleason, 1957) (see Piron, 1976, for the general case of a family of Hilbert spaces, for which there is a nontrivial set of minimal propositions):

Theorem: *Given a propositional system $L = P(H)$, where H is a Hilbert space (of dimension ≥ 3) over the reals, complex numbers or quaternions, there exists a unique function $w(p,b)$ defined on the atoms p (corresponding to the one-dimensional subspaces of H) and the propositions b of L which satisfies (as in Eq. (1.2) and Eq. (1.3))*

$$\begin{aligned}
&(1) \quad 0 \leq w(p,b) \leq 1 \\
&(2) \quad p \subset b \Leftrightarrow w(p,b) = 1 \\
&(3) \quad b \perp c \Rightarrow w(p,b) + w(p,c) = w(p, b \cup c) .
\end{aligned} \tag{1A.1}$$

We begin the proof by noting that there is a vector f_p in H, associated with the atom p, satisfying

$$\langle f_p \mid f_p \rangle = \| f_p \|^2 = 1.$$

Each proposition b in $P(H)$ can be represented by a projection operator Q into a linear closed subspace of H. Then

$$w(p,b) = \langle f_p | Q | f_p \rangle$$

satisfies the conditions of the theorem.

Our principal task is then to show uniqueness. If there were another function $w(p,b)$ satisfying these conditions, it would have to have a different value on some pair p, b. For such functions, there would be another proposition q (an atom) for which, in this case, $w(p,q)$ has a different value. However, if the function were unique, the value would necessarily be the same. Such a q can be constructed as follows. Note that

$$[(p \cup b') \cap b] \cup (p' \cap b) = b$$

and that, since p and p' are orthogonal,

$$[(p \cup b') \cap b] \perp p' \cap b.$$

However, $w\left(p, p' \cap b\right) = 0$, so

$$q = \left(p \cup b'\right) \cap b \tag{1A.2}$$

for an atom. The other function would, by construction, have a different value for $w\,(p, q)$. We choose the two vectors f_p and f_q in such a way that $\left\langle f_p \mid f_q \right\rangle$ is real. We may then consider just three vectors associated with the atoms p, q, i.e. f_p, f_q and a vector (real) orthogonal to these. The restriction of $w\,(p, b)$ to the three-dimensional real Hilbert subspace generated by f_p, f_q and a third vector orthogonal to these still satisfies the conditions of the theorem. To complete the proof, it is then sufficient to prove the uniqueness of w in the case of the real three-dimensional Hilbert space $\left(R^3\right)$. This construction, therefore, has the minimum dimension necessary to carry out a proof of uniqueness.

To carry out the proof, let us assume p in $w\,(p, b)$ to be fixed. The lattice of subspaces of R^3 is then the points and lines of the projective plane realized as the intersection of R^3 with the tangent plane at p to the unit sphere. In the same way as the complex plane is mapped onto the unit sphere including the point at infinity, we are considering the plane as a (projective) representation of the sphere of unit vectors in R^3. (It may be helpful for the reader to draw his own diagrams for the construction described here.)

We seek a unique function w(q), where we drop reference to p, now fixed, defined at the points q of the plane which has the value 1 at p and 0 at the point(s) at infinity.

If q lies on some arbitrary line L in the plane, then $w\,(q)$ takes on a maximal value at a point q_0 where the line pq_0 is perpendicular to the line L. This follows from the fact that if q is a point on L, and q' is its orthogonal complement on L, $q \cup q'$ on the line is just q_0. Hence, by (3) of Eq. (1A.1),

$$w\,(q) + w\left(q'\right) = w\,(q_0)$$
$$\text{or} \quad w\,(q_0) \geq w\,(q)\,.$$

We now note that $w\,(q)$ decreases along the line L. To see this, consider a point at q and a line L_q perpendicular to pq. Move along this line to q_1; we know by the foregoing argument that

$$w\,(q) \geq w\,(q_1)\,.$$

Now erect a line at q_1 perpendicular to pq_1 and move to a point on this new line, r. Clearly,

$$w\,(q_1) \geq w\,(r)\,.$$

Now put another line at this point r, and connect it back to L_q at the point q_2. Since

$$w\,(r) \geq w\,(q_2)$$

along L_q, it follows that

$$w(q) \geq w(q_1) \geq w(q_2), \tag{1A.3}$$

forming a decreasing sequence.

We prove now the first lemma of four leading to the uniqueness of the function $w(p, q)$. The method we follow is to prove each lemma making some crucial assumptions, and each succeeding lemma proves those assumptions. In the fourth lemma the proof is complete.

Lemma 1: *If the value of $w(p, q)$ depends only on the angle θ between the rays p and q, then it is unique and given by*

$$w(q) = \cos^2\theta. \tag{1A.4}$$

To prove this lemma, we work as before in the plane tangent to R^3 at the point p and erect another point q at a "distance" λ (corresponding to the square of the actual distance), say, below p. We then erect another point q' at an equal distance λ from p, labeling the midpoint of the line qq' by q_1. By the rules of ordinary geometry, the line pq_1 is orthogonal to the line qq'; it is the closest point on that line to p. It then follows from our previous arguments (q' is the orthogonal complement of q on this line) that

$$w(q') + w(q) = w(q_1).$$

But the angles $q'q_1$ and q_1q are equal, and by the assumptions of our lemma, it then follows that

$$2w(q) = w(q_1).$$

There is a line L_q, perpendicular to pq at a point r, passing through q', and a right triangle that can be constructed from r to the apex q_2 to q, with the line rpq as hypotenuse. To satisfy Pythagoras's theorem, we see that the distance pr is $\frac{1}{\lambda}$. pq_2 is unity (this line is orthogonal to qp). The distance qq_2 is $1+\lambda$, and the distance rq_2 is $1+\frac{1}{\lambda}$. Finally, $q'r$ is $\lambda - \frac{1}{\lambda}$. Now we denote the total length of $q'q$ as $2y$ (this line is bisected by q_1). Again, by Pythagoras, the length of qr is $1+\lambda+1+\frac{1}{\lambda}$. Adding this to $q'r$, which is $\lambda - \frac{1}{\lambda}$, we find the simple result that $4y = 2(1+\lambda)$. Finally, using the fact that pq has length (squared) λ, the length of pq_1, which we call z, is

$$z = \lambda - y = \lambda - \frac{1}{2}(1+\lambda) = \frac{1}{2}(\lambda - 1).$$

We now rewrite the relation previously obtained, $2w(q) = w(q_1)$, as

$$2w(\lambda) = w\left(\frac{1}{2}(\lambda - 1)\right)$$

for $\lambda > 1$. Since by our construction, $r \perp q$,

$$w(\lambda) + w\left(\frac{1}{\lambda}\right) = w(p) = 1,$$

we have that

$$1 - w(\lambda) = w\left(\frac{1}{\lambda}\right).$$

If we now define

$$x = (1+\lambda)^{-1} = \cos^2\theta,$$

the rest of the demonstration follows by simple algebra.

Since $\lambda = \frac{1}{x} - 1$, by defining

$$f(x) = w(\lambda) = w\left(\frac{1-x}{x}\right),$$

one easily finds that

$$2f(x) = f(2x) \tag{1A.5}$$

for $0 \leq x \leq \frac{1}{2}$ (i.e. $\lambda > 1$), and for a second relation,

$$1 - f(x) = f(1-x). \tag{1A.6}$$

To see this, set $y = \frac{2}{\lambda+1} = 2x$; then, using the definition,

$$f(y) = w\left(\frac{1-y}{y}\right) = w\left(\frac{1}{2}(\lambda-1)\right) = 2w(\lambda),$$

it follows that $f(y) = f(2x) = 2f(x)$.

The second relation follows from the fact that

$$f(1-x) = w\left(\frac{x}{1-x}\right) = w\left(\frac{1}{\lambda}\right),$$

so that $1 - \mathrm{f}(x) = f(1-x)$, for $0 \leq x \leq 1$.

The identification $f(x) = x$ with $x = \cos^2\theta$ for some θ satisfies both these relations and satisfies the statement of the lemma. To see that this solution is the only solution which increases, we may expand both sides of the equation $2f(x) = f(2x)$ in Taylor series about $x = 0$. The condition $f(0) = 0$ follows from the requirement that $w \to 0$ at ∞; it follows that all derivatives equal to or higher than second order must vanish, and the function must therefore be linear. Substituting $f(x) = \alpha x$ into the second relation, Eq. (1A.6), we see that $1 - \alpha x = \alpha(1-x)$ so that α must be unity. The solution is therefore unique.

We now prove one of the assumptions of Lemma 1.

Lemma 2: *If $w(q)$ is continuous, then its value depends only on the angle between the rays p and q.*

The remaining two lemmas (lemmas 3 and 4) prove continuity.

To prove this lemma, let q and r be two points on the projective plane situated at the same distance from p. To prove that $w(q) = w(r)$, we start by proving that for any $q_0 \in qp$ sufficiently close to q, the signs of $w(q_0) - w(r)$ and $\lambda - \lambda_0$, where λ and λ_0 are the distances pq and pq_0 respectively, are the same. If $\lambda > \lambda_0$, we can join q_0 to r by a sequence $q_0,\ q_1,\ q_2, ...$ of sequentially perpendicular steps, since at each step $\lambda_1 \geq \lambda_0,\ \lambda_2 \geq \lambda_1, ...$ up to r, which reaches λ, by construction (note that we started with $\lambda_0 < \lambda$). Then

$$w(q_0) \geq w(q_1) \geq w(q_2) \geq ... \geq w(r), \tag{1A.7}$$

since the lengths increase at every step. But we can take q_0 arbitrarily close to q. The same set of inequalities can be established in the other direction, starting with a point r_0 on pr, and hence $w(q) = w(r)$; i.e. the value of $w(q)$ depends only on the distance between p and q (the angle).

Lemma 3: *If $w(q)$ is continuous at some point q_0, then it is continuous at every point.*

We first show that if $w(q)$ is continuous at q_0, it is continuous at each point q_1 orthogonal to q_0. Then q_0 and q_1 lie symmetrically on both sides of the point of a line from p perpendicular to q_0q_1. Denote an ε neighborhood of q_0 by U, and take a point q' on the line q_0q_1 in U; further, consider the point q on the line q_0q_1 orthogonal to q'. As we have done before, we use the relations

$$\begin{aligned} w(q) + w(q') &= w(q_0) + w(q_1) \\ w(r_0) + w(r_1) &= w(r') + w(q_0), \end{aligned}$$

where $r_0,\ r_1$ and r' are defined in a similar way on a line passing at some angle through q, for which q and r' are orthogonal and $r_0 \in U$ and r_1 are orthogonal. It follows from these relations that

$$\begin{aligned} |w(r_1) - w(q_1)| &= |w(q_0) \quad w(r_0) + w(r') - w(q')| \\ &= |w(q_0) - w(r_0) + w(r') - w(q_0) + w(q_0) - w(q')| \\ &\leq |w(q_0) - w(r_0)| + |w(r') - w(q_0)| + |w(q_0) - w(q')| \\ &\leq 3\varepsilon, \end{aligned}$$

where we have used the bounding inequalities between the relation between the $w(q)$'s and the distances. Our construction, furthermore, requires $r', q' \in U_{q_0}$. The subset $r_0 \ni r_1 \in U$ then forms an ε neighborhood of q_1 and is therefore continuous at q_1. We finally note that there always exists a point $q^{\perp}$ perpendicular to two arbitrary points q', r'.

Lemma 4: *The function $w(q)$ is continuous at some point q_0.*

On a line L through p, $w(q)$ is a decreasing function of λ (distance from p). A decreasing bounded function is continuous almost everywhere. Hence $w(q)$ is continuous on L at some point q_0. Finally, if $w(q_2) - w(q_1) < \varepsilon$, then $|w(q) - w(q_0)| < \varepsilon$ at every point in the triangle formed by $rr'q_1$ (all points in this triangle are farther away from p than the distance λ at q_2, in the ε neighborhood of q_0).

This completes the lemmas for the proof of the Gleason theorem, in general.

References

Balescu, R. (1975). *Equilibrium and Non-equilibrium Statistical Mechanics* (New York, John Wiley), revised 1999 as *Matter out of Equilibrium* (London, Imperial College Press).

Birkhoff, G. (1961). *Lattice Theory* (Providence, American Mathematical Society).

Birkhoff, G. and von Neumann, J. (1936). *Ann. Math.* **37**, 823.

Dirac, P. A. M. (1958). *Quantum Mechanics*, 4th edn. (London, Oxford University Press).

Dvurecenskij, A. (1993). *Gleason's Theorem and Its Applications* (Dordrecht, Kluwer).

Farquahar, I. E. (1964). *Ergodic Theory in Statistical Mechanics* (London, Interscience).

Gleason, A. M. (1957). *J. Math. Mech.* **6**, 885.

Huang, K. (1987). *Statistical Mechanics*, 2nd edn. (New York, John Wiley).

Jauch, J. M. (1968). *Foundations of Quantum Mechanics* (Reading, Addison-Wesley).

Landau, L. D. and Lifshitz, E. M. (1970). *Statistical Mechanics* (Reading, Addison-Wesley).

Mackey, G. W. (1963). *Mathematical Foundations of Quantum Mechanics* (Reading, Benjamin).

Piron, C. (1976). *Foundations of Quantum Physics* (Reading, Benjamin).

Tolman, R. C. (1938). *The Principles of Statistical Mechanics* (London, Oxford).

von Neumann, J. (1955). *Mathematical Foundations of Quantum Mechanics* (Princeton, Princeton University Press).

Wigner, E. (1936). *Phys. Rev.* **40**, 749.

2
Elementary examples

2.1 Introduction

Now we will turn to some elementary and familiar examples of quantum mechanics to remind us of matters which will be used in the subsequent discussions. The focus will be the harmonic oscillator and also the two-level atom and spin $\frac{1}{2}$ systems (Dirac, 1958; Louisell, 1973; Cohen-Tannoudji *et al.,* 1977; Jordan, 1986; Liboff, 1998).

2.2 Harmonic oscillator

The Hamiltonian operator is

$$\hat{H} = \frac{1}{2}\left(\hat{p}^2 + \omega^2 \hat{q}^2\right) = \hat{H}^{\dagger}. \tag{2.1}$$

The classical equations of motion are

$$\begin{aligned} \frac{dq}{dt} &= \frac{\partial H}{\partial p} = p \\ \frac{dp}{dt} &= -\frac{\partial H}{\partial q} = -\omega^2 q. \end{aligned} \tag{2.2}$$

In quantum mechanics,

$$\left[\hat{q}, \hat{p}\right] = i\hbar. \tag{2.3}$$

The "hat" denotes operator.

The time-dependent Heisenberg equations are of the same form as the classical counterpart:

$$\begin{aligned} \frac{d\hat{q}(t)}{dt} &= \hat{p}(t) \\ \frac{d\hat{p}\,(t)}{dt} &= -\omega^2 \hat{q}\,(t)\,. \end{aligned} \tag{2.4}$$

This is generally true in one dimension, where we have

$$\frac{d\hat{q}(t)}{dt} = \frac{1}{i}\left[\hat{q}(t), \hat{H}\left(\hat{p}(t), \hat{q}(t)\right)\right] = \frac{\partial \hat{H}(t)}{\partial \hat{p}(t)}$$
$$\frac{d\hat{p}(t)}{dt} = \frac{1}{i}\left[\hat{p}(t), \hat{H}\left(\hat{p}(t), \hat{q}(t)\right)\right] = -\frac{\partial H(t)}{\partial \hat{q}(t)},$$

where $\hat{H}\left(\hat{p}(t), \hat{q}(t)\right)$ is the Heisenberg Hamiltonian operator. This, of course, is the classical correspondence rule

$$\{A, B\} \to \frac{1}{\hbar i}[A, B]$$

$$\left[\hat{q}(t), \hat{p}(t)\right] = i\hbar,$$

where the Heisenberg operators $\hat{q}(t)$, $\hat{p}(t)$ are related to the Schrödinger $\hat{q}$, $\hat{p}$ by

$$\hat{q}(t) = U^{\dagger}(t, 0)\,\hat{q}U(t, 0) \qquad (2.5)$$
$$\hat{p}(t) = U^{\dagger}(t, 0)\,\hat{p}U(t, 0).$$

Here $U(t) = \exp\left(-i\hat{H}t\right)$, $\hbar = 1$. Utilizing this, we obtain the solutions to Eq. (2.4):

$$\hat{q}(t) = \hat{q}\cos\omega t + \frac{\hat{p}}{\omega}\sin\omega t \qquad (2.6)$$
$$\hat{p}(t) = -\omega\hat{q}\sin\omega t + \hat{p}\cos\omega t.$$

These operator equations have exactly the same form as the solutions to the classical equations. For this reason, this is one of the few cases in which an exact Heisenberg operator solution may be obtained. It is easily shown that the time-dependent commutation laws follow.

The Schrödinger equation is

$$i\frac{\partial}{\partial t}\left|\psi(t)\right\rangle = \hat{H}\left|\psi(t)\right\rangle. \qquad (2.7)$$

In this "picture" the operators, $\hat{H}$ etc., are time independent. From this the von Neumann equation for $\hat{\rho}(t)$ is obtained (see the previous chapter):

$$i\frac{d\hat{\rho}}{dt} = \left[\hat{H}, \hat{\rho}\right] \qquad (2.8)$$

Keep in mind that we are working in the Schrödinger picture. For the harmonic oscillator,

$$\psi(t) = \exp(-iHt)\left|\psi(0)\right\rangle = U(t, 0)\left|\psi(0)\right\rangle$$
$$= -i\left[\cos\hat{H}t + i\sin\hat{H}(t)\right]\left|\psi(0)\right\rangle. \qquad (2.9)$$

To reduce this further, let us introduce the well-known creation ($a^\dagger$) and annihilation (a) operators. (Both are non-Hermitian.)

$$\hat{a} = \frac{1}{\sqrt{2\omega}} \left(\omega \hat{q} + i \hat{p} \right) \tag{2.10}$$

$$\hat{a}^\dagger = \frac{1}{\sqrt{2\omega}} \left(\omega \hat{q} - i \hat{p} \right) \tag{2.11}$$

From the commutation law, Eq. (2.3), we obtain

$$\left[\hat{a}, \hat{a}^\dagger \right] = 1. \tag{2.12}$$

Also important are

$$\begin{aligned} \left[\hat{a}, \hat{a}^\dagger \hat{a} \right] &= \hat{a} \\ \left[\hat{a}^\dagger, \hat{a}^\dagger \hat{a} \right] &= -\hat{a}^\dagger. \end{aligned} \tag{2.13}$$

In this representation,

$$\hat{H} = \hbar \omega \left(\hat{a}^\dagger \hat{a} + \frac{1}{2} \right). \tag{2.14}$$

These relations are true in the Heisenberg as well as the Schrödinger picture.

Now, for the harmonic oscillator,

$$U(t, 0) = \exp \left(-i \omega \hat{a}^\dagger \hat{a} t \right) \exp \left(\frac{-i \omega t}{2} \right).$$

Let us introduce the number representation

$$\hat{N} \left| n \right\rangle = n \left| n \right\rangle, \tag{2.15}$$

equivalent to the energy representation

$$\hat{H} \left| E \right\rangle - E \left| E \right\rangle$$

$$\hat{N} = \hat{a}^\dagger \hat{a} = \hat{N}^\dagger.$$

From Eq. (2.13),

$$\begin{aligned} aN - Na &= a \\ a^\dagger N - N a^\dagger &= a^\dagger. \end{aligned} \tag{2.16}$$

With these raising and lowering operators, we may construct a complete set of states (Dirac, 1958). For normalized states we have

$$\begin{aligned} \hat{N}\,|n\rangle &= n\,|n\rangle \qquad \text{n integer and positive} \qquad (2.17)\\ &< n \mid n' > = \delta_{nn'} \\ \hat{a}\dagger\,|n\rangle &= \sqrt{n+1}\,|n+1\rangle \\ \hat{a}\,|n\rangle &= \sqrt{n}\,|n-1\rangle \\ a\,|0\rangle &= 0 \\ |n\rangle &= \frac{\hat{a}^{\dagger n}\,|0\rangle}{\sqrt{n!}} \end{aligned}$$

and completeness

$$\sum_{n=0}^{\infty} |n\rangle\,\langle n| = I.$$

The energy is

$$E_n = \omega\left(n + \frac{1}{2}\right).$$

In the number states, the harmonic oscillator von Neumann equation is

$$\begin{aligned} i\dot{\rho}_{nn'} &= (E_n - E_{n'})\,\rho_{nn'} \\ &= \omega\left(n - n'\right)\rho_{nn'}. \end{aligned}$$

The solution is simply

$$\rho_{nn'}(t) = \exp - \left(i\omega\left(n - n'\right)t\right)\rho_{nn'}(0). \qquad (2.18)$$

The diagonal and off-diagonal elements are uncoupled. Diagonal elements are constant, and the off-diagonal elements oscillate, and

$$\sum_n \rho_{nn}(t) = \sum_n \rho_{nn}(0) = 1. \qquad (2.19)$$

In the so-called random phase approximation, we replace $\rho_{nn'}(t)$ by its average over $n - n'$. Then the oscillations cancel, and $\bar{\rho}_{nn'}(t) = \rho_{nn'}(0)$ is time independent. The comments made are also true for any exact diagonal representation, not just the harmonic oscillator being discussed here. We may write the coordinate representation $u_n(q)$. From

$$a\,|0\rangle = 0 = (q + ip)\,|0\rangle,$$

we have

$$\left(\omega q' + \frac{d}{dq'}\right)u_0\left(q'\right) = 0, \qquad (2.20)$$

whose normalized solution is the Gaussian

$$u_0(q) \equiv < q \mid 0 > = \left(\frac{\omega}{\pi}\right)^{\frac{1}{4}} \exp\left(\frac{-\omega q^2}{2}\right). \tag{2.21}$$

The time-dependent solution is

$$u_0(q,t) = \exp\left(-i\frac{\omega}{2}t\right) u_0(q).$$

It is easily seen that the ground state is a minimum uncertainty state $\Delta q \Delta p = \frac{1}{2}\hbar$.

Let us now consider the coherent state representation. We introduce the non-Hermitian eigenvalue problem,

$$a|\alpha\rangle = \alpha|\alpha\rangle. \tag{2.22}$$

The eigenvalues are not real, nor are they orthogonal.

To solve this, we use the completeness of the number representation $|\alpha\rangle = \sum_{n=0}^{\infty} c_n(\alpha)|n\rangle$. Next, we form

$$a|\alpha\rangle = \sum_{n=1}^{\infty} c_n(\alpha)\sqrt{n}\,|n-1\rangle = \sum_{n=0}^{\infty} \alpha c_n(\alpha)|n\rangle \tag{2.23}$$

and shift indices $n \to n+1$. Take the scalar product with $|m\rangle$. We obtain the recursion relation

$$c_{n+1}(\alpha)\sqrt{n+1} = \alpha c_n(\alpha). \tag{2.24}$$

This gives

$$c_n(\alpha) = \frac{\alpha^n}{\sqrt{n!}} c_0.$$

Thus,

$$|\alpha\rangle = c_0 \sum_{n=0}^{\infty} \frac{\alpha^n}{\sqrt{n!}}|n\rangle.$$

It is easy to show

$$|\langle n \mid \alpha\rangle|^2 = \frac{\alpha^{2n}\exp\left(-\frac{\alpha^2}{2}\right)}{n!},$$

a Poisson distribution. From this $\langle n\rangle = \alpha^*\alpha$, and

$$\frac{\left\langle (n-\langle n\rangle)^2\right\rangle^{\frac{1}{2}}}{\langle n\rangle} = \frac{1}{|\alpha|} = \frac{1}{\langle n\rangle^{\frac{1}{2}}}.$$

We take $\langle \alpha \mid \alpha \rangle = 1$ and obtain

$$\langle \alpha \mid \alpha \rangle = |c_0|^2 \exp |\alpha|^2 ,$$

so

$$|\alpha\rangle = \exp \frac{(-|\alpha|^2)}{2} \exp \alpha \hat{a}^\dagger \exp\left(-\alpha^* \hat{a}\right) |0\rangle , \tag{2.25}$$

taking α to be complex. The completeness relation is

$$\int d^2\alpha \, |\alpha\rangle \langle \alpha| = 1 = \sum_0^\infty |n\rangle \langle n| , \tag{2.26}$$

where $d^2\alpha = r dr d\theta$, and the non-orthogonality is seen by

$$|\langle \beta \mid \alpha \rangle|^2 = \exp\left(-|\alpha - \beta|^2\right) . \tag{2.27}$$

The expansion in terms of coherent states is not unique (Nussenzweig, 1973). They are overcomplete and non-orthogonal. In spite of this, one may expand an arbitrary vector in Hilbert space in terms of them. If we assume that the expansion is an entire function, $f(\alpha\alpha^*)$, of the complex α plane, then the representation is unique.

We may show

$$\langle q \rangle = \sqrt{\frac{1}{2\omega}} \left(\alpha + \alpha^*\right) \tag{2.28}$$

$$\langle p \rangle = i \sqrt{\frac{\omega}{2}} \left(\alpha^* - \alpha\right)$$

$$\langle q^2 \rangle = \frac{1}{2\omega} \left(\alpha^{*2} + \alpha^2 + 2\alpha^*\alpha + 1\right) \tag{2.29}$$

$$\langle p^2 \rangle = \frac{-\omega}{2} \left(\alpha^{*2} + \alpha^2 - 2\alpha^*\alpha - 1\right) .$$

Thus, $\Delta p \Delta q = \frac{1}{2}$, since $(\Delta q)^2 = \frac{1}{2\omega}$ and $(\Delta p)^2 = \frac{\omega}{2}$. All the coherent states are minimum uncertainty. They are quasi-classical. We may obtain $\langle q \mid \alpha \rangle$ to verify this. It is the generalized Gaussian

$$\langle q \mid \alpha \rangle = \left(\frac{\omega}{\pi}\right)^{\frac{1}{4}} \exp\left[\frac{-\omega}{2} \left(q - \langle \hat{q} \rangle\right)^2 + i \langle \hat{p} \rangle q + iu\right] , \tag{2.30}$$

where u is an arbitrary phase and as above,

$$\langle \Delta q \rangle^2 = \frac{1}{2\omega}$$

$$\langle \Delta p \rangle^2 = \frac{\omega}{2} .$$

Now we introduce the first example met here of a phase space distribution function, $P(\alpha\alpha^*, t)$, of Glauber (1963) and Sudarshan (1963). Here the "phase space" is α, α^*. Now

$$\int d^2\alpha P\left(\alpha\alpha^*, t\right) = 1. \tag{2.31}$$

$P\left(\alpha\alpha^*\right)$ is a "diagonal" representation of the density operator in coherent states

$$\rho = \int d^2\alpha P\left(\alpha\alpha^*\right) |\alpha\rangle \langle\alpha| .$$

It has the important property

$$tr\hat{\rho}\hat{O} = \left\langle O\left(\hat{a}, \hat{a}^\dagger\right)\right\rangle = \int d^2\alpha O_{cl}\left(\alpha a^*\right) P\left(\alpha\alpha^*\right). \tag{2.32}$$

Quantum averages are calculated quasi-classically. There is a correspondence rule, the normal ordering rule. In $\hat{O}$ the $\hat{a}$ is placed to the right of the $\hat{a}^\dagger$. For instance, by commutation, $aa^\dagger \rightarrow a^\dagger a + 1$. Phase space distribution functions, such as the Wigner function, will be discussed in greater detail in subsequent chapters. We must remark $P\left(\alpha\alpha^*, t\right) \not> 0$. It is real and normalizable. Let

$$P\left(\alpha\alpha^*, t\right) = tr\rho\left(t\right)\delta\left(\alpha^* - a^\dagger\right)\delta\left(\alpha - a\right). \tag{2.33}$$

This is a somewhat sophisticated statement because of the operator δ functions. Utilizing this definition and the von Neumann equation, we may write for the harmonic oscillator

$$i\frac{\partial P}{\partial t} = \mathrm{Tr}\left[\rho\left(t\right)\left[\delta\left(\alpha^\dagger - a^*\right)\delta\left(\alpha - a\right), \omega a^\dagger a\right]\right].$$

We will evaluate this in the appendix to this chapter. We obtain a Fokker–Planck equation for $P\left(\alpha\alpha^*, t\right)$ (Gardiner, 1991).

$$\frac{\partial P\left(\alpha\alpha^*, t\right)}{\partial t} = i\omega\left[\alpha\frac{\partial P}{\partial\alpha} - \alpha^*\frac{\partial P}{\partial\alpha^*}\right]. \tag{2.34}$$

It is a first-order partial differential equation in t, α, α^*. The general solution may be obtained from the characteristic equations

$$dt = \frac{d\alpha}{-i\omega\alpha} = \frac{d\alpha^*}{i\omega\alpha^*}, \tag{2.35}$$

which are the "Hamilton equations" of the α, α^* "phase space." The solution is

$$\begin{aligned} \alpha\left(t\right) &= \alpha_0 \exp\left(-i\omega t\right) \\ \alpha^*\left(t\right) &= \alpha_0^* \exp\left(i\omega t\right). \end{aligned} \tag{2.36}$$

The general solution is an arbitrary function $f(\alpha(t), \alpha^*(t))$. If the initial value is Gaussian in α, i.e.

$$P\left(\alpha, \alpha^*, 0\right) = N \exp\left(-|\alpha - \alpha_0|^2\right),$$

then

$$P\left(\alpha, \alpha^*, t\right) = N \exp\left(-|\alpha(t) - \alpha_0|^2\right).$$

For

$$P\left(\alpha\alpha^{*,} t\right) = \delta^2\left(\alpha(t) - \alpha_0\right),$$

the coherent state propagates in time as $\exp i\omega t$. This was first seen by Schrödinger (1926).

Let us consider an extension of the harmonic oscillator by including a damping term. A particularly simple example is the phase damped oscillator with the interaction

$$V = \Gamma a^\dagger a + \Gamma^\dagger a^\dagger a \tag{2.37}$$

(Walls and Milburn, 1985; Gardiner, 1991). The von Neumann equation may be written

$$\dot{\rho} = -i\omega\left[a^\dagger a, \rho\right] + \frac{1}{2}K\left(\bar{N}+1\right)2a^\dagger a\rho a^\dagger a - \left(a^\dagger a\right)^2\rho - \rho\left(a^\dagger a\right)^2. \tag{2.38}$$

This is the Lindblad form and is discussed in detail in Chapters 4, 5 and 6. Here $\bar{N} = \frac{1}{\exp\left(\frac{\omega}{kT}\right)-1}$, and K is a damping constant. In the number representation,

$$\langle n|\dot{\rho}|m\rangle = \left\{-i\omega(n-m) - \frac{1}{2}K\left(2\bar{N}+1\right)(n-m)^2\right\}\langle n|\rho|m\rangle.$$

The diagonal and off-diagonal elements $\langle n|\rho|m\rangle$ are still uncoupled. The solution is immediate:

$$\langle n|\rho(t)|m\rangle = \exp\left(-i\omega(n-m)t\right)\exp - \left[(2\bar{N}+1)K(n-m)^2\frac{t}{2}\right]\langle n|\rho(0)|m\rangle.$$

The off-diagonal elements decay as $(n-m)^2 K\left(2\bar{N}+1\right)$ to the constant diagonal initial state $\langle n|\rho(0)|m\rangle$. More will be said of this in the discussion of decoherence in Chapter 12.

To obtain the equation for $P(\alpha)$, we use the operator correspondence discussed in the appendix:

$$\begin{aligned} a\rho &\to \alpha P\left(\alpha\alpha^*\right) \\ a^\dagger\rho &\to \left(\alpha^* - \frac{\partial}{\partial\alpha}\right) P\left(\alpha\alpha^*\right) \\ \rho a &\to \left(\alpha - \frac{\partial}{\partial\alpha^*}\right) P\left(\alpha\alpha^*\right) \\ \rho a^\dagger &\to \alpha^* P\left(\alpha\alpha^*\right) \end{aligned} \tag{2.39}$$

to obtain the Fokker–Planck equation,

$$\frac{\partial P}{\partial t} = \left\{\frac{1}{2}K\left(\frac{\partial}{\partial\alpha}\alpha + \frac{\partial}{\partial\alpha^*}\alpha^*\right) - i\omega\left(\frac{\partial}{\partial\alpha}\alpha - \frac{\partial}{\partial\alpha^*}\alpha^*\right) + K\bar{N}\frac{\partial^2}{\partial\alpha\partial\alpha^*}\right\} P. \tag{2.40}$$

By introducing $\alpha = x + iy$ (Scully and Zubairy, 1997), we find the average:

$$\langle\alpha(t)\rangle = \alpha(0)\exp\left[-\left(\frac{K}{2}\right) - i\omega\right] t. \tag{2.41}$$

In the coherent state, we obtain a classical damped oscillator solution.

$P(\alpha\alpha^*, t)$ need not be positive. If it is, then the state of the system is classical, $P(\alpha\alpha^*)$ being a true probability distribution. $P(\alpha\alpha^*)$ may exist for nonclassical or truly quantum states. However, if $\alpha = x + iy$, we obtain a Fokker–Planck equation in x, y with positive diffusion coefficient, so $P(\alpha a^*, t) > 0$.

2.3 Spin one-half and two-level atoms

The spin of the electron is

$$\mathbf{S} = \frac{1}{2}\hbar\sigma \qquad \text{Let } \hbar = 1 \tag{2.42}$$

(Cohen-Tannoudji *et al.*, 1977). σ obeys $\mathbf{m}_S = -\frac{e}{2\mu}\boldsymbol{\sigma}$, and $\mathbf{m}_s$ is the spin magnetic moment. σ_j has the properties

$$\begin{gathered} \left[\sigma_i, \sigma_j\right]_- = 2i\sigma_k \\ i, j = 1, 2, 3. \end{gathered} \tag{2.43}$$

These are angular momentum commutation laws for half integer **I**. Now $\sigma_i^2 = 1$, so

$$\sigma_i\sigma_j = i\sigma_k. \tag{2.44}$$

We define (analogous to a in Eq. (2.10))

$$\begin{aligned}\sigma_\pm &= \frac{1}{2}(\sigma_1 \pm \sigma_2) \\ \sigma_+ &= \sigma_-^\dagger .\end{aligned} \tag{2.45}$$

They are not themselves Hermitian. Now we find the commutation laws,

$$\begin{aligned}\left[\sigma_\pm, \sigma_1\right]_- &= \pm\sigma_3 \\ \left[\sigma_\pm, \sigma_2\right]_- &= i\sigma_3 \\ \left[\sigma_\pm, \sigma_3\right]_- &= \mp\sigma_2 \\ \left[\sigma_+, \sigma_-\right]_- &= \sigma_3,\end{aligned} \tag{2.46}$$

as well as *anti-commutation* laws,

$$\begin{aligned}\left[\sigma_\pm, \sigma_1\right]_+ &= 1 \\ \left[\sigma_\pm, \sigma_2\right]_+ &= \pm i \\ \left[\sigma_\pm, \sigma_3\right]_+ &= 0 \\ \left[\sigma_+, \sigma_-\right]_+ &= 1\end{aligned} \tag{2.47}$$

and

$$\begin{aligned}\sigma_{1.}^2 &= \sigma_2^2 = \sigma_3^2 \qquad \boldsymbol{\sigma}^2 = 3 \\ \sigma_+^2 &= \sigma_-^2 = 0.\end{aligned} \tag{2.48}$$

For spin $\frac{1}{2}$ and the general properties of angular momentum, the wave function for the basis states $\left|\frac{1}{2}\right\rangle$, $\left|-\frac{1}{2}\right\rangle$ are

$$\begin{aligned}\alpha &\equiv \binom{1}{0} \quad \beta \equiv \binom{0}{1} \\ &\equiv |+1\rangle \qquad \equiv |-1\rangle .\end{aligned} \tag{2.49}$$

The α state is spin positive ($m_s = +1$) along the "3" direction, and β spin down ($m_s = -1$). Generally,

$$\begin{aligned}|\psi\rangle &= a\alpha + b\beta = a\,|+1\rangle + b\,|-1\rangle \\ a^2 + b^2 &= 1.\end{aligned}$$

In this representation,

$$\sigma_1 = \begin{pmatrix} 0 & 1 \\ 1 & 0 \end{pmatrix}, \; \sigma_2 = \begin{pmatrix} 0 & -i \\ i & 0 \end{pmatrix}, \; \sigma_3 = \begin{pmatrix} 1 & 0 \\ 0 & -1 \end{pmatrix}, \tag{2.50}$$

the familiar Pauli matrices. Continuing, we find

$$\sigma_+ = \begin{pmatrix} 0 & 1 \\ 0 & 0 \end{pmatrix}, \ \sigma_- = \begin{pmatrix} 0 & 0 \\ 1 & 0 \end{pmatrix}. \tag{2.51}$$

Finally, σ_3 has obvious eigenvalues, ± 1, and σ_1, σ_2 raise and lower states:

$$\begin{aligned} \sigma_1 \left|\pm 1\right\rangle &= \left|\mp 1\right\rangle \\ \sigma_2 \left|\pm 1\right\rangle &= \pm i \left|\mp 1\right\rangle \end{aligned} \tag{2.52}$$

and we also have

$$\begin{aligned} \sigma_+ \left|+1\right\rangle &= 0 \\ \sigma_+ \left|-1\right\rangle &- \left|+1\right\rangle \\ \sigma_- \left|+1\right\rangle &= \left|-1\right\rangle \\ \sigma_- \left|-\right\rangle &= 0. \end{aligned} \tag{2.53}$$

σ_+ defines the $\left|+1\right\rangle$ "vacuum," and σ_- the $\left|-1\right\rangle$ "vacuum." Recall that $\boldsymbol{\sigma}$ and I form a complete set of 2×2 matrices. Because of this completeness, we may write any 2×2 density matrix in these terms, i.e.

$$\rho = \frac{1}{2}\left[a_0 I + \mathbf{r} \cdot \boldsymbol{\sigma}\right]. \tag{2.54}$$

The coefficients may be written

$$a_0 = \mathrm{Tr}\rho$$

$$r_i = \mathrm{Tr}\rho\sigma_i.$$

The above operators have been written in the Schrödinger picture. If $\rho^2 = \rho$, it is a pure state. If

$$\rho = \begin{pmatrix} \frac{1}{2} & 0 \\ 0 & \frac{1}{2} \end{pmatrix},$$

then $\rho^2 = \frac{\rho}{2}$, and in this case, we have a mixture. We find $\langle s_i \rangle = 0$. The spin is unpolarized, since all directions are equivalent. A pure polarization state is

$$\rho(\theta, \phi) = \begin{pmatrix} \cos^2 \frac{\theta}{2} & \sin \frac{\theta}{2} \cos \frac{\theta}{2} \exp(-i\theta) \\ \sin \frac{\theta}{2} \cos \frac{\theta}{2} \exp(i\theta) & \sin^2 \frac{\theta}{2} \end{pmatrix}.$$

Here $\langle \mathbf{s} \rangle = \frac{1}{2}\boldsymbol{\mu}$, $\boldsymbol{\mu}$ being a classical vector whose polar angles are θ, ϕ. Remember that the mixture state *is not* a unique state $\left|\psi\right\rangle$.

The unperturbed spin Hamiltonian is

$$H = \frac{\hbar\omega}{2}\sigma_z, \tag{2.55}$$

so

$$U(t,0) = \exp\left(-i\frac{\omega}{2}\sigma_z t\right). \tag{2.56}$$

The Heisenberg equations are

$$\begin{aligned}\frac{d\sigma_z(t)}{dt} &= 0 \\ \frac{d\sigma_+(t)}{dt} &= \frac{i\omega}{2}\sigma_+(t) \\ \frac{d\sigma_-}{dt} &= -\frac{i\omega}{2}\sigma_-(t).\end{aligned} \tag{2.57}$$

Now let us turn to the quantum dynamics of the two-level system, or one of spin (Nussenzweig, 1973). We have

$$|\psi(t)\rangle = a(t)|+1\rangle + b(t)|-1\rangle, \tag{2.58}$$

and the density matrix is again

$$\rho(t) = |\psi(t)\rangle\langle\psi(t)| \to \begin{pmatrix} a^2 & ab^* \\ a^*b & b^2 \end{pmatrix}. \tag{2.59}$$

We will choose a semi-phenomenological Hamiltonian including damping:

$$H = H_0 + V = \begin{pmatrix} E_+ & 0 \\ 0 & E_- \end{pmatrix} + \begin{pmatrix} 0 & V_{+-} \\ V_{+-} & 0 \end{pmatrix} \tag{2.60}$$

$$V_{+-} = V^*_{-+},\ E_+ - E_- = \omega_0.$$

If

$$\hat{V} = -e\hat{x}\cdot\mathbf{E}(\mathbf{r}_1 t), \tag{2.61}$$

$\mathbf{E}(\mathbf{r}_1 t)$ being the *classical* electric field, then the dipole moment is $\mu_{+-} = e\langle x_{+-}\rangle$, and $V_{++} = V_{--} = 0$. The polarization is $\langle P\rangle = \mu_{+-}\left(\rho_{+-} + \rho^*_{+-}\right)$.

We introduce now a phenomenological damping term Γ and write

$$\frac{id\rho}{dt} = [H,\rho]_- + \frac{-i}{2}[\Gamma,\rho]_+, \tag{2.62}$$

where $\Gamma = \begin{pmatrix} \gamma_+ & 0 \\ 0 & \gamma_- \end{pmatrix}$. This, of course, leads to exponential decay in time

$$\left|\psi_+(t)\right\rangle = \left|\psi_+(0)\right\rangle \exp\left(-i\left(\omega - i\frac{\gamma_+}{2}\right)t\right).$$

It has its origins most simply in the Weisskopf–Wigner theory of spontaneous emission, which will be discussed in detail in later chapters (Weisskopf and

Wigner, 1930). Without damping, we may give Eq. (2.62) a geometric interpretation. Using Eq. (2.54) we have

$$\rho = \frac{1}{2}\begin{pmatrix} \rho_0 + r_3 & r_1 - ir_2 \\ r_1 + ir_2 & \rho_0 - r_3 \end{pmatrix}. \tag{2.63}$$

Therefore,

$$\rho_0 = \mathrm{Tr}\rho = 1 = |a|^2 + |b|^2$$

$$\begin{aligned} r_1 &= ab^* + a^*b \\ r_2 &= i\left(ab^* - a^*b\right) \\ r_3 &= |a|^2 - |b|^2, \end{aligned} \tag{2.64}$$

and now Eq. (2.60) is in terms of the Pauli matrix representation:

$$H = \frac{1}{2}\left(V_1\sigma_1 + V_2\sigma_2 + \omega_0\sigma_3\right), \tag{2.65}$$

and $V_{+-} \equiv \frac{1}{2}(V_1 - V_2)$. Utilizing $\left[\sigma_{1,}\sigma_2\right] = 2i\sigma_3$, the von Neumann equation is

$$\begin{aligned} \dot{\rho} &= 0 \\ \dot{r}_1 &= V_2 r_3 - \omega_0 r_2 \\ \dot{r}_2 &= \omega_0 r_1 - V_1 r_3 \\ \dot{r}_3 &= V_1 r_2 - V_2 r_{1.} \end{aligned} \tag{2.66}$$

For a pure state, the vector $\mathbf{r}$ has unit length. We may write

$$\frac{d\mathbf{r}}{dt} = \boldsymbol{\omega} \times \mathbf{r}, \tag{2.67}$$

where

$$\begin{aligned} \omega_1 &= V_1 \\ \omega_2 &= V_2 \\ \omega_3 &= \omega_0. \end{aligned} \tag{2.68}$$

These are the optical Bloch equations written by Feynman, Vernon and Hellwarth (Feynman *et al.*, 1957). The physical picture is that $\mathbf{r}$ precesses around $\boldsymbol{\omega}$. In the case of spin $\frac{1}{2}$, $\mathbf{r}$ is proportional to $\langle\mu\rangle$, the average magnetic moment, and $\boldsymbol{\omega}$ proportional to the magnetic field. Then $\mathbf{r}$ is truly a physical space with $\langle\mu\rangle$ precessing in this space about the magnetic field. We will discuss this later in this chapter.

For the electromagnetic field, the geometry is more abstract. If $E(t)$ is also sinusoidal, then

$$V_1 = (V_{+-} + V_{-+})$$
$$V_2 = \frac{1}{i}(V_{+-} - V_{-+}),$$

where $V_{+-} = -\frac{1}{2}\mu_{+-}(\mathfrak{E}\exp(i\omega t) + \mathfrak{E}\exp(-i\omega t))$. The optical field perturbation is "rotating" in the 1,2 plane. There is a $\pm$ rotation. For positive ω_0 we ignore the $-\omega$ rotation, since it may not add in phase. This is the rotating wave approximation. To solve we go to a rotating frame in $+\omega$. In this rotating frame, $|V - \omega|$ precesses about $\mathbf{V} - \boldsymbol{\omega}$. The angular rotation velocity is the nutation frequency, Ω:

$$\Omega \equiv |V - \omega| = \sqrt{\left|\frac{\mu_{+-}\mathfrak{E}}{\hbar}\right|^2 + (\omega_0 - \omega)^2}. \tag{2.69}$$

This is the Rabi formula (Rabi, 1937), leading to a population inversion,

$$p_+(t) = |a(t)|^2 = \frac{|\mu_{+-}|^2}{\Omega^2}\mathfrak{E}^2 \sin^2\frac{\Omega t}{2}. \tag{2.70}$$

The above calculation is a geometric interpretation of that which may be done in other ways (Scully and Zubairy, 1997).

This result may also be obtained immediately from the von Neumann equation, Eq. (2.66), assuming

$$\begin{aligned} \rho_{++} &= \rho^0_{++}\exp(\lambda t) \\ \rho_{--} &= \rho^0_{--}\exp(\lambda t) \\ \rho_{+-} &= \rho^*_{-+} = \rho^0_{+-}\exp(-i(\omega_0 - \omega)t)\exp\lambda t. \end{aligned} \tag{2.71}$$

The determinant of the coefficients gives

$$\lambda^2\left\{\lambda^2 + (\omega_0 - \omega)^2 + \left|\frac{\mu_{+-}\mathfrak{E}}{\hbar}\right|^2\right\} = 0,$$

having roots $\lambda_1 = 0$ and $\lambda_2 = i\Omega = \lambda_3^*$, where Ω is given in Eq. (2.69).

Semi-classical electron spin resonance is another example of two-level system dynamics. Here we treat electron spin resonance briefly. An electric dipole moment interacts with a radio frequency field. We take

$$H = -\boldsymbol{\mu}\cdot\mathfrak{H}, \tag{2.72}$$

$\mathfrak{H}$ being the classical magnetic field with

$$\boldsymbol{\mu} = -\frac{\gamma}{2}\boldsymbol{\sigma}. \tag{2.73}$$

We also take $\mathfrak{H}_1$, $\mathfrak{H}_2$ rotating and $\mathfrak{H}_0$ being constant in the z direction. We have

$$H = \frac{\gamma}{2}\left[\mathfrak{H}_0\sigma_z + \mathfrak{H}_1\left(\sigma_+ \exp\left(-i\omega t\right) + \sigma_- \exp\left(+i\omega t\right)\right)\right]. \tag{2.74}$$

For the spin $\frac{1}{2}$ states already discussed in detail,

$$E_+ - E_- = \omega_0 = \gamma\mathfrak{H}_0. \tag{2.75}$$

We may show

$$\begin{aligned} U(t) &= \exp\left(-iHt\right) \\ &= \exp\left(-i\omega t\frac{\sigma_z}{2}\right)\left[\cos\frac{1}{2}\Omega t - i\sin\frac{1}{2}\Omega t\left(\cos\theta\sigma_z + \sin\theta\sigma_x\right)\right], \end{aligned} \tag{2.76}$$

where

$$\Omega^2 = \left(\omega - \omega_0\right)^2 + \left(\gamma\mathfrak{H}\right)^2,$$

since

$$\begin{aligned} \Omega\cos\theta &= \omega_0 - \omega \\ \Omega\sin\theta &= \gamma\mathfrak{H}_1. \end{aligned}$$

From this we may obtain $|\psi(t)\rangle$. If $|\psi(0)\rangle = |+\rangle$, then

$$|c_-(t)|^2 = |\langle -|\psi(t)\rangle|^2 = \sin^2\theta\sin^2\frac{1}{2}\Omega t,$$

and we may write it as

$$\begin{aligned} |\langle -|\psi(t)\rangle|^2 = & \frac{\left(\gamma\mathfrak{H}_1\right)^2}{\left(\omega-\omega_0\right)^2 + \left(\gamma\mathfrak{H}_1\right)^2} \\ & \times \sin^2\frac{1}{2}t\sqrt{\left(\omega-\omega_0\right)^2 + \left(\gamma\mathfrak{H}_1\right)^2}. \end{aligned} \tag{2.77}$$

These are the Rabi oscillations in their earliest example of spin resonance. At resonance $\omega = \omega_0$,

$$\begin{aligned} \langle\sigma_x\rangle &= \sin\omega_0 t\sin\gamma\mathfrak{H}_1 t \\ \langle\sigma_z\rangle &= \cos\gamma\mathfrak{H}_1 t \\ \langle\sigma_z\rangle &= \cos\gamma\mathfrak{H}_{1t}. \end{aligned} \tag{2.78}$$

$\langle\sigma_x\rangle$ and $\langle\sigma_y\rangle$ precess at ω_0, and $\langle\sigma_z\rangle$ nutates at frequency $\gamma\mathfrak{H}_1$.

Appendix 2A: the Fokker–Planck equation

We will here derive the Fokker–Planck equation for $P(\alpha a^*, t)$ for the case of the harmonic oscillator, Eq. (2.34) (Gardiner, 1991). We use Bargman states (Bargman, 1961, 1962), defined as

$$\|\alpha\rangle = \exp\left(-\frac{1}{2}|\alpha|^2\right)|\alpha\rangle, \tag{2A.1}$$

which because of the Gaussian prefactor are analytic functions of $|\alpha\rangle$. Then

$$|f\rangle = \frac{1}{\pi}\int d^2\alpha\, f(\alpha^*)\exp\left(-\frac{1}{2}|\alpha^2|\right)|\alpha\rangle \tag{2A.2}$$

is unique. Also, for operator in Hilbert space

$$O\left(\alpha^*\beta\right) = \langle\alpha\,\|O\|\,\beta\rangle, \tag{2A.3}$$

the matrix elements in Bargman states are well defined.

For Bargman states,

$$\begin{aligned} a^+\|\alpha\rangle &= \frac{\partial}{\partial\alpha}\|\alpha\rangle \\ \langle\alpha\|\,a &= \frac{\partial}{\partial\alpha^*}\langle\alpha\|\,. \end{aligned} \tag{2A.4}$$

In these states, the $P(\alpha a^*)$ representation becomes

$$\hat{\rho} = \int d^2\alpha\,\|\alpha\rangle\langle\alpha\|\exp\left(-\alpha a^*\right)P\left(\alpha\alpha^*\right). \tag{2A.5}$$

Upon using Eq. (2A.4) and integrating by parts, we obtain

$$\hat{a}^\dagger\hat{\rho} = \int d^2\alpha\,\|\alpha\rangle\langle\alpha\|\exp\left(-\alpha\alpha^*\right)\left(\alpha^* - \frac{\partial}{\partial\alpha}\right)P\left(\alpha\alpha^*\right).$$

This is an operator rule for $a^+\rho$ on $P(\alpha\alpha^*)$. We easily obtain the rules

$$\begin{aligned} \hat{a}\hat{\rho} &\to \alpha P\left(\alpha\alpha^*\right) \\ \hat{a}^\dagger\hat{\rho} &\to \left(\alpha^* - \frac{\partial}{\partial\alpha}\right)P\left(\alpha\alpha^*\right) \\ \hat{\rho}\hat{a} &\to \left(\alpha - \frac{\partial}{\partial\alpha^*}\right)P\left(\alpha\alpha^*\right) \\ \hat{\rho}\hat{a}^\dagger &\to \alpha^* P\left(\alpha\alpha^*\right), \end{aligned} \tag{2A.6}$$

where the right sides are the complex functions α, α^* and derivatives under the integral as above. This correspondence is discussed in much more detail in later chapters.

We now consider the harmonic oscillator in normal ordered form:

$$H = \omega \left(a^{\dagger} a + \frac{1}{2} \right).$$

This will be the source of the correspondence rule to follow.

The von Neumann equation is, in this simple example,

$$i \frac{\partial \hat{\rho}}{\partial t} = \omega \left[a^{\dagger} a, \rho \right].$$

Using the preceding operator correspondence, maintaining the order

$$\hat{a}^{\dagger} \hat{a} \to \left(\alpha^* - \frac{\partial}{\partial \alpha} \right) \alpha P$$
$$\rho \hat{a}^{\dagger} \hat{a} \to \left(\alpha - \frac{\partial}{\partial \alpha^*} \right) \alpha^* P,$$

and using Eq. (2A.6) with the von Neumann equation, we find the integrand to be

$$\frac{\partial P}{\partial t} = i \left(-\omega \frac{\partial}{\partial \alpha} + \omega \frac{\partial}{\partial \alpha^*} \alpha^* \right) P, \tag{2A.7}$$

a complex Fokker–Planck equation for $P\left(\alpha \alpha^*\right)$.

The real variables may be introduced with

$$\alpha = x + iy$$
$$\alpha^* = x - iy.$$

We obtain

$$\frac{\partial P}{\partial t} = \omega \left[\frac{\partial}{\partial x} y - \frac{\partial}{\partial y} x \right] P,$$

which is a classical Liouville equation in the phase space x, y. The method of characteristics has already given the solution used in this chapter,

$$P\left(\alpha \alpha^*, t\right) = \delta^2 \left(\alpha - \alpha\left(t\right)\right).$$

References

Bargman, V. (1961). *Comm. Pure and Applied Math.* **14**, 187.

Bargman, V. (1962). *Proc. Natl. Acad. Sci. (USA)* **48**, 199.

Cohen-Tannoudji, C., Diu, B. and Laloé, F. (1977). *Quantum Mechanics,* vol. 1 (New York, Wiley).

Dirac, P. A. M. (1958). *Quantum Mechanics,* 4th edn. (London, Oxford University Press).

Feynman, R. P., Vernon F. L. Jr. and Hellwarth, R. W. (1957). *J. Appl. Phys.* **28**, 49.

Gardiner, C. W. (1991). *Quantum Noise* (New York, Springer).

Glauber, R. J. (1963). *Phys. Rev.* **131**, 2766.

Jordan, T. F. (1986). *Quantum Mechanics in Simple Matrix Form* (New York, Wiley).
Liboff, R. (1998). *Quantum Mechanics,* 3rd edn. (New York, Addison-Wesley).
Louisell, W. (1973). *Quantum Statistical Properties of Radiation* (New York, Wiley).
Nussenzweig, H. M. (1973). *Introduction to Quantum Optics* (New York, Gordon and Breach).
Rabi, I. J. (1937). *Phys. Rev.* **51**, 652.
Schrödinger, E. (1926). *Naturwissenschaften* **14**, 664.
Scully, M. O. and Zubairy, M. S. (1997). *Quantum Optics* (New York, Cambridge University Press).
Sudarshan, E. C. G. (1963). *Phys. Rev. Lett.* **10,** 277.
Walls, D. F. and Milburn, G. W. (1985). *Phys. Rev.* **A 31**, 2403.
Weisskopf, V. F. and Wigner, E. P. (1930). *Z. Physik* **54**, 63.

3

Quantum statistical master equation

3.1 Reduced observables

The fundamental density operator ρ having the properties

$$\langle A\rangle = \mathrm{Tr}\rho A \tag{3.1}$$

$$\mathrm{Tr}\rho = 1 \tag{3.2}$$

$$\rho^{\dagger} = \rho \tag{3.3}$$

was introduced in Chapter 1.

Here A is the observable. $\rho(t)$ obeys von Neumann's (Liouville) equation,

$$i\dot{\rho}(t) = [H, \rho(t)] \equiv L\rho(t) \quad (-\infty \leq t \leq \infty), \tag{3.4}$$

and here $\hbar = 1$. It might be the case that A is diagonal in a discrete representation $|m\rangle$, where

$$A\,|m\rangle = a_m\;|m\rangle\,. \tag{3.5}$$

Thus,

$$\langle A\rangle = \sum_m a_m\;\rho_{mn}(t)\delta_{nm},$$

and only diagonal elements of ρ are important.

$$\rho_{mm} \geqslant 0$$

$$\sum_m \rho_{mm} = 1.$$

This is the case in elementary applications of equilibrium statistical mechanics, as in the text of Reif (1965). Of course, $\rho_{mm}(t) = P_m(t)$, the probability that the system is in state $|m\rangle$ at time t. For this average the off-diagonal elements of $\rho(t)$

do not enter. This "reduction" clearly depends upon what is being observed. It is important in that it simplifies the description. The full density operator is no longer necessary to the calculation of such averages. This is also true classically when we are considering hydrodynamic observables such as $n(\mathbf{r}, \mathbf{p}, t)$, the local number density in the spacially inhomogenous fluid. Then the N-particle distribution function $f_N(\mathbf{r}_1\mathbf{p}_1, \mathbf{r}_2\mathbf{p}_2, \ldots, \mathbf{r}_N\mathbf{p}_N, t)$ is not necessary, and we may use one-body distributions, $f_1(\mathbf{r}_1\mathbf{p}_1, t)$. For the details of this, the reader should see the texts of Balescu (1975) and Huang (1987).

Quantum reduced distribution functions may also be introduced. The Wigner function (Wigner, 1932; Balescu, 1975) is one. It is defined as

$$w(x, p) = \frac{1}{2\pi} \int_{-\infty}^{+\infty} d\xi \exp(-ip\xi) \left\langle x + \frac{1}{2}\xi \left| \hat{\rho} \right| x - \frac{1}{2}\xi \right\rangle \tag{3.6}$$

(Schleich, 2001). It is not a probability distribution, since $w(x, \rho) \not\geq 0$. More will be said in the next chapter, where we discuss the quantum Boltzmann equation and its derivation.

For the purpose of obtaining reduced forms of the density operator and its matrix elements, we will introduce here a projection operator, P, and its realizations. This simple approach is due to Nakajima (1958) and Zwanzig (1960a). The equations are called master equations.

For the reduction we will use a tetradic representation of operators. The fundamental operator is $L \equiv [H,]$, written

$$L_{mm'nn'} = H_{mn}\delta_{m'n'} - \delta_{mn}H_{n'm'}, \tag{3.7}$$

where the mapping of "ordinary" observables in Hilbert space (A) $C = LA$ is written

$$C_{mn} = \sum_{m'n'} L_{mnm'n'} A_{m'n'}. \tag{3.8}$$

This is discussed further in Section 3.3.

For a simple reduction of ρ to its diagonal elements, we have

$$(P\rho)_{mn} = \rho_{nm}\delta_{mn}. \tag{3.9}$$

In tetradic representation, the projection operator is

$$P_{mnm'n'} = \delta_{mn}\delta_{mm'}\delta_{nn'}. \tag{3.10}$$

P has the properties

$$P^2 = P \tag{3.11}$$

$$P^\dagger = P. \tag{3.12}$$

The latter property is not necessary but assures an orthogonal projection. It is true in the case of Eq. (3.10). The projection operator method is quite general, and with it we may obtain an "intermediate" equation, the generalized master equation. From Eq. (3.4) we have

$$iP\dot{\rho} = PL(P\rho + (1-P)\rho) \tag{3.13}$$

$$i(1-P)\dot{\rho} = (1-P)L(P\rho + (1-P)\rho). \tag{3.14}$$

Writing a formal solution to Eq. (3.14), we have

$$\begin{aligned}(1-P)\rho = &-i\int_0^t dt'[\exp(-i(1-P)L(1-P)t')(1-P) \times LP\rho(t-t')] \\ &+ \exp(-i(1-P)L(1-P)t)(1-P)\rho(0); \quad t > 0.\end{aligned} \tag{3.15}$$

Here a time initial value, $\rho(0)$, has been assumed, with $0 \leq t \leq \infty$. Thus, Eq. (3.15) is not equivalent to the von Neumann equation, where $-\infty \leq t \leq \infty$. Putting Eq. (3.15) into Eq. (3.13), a closed equation for $P\rho(t)$ may be obtained. It is non-Markovian (see the appendix to this chapter) in the sense that it depends on $P\rho(t-t')$. This is the so-called generalized master equation of Montroll, Zwanzig, Prigogine and Résebois (Montroll, 1960; Zwanzig, 1960a; Prigogine and Résebois, 1961; Prigogine, 1963). It represents a starting point for further discussion but by itself is too general and unwieldy.

The point of this chapter is to develop, physically, useful master equations of a Markovian nature. Such a generalized master equation was first obtained by Van Hove (1957), using diagrammatic perturbation theory. Its form is difficult to compare with that obtained by Eq. (3.13) and Eq. (3.15). We shall not try, but refer the reader to the work of Swenson (1962). He showed that perturbation theory is not necessary.

3.2 The Pauli equation

We will now turn to the simplest example of a quantum master equation first introduced by Pauli (1928). We repeat the original derivation of Pauli and discuss its weakness. Also, we will consider the structure of this original quantum master equation as a prototype example.

We take the Hamiltonian as

$$H = H^0 + \lambda V, \tag{3.16}$$

where the unperturbed contribution is

$$H^0 \left|\alpha\right\rangle = E^0_\alpha \left|\alpha\right\rangle , \tag{3.17}$$

with $|\alpha\rangle$ being the unperturbed discrete eigenstates. The perturbation λV, here assumed small, is characterized by the parameter λ. A simple example would be in a cubic anharmonic oscillator, the harmonic approximation being most important. In perturbation, the states $|\alpha\rangle$ are the basis set and the "language" of the discussion.

The state at time t is

$$\phi(t) = \sum_\alpha c(\alpha, t) \left|\alpha\right\rangle . \tag{3.18}$$

Now

$$P(\alpha, t) = |c(\alpha, t)|^2 \tag{3.19}$$

the probability at the time t that the system is in state $|\alpha\rangle$. Utilizing second-order (in λ) time-dependent perturbation theory, the transition rate is

$$W\alpha\alpha' = 2\pi\lambda^2\delta(E^0_{\alpha'} - E^0_\alpha) \left|\left\langle\alpha\right| V \left|\alpha'\right\rangle\right|^2 . \tag{3.20}$$

This is, of course, the "golden rule" (Dirac, 1958). The energy-conserving delta function is the continuum limit of the discrete state index α. For instance, for a lattice in three dimensions with periodic boundary conditions in the infinite volume limit,

$$\sum_\alpha \Rightarrow \frac{V}{8\pi^3} \int d_3\alpha .$$

The Pauli equation may now be obtained. Using the unitary time evolution

$$\phi(t + \Delta t) = \exp(-i(H^0 + \lambda V)\Delta t)\phi(t), \tag{3.21}$$

we have

$$\begin{aligned} P(\alpha, t + \Delta t) = \sum_{\alpha'\alpha''} & c^*(\alpha'' t) \left\langle\alpha''\right| \exp(i(H_0 + \lambda V)\Delta t \left|\alpha\right\rangle \\ & \times \left\langle\alpha\right| \exp -i(H_0 + \lambda V)\Delta t \left|\alpha'\right\rangle c(\alpha', t). \end{aligned} \tag{3.22}$$

The continuous-in-time random phase approximation is now made. The $\alpha \neq \alpha'$ contributions rapidly oscillate and cancel, leaving only the $\alpha' = \alpha''$ contributions to the summation. Eq. (3.22) becomes

$$P(\alpha, t + \Delta t) = \sum_{\alpha'} | \left\langle\alpha\right| \exp(-i(H^0 + \lambda V)\Delta t) \left|\alpha'\right\rangle |^2 \; P(\alpha', t), \tag{3.23}$$

where $\Delta t \geqslant 0$. To second order in λ, Eq. (3.23) becomes

$$P(\alpha, t+\Delta t) = P(\alpha, t)\delta_{\alpha a'} + 2\pi\lambda^2\Delta t \sum_{\alpha'} \delta(E^0_{\alpha'} - E^0_\alpha) \mid \langle \alpha' | V | \alpha \rangle \mid^2 [P(\alpha', t) - P(\alpha, t)]. \tag{3.24}$$

Thus, to this order,

$$\frac{d}{dt} P(\alpha, t) = \sum_{\alpha'} [W_{\alpha\alpha'} P(\alpha', t) - W_{\alpha'\alpha} P(\alpha, t)]. \tag{3.25}$$

This is Pauli's argument. $W_{\alpha a'}$ is given by Eq. (3.20). It is a gain–loss (birth–death!) equation between states $|\alpha\rangle$, $|\alpha'\rangle$. It is Markovian, being an equation for $P(\alpha, t)$ in terms of $P(\alpha', t)$. This is a continuous-in-time stochastic Kolmogorov equation (Kolmogorov, 1950). For this the reader should note the appendix to this chapter. The name "master" is derived from this.

The validity of perturbation theory must be examined for a given problem. The reader can consult any good book on applied quantum mechanics to see examples.

The limit of continuous spectrum for $|\alpha\rangle$ is more subtle and is discussed in detail in Chapter 18. It depends on the level spacing, which depends on V for free particles with periodic boundary conditions in one dimension. This is one aspect of the *thermodynamic limit* as the volume V approaches infinity,

$$V \to \infty, \tag{3.26}$$

such that $\frac{\mathfrak{N}}{V} = c =$ constant as $\mathfrak{N} \to \infty$. $\mathfrak{N}$ is the number of particles. This will be used in many applications in later chapters. We note, however, that this is not true for harmonic oscillators in a container. They have no V dependence to the spectrum.

The repeated random phase assumption *at all time* has a flaw. It is inconsistent, as was first pointed out by Van Hove (1962).

From Eq. (3.23) we may also show

$$P(\alpha, t-\Delta t) = P(\alpha, t)\delta_{\alpha a'} - 2\pi\lambda^2\Delta t \sum_{\alpha} W_{\alpha a'} P(\alpha', t) - W_{\alpha' a} P_\alpha(\alpha, t).$$

Thus,

$$\lim_{\Delta t \to 0^+} \frac{\Delta P(\alpha, t)}{\Delta t} = - \lim_{\Delta t \to 0^-} \frac{\Delta P(\alpha, t)}{\Delta t}. \tag{3.27}$$

The only solution is $\dot{P}(\alpha, t) = 0$ for all time. In a sense this is the "watched pot" difficulty. Repeated continuous random phase leads to no change in the equilibrium state. To remove this difficulty, we must random phase *initially* only (Van Hove, 1962; Prigogine, 1963).

3.3 The weak coupling master equation for open systems

Let us consider open systems, which are a central theme of this book. We will take the Hamiltonian to be

$$H = H^0 + \lambda V, \tag{3.28}$$

where

$$H^0 = H_S + H_R \,. \tag{3.29}$$

The system of interest is S, which is in contact with a "reservoir" R through the interaction λV. The reservoir R may be a very large system in approximate thermodynamic equilibrium. However, this need not be the case. The two systems together are isolated. H is a conserved Hamiltonian. The unit operator in Eq. (3.28) is understood. We will take R to be macroscopic. By tracing over the R states (Tr_R), we will obtain a reduced density operator ρ_S for the system of interest. Now

$$i\dot{\rho}(t) = L\rho(t) \equiv [H, \rho], \tag{3.30}$$

and

$$\rho_S\,(t) = \mathrm{Tr}_R \rho(t). \tag{3.31}$$

We assume initially that the two systems are uncorrelated.

We will choose the relevant projection operator to be

$$P\rho = \rho_R\,(0)\mathrm{Tr}_R\,\rho. \tag{3.32}$$

This was first introduced by Argyres and Kelley (1964) in the discussion of spin resonance (see also Peier and Thellung, 1970; Peier, 1972; Agarwal, 1973; Haake, 1973; Louisell, 1973). ρ and A are assumed to have a finite trace in R, that is, trace class in the Hilbert space $\mathfrak{L}_R$. P is idempotent, since $\mathrm{Tr}_R \rho_R(0) = 1$. It is not necessarily Hermitian. We form (A, PB) and examine (PA, B). Let $A = A_s A_R$ and $B = B_s B_R$. We have the condition for hermiticity,

$$\mathrm{Tr}_R\, B_R\, \mathrm{Tr}_R\, A^{\dagger}_R \rho_R\,(0) = \mathrm{Tr}_R\,(B_R\,\rho_R\,(0))\mathrm{Tr}_R\, A^{\dagger}_R,$$

which is not necessarily true.

We assume

$$\begin{aligned} \rho(0) &= \rho_R(0)\rho_S(0) \\ [H_R, \rho_R(t)] &= 0 \\ PL_S &= L_S P \\ PL'P &= 0, \end{aligned} \tag{3.33}$$

where $L' = [\lambda V,]$. The latter follows by incorporating the diagonal part of λV into H^0. Now, following Eqs. (3.13), (3.14) and (3.15), we have

$$i\dot{\rho}_S(t) = L_S\rho_S(t) - i\int_0^t dt' G(t-t')\rho_S(t'), \tag{3.34}$$

where the kernel is

$$G(t-t') = \lambda^2 \mathrm{Tr}_R[L' \exp(-i(t-t')(L^0 + (1-P)\lambda L'))\, L'\rho_R(0)]. \tag{3.35}$$

Here the reduced system density operator is

$$\rho_S(t) = \mathrm{Tr}_R P\rho(t). \tag{3.36}$$

Since we are interested in obtaining the Pauli equation, we will form an equation for ρ_{Sd}, the diagonal part of $\rho_S(t)$, introducing a further projection $D\rho_S(t)$, where

$$DA_S A_R = A_{Sd} A_R. \tag{3.37}$$

Assume also

$$\rho_S(0) = \rho_{Sd}(0). \tag{3.38}$$

Eq. (3.35) becomes

$$G(\tau) = D\mathrm{Tr}_R\{\lambda^2 L' \exp(-i\tau[L^0 + \lambda(1-P)L'])]L'\rho_R(0)\} \tag{3.39}$$

in the equation for $\rho_{Sd}(t)$.

Let us rescale the time, since we are interested in the singular limit, $\lambda \to 0$, $t \to \infty$; $\lambda^2 t =$ constant (Van Hove, 1962). Eq. 3.34 becomes

$$\frac{d\tilde{\rho}_{sd}(\tilde{t})}{d\tilde{t}} = -\int_0^{\frac{\tilde{t}}{\lambda^2}} dt' G(t')\tilde{\rho}_{Sd}(\tilde{t} - \lambda^2 t'), \tag{3.40}$$

where

$$\tilde{t} = \lambda^2 t.$$

Now

$$\tilde{\rho}_{Sd}(\tilde{t}) = \rho_{Sd}(t).$$

In the limit $\lambda \to 0$,

$$\frac{d\tilde{\rho}_{Sd}(\tilde{t})}{d\tilde{t}} = -\int_0^{\infty} dt' G(t')\tilde{\rho}_{Sd}(\tilde{t}); \tag{3.41}$$

we obtain a general Markovian equation. Later we will make some further comments on Eq. (3.40) and Eq. (3.41).

To lowest order in λ (sometimes here called a Born approximation after scattering theory), Eq. (3.39) becomes

$$G^0(t') = \lambda^2 D\mathrm{Tr}_R[L' \exp(-iL^0 t')L'\rho_R(0)]. \tag{3.42}$$

Let us first evaluate Eq. (3.41) and Eq. (3.42); later in this chapter, we will comment on what time scale we expect the Van Hove limit to hold. We take the Laplace transform and obtain

$$\hat{\rho}_{Sd}(0) = -i\tilde{z}\hat{\rho}_{Sd}(\tilde{z}) - iG^0(\lambda^2\tilde{z})\hat{\rho}_{Sd}(\tilde{z}) \tag{3.43}$$

$$\text{with} \quad z = \lambda^2\tilde{z}.$$

$$\rho_{Sd}(\tilde{z}) = \frac{1}{\lambda^2}\hat{\rho}(\tilde{z}). \tag{3.44}$$

The Laplace transform of Eq. (3.42) is

$$G^0(\lambda^2\tilde{z}) = -iD\text{Tr}L'\frac{1}{\lambda^2\tilde{z} - (L^0 + \lambda(1-P)L')}L'\rho_R(0). \tag{3.45}$$

As $\lambda \to 0$, we write formally

$$G^0(0+) \equiv \lim_{\varepsilon \to 0^+} G^0(0 + i\varepsilon).$$

The limit is obtained because we have already made a causality assumption in the derivation of the generalized master equation. We write the result of the limit as

$$G^0(0+) = +i\mathfrak{P}D\text{Tr}\left[L'\frac{1}{L^0}L'\rho_R(0)\right] - \pi D\text{Tr}_R[L'\delta(L^0)L'\rho_R(0)], \tag{3.46}$$

where the distributions

$$\lim_{\varepsilon \to 0^+}\frac{1}{x + i\varepsilon} = \mathfrak{P}\frac{1}{x} + i\delta(x),$$

$\mathfrak{P}(x)$ being the principal part function and $\delta(x)$ the Dirac delta function. Eq. (3.46) is just a formal statement with operators indicating what must be evaluated after the tetradic operations in that representation have been done.

As an example, we may use the simplest representation of tetradic $L_{mn,m'n'}$, due to Résebois (Résebois, 1961; Prigogine, 1963). Let

$$\nu = n - m \tag{3.47}$$

$$N = \frac{n+m}{2}.$$

Then $\langle n|\, A\, |m\rangle = A_{n-m}(\frac{n+m}{2}) \equiv A_\nu(N)$, and Eq. (3.7) becomes

$$\langle\nu|\, L\, |\nu'\rangle = \eta^{+\nu'}H_{\nu-\nu'}(N)\eta^{-\nu} - \eta^{-\nu'}H_{\nu-\nu'}(N)\eta^{+\nu} \tag{3.48}$$

with the shift operator

$$\eta^\nu A_{\nu'}(N) = A_{\nu'}\left(N + \frac{\nu}{2}\right). \tag{3.49}$$

This is a "classical-like" representation. Furthermore, since

$$H^0 \left|n'\right\rangle = E_n^0 \delta_{nn'},$$

we have $H^0_{\nu-n'}(N) = E^0(N)\delta_{\nu-\nu'}$ and $\langle\nu| L^0 \left|\nu'\right\rangle = E^0 \cdot \nu\delta_{\nu\nu'}$. Now assume the basis states are those which diagonalize $\rho_R(0)$. We also assume the thermodynamic limit, $\mathfrak{N} \to \infty,\ V \to \infty\ ;\ \frac{\mathfrak{N}}{V} = c =$ constant. For particular systems to be discussed, $n,\ m$ and ν become continuous. ν may be viewed as a frequency and has the range $-\infty$ to $+\infty$. Thus, the singular operators in Eq. (3.46) have a meaning. More will be said concerning the limit in Chapter 18.

The first term in Eq. (3.46) is proportional to $\frac{\mathfrak{P}}{E^0 \cdot \nu}$ and vanishes. We are left with

$$G^0(0+) = \pi D \mathrm{Tr}_R[L'\delta(L^0)L'\rho_R(0)].$$

Now

$$L' = L'_S + L'_{SR} \quad and \quad \mathrm{Tr}_R[L'_{SR}\rho_R(0)] = 0.$$

Thus,

$$G^0(0+) = \pi D L'_S\delta(L^0_S)L'_S + \pi D \mathrm{Tr}_R L'_{SR}\delta(L^0)L'_{SR}\rho_R(0).$$

From this, using Eq. (3.48) or Eq. (3.7), we obtain

$$\begin{aligned}\frac{d\tilde{\rho}_{Snn}(\tilde{t})}{d\tilde{t}} = &-2\pi \sum_m \left|H'_{Snm}\right|^2 \delta(E_n^0 - E_m^0)[\tilde{\rho}_{Snn}(\tilde{t}) - \tilde{\rho}_{Smm}(\tilde{t})] \\ &- 2\pi \sum_m \sum_{\alpha\beta}[\left|H_{SRn\alpha m\beta}\right|^2 \delta(E_n^0 + E_\alpha^0 - E_m^0 - E_\beta^0)] \\ &\times [\rho_{R\alpha\alpha}(0)\tilde{\rho}_{Snn}(\tilde{t}) - \rho_{R\beta\beta}(0)\tilde{\rho}_{Smm}(\tilde{t})].\end{aligned} \tag{3.50}$$

Here E_m^0 and E_α^0 are the system and reservoir eigenstates respectively. This is a Pauli master equation for the system in interaction with a reservoir. The two terms have the apparent meaning of a system gain–loss dynamics due to the interaction within the system and due also to the system interaction with the reservoir. It is characterized by the initial reservoir probability, $\rho_{R\alpha\alpha}(0)$. The most important result (following Van Hove) is to obtain this in the singular limit $\lambda \to 0,\ t \to \infty;\ \lambda^2 t$ finite. The random phase assumption is made at $t = 0$, only with $\rho(0) = \rho_S(0)\rho_R(0)$.

In the subsequent sections we will examine the validity of this and in particular ask on what time scale we may expect the dynamics to be obeyed by $\tilde{\rho}_{Sd}(t)$. A similar equation may be obtained for the off-diagonal elements of $\rho(t)$. For this see Peier (1972) and Louisell (1973). Eq. (3.50) contains $\rho_{R\alpha\alpha}(0)$, which may be taken as a thermodynamic equilibrium state for a large system. This then introduces a temperature as a parameter in the reservoir.

General results of this Pauli equation will be discussed in Chapter 5. The principal applications will be seen in later sections of this book, particularly in the discussion of quantum optics in Chapter 11. The reader should consult the fine book of Louisell (1973) and the early reviews of Agarwal (1973) and Haake (1973). We will use this in Chapter 19 to discuss boundary scattering and the Landauer theory.

In the more chemically oriented area, the book of Oppenheim (Oppenheim *et al.*, 1977) is a must. This reprint volume contains many valuable articles, including those of Zwanzig and Van Hove, as well as others contributing to chemical physics relaxation phenomena. Of particular interest is the discussion by Oppenheim of the formal solutions to finite dimensional master equations. For this, the more recent book of Gardiner (1985) also should be consulted. Gardiner's handbook is extremely useful to anyone working in stochastic processes, no matter the topic. It is not our purpose to turn to this arena but rather to continue the discussion of the derivation of the quantum Pauli equation.

3.4 Pauli equation: time scaling

The Van Hove $\lambda^2 t$ limit leads from the "exact" generalized master equation, Eq. (3.34), to the weak coupling Pauli equation. This is similar to the singular Grad limit (Grad, 1958) in the derivation of the Boltzmann equation from the classical hierarchy. Some comments will be made on this in the next chapter. Here it is important to ask on what real physical time scale the Pauli equation holds (Peier, 1972; Davies, 1974; Davies, 1976; Middleton and Schieve, 1977).

The considerations of simple decay models such as that of Friedrichs (1948) and other exact results (Goldberger and Watson, 1964; Horwitz and Marchand, 1967; Middleton and Schieve, 1973) make it clear that the decay of $G(t - t')$ in the generalized master equation cannot be only exponential in $0 \leq t \leq \infty$, thus guaranteeing its Markovianization. It is at least not exponential. Two time scales exist: one τ_B as $t \to \infty$, and the other τ_c as $t \to 0$. The lower limit was examined by Horwitz and Marchand (1967). They argue that near $t = 0$, we may neglect the time integral in Eq. (3.34). In the remaining term, $\rho_S(0)$ is diagonal, and $PL'P = 0$. Thus $\rho_S(t)$ near $t = 0$ is time independent, and there can be no exponential decay.

The long-time behavior is difficult to treat and subject to much consideration. Qualitatively, in decay-scattering models, the energy E^0 is bounded from below ($E^0_m = 0$), and a branch cut must appear in the Laplace transform space of the resolvent (Goldberger and Watson, 1964). A power law decay $t^{-\frac{3}{2}}$ results as $t \to \infty$. This is a manifestation of the Paley–Weiner theorem (see Chapter 17).

To understand this in more detail, let us consider the generalized master equation for a simple Friedrich's model (Middleton and Schieve, 1973). We have then a discrete state $|E\rangle$ and a continuum $|\omega\rangle$, such that

$$\langle E \mid E\rangle = 1, \langle \omega \mid \omega' \rangle = \delta(\omega - \omega') \qquad \langle E \mid \omega\rangle = 0. \tag{3.51}$$

In this basis and for an isolated system here considered, the eigen representation is $H = H^0 + H'$, where

$$H^0 = E\,|E\rangle\,\langle E| + \int d\omega\, \omega\,|\omega\rangle\,\langle\omega| \tag{3.52}$$

and $H' = V(\omega)\,|\omega\rangle\,\langle E|$. The tetradic representation (see Eq. (3.7)) of $L' = [H']$ is

$$\begin{aligned} L'_{\omega\mu E\nu} &= V(\omega)\delta(\mu-\nu). \\ L'_{\nu E\mu\omega} &= -V(\omega)\delta(\mu-\nu) \\ L'_{\omega EEE} &= V(\omega) \\ L'_{EEE\omega} &= -V(\omega) \\ L_{abcd} &= L^*_{cdab.} \end{aligned} \tag{3.53}$$

For the isolated system, the relevant diagonal projection tetradic operator is

$$\begin{aligned} P_{EEEE} &= 1 \\ &= 0 \ \text{ otherwise.} \end{aligned}$$

This projects to the sole diagonal density matrix ρ_{EE} for the discrete state $|E\rangle$, and thus it is the probability to be in $|E\rangle$.

We write the Laplace transform of the kernel of the generalized master equation as

$$\hat{G}(z) = PL\mathfrak{P}(z)LP \tag{3.54}$$

where

$$\mathfrak{P}(z) = Q(z - QLQ)^{-1}Q$$

and

$$Q = 1 - P. \tag{3.55}$$

We easily obtain coupled equations for the tetradic matrix elements of $\mathfrak{P}(z)$. This is the merit of this simple model.

The relevant matrix element equations are

$$\mathfrak{P}_{E\mu\nu E} = -V^*(\mu)\Delta_1(\mu)\left[\frac{V^*(\nu)\Delta_2(\nu)}{z+\mu-\nu} - \int d\omega K_2(\mu,\omega)V(\omega)\mathfrak{P}_{E\omega\nu E}\right] \quad (3.56)$$

$$\mathfrak{P}_{\mu EE\nu} = -V(\mu)\Delta_2(\mu)\left[\frac{V(\nu)\Delta_1(\nu)}{z+\nu-\mu} - \int d\omega K_1(\mu,\omega)V^*(\omega)\mathfrak{P}_{\omega EE\nu}\right], \quad (3.57)$$

where

$$K_1(\mu,\omega,z) \equiv \int d\eta \frac{\mid V(\eta)\mid^2 \Delta_1(\eta,z)}{(z+\eta-\mu)(z+\eta-\omega)} \quad (3.58)$$

$$K_2(\mu,\omega,z) \equiv \int d\eta \frac{\mid V(\eta)\mid^2 \Delta_2(\eta,z)}{(z+\mu-\eta)(z+\omega-\eta)} \quad (3.59)$$

and

$$\Delta_1(\mu,z) \equiv \left[z+\mu-E-\int d\omega \frac{\mid V(\omega)\mid^2}{z+\mu-\omega}\right]^{-1} \quad (3.60)$$

$$\Delta_2(\mu,z) \equiv \left[z+E-\mu-\int d\omega \frac{\mid V(\omega)\mid^2}{z+\omega-\mu}\right]^{-1}. \quad (3.61)$$

Now the other tetradic matrix elements of $\mathfrak{P}(z)$ are simply related to the previous six equations. We have, for instance,

$$P_{E\xi\eta\mu} = (z+\mu-\eta)^{-1}[V^*(\eta)\mathfrak{P}_{E\xi E\mu} - V(\mu)\mathfrak{P}_{E\xi\eta E}].$$

We may obtain solutions to the integral equations (3.57) and (3.58) if we factorize the kernel, K. To proceed further, we simplify the spectrum of $|\omega\rangle$. We take

$$H^0 = E\,|E\rangle\,\langle E| + \int_{-\infty}^{+\infty} d\omega\omega\,|\omega\rangle\,\langle\omega|\,. \quad (3.62)$$

This assumption from the point of view of our earlier discussion removes the branch cut and power law decay as $t \to \infty$ (Goldberger and Watson, 1964). However, certain features remaining in the calculation will still play a similar role as $t \to \infty$. Assume further that $|V(\omega)|^2$ is a Lorentzian:

$$|V(\omega)|^2 = \frac{g^2\gamma^3}{4\pi}\;\frac{1}{(\omega-E)^2+\gamma^2}, \quad (3.63)$$

where

$$g^2 \equiv 4\pi\frac{\lambda}{\gamma} \quad (3.64)$$

is the dimensionless height-to-width ratio of the interaction of the single-level $|E\rangle$ with the continuum "field" $|\omega\rangle$. These parameters will scale the time dependence.

Eq. (3.60) and Eq. (3.61) become

$$\Delta_1(\omega, z) = \frac{z + \omega - E + i\gamma}{(z + \omega - E + i\alpha)(z + \omega - E + i\beta)} \tag{3.65}$$

$$\Delta_2(\omega, z) = \frac{z + E - \omega + i\gamma}{(z + E - \omega + i\alpha)(z + E - \omega + i\beta)}, \tag{3.66}$$

where

$$2\alpha \equiv \gamma \left[1 - (1 - g^2)^{\frac{1}{2}}\right] \tag{3.67}$$

and

$$2\beta \equiv \gamma \left[1 + (1 - g^2)^{\frac{1}{2}}\right].$$

Now

$$K_1(\omega, \nu, z) \equiv (z + 2i\gamma)h(z) f_1(\omega, z) f_1(\nu, z) \tag{3.68}$$

$$K_2(\omega, \nu, z) \equiv (z + 2i\gamma)h(z) f_2(\omega, z) f_2(\nu, z),$$

where

$$f_1(\omega, z) \equiv (z + i\gamma + E - \omega)^{-1} \tag{3.69}$$

$$f_2(\omega, z) \equiv (z + i\gamma + \omega - E)^{-1}$$

$$h((z) \equiv \alpha\beta(z + i\gamma + i\alpha)^{-1}(z + i\gamma + i\beta)^{-1}.$$

The solution of the integral equation for $\mathfrak{P}_{\mu EE\nu}$ is now found with the factored kernel $K_1 K_2$. We multiply Eq. (3.57) by $V^*(\mu) f_1(\mu z)$ and integrate on μ to obtain $\int_{-\infty}^{+\infty} d\mu V^*(\mu) f_1(\mu z)\mathfrak{P}_{\mu EE\nu}$. Substituting again in Eq. (3.57), we obtain the solution

$$\mathfrak{P}_{\mu EE\nu} = -V(\mu)V(\nu)\Delta_1(\nu)\Delta_2(\mu) \times \left[\frac{1}{z + \nu - \mu} + (2 + i\gamma)\frac{f_2(\nu) f_1(\mu)h^2(z)}{1 - h^2(z)}\right]. \tag{3.70}$$

Similarly,

$$\mathfrak{P}_{E\mu\nu E} = -V^*(\nu)V^*(\mu)\Delta_2(\nu)\Delta_1(\mu) \times \left[\frac{1}{z + \mu - \nu} + (z + 2i\gamma)\frac{f_1(\nu) f_2(\mu)h^2(z)}{1 - h^2(z)}\right]. \tag{3.71}$$

From Eq. (3.70) and Eq. (3.71), Eq. (3.54) becomes

$$G(z)_{EEEE} = 2\phi(z), \tag{3.72}$$

where

$$\phi(z) = \frac{\alpha\beta(z + i\gamma)}{(z + i\mu_+)(z + i\mu_-)} \tag{3.73}$$

and

$$2\mu_\pm = 3\gamma \pm \gamma(1 - 2g^2)^{\frac{1}{2}}.$$

Eq. (3.72) and Eq. (3.73) are the important results. We hope the reader has followed this solution for this simple model. It is one of the few.

Let us comment on the analytic results. $G(z)$ is analytic in the half plane Im (z) > Re $(\mu) \geq 0$. $G(t)$ will decay to zero as $t \to \infty$ unless the interaction amplitude $\gamma = 0$. We also note that

$$G(z = 0) = -2i\pi\lambda\gamma(\gamma + \pi\lambda)^{-1}. \tag{3.74}$$

The time-dependent decay may now be examined. Assume

$$\rho_{EE}(0) = 1, \;\; Q\rho(0) = 0. \tag{3.75}$$

Then, from Eq. (3.72),

$$\begin{aligned} \rho_{EE}(t) &= -\frac{1}{2\pi i}\int_c \frac{dz \exp(-izt)}{z - G(z)} \\ &= (\beta - \alpha)^{-2}[\beta^2 \exp(-2\alpha t) + \alpha \exp(-2\beta t) - 2\alpha\beta \exp(-\gamma t)]. \end{aligned} \tag{3.76}$$

For $g^2 < 1$ (weak coupling!), α, β are then positive real numbers, and for $t >> \gamma^{-1}$ the solution approximates

$$\rho_{EE}(t) = \beta^2(\beta - \alpha)^{-2} \exp(-2\alpha t), \tag{3.77}$$

a simple exponential decay.

Now we note that in the constant interaction limit $\gamma \to \infty$,

$$G(z) = -2\pi i\lambda \, |V(\omega = E)|^2 . \tag{3.78}$$

The exact dynamics at all time become the Pauli master equation dynamics at all time. This suggests further introducing time τ_c of collision duration. We Fourier transform $|V(\omega)|^2$:

$$\int_{-\infty}^{+\infty} dk \, |V(k)|^2 \exp(-ikt) = 2\pi\gamma \exp(-\gamma t)$$

and define $\tau_c = \gamma^{-1}$ as the interaction duration time scale. Take $\lambda^2\Gamma \equiv 2\pi\lambda^2 \, |V(\omega)|^2$ as the transition rate of the Pauli equation. Call this time scale $\tau_B = (\lambda^2\Gamma)^{-1}$ the relaxation time scale. It is apparent that we expect the Pauli-like dynamics for $2\tau_c/\tau_B << 1$ as $\gamma \to \infty$, $\tau_c \to 0$.

For $g^2 > 1$ the time dependence is more complicated:

$$\rho_{EE}(t) = \frac{\exp(-\gamma t)}{G^2}\left[\gamma G \sin Gt + (G^2 - \gamma^2)\cos Gt + \frac{1}{2}(G^2 + \gamma^2)\right]. \quad (3.79)$$

Here $G = \gamma(g^2 - 1)^{\frac{1}{2}}$. The solution is damped oscillations, not simple exponential, as in Eq. (3.77). If $\lambda \to \infty$ or $\gamma \to \infty$, such that $\gamma\lambda$ = constant, $\rho_{EE}(t) = \cos^2(\pi\lambda\gamma t)$. For $g^2 > 1$ there is no simple decay at long time. This is a strong coupling manifestation of a new behavior analogous to branch cut time dependence for this model.

We now return to the discussion of open systems where the projection is that introduced by Eq. (3.32) (Middleton and Schieve, 1977). Let us focus on a time asymptotic equation of the form

$$\frac{d\tilde{\rho}_S(t)}{dt} = -iL_S\tilde{\rho}_S(t) + \int_0^\infty d\tau G(\tau)\tilde{\rho}_S(t-\tau). \quad (3.80)$$

Here we have simply made the assumption $t \to \infty$ in the limit of the integral. We are also considering $\tilde{\rho}_S$ rather than $\tilde{\rho}_{Sd}$. Assume formally

$$\tilde{\rho}_S(t) = \exp(-i\Theta t)\tilde{\rho}_S(0) \quad (3.81)$$

and an operator equation for Θ results,

$$\Theta = L_S + \int_0^\infty d\tau \exp(-i\Theta\tau). \quad (3.82)$$

An iterated equation for Θ was first obtained by Résebois (Prigogine and Résebois, 1961). If Θ is a scalar, then this expansion is the Lagrange expansion

$$\Theta = \sum_{n=1}^{\infty} \frac{1}{n!}\frac{d^{n-1}}{dz^{n-1}} G(z)\,|_{z=0}\,. \quad (3.83)$$

This is so for the Friedrich's model already discussed, since Θ there commutes with its derivatives with respect to z. We may define

$$i\frac{d\tilde{\rho}_S(t)}{dt} = \Theta\tilde{\rho}_S. \quad (3.84)$$

Claude George (Prigogine *et al.*, 1969) showed that Eq. (3.54) is an exact projection, $\Pi\rho$, of the von Neumann equation, Eq. (3.4), where

$$\Pi^2 = \Pi$$
$$\Pi L = L\Pi.$$

(For the details, see Prigogine *et al.*, 1973.) See also Balescu (1975) and Schieve (1974). This will also be discussed in Chapter 18.

We may write a perturbation expansion of Θ, $\Theta = \sum_{n=1}^{\infty} \lambda^2 \Theta_{2n}$. Let us assume the reservoir time scale to be simply

$$\begin{aligned} \delta_{\tau_c} &= \frac{1}{\tau_c} \quad \text{for } t \leq \frac{\tau c}{2} \\ &= 0 \quad \text{otherwise.} \end{aligned} \tag{3.85}$$

We take the reservoir to be a free field correlation function:

$$\langle F_1(t_1) F_2(t_2) \rangle = \mathrm{Tr} F(t_1) F(t_2) \rho_R(0) = F^2 \delta_{\tau_c}(t_1 - t_2).$$

We further assume a Gaussian factorization of the higher order correlation functions. These (and also $< \Theta >$) are completely determined by the two-point-in time correlation function above. This is discussed in detail by Middleton (Middleton and Schieve, 1977). It is found that

$$\left\| \Theta_{2n}^{\tau_c} \right\| \leq < F >^2 \|[V_S,]\|^{2n} \left(\frac{\tau_c}{2} \right)^{n-1} d_n; \qquad n \geq 1. \tag{3.86}$$

Here d_n is a numerical factor close to unity. We define τ_B, the relaxation time, as

$$\tau_B^{-1} = \lambda^2 < F^2 > \|[V_S,]\|^2 . \tag{3.87}$$

This is the Pauli equation relaxation time estimate. Then we obtain the inequality

$$\lambda^{2n} \left\| \Theta_{2n}^{\tau_c} \right\| \leq \tau_B^{-1} (\frac{\tau_c}{\tau_B})^{n-1} d_n; \qquad n > 1. \tag{3.88}$$

The $n = 1$ term is the Pauli answer. We see that $\tau_c \to 0$ leads to this as an **exact** result. Also, if $\tau_B \to \infty$, we obtain the Van Hove limit. No proof has been possible concerning the convergence of the series for Θ in general. For the simple Friedrichs model, the convergence has been shown for the Lagrange expansion.

Finally, let us comment on the difficulty of the lower bound energy in decay scattering (Goldberger and Watson, 1964). Levy (1959) was the first to point out the existence of a power law non-exponential decay (Riley and Wiener, 1934). This does not exist in the Friedrichs model of this section, since we assumed $E^0 \to \infty$. In other cases an estimate of the time T, when the power law becomes comparable to the exponential decay, is

$$T \approx \frac{5}{\gamma} \ln \frac{E_0}{\gamma},$$

where $\gamma = \tau_B^{-1}$, the exponential time constant. For common values of γ, $T \approx 10 - 100$, and the power law is apparently unobservable. Hawker and Schieve have argued that at this time the amplitude is so small that it plays no role in the physical results of master equations and kinetic equations. This follows the 1975 unpublished University of Texas Ph.D. thesis of my student Kenneth Hawker, entitled "Contributions to Quantum Kinetic Theory."

3.5 Reservoir states: rigorous results and models

In Section 3.4 we suggested that the Pauli equation is obtained from the generalized master equation in the singular limits:

(i) $\tau_c \to 0$ zero memory
(ii) $\tau_B \to \infty$ Van Hove limit.

To a large measure, the difficulty remaining is to put reasonable conditions on the reservoir state $\rho_R(0) = 0$ to carry through a vigorous development of the argument. In his thesis, Middleton outlined and discussed a number of possible avenues. For one, the Gaussian factorization of the reservoir multitime correlation functions to $\langle F(t_1)F(t_2)\rangle$ may be obtained by the assumption of chaotic initial reservoir states for bosons:

$$\rho_R(0) = \rho_K, \tag{3.89}$$

where

$$\rho_K = \sum_{k=0}^{\infty} \left(\frac{\overline{n}k}{1 - \overline{n}k} \right) |nk >< nk| \,.$$

In quantum optics these represent states of a thermal source. With this in the infinite volume limit ($V \to \infty$), the argument carried through at the end of the previous section may be done. The zero memory limit is independent of the weak coupling assumption. It is thus true independent coupling strength between the system and reservoir. In the Friedrichs model, discussed in Section 3.4, the constant coupling limit corresponds to a white noise (zero memory) reservoir limit.

In Chapter 17 we will discuss the Friedrichs model formulated as an open system of two-level atoms interacting in the rotating wave approximation with the reservoir in the vacuum state.

Davies (1974, 1976) has given an impressive, rigorous proof of the $\lambda^2 t$ limit (case 2 at the beginning of this section). He assumed that the reservoir state is represented by correlation functions for an infinite free Fermi system in equilibrium. This requires the existence of certain integrals over time correlation functions, where he exploited the properties of the Volterra integral equation. These methods were adopted by Middleton (Middleton and Schieve, 1977). Frigerio and coauthors (Frigerio *et al.*, 1976) have obtained the zero memory limit (case 1) using a weak coupling argument similar to Davies's.

3.6 The completely positive evolution

It has been suggested that there is a class of physically relevant and mathematically interesting semigroup transformations termed *completely positive*. This was introduced by Kraus (1971) and later developed by others (Davies, 1976; Gorini *et al.*, 1976; Lindblad, 1976). The focus of our discussion will be to show how the Lindblad or Kossakowski quantum master equation is obtained. We will also discuss some of its properties and recent applications. We will follow a very readable and nonrigorous discussion by George Sudarshan (1991). There is also a review by Gorini (Gorini *et al.*, 1978).

Consider the dynamic linear map $\rho \to \rho'$, where

$$\begin{aligned} \rho' &= \int d\alpha \varepsilon(\alpha) B(\alpha) \rho B^{\dagger}(\alpha) \\ B^{\dagger} &= B \\ \varepsilon^2(\alpha) &= 1 \\ \text{and} \quad \int d\alpha \varepsilon(\alpha) B(\alpha) B^{\dagger}(\alpha) &= 1. \end{aligned} \tag{3.90}$$

This utilizes the diagonal representation of the operator B in the Stieltje's integral.

Complete positivity is defined as

$$\varepsilon(\alpha) = 1 \qquad \textbf{all } \alpha. \tag{3.91}$$

For a discrete spectrum then,

$$\rho' = \sum_{\alpha} B^{+}(\alpha) \rho B(\alpha). \tag{3.92}$$

A tetradic representation is

$$\rho_{rs} = \sum_{r's'} B_{rr',ss'} \rho_{r's'} \tag{3.93}$$

with

$$B_{rr',s's} = \int d\alpha B_{rr'}(\alpha) B^{*}_{s's}. \tag{3.94}$$

Complete positivity is a stronger condition than positivity. Not many physical examples have been obtained, but considerable attention is being given to this now, since it represents a method of quantizing dissipative systems. The simplest mathematical example is

$$\begin{aligned} \rho' &= V \rho V^{\dagger} \\ V^{\dagger} V &= 1, \end{aligned} \tag{3.95}$$

where

$$V V^{\dagger} = 1 - \Delta.$$

The V is an isometry familiar in scattering theory as the Möller wave operator (Goldberger and Watson, 1964).

Let us consider an extended $\mathfrak{H}$ built of the product $\mathfrak{H} \otimes \mathfrak{H}_R$, $\mathfrak{H}$ being the system Hilbert space and $\mathfrak{H}_R$ that of the reservoir. In the extended space, we assume unitary time evolution:

$$\rho' = V\rho V^\dagger \qquad (3.96)$$
$$\dot{\rho} = [\rho, H].$$

In $\mathfrak{H}$ consider the isometric map

$$(\rho \times \sigma)' = V(\rho \times \sigma)V^\dagger; \qquad V^\dagger V = 1. \qquad (3.97)$$

σ is the density operator of the reservoir in $\mathfrak{H}_R$. Now we trace over $\mathfrak{H}_R$ and assume the reservoir is diagonal in its ground state:

$$\sigma_1 = 1$$
$$\sigma_n = 0; \qquad n \neq 1.$$

Then

$$\rho' = \mathrm{Tr}_R(V(\rho \times \sigma)V^\dagger) = \sum_{\alpha B} V_{\alpha B}\rho\sigma_{\alpha B}(V_{\alpha B})^\dagger = \sum_\alpha V_{\alpha 1}\rho V_{\alpha 1}^\dagger. \qquad (3.98)$$

Here $V(\alpha) = V_{\alpha 1}$.

This is a completely positive map of ρ in $\mathfrak{H}$ and the result of the assumption on the reservoir state. We time evolve $V(\alpha)$ in $\mathfrak{H}$:

$$V_{\alpha 1}(t) = \exp(-itH)V_{\alpha 1}(0), \quad t \geq 0, \qquad (3.99)$$

a Heisenberg semigroup evolution. Then, to second order in t,

$$\rho'(t) = \rho(0) - itH_{11}\rho + it(H_{11})^\dagger \qquad (3.100)$$
$$+ \frac{(it)^2}{2!}\left[\sum_\beta H_{1\beta}H_{\beta 1}\rho + \sum_\beta \rho H_{\beta 1}^\dagger H_{1\beta}^\dagger\right] - it^2 \sum_\alpha H_{\alpha 1}\rho H_{\alpha 1}^\dagger.$$

We can rewrite Eq. (3.100) as

$$\rho' = \rho - it[h, \rho] + \frac{(it)^2}{2!}[h, [h, \rho]] - t\sum_\alpha [L_\alpha^\dagger L_\alpha \rho + \rho L_\alpha^\dagger L_\alpha - 2L_\alpha^\dagger \rho L_\alpha], \quad (3.101)$$

defining

$$h = H_{11} \qquad (3.102)$$
$$L_\alpha = t^{\frac{1}{2}} H_{\alpha 1}; \qquad \alpha > 1.$$

We assume L_α to be defined as $t \to 0$. The remaining t^2 term is neglected in this limit. Eq. (3.102) gives

$$\dot{\rho} = -i[h, \rho] + \sum_\alpha [L_\alpha^\dagger, \rho]L_\alpha + L_\alpha^\dagger[\rho, L_\alpha]. \tag{3.103}$$

This is the Lindblad–Kossakowski equation for the completely positive semigroup time evolution (Lindblad, 1976; Gorini and Kossakowski, 1976). In this heuristic derivation, much has been assumed. The reader should consult the references for more complete treatment.

If a system obeys completely positive semigroup, then it will follow Eq. (3.103). Gorini (Gorini *et al.,* 1978) discussed the two-level atom system. We will not write down the map. They defined a polarization vector M_i and derived Bloch equations for these quantities:

$$\dot{M}_i(t) = \sum_{j,k=1}^{3} \varepsilon_{ijk} h_j (M_k(t) - M_k(0)) - \gamma_i (M_i(t) - M_i(0)). \tag{3.104}$$

$M_i(0)$ is the equilibrium state if $\gamma_1\gamma_2\gamma_3 > 0$. γ_i^{-1} are, of course, the relaxation times. The conditions for complete positivity imply

$$\gamma_1 + \gamma_2 \geq \gamma_3; \quad \gamma_2 + \gamma_3 \geq \gamma_1; \quad \gamma_3 + \gamma_1 \geq \gamma_2. \tag{3.105}$$

Take the magnetic field in the "3" direction. Then $M_1(0) = M_2(0) = 0$, and

$$\gamma_1^{-1} = \gamma_2^{-1} \equiv \gamma_\perp^{-1} = \mathfrak{T}_\perp$$
$$\gamma_3^{-1} = \gamma_\parallel^{-1} = \mathfrak{T}_\parallel,$$

defining the parallel and perpendicular relaxation times. The necessary and sufficient condition for complete positivity is

$$2\mathfrak{T}_\parallel \geq \mathfrak{T}_\perp. \tag{3.106}$$

This seems to be true experimentally (Haake, 1973). However, there is a recent exception (Weinstein *et al.,* 2004).

The alert student will note that the von Neumann equation for phase damping, Eq. (2.38), is of the completely positive type, having the positive solutions given there. We remark that the Fokker–Planck equation, Eq. (2.40), has a positive definite diffusion coefficient. This is as expected. The phase-damping model is a good example of the Lindblad dynamics. More will be said about completely positive dynamical evolution in later chapters on dissipative evolution, particularly in Chapter 6.

Appendix 3A: Chapman–Kolmogorov master equation

The Kolmogorov master equation is the stochastic mathematical basis of Pauli-like non-Markovian master equations. We will discuss this here briefly (Kolmogorov, 1950; Gardiner, 1985).

Consider particles in a state $|l\rangle$. l is a continuum. We introduce the conditional probability

$$h_1^{(l)}(12\ldots l)dl, \tag{3A.1}$$

being the probability that a particle is in "dl" near 1, given that 2 is in $|2\rangle$, 3 is in $|3\rangle$, etc. If the events are independent,

$$f_l(1,2\ldots l) = f_1(1)f_2(2)\ldots \text{where} \int f_1(i)di = 1,$$

then

$$h_1^l(1 \mid 2\ldots l) = f_1(1).$$

Other conditional probabilities are

$$h_2^l(1,2 \mid 3\ldots l) = \frac{f_l(1\ldots l)}{f_{l2}(3\ldots l)}.$$

Now we write

$$f_s dl\ldots ds = \prod_s(1,2\ldots t \mid 1_0\ldots s_{0,}t_0)dl\ldots ds. \tag{3A.2}$$

Let $z \equiv (1,2\ldots s)$. $\Pi(zt \mid z_0t_0)dz$ is the conditional probability of being in dz around z at t, given it was in dz_0 at t_0 with

$$\int dz\Pi(zt \mid z_0t_0) = 1 \tag{3A.3}$$

for an intermediate time t^1, $t_0 < t^1$, t.

$$\Pi(zt \mid z_0t_0) = \int \Pi(zt \mid z^1t^1; z_0t_0)\Pi(z^1t^1 \mid z_0t_0)dz^1. \tag{3A.4}$$

The Chapman–Kolmogorov equation, Eq. (3A.4), is rather subtle. The convolution depends on z_0, t_0. This is nonlincar and has memory of z_0, t_0 in $\Pi(zt \mid z^1t^1; z_0t_0)$ for all t^1. We neglect the memory and write

$$\Pi(zt \mid z_0t_0) = \int dz^1\Pi(zt \mid z^1t^1)\Pi(z^1t^1 \mid z_0t_0). \tag{3A.5}$$

This is the Markovian Chapman–Kolmogorov equation. It has many solutions. For a discrete basis, we let

$$\int dz \to \sum_l,$$

and then

$$\sum_{l} \Pi(z \mid z^{1}; t) = 1 \tag{3A.6}$$

$$\Pi(l \mid l_0, 0) = \delta_{ll^0}.$$

We have

$$\Pi(l \mid l_0; \tau + \Delta t) = \sum_{j} \Pi(l \mid j; \Delta t)\Pi(j \mid l_0; \tau). \tag{3A.7}$$

Now we assume near $\tau = 0$ that Δ is small and introduce the transition probability a_{ij}, $j \neq l$. Then $\Pi(l, j, \Delta t) = a_{lj}\Delta t, \quad l \neq j$, would be

$$\Pi(l \mid l; \Delta t) = 1 - \text{ probability that } l \nrightarrow l \text{ in } \Delta t = 1 - \Delta t \sum_{l \neq l'} a_{l'l}. \tag{3A.8}$$

Substituting this in Eq. (3A.7), we write

$$\frac{\Pi(l \mid l_0; t + \Delta t) - \Pi(l \mid l_0; t)}{\Delta t} = \sum_{j} [a_{lj}\Pi(j \mid l_0; t) - a_{jl}\Pi(l \mid l_0; t)],$$

or as $\Delta t \to 0$,

$$\frac{d\Pi(l \mid l_0; t)}{dt} = \sum_{j} [a_{lj}\Pi(j \mid l; t) - a_{jl}\Pi(l \mid l_0; t)]. \tag{3A.9}$$

This is the differential Kolmogorov equation in a discrete space l. Pauli identified a_{ij} in the quantum case with the "golden rule" transition rate, as we have discussed.

A simple example is a Poisson process. Let

$$\begin{aligned} a_{ij} &= \lambda \qquad j = i + 1 \quad i = 0, 1, 2, \ldots \\ \alpha_{ij} &= 0 \qquad \text{otherwise.} \end{aligned}$$

We obtain

$$\frac{d\Pi_{ij}(t)}{dt} = -\lambda\Pi_{ij}(t) + \lambda\Pi_{ij-1}(t)$$

$$\text{or} \qquad \frac{d\Pi_{ij}(t)}{dt} = -\lambda\Pi_{ij}(t) + \lambda\Pi_{i+1,j}(t).$$

The solution to this set of Kolomogorov equations is

$$\begin{aligned} \Pi_{ij}(t) &= \frac{(\lambda t)^{j-1}}{(j-1)!}\exp(-\lambda t) \qquad j \geq i \\ \Pi_{ij}(t) &= 0 \qquad j < i \end{aligned}$$

with $\Pi_{ij}(0) = \delta_{ij}$.

References

Agarwal, G. (1973). *Progress in Optics* **11**, ed. E. Wolf (Amsterdom, North–Holland).
Argyres, P. N. and Kelley, P. L. (1964). *Phys. Rev.* **134**, 99.
Balescu, R. (1975). *Equilibrium and Non-equilibrium Statistical Mechanics* (New York, Wiley), rev. 1999 as *Matter out of Equilibrium* (London, Imperial College Press).
Davies, E. B. (1974). *Commun. Math. Phys.* **39**, 91.
Davies, E. B. (1976). *Quantum Theory of Open Systems* (New York, Academic Press).
Dirac, P. A. M. (1958). *The Principles of Quantum Mechanics*, 4th edn. (London, Oxford University Press).
Friedrichs, K. (1948). *Common, Pure Appl. Math* .**1**, 361.
Frigerio, A., Novellone, C., and Verri, M. (1976). Master equation treatment of the singular reservoir limit. In *Institute de Fisica dell Universita Milano IFUM* **183/ FT** (Milan).
Gardiner, C. W. (1985). *Handbook of Stochastic Methods*, 2nd edn. (Berlin, Springer).
Goldberger, M. L. and Watson, K. M. (1964). *Collision Theory* (New York, Wiley).
Gorini, V. and Kossakowski, A. (1976). *J. Math. Phys.* **17**, 1298.
Gorini, V., Kossakowski, A. and Sudarshan, E. C. G. (1976). *J. Math. Phys.* **17**, 821.
Gorini, V., Frigerio, A., Verri, M., Kossakowski, A. and Sudarshan, E. C. G. (1978). *Rep. Math. Phys.* **13**, 149.
Grad, H. (1958). Principles of the kinetic theory of gases. In *Encyclopedia of Physics,* ed. S. Flugge (New York, Springer).
Haake, F. (1973). *Tracts in Modern Physics* **66** (Berlin, Springer).
Horwitz, L. and Marchand, J. (1967). *Helv. Phys. Acta.*
Huang, K. (1987), *Statistical Mechanics*, 2nd edn. (New York, Wiley).
Kolmogorov, A. N. (1950). *Foundations of the Theory of Probability* (New York, Chelsea).
Kraus, K. (1971). *Ann. Phys. N.Y.* **64**, 311.
Levy, M., (1959). *Nuovo C.M.* **14**, 3516.
Lindblad, G. (1976). *Commun. Math. Phys.* **48**, 119.
Louisell, W. H. (1973). *Quantum Statistical Properties of Radiation* (New York, Wiley).
Middleton, J. and Schieve, W. C. (1973). *Physica* **64**, 139.
Middleton, J. and Schieve, W. C. (1977). *Int. J. Quantum Chem.* **11**, 625.
Montroll, E. W. (1960). *Lectures in Theoretical Physics* **3** (Boulder), 221. New York; Interscience.
Nakajima, S. (1958). *Prog. Theor. Phys.* **20**, 948.
Oppenheim, K., Shuler, K. and Weiss, G. (1977). *Stochastic Processes in Chemical Physics* (Cambridge, Mass., MIT Press).
Pauli, W. (1928). *Festschr. zum 60 Geburtstag A. Sommerfeld* (Leipzig, Hirzl).
Peier, W. (1972). *Physica* **57**, 229 and 565.
Peier, W. and Thellung, A. (1970). *Physica* **46**, 577.
Prigogine, I. (1963). *Non-equilibrium Statistical Mechanics* (New York, Interscience).
Prigogine, I. and Résebois, P. (1961). *Physica* **27**, 629.
Prigogine, I., George, C. and Henin, F. (1969). *Physica* **45**, 418.
Prigogine, I., George, C., Henin, F. and Rosenfeld. (1973). *Chemica Scripta* **4**, 5.
Reif, F. (1965). *Fundamentals of Statistical and Thermal Physics.* (New York, McGraw-Hill).
Résebois, P. (1961). *Physica* **27**, 541.
Riley, R. and Wiener, N. (1934). *Fourier Transform in the Complex Domain* (Providence, University of Rhode Island Press).

Schieve, W. C. (1974). Aspects of non-equilibrium quantum statistical mechanics: an introduction, in *Lectures in Statistical Physics*, ed. W. C. Schieve and J. S. Turner, in *Lecture Notes in Physics* **28** (Berlin–Heidelberg–New York, Springer).
Schleich, W. (2001). *Quantum Optics in Phase Space* (Berlin, Wiley–VCH).
Sudarshan, E. C. G. (1991). *J. Math. Phys. Sci.* **25**, 573.
Swenson, R. (1962). *J. Math. Phys.* **3**,1017.
Van Hove, L. (1955). *Physica* **21**, 517.
Van Hove, L. (1957). *Physica* **23**, 441.
Van Hove, L. (1962). *Fundamental Problems in Statistical Mechanics,* ed. E. G. D. Cohen (Amsterdam, North-Holland).
Weinstein, Y. S., Havel, T. F., Emerson, J., Boulant, N., Saraceno, M., Lloyd, S., and Cory, D. J. (2004). *J. Chem. Phys.* **121**, no. 13, 6117.
Wigner, E. P. (1932). *Phys. Rev.* **40**, 749.
Zwanzig, R. (1960a). *J. Chem. Phys.* **33**. 1338
Zwanzig, R. (1960b). *Lectures in Theoretical Physics* **3**, 106 (Boulder, Colo., New York, Interscience).
Zwanzig, R. (1965). *Physica* **30**, 1109.

4
Quantum kinetic equations

4.1 Introduction

Now let us discuss the fundamental jewel of non-equilibrium statistical dynamics, the Boltzmann equation (Boltzmann, 1872). Of course, we will be discussing the quantum version of this equation and its structure, which shows a remarkable similarity to the original classical example. This fact alone would speak of the genius of the founder of statistical mechanics. We will also touch on the other fundamental equation of plasma, the Vlasov equation (Vlasov, 1938; Balescu, 1975), which again will be the quantum version.

What distinguishes these from the Markov master equations of the previous chapter? They are spacially inhomogeneous, thus necessitating the introduction of phase space distribution functions, $w(xp,t)$, into quantum mechanics. We shall do this in some detail in this chapter. This could also have been done earlier. Phase space distribution functions seem not to be a natural thing in quantum mechanics because of the noncommutivity of the position, x, and momentum, p. We will see that this is not necessarily the case.

4.2 Reduced density matrices and the B.B.G.Y.K. hierarchy

The method of presentation of the following is similar to that of K. Hawker in his unpublished 1975 University of Texas doctoral thesis, "Contributions to Quantum Kinetic Theory." The von Neumann equation for the density operator is

$$i\dot{\rho}_N = [H_{N_1}\rho_N] \equiv L_N\rho_N; \qquad \infty \leq t \leq \infty. \tag{4.1}$$

The N emphasizes that we have an N-body system where we assume

$$H_N = \sum_{i=1}^{N} H_i^0 + \sum_{i<j}^{N} V_j = \sum_{i=1}^{N} \frac{p_i^2}{2m} + \sum_{i<j}^{N} V_{ij}. \tag{4.2}$$

We make the simple assumption of structurally simple identical particles. At $t = 0$,

$$\rho(1, 2, \ldots, r, s, z, \ldots, N, 0) = \rho(1, 2, \ldots, s, z, r, \ldots, N, 0),$$

and we assume that there are no particles with internal structure. Because of H_N chosen here, this symmetry is propagated in time. We will discuss quantum exchange symmetries in a later section and show that the main features of the following discussion go unchanged. In a sense, we are dealing with quantum particles which are "Boltzons," to use a rather cryptic title.

The interaction potential is assumed to be a sum of pair potentials, V_{ij}, which depend on the scalar distance between the particle pairs. All masses are taken to be equal, although this is not necessary. The reduced density operator for a set of $s < N$ particles is $\rho_s = V^s \operatorname*{Tr}_{s+1\ldots N} \rho_N$. This is, of course, a reduction similar to that discussed in the chapter on master equations. The main point is that N-body observables do not depend on ρ_N but rather on simpler objects such as ρ_1, ρ_2 etc. It has not been possible to introduce projection operators to achieve Eq. (4.12). However, it is not really necessary here.

We shall form a hierarchy of the ρ_s. We trace over $(2 \ldots N)$ variables in Eq. (4.1) and obtain

$$i\dot{\rho}_1 = \left[H_1^0, \rho_1\right] + V \sum_{i<j}^{N} \operatorname*{Tr}_{(2\ldots N)} \left[V_{ij}, \rho_N\right].$$

We may show by the identical particle assumption

$$\sum_{1\leq i\leq j\leq N} \operatorname*{Tr}_{(2\ldots N)} \left[V_{ij}, \rho_N\right] = \sum_{2\leq j<N}^{N} \operatorname*{Tr}_{(2\ldots N)} \left[V_{ij}, \rho_N\right] = \frac{N-1}{V} \operatorname*{Tr}_{(2)} \left[V_{12}, \rho_{12}\right]. \quad (4.3)$$

We obtain

$$i\dot{\rho}_1 = L_1^0 \rho_1 + \frac{N-1}{V} \operatorname*{Tr}_{(2)} L'_{12} \rho_{12}, \quad (4.4)$$

where

$$\begin{aligned} L_1^0 &= \left[H_1^0, \right] \\ L'_{12} &= \left[V_{12}, \right] \end{aligned} \quad (4.5)$$

and

$$\rho_{12} = V^2 \operatorname*{Tr}_{(3\ldots N)} \rho_N, \quad (4.6)$$

being the next more complicated two-particle density operator. This may be carried on successively. Operating with $V^2 \operatorname*{Tr}_{(3\ldots N)} \rho_N$ on Eq. (4.1), we find

$$i\dot{\rho}_{12} = \left[H_{12}, \rho_{12}\right] + \frac{N-2}{V} \sum_{(3)} \operatorname*{Tr}_{(3)} \left[\left(V_{12} + V_{23}\right), \rho_{123}\right] \tag{4.7}$$

where $\rho_{123} = V^3 \operatorname*{Tr}_{(4\ldots N)} \rho_N$. This is the B.B.G.Y.K. (Born, Bogoliubov, Green, Yvon, Kirkwood) hierarchy first written by Yvon (1935). The general formula is obvious, and we do not need to write it here. The final equation is Eq. (4.1). As with the generalized master equation, this hierarchy is a beginning. Further analysis must be done to obtain closed kinetic equations.

At this point we will take the thermodynamic limit

$$N \to \infty,\ V \to \infty,\ \frac{N}{V} = n_0 = c < \infty, \tag{4.8}$$

obtaining to low order

$$i\dot{\rho}_1 = L_1^0 \rho_1 + n_0 \operatorname*{Tr}_{(2)} L'_{12} \rho_{12} \tag{4.9a}$$

$$i\dot{\rho}_{12} = L_{12} \rho_{12} + n_0 \operatorname*{Tr}_{(3)} \left[\left(L'_{13} + L'_{23}\right) \rho_{123}\right] \tag{4.9b}$$

$$i\dot{\rho}_{123} = L_{123} \rho_{123} + n_0 \operatorname*{Tr}_{(4)} \left[\left(L'_{14} + L'_{24} + L'_{34}\right) \rho_{1234}\right] \tag{4.9c}$$

and so on, with

$$L'_{ij} = \left[V_{ij}, \right] \qquad L_1^0 = \left[H_1^0, \right]$$
$$L_{12} = \left[H_1^0 + H_2^0 + V_{12}, \right] = L_{12}^0 + L'_{12}.$$

Taking the thermodynamic limit here is merely a convenience. We do not imply that such an infinite set of operator equations has been derived in a rigorous fashion, a formidable and daunting task. However, Gallavotti has proved the existence of time-dependent correlation functions for a one-dimensional classical B.B.G.Y.K. hierarchy (Gallavotti *et al.*, 1970). There are several subsequent reasons for this limit, such as the elimination of Poincaré recurrences still present in finite quantum systems with the use of almost periodic functions (Bocchieri and Loingier, 1957) and the introduction of asymptotic scattering states in the derivation of the quantum Boltzmann equation, discussed in the next section. The recurrence proof is given in the appendix to Chapter 6.

4.3 Derivation of the quantum Boltzmann equation

We will now use the first two elements, Eqs. (4.9a) and (4.9b), and by an argument similar to that made by Bogoliubov and Green in the classical case, obtain the Boltzmann operator equation for $\rho(t)$ (Born and Green, 1949; Green, 1961; Bogoliubov, 1962; McLennan, 1989). First, we define the two-body correlation function g_{12} by

$$\rho_{12} = \rho_1 \rho_2 + g_{12}. \tag{4.10}$$

We write Eq. (4.9b) as

$$i\dot{g}_{12} = L'_{12} g_{12} + L_{12} \rho_1 \rho_2 - i \frac{d}{dt} (\rho_1 \rho_2) + 0(n_0).$$

With Eq. (4.9a) we obtain

$$i\dot{g}_{12} = L_{12} g_n + L'_{12} \rho_1 \rho_2 + 0(n_0).$$

We formally solve this equation to obtain

$$\begin{aligned} \rho_{12} &= \rho_1 \rho_2 \\ &\quad + \exp(-iL_{12}t) g_{12}(0) \\ &\quad - i \int_0^t d\tau \exp(-iL_{12}\tau) L'_{12} \rho_1 (t-\tau) \rho_2 (t-\tau). \\ 0 &< t < \infty \end{aligned} \tag{4.11}$$

Eq. (4.11) is not of definite order in n_0 because of the convolution term. Note the causality assumption and $0 < t < \infty$.

To obtain a Markovian result to lowest order in n_0, we may show in Eq. (4.11) that

$$\rho_1 (t-\tau) \rho_2 (t-\tau) = \exp\left(iL^0_{12}\tau\right) \rho_1(t) \rho_2(t). \tag{4.12}$$

We note that the density expansion has time localized, $\rho_1(t-\tau) \to \rho_1(t)$ in the time integration to order n_0. Now we also take the asymptotic limit $t \to \infty$. The justification for this is similar to the time scaling discussed in the previous chapter. Here the time scales are the duration of the two-body scattering collision (τ_c above) and τ_B, the Boltzmann relaxation time. Also, there is an additional longer time, a hydrodynamic time τ_H, which governs the rate of spacial homogeneity. This is not yet explicit. We assume $\tau_c < t \approx \tau_B < \tau_H$. These qualitative remarks are not at all rigorous. There are some mathematical results, and the interested reader should consult the mathematical literature.

We recognize a perfect differential in τ. After integration we obtain

$$\rho_{12}(t) = \exp(-iL_{12}t) g_{12}(0) + \exp(-iL_{12}t) \exp\left(iL^0_{12}t\right) \rho_1(t) \rho_2(t) + 0(n_0). \tag{4.13}$$

Using this in Eq. (4.9a), we obtain a closed nonlinear equation for $\rho_1(t)$:

$$i\dot{\rho}_1(t) = L_1^0\rho_1(t) + n_0 \underset{(2)}{\mathrm{Tr}}\left[L_{12}\lim_{t\to\infty}\exp(-iL_{12}t)\exp\left(+iL_{12}^0\right)\rho_1\rho_2\right] \quad (4.14)$$
$$+\, n_0 \underset{(2)}{\mathrm{Tr}}\left[L'_{12}\lim_{t\to\infty}\exp\left(-iL_{12}t\right)g_{12}(0)\right] + 0\left(n_0^2\right).$$

Eq. (4.14) is an operator form of the Boltzmann equation for $\rho_1(t)$ if we choose $g_{12}(0) = 0$. It is the lowest order in n_0 and nonlinear. If $g_{12}(0) = 0$, the initial correlations are zero and do not influence $\rho_1(t)$ at a later time. No assumption of $g_{12}(t) = 0$ at $t > 0$ is made in Eq. (4.14). This argument is a proof of Boltzmann's "*Stosszahlansatz*" classically and follows naturally from the methods of Bogoliubov and Green (Tolman, 1938; McLennan, 1989). Rigorous proof of this result has been made classically by Lanford (1969). A quantum proof of the *Stosszahlansatz* has been given by Petrosky and Schieve (1982) using the two-time resolvent approach of Van Hove (1955, 1957, 1962).

We must now interpret the remaining order n_0 term in Eq. (4.14). We let

$$K_{12}(t) = \exp\left(-iL_{12}t\right)\exp\left(iL_{12}^0 t\right)$$

and consider $\lim_{t\to\infty} K_n(t) \equiv K_{12}(\infty)$. We have in Eq. (4.14) $K_{12}(\infty)$. Now

$$K_{12}(t)\,\rho_1\rho_2 = \mathfrak{S}_{12}(t)\,G_{12}^\dagger(t)\,\rho_1\rho_2 G_{12}(t)\,\mathfrak{S}_{12}^\dagger(t), \quad (4.15)$$

where

$$\exp\left(-iL_{12}t\right)A = \exp\left(-iH_{12}t\right)A\exp\left(iH_{12}t\right),$$

where the Green's functions are

$$\mathfrak{S}_{12}(t) = \exp\left(-iH_{12}t\right) \quad (4.16)$$
$$G_{12}(t) = \exp\left(-iH_{12}^0 t\right).$$

Now we take

$$\lim_{t'\to\infty}\mathfrak{S}\left(t'\right)G_{12}^\dagger\left(t'\right)\rho_1\rho_2 G_{12}\left(t'\right)\mathfrak{S}_{12}^\dagger\left(t'\right)$$
$$= \left[\lim_{t'\to\infty}\mathfrak{S}\left(t'\right)G_{12}^\dagger\left(t'\right)\right]\rho_{'}\rho_2\lim_{t''\to\infty}\left[G\left(t''\right)\mathfrak{S}_{12}^\dagger\left(t''\right)\right],$$

provided the limits are independent on both sides of $\rho_1\rho_2$. Now the Möller scattering operators are

$$\Omega_{12} = \lim_{t\to\infty}\mathfrak{S}_{12}(t)\,G_{12}^\dagger(t) \quad (4.17)$$
$$\Omega_{12}^\dagger = \lim_{t\to\infty}G_{12}(t)\,\mathfrak{S}_{12}^\dagger(t),$$

where

$$\Omega^\dagger\Omega = 1 \quad \neq \Omega\Omega^\dagger \tag{4.18}$$
$$\Omega^\dagger H\Omega = H^0$$

(Taylor, 1972). The first emphasizes the isometric property of the Möller operator. Isometrics were introduced at the end of Chapter 3. We do not distinguish $\Omega_\pm = \lim_{t\to\pm\infty} \mathfrak{S}_{12}(t)\, G^\dagger_{12}(t)$ here, since we need only the positive time limit.

We have

$$K_{12}(\infty)\,\rho_1\rho_2 = \Omega_{12}\rho_1\rho_2\Omega^\dagger_{12}.$$

Eq. (4.14) becomes the *quantum operator* Boltzmann equation in the scattering theory form:

$$i\dot{\rho}_1 = L^0_1\rho_1 + n_0\mathrm{Tr}_{(2)}\left[V_{12}, \Omega_{12}\rho_1\rho_2\Omega^\dagger_{12}\right]. \tag{4.19}$$

We could also have derived it to n_0^2 including three-body effects, but that is not our purpose here.

In obtaining this result, we have used an implicit density expansion. The density expansion has been known for some time to have difficulties (McLennan, 1989), but not at the order at which we are using it here. Persistent time correlations appear in the n_0^2 order of the form $t^{-\frac{3}{2}}$, and this results in the transport coefficients not being analytic functions of the density. Re-summation of the n_0 expansion has to some extent alleviated the problem, but not entirely. The result is that the consistent theory of transport coefficients (a principal object of general solutions of the Boltzmann equation) has not yet been successful. The reader should see the text of McLennan (1989) for a clear discussion of this in the classical work.

4.4 Phase space quantum Boltzmann equation

Because many systems such as gases and fluids are spacially inhomogeneous, it is necessary, as in the classical approximation, to use a distribution function (like $F(Rp, t)$) language which is somewhat alien to quantum mechanics. This is a phase space distribution function which serves the purpose of the classical counterpart. Phase space distribution functions are defined by the requirement

$$\mathrm{Tr}O^{op}\rho = \int dRdpO(Rp)\,w(R, p). \tag{4.20}$$

Here $O(Rp)$ is the classical counterpart of O^{op}. We will discuss the phase space distribution functions quite generally in the appendix to this chapter, following the work of Cohen (Cohen, 1966; Margenau and Cohen, 1967). The Wigner function has been used almost exclusively in Eq. (4.20), and we shall do so also in this section (Wigner, 1932). Recently, Schleich (2001) has made extensive use of the Wigner phase space distribution function in phenomena of quantum optics. There the discussion is centered on quantum states in phase space, not on kinetic theory, which is the topic of interest here. $P(\alpha\alpha^*)$, which we already met in Chapter 2, plays a similar role to $w(R, p)$.

The relation of the Wigner function $w(\mathbf{R}, \mathbf{p})$ to the matrix elements of the density operator is the transform

$$w(\mathbf{R}, \mathbf{p}) = (2\pi)^{-3} \int dk' \exp\left(i\, \mathbf{R} \cdot \mathbf{k}'\right) \times < p + \frac{k'}{2} |\rho| \, p - \frac{k'}{2} >; \qquad \hbar = 1. \tag{4.21}$$

Eq. (4.21) is a Fourier transform of $\rho(p)$ with respect to a parameter k'. See the similarity to the ν, N representation mentioned in Chapter 3. From the general discussion in this chapter's appendix, notice that the marginal distributions are

$$\phi(p) = \int dRw(Rp), \text{ the momentum distribution function} \tag{4.22}$$

$$n(R) = \int dpw(Rp), \text{ the number density.}$$

We have dropped explicit vector notation for simplicity. As with other phase space distribution functions (see appendix),

$$w(Rp) \not\geq 0 \qquad \int dRdpw(Rp) = 1. \tag{4.23}$$

Thus, it is *not* a probability distribution in the classical sense. The fact that $w(R, p)$ is not a true classical distribution function is to be expected, as was noted in Chapter 1. It is a tool for calculating averages in phase space via the rule of Eq. (4.20), of which more is said in the appendix and in Chapter 6. There are many old and new references to the Wigner function in quantum mechanics. For a nice bibliography, see Schleich (2001).

There is an additional interesting theorem for a pure state:

$$\operatorname{Tr}(\rho_1\rho_2) = 2\pi \int_{-\infty}^{+\infty} dx \int_{-\infty}^{+\infty} dp \;\; w_{\rho_1}(xp)\, w_{\rho_2}(xp) = |< \Psi_1 | \, \Psi_2 >|^2 . \tag{4.24}$$

We recognize that the right side is the transition rate $1 \to 2$.

$$w_{\rho_1}(xp) = \frac{1}{2\pi} \int_{-\infty}^{=\infty} d\psi \exp(-ip\psi) \left\langle x + \frac{\psi}{2} \middle| \rho_1 \middle| x - \frac{\psi}{2} \right\rangle .$$

The proof is left as a problem.

Now we observe that

$$\mathrm{Tr}\,(\rho_1\rho_2) = 0$$

$$\int_{-\infty}^{+\infty} dx \int_{-\infty}^{+\infty} dp \;\; w_{\rho_1}(xp)\, w_{\rho_2}(xp) = 0,$$

and $w_{\rho_1}(xp)$ must take on negative values. We should mention that the Hudson–Piquet theorem states that the only nonnegative Wigner function is a Gaussian (Hudson, 1974; Piquet, 1974). We will use this fact in Chapter 6.

We also note from Eq. (4.23) that if $\rho_1 = \rho_2 = \rho$,

$$2\pi \int_{-\infty}^{+\infty} dx \int_{-\infty}^{+\infty} dp \quad w_\rho(xp)^2 \leq 1,$$

since $\mathrm{Tr}\rho^2 \leq 1$.

One of the first uses of the Wigner function to discuss hydrodynamic systems was in the papers of Irving and Zwanzig (1951) and Born and Green (1949). An early review of applications to the kinetic theory of gases is that of Mori, Oppenheim and Ross (1962). The algebra to transform the operator Boltzmann equation to one for the Wigner function is awkward. First, we write Eq. (4.19) in the momentum representation, $\rho_{\alpha\beta}$. Then we transform it to the Wigner function using

$$w\left(R, \frac{\alpha+\beta}{2}\right) = (2\pi)^{-3} \int d(\alpha-\beta) \exp\left(iR(\alpha-\beta)\right) \rho_{\alpha\beta}. \tag{4.25}$$

We change variables to

$$p = \frac{\alpha+\beta}{2}$$
$$p_1 = (\alpha - \beta)$$

and obtain

$$i\left[\frac{\partial w}{\partial t} + m^{-1}\mathbf{p}\cdot\nabla w\right] = (2\pi)^{-3}\int dp_1 d\gamma d\mu d\upsilon'$$
$$\times \exp\left(i\mathbf{p}_1\cdot\mathbf{R}\right)\begin{bmatrix} < p+\frac{p_1}{2},\gamma\left|V\Omega\right|\mu\upsilon >< \mu'\upsilon'\left|\Omega^\dagger\right| p-\frac{p_1}{2},\gamma > \\ - < p+\frac{p_1}{2},\gamma\left|\Omega\right|\mu\upsilon >< \mu'\upsilon'\left|\Omega^\dagger V\right| p-\frac{p_1}{2},\gamma > \rho_{\mu'\mu}\rho_{\upsilon\upsilon'}\end{bmatrix}. \tag{4.26}$$

We may eliminate $\rho_{\mu'm'}$ and $\rho_{\upsilon\upsilon'}$ by the inverse of Eq. (4.25), obtaining a nonlinear equation for w.

Next, we write

$$i\left[\partial_t w + m^{-1} p \cdot \nabla w\right] = 2\pi^{-3} \int dp_1 d\gamma d\mu \ldots d\nu' dr_1 dr_2$$
$$\times \left\{ \begin{array}{c} \exp\left(ip_1 \cdot R\right) \exp\left(-ir_1\left(\mu - \mu'\right)\right) \exp\left(-ir_2\left(\nu - \nu'\right)\right) \\ \left[\begin{array}{c} \left\langle p + \frac{p_1}{2},\ \gamma \left|V\Omega\right| \mu\nu\right\rangle \left\langle \mu'\nu' \left|\Omega^{\dagger}\right| p - \frac{p_1}{2},\ \gamma\right\rangle \\ -\left\langle p + \frac{p_1}{2},\ \gamma \left|\Omega\right| \mu\nu\right\rangle \left\langle \mu'\nu' \left|\Omega^{\dagger} V\right| p - \frac{p_1}{2},\ \gamma\right\rangle \end{array} \right] \\ \times w\left(r_1, \frac{\mu+\mu'}{2}\right) w\left(r_2, \frac{\nu+\nu'}{2}\right) \end{array} \right\}.$$

We change variables:

$$r_1 = R + \frac{x+y}{2} \qquad \mu = \frac{k'}{2} + k_1 \qquad \mu' = \frac{k_2'}{2} + k_2$$
$$r_2 = R + \frac{x-y}{2} \qquad \nu = \frac{k_1'}{2} - k_1 \qquad \nu' = \frac{k_2'}{2} - k_2\ .$$

We use the translational invariance of V and Ω. We then recognize the δ functions

$$\delta\left(p + \frac{p_1}{2} - \gamma - k_1'\right), \qquad \delta\left(p - \frac{p_1}{2} + \gamma - k_2'\right).$$

After doing the integrals on k_1', k_2', we again change variables:

$$q = \frac{p_1}{2}, \qquad k = \frac{k_1 + k_2}{2}, \qquad q_2 = k_1 - k_2$$

and obtain

$$i\left[\partial_t w + m^{-1} p_1 \nabla w_1\right] = 2\pi^{-3} \int d\gamma dq_1 dq_2 dk dx dy\ \exp -i\left(xq_1 + yq_2\right)$$
$$\left\{ \begin{array}{c} \left\langle \frac{p-\gamma+q_1}{2} \right| V\Omega \left| k + \frac{q_2}{2} \right\rangle \left\langle k - \frac{q_2}{2} \right| \Omega^{\dagger} \left| \frac{p-\gamma+q_1}{2} \right\rangle \\ -\left\langle \frac{p-\gamma+q_1}{2} \right| \Omega \left| k + \frac{q_2}{2} \right\rangle \left\langle k - \frac{q_2}{2} \right| \Omega^{\dagger} V \left| \frac{p-\gamma-q_1}{2} \right\rangle \end{array} \right\}$$
$$\times w\left(R + \frac{x+y}{2}, \frac{p+\gamma}{2} + k\right) w\left(R + \frac{x-y}{2}, \frac{p+\gamma}{2} - k\right).$$

Now $V^{\dagger} = V$, so the bracket {} is the difference between the term and its complex conjugate. Defining $(P - \gamma)/2 = q$, we finally have

$$\partial_t w + m^{-1}\mathbf{p} \cdot \nabla w = \mathrm{Im} \int \ldots \int dq dq_1 dq_2 dk dx dy$$
$$\times \left\{ \begin{array}{c} \exp\left(-i\left(xq_1 + yq_2\right)\right) J\left(q_1 k_1 \frac{q_1}{2} \frac{q_2}{2}\right) \\ \times w\left(R + \frac{x+y}{2},\ p - q + k\right) w\left(R + \frac{x-y}{2},\ p - q - k\right) \end{array} \right\}, \tag{4.27}$$

where

$$J\left(qk\frac{q_1}{2}\frac{q_2}{2}\right) \equiv \left\langle q + \frac{q_1}{2} \right| V\Omega \left| k + \frac{q_2}{2} \right\rangle \left\langle k - \frac{q_2}{2} \right| \Omega^{\dagger} \left| q - \frac{q_1}{2} \right\rangle. \tag{4.28}$$

Eq. (4.27) is not yet the spacial form of the quantum Boltzmann equation but rather a generalization. It is nonlocal, and the J is not the collision cross section. This equation has been treated and transport effects discussed by Kenneth Hawker in his doctoral thesis.

Assume now that $w(R, p, t)$ varies negligibly over the range of particle interaction. x and y are connected in J by a Fourier transform, and the interaction is short range, rather like a Yukawa-type potential, $\exp\left(\frac{-\mu r}{r}\right)$. This makes possible the local homogeneity approximation. In Eq. (4.27) we have by a Taylor expansion

$$w\left(R + \frac{x+y}{2}, p - q + k\right) w\left(R + \frac{x-y}{2}, p - q - k\right)$$
$$\approx w(R, p - q + k)\, w(R, p - q - k).$$

Now the x and y integrations are done with the help of $\delta(q_1)\,\delta(q_2)$. We obtain

$$\frac{\partial w(R\ p)}{\partial t} + m^{-1}\mathbf{p}\cdot\nabla w(R\ p)$$
$$= \operatorname{Im} 2 \int \ldots \int dq dk \left[J(qk00)\, w(R\ p - q + k)\, w(R\ p - q - k)\right]. \tag{4.29}$$

It remains to relate $\operatorname{Im} J(qk00)$ to the scattering cross section. Considering scattering theory (Taylor, 1972), we may define the T operator, $T^{\dagger} = V\Omega^{\dagger}$, where the $\dagger$ means outgoing scattered wave $\left|\beta^{\dagger}\right\rangle = \Omega^{\dagger}|\beta\rangle$, and the operator Lippmann–Schwinger equation gives

$$\Omega^{\dagger} = 1 + G^{\dagger}T^{\dagger}. \tag{4.30}$$

$G^{\dagger}$ is the free particle Green operator. Thus, $\left\langle\alpha\left|\Omega^{\dagger}\right|\beta\right\rangle = \langle\alpha\mid\beta\rangle + \left(E_{\beta} - E_{\alpha} + i\varepsilon\right)^{-1}\left\langle\alpha\left|T^{\dagger}\left(E_{\beta}\right)\right|\beta\right\rangle$. The convention is that the T operator is in the same energy state as that to its right. Using Eq. (4.29), Eq. (4.30) and the adjoints, we have

$$J(qk00) = T_{qk}\delta(q - k) + T_{qk}T_{qk}^{*}\left(E_k - E_q - i\varepsilon\right)^{-1}.$$

Here $T_{qk} \equiv \langle q\,|T|\,k\rangle$. We may use the optical theorem (Taylor, 1972)

$$T_{qq} - T_{qq}^{*} = -2\pi i \int dk' \left|T_{qk^1}\right|^2 \delta(E_q - E_{k^1})$$

and

$$\lim_{\varepsilon\to 0}\left[\left(E_k - E_q - i\varepsilon\right)^{-1} - \left(E_k - E_q + i\varepsilon\right)^{-1}\right] = 2\pi i\delta\left(E_k - E_q\right)$$

to obtain

$$\operatorname{Im} J(qk00) = \pi\left[\left|T_{qk}\right|^2\delta\left(E_q - E_k\right) - \delta(q - k)\int dk^1 \left|T_{qk^1}\right|^2\delta\left(E_q - E_{k^1}\right)\right].$$

Eq. (4.30) becomes

$$\partial_t w + m^{-1} p \cdot \nabla w = 8\,(2\pi)^4 \int dq dk \tag{4.31}$$

$$\left\{ \begin{array}{c} |T_{qk}|^2 \,\delta\left(E_q - E_k\right) \\ \times \left[w\,(p-q+k)\,w\,(p-q-k) - w\,(p)\,w\,(p-2q)\right] \end{array} \right\} \quad t > 0.$$

When we introduce the quantum differential cross section

$$|T_{qk}|^2 = [m^2\,(2\pi)^4]^{-1} \frac{d\sigma}{d\Omega}\,(k \to q)$$

(Taylor, 1972) and change integration variables

$$\int d\mathbf{k} \to \int_0^{\infty} dk\; k^2 \int d\Omega_k,$$

we obtain a result which has exactly the same form as the classical Boltzmann equation with the replacement of the classical differential cross section by the quantum one. We can only be amazed at Boltzmann's genius in writing this equation. The later methods of calculating the transport coefficients, though difficult indeed, go through here in quantum mechanics (Chapman and Cowling, 1960). However, the question of the positivity of the Wigner function does enter, and this will be discussed in some detail in Chapter 6 when dissipation is considered. Uehling and Uhlenbeck (1933) first derived this equation, including exchange statistics, which we have not done here for simplicity, preferring to emphasize the connection to statistical dynamics and the von Neumann equation. The properties of Eq. (4.31) are identical to the classical case. The exchange scattering will be discussed in the next sections.

The equation is nonlinear and of the birth–death (gain–loss) form. The principal quantum effect is wave diffraction in the cross section. In the following table, we give a few numerical estimates of parameters for the case of helium and argon at 30 K at 0.1 atm:

	He	Ar
R	4×10^{-8} cm	6×10^{-8} cm
nR^3	1.2×10^{-3}	5×10^{-3}
$n\lambda_T^3$	8.34×10^{-5}	3.48×10^{-6}
$\frac{\lambda_T}{\lambda_{MFP}}$	4.9×10^{-3}	4.5×10^{-4}
$\frac{\lambda_T}{R}$	4.1×10^{-1}	8.8×10^{-2}

λ_T is the thermal DeBroglie wavelength $\frac{2\pi\hbar^2}{mkT}$, and $\lambda_{MFP} = \left(nR^2\right)^{-1}$ the mean free scattering distance. $\frac{\lambda_T}{\lambda_{MFP}}$ estimates the importance of diffraction in scattering, important in the two cases shown. $n\lambda_T^3$ estimates the importance of exchange scattering at these low densities and for heavy masses. nR^3 small validates the truncation of the hierarchy, leaving only the important binary collision effects. The last rows show that quantum dynamic diffraction, $\frac{\lambda_T}{R}$, is more important than statistical exchange in these cases. The values of R are taken from Farrar *et al.* (1973). A final comment here concerning phase space distribution functions is that it is possible to transform the exact von Neumann equation to the Wigner representation. The result is called the quantum Liouville equation.

Let us take the nth momentum moment of Eq. (4.31), forming $\langle p^n \rangle$ using Eq. (4.20):

$$\langle p^n \rangle = \int d_3 p^n w\,(R\;pt)\,.$$

We then have an equation in a symmetric form:

$$\begin{aligned}\partial_t \langle p^n \rangle + m^{-1}\nabla \int d_3 p\mathbf{p}\mathbf{p}^n w = 8\,(2\pi)^4 \int\int\int dqdkdp'\delta\left(E_q - E_k\right) \\ \times\left[\left|T_{qk}\right|^2 \left(p' - q\right)^n - \left|T_{kq}\right|^2 \left(p' - k\right)^n\right] \\ \times\left[w\left(p' + k\right) w\left(p' - k\right) - w\left(p' + q\right) w\left(p' - q\right)\right].\end{aligned} \tag{4.32}$$

Now the R dependence will be implicit in w. The right side vanishes for $n = 0, 1$. The latter case is true because of parity $\left|T_{-q-k}\right|^2 = \left|T_{qk}\right|^2$. The $n = 2$ case also vanishes because of this, and $q^2 = k^2$ from the kinetic-energy-conserving delta function. This is the well-known result that the Boltzmann equation conserves particle number, momentum, and one-particle energy. Thus, we can write, from the left side in these cases, the macroscopic hydrodynamic conservation laws (in a common notation):

$$\begin{aligned}&\partial_t \rho + \nabla \cdot (\rho \mathbf{u}) = 0 \\ &\rho \partial_t \mathbf{u} + \rho \mathbf{u} \cdot \nabla \mathbf{u} = \nabla \cdot \mathfrak{P} \\ &\rho \partial_t e + \rho \mathbf{u} \cdot \nabla \mathbf{e} = \nabla \cdot \mathbf{q} - \sum_{ij} \mathfrak{P}_{ij} D_{ij}.\end{aligned} \tag{4.33}$$

Here the strain tensor D_{ij} and pressure tensor $\mathfrak{P}_{ij}$ appear. Also $\rho = mn$, $\mathbf{u} = \rho^{-1} \langle \mathbf{p} \rangle$, e is the energy density, and $\mathbf{q}$ is the heat flux. These are not a complete set of equations unless we introduce the phenomenological (or derived) transport laws. $\mathbf{q} = -\Lambda \nabla T$, and

$$\mathfrak{P}_{ij} = 2\eta\left(Dij - \frac{1}{3} D_{ij}\delta_{ij}\right) + \xi D_{ij}\delta_{ij}. \tag{4.34}$$

Here Λ is the thermal conductivity, η the shear viscosity, and ξ the bulk viscosity. It is the main object of the Boltzmann kinetic theory (quantum or classical) to calculate these coefficients (Chapman and Cowling, 1960; Balescu, 1975; K. Hawker in his 1975 unpublished Ph.D. thesis; McLennan, 1989).

We will not undertake this in detail here, but rather consider a simple case as an example. Let us expand the Boltzmann equation around the local Maxwellian solution (to be discussed further in Chapter 6):

$$f^0(1) = \frac{n_0}{(2\pi mkT)^{\frac{3}{2}}} \exp\left[-\frac{m}{2kT}(\mathbf{V}-\mathbf{u})^2\right] \tag{4.35}$$

$\mathbf{u}$ is the mean particle velocity at R, and T the absolute temperature at R. This classical assumption linearizes the equation, which we write in the steady state, $\frac{\partial w}{\partial t} = 0$, as for the first correction w' :

$$\mathbf{v}\cdot\frac{\partial}{\partial \mathbf{r}_1} f^0(E_v) = \left[J f^0(1)\, w'(2) - J w'(2)\, f^0(1)\right]. \tag{4.36}$$

This is a linear inhomogeneous integral equation for $w'(1) \equiv f^0(1)(1+\Phi(1))$. The kernel contains the differential cross section. We wish to calculate the heat flow, $\mathbf{q}$:

$$\begin{aligned} \mathbf{q} &= \int w(\mathbf{v})\, \mathbf{v}\frac{m}{2} v^2 d_3 v \\ &= \int f^0(E_v)\, \Phi(v)\, \frac{m}{2} v^2 \mathbf{v} d_3 v. \end{aligned} \tag{4.37}$$

Now we may write to this order by means of Eq. (4.33)

$$\frac{\partial}{\partial r} f^0(E_v) = f^0(E_v)\, \frac{\frac{mv^2}{2} - h\nabla T}{k_B T^2}$$

(Huang, 1987), where h is the enthalpy per particle. Now $\langle v_z \rangle = 0$ and $\nabla T = (\nabla T)\, e_z$.

Let us make the Bhatnager, Krook, Gross (Bhatnager *et al.*, 1954) approximation to the Boltzmann collision kernel:

$$J_{BKG} = \nu(f^0 - w). \tag{4.38}$$

When $\nu \equiv \frac{1}{\tau^c}$, the collision frequency is $\nu = \int f^0 v \sigma d\Omega d_3 v$, σ being the differential cross section. Note that f^0 is the local Maxwellian. This approximation is deceptively simple and still contains many aspects of the Boltzmann equation itself. It implies, for instance, the proper relaxation to the equilibrium state and the preceding continuity equations. With this approximation we may easily solve for w (and Φ), thus obtaining the transport law

$$q_z = -\int d_3 v f^0 \tau \frac{1}{2} m \mathbf{v}^2 v_z \left(\frac{mv^2}{2kT} - \frac{5}{2} \right) \frac{v_z}{T} \frac{\partial T}{\partial z}. \tag{4.39}$$

Thus, the thermal conductivity is

$$\lambda = \frac{m^5 \tau}{6kT} \int d_3 v v^4 \left(\frac{m}{2kT} - \frac{5}{2} \right) f^0 = \frac{5}{2} \tau k T n_0. \tag{4.40}$$

Similarly, for the viscosity, we have

$$\eta = \frac{\tau m^5}{kT} \int d_3 v v_i^2 v_j^2 f^0 = \tau n_0 k T; \qquad \text{any } i, j. \tag{4.41}$$

In Eqs. (4.39) and (4.40), n is the number density. Immediately we see $\frac{\lambda}{\eta c_v} = \frac{5}{2}$. This is also true of the low-order approximation to the Chapman–Enskog solution, where J not J_{BKG} is used.

The above illustrates two important points. In the steady solution to the transport flux, we obtain from the Boltzmann equation (quantum and classical) the transport law (Eq. 4.39) and numerical estimates of λ and η. Through τ these are related to the quantum binary scattering cross section, σ. We will not discuss the full details of the Chapman–Enskog (or other procedures) for obtaining more exact results. See Chapter 6 for further discussion of the role of transport coefficients in irreversible thermodynamics.

Now consider exchange scattering, already mentioned (Taylor, 1972; McLennan, 1989). For identical particles, the proper Hilbert space is $\mathfrak{H}_\varepsilon$ of functions with the proper exchange symmetry. Let $\Pi\,(\alpha, \beta)$ be the exchange operator for particles α, β:

$$\Pi\,(\alpha\beta)\,\Psi = \varepsilon \Psi \tag{4.42}$$

$$\varepsilon = 1\,(-1) \qquad \text{for fermions } -1, \text{ bosons } +1.$$

Instead of space $\mathfrak{H}_\varepsilon$, it is more convenient to utilize the full Hilbert space $\mathfrak{H}$ and the projector

$$\mathcal{E} = \frac{1}{N!} \sum_{\Pi} \varepsilon^{\Pi} \Pi \tag{4.43}$$

$$\mathcal{E} = \mathcal{E}^2 = \mathcal{E}^\dagger,$$

Π being the permutation operator on N particles such that

$$\begin{aligned}\varepsilon^{\Pi} &= 1 \text{ boson} \\ &= \pm 1 \text{ fermion} \\ &\qquad \text{(even or odd permutations).}\end{aligned}$$

The density matrix on $\mathfrak{H}$ is now $\rho = \varepsilon \rho_e \varepsilon$ and $\rho\varepsilon = \varepsilon\rho$. The useful relation is that for any A,

$$\mathrm{Tr}_{\varepsilon} A = \mathrm{Tr} A \varepsilon. \tag{4.44}$$

The exchange symmetry requires a modification of the initial factorization assumption. Eq. (4.10) is now taken as

$$\rho_{12}(t_0) = \rho_1(1, t_0)\, \rho_1(2, t_0)\, [1 \pm \Pi(12)]. \tag{4.45}$$

In matrix elements, in the two-particle momentum representation, this is

$$\rho_{12}\left(p_1 p_2 \mid p_1' p_2'\right) = \rho_1\left(p_1 \mid p_1'\right) \rho_1\left(p_2 \mid p_2'\right) \pm \rho_1\left(p_1 \mid p_2'\right) \rho_1\left(p_2 \mid p_1'\right).$$

The arguments in the derivation of the operator or Wigner function Boltzmann equation are now as earlier. The collision part of Eq. (4.31) is now the same form,

$$J = 4\pi^2 \int d_3 p_1' d_3 p_2 \delta\left(E - E'\right) |T_{\varepsilon}|^2 \left[w\left(p_1'\right) w\left(p_2'\right) - w\left(p_1\right) w\left(p_2\right)\right], \tag{4.46}$$

where the scattering matrix is $T_{\varepsilon}\left(p \mid p'\right) = T\left(p \mid p'\right) \pm T\left(-p \mid p'\right)$. Here the optical theorem for T_{ε} has been used. The cross section $\sigma = f^2(\theta)$ is replaced by $\frac{1}{2}|f(\theta) \pm f(\pi - \theta)|^2$, $f(\theta)$ being the scattering amplitude for spherically symmetric scattering. This leads to characteristically quantum interference effects. In addition, the steady state (equilibrium, $J = 0$) must be invariant under Π, so w^0 is then the Bose–Einstein or Fermi distribution, rather than Boltzmann. The former give the well-known equilibrium results, which will be discussed later. However, we must note here that the Wigner function Boltzmann equation, Eq. (4.46), conserves single-particle energy, and thus one obtains Bose, Fermi and, in the limit, the Maxwellian distribution for equilibrium. This is not true for the exact hierarchy expansion, Eq. (4.19), where no spacial localization has been introduced. There we obtain $p = nkT\,[1 - nB\,(T)]$. $B\,(T)$ is the quantum second virial coefficient. This effect is properly taken into account by systematically treating the spacial delocalization or collisional transfer corrections (Thomas and Snider, 1970; K. Hawker, unpublished Ph.D. thesis, 1975).

The comments above are strictly for repulsive potentials where no bound states are present. If there are bound states, then a naive examination of collisional memory in the derivation of the Boltzmann equation is not possible. Both collisional memory and initial correlations (bound states) must be considered.

4.5 Memory of initial correlations

The evolution of initial correlations are given by the operator

$$D_1(t) = \mathrm{Tr}_{(2)}\left[V_{12,} \exp\left(-iH_{12}t\right) g_{12}(0) \exp\left(iH_{12}t\right)\right].$$

See Eq. (4.10) and the following material. Eq. (4.10) may be written (after considerable calculation) in terms of Wigner functions as

$$\begin{aligned} D(Rpt) = {} & 2i\,(4\pi)^{-3}\,\mathrm{Im}\int\cdots\int dp_1 dk_2 dk_3 d\gamma dx dy \\ & \times \exp\left(-ip_1(p+\gamma)\frac{t}{2m}\right)\exp\left(-ip_1\frac{x}{2}\right) \\ & \left[\left\langle\frac{p+\frac{p_1}{2}-\gamma}{2}\,|V\mathfrak{G}|\,k_2\right\rangle\left\langle k_3\,|\mathfrak{G}^\dagger|\,\frac{p-\frac{1}{2}-\gamma}{2}\right\rangle \exp -iy(k_2-k_3)\right. \\ & \left.\times\, G\left(R+\frac{x+y}{2}, R+\frac{x-y}{2}, \frac{p+\gamma+k_2+k_3}{2}, \frac{p+\gamma-k_2-k_3}{2}; 0\right)\right]. \end{aligned} \tag{4.47}$$

$\mathfrak{G}$ is the two-particle Green's function, and G the Fourier transform,

$$\begin{aligned} \langle ab\,|g(0)|\,cd\rangle = {} & \int dr_1 dr_2 \exp(-ir_1(a-c))\exp(-ir_2(b-d)) \\ & \times G\left(r_1r_2, \frac{a+c}{2}\frac{b+d}{2}; 0\right). \end{aligned}$$

In the preceding equations, r_1, r_2 are the spacial coordinates of two statistical particles, g their relative separation, and x the difference between the center of mass position and the point R. These expressions may be somewhat simplified in the spacially homogeneous approximation.

The weak coupling (Born) approximation for homogeneous systems to Eq. (4.47) is

$$\begin{aligned} D(Rp,t) = {} & 16\,\mathrm{Im}\int\cdots\int dy dk dk' \\ & \left\{\exp\left(-iy\cdot k'\right)\exp\left(-2i\left(k\cdot k'\right)\frac{t}{m}\right)V\left(k'\right)\right. \\ & \left.\times\, G\left(y_1 p+\frac{k'}{2}, p-2k+\frac{k'}{2}\right)\right\}. \end{aligned} \tag{4.48}$$

We again make the equilibrium approximation to $g(0)$. Assuming high-temperature (Born) approximation and retaining the lowest-order terms,

$$G_{eq}\left(y, p+\frac{k^1}{2}, p+\frac{k^1}{2}-2k\right)$$
$$= -\beta^4\pi^2(2\pi m)^{-3}(2\pi)^3 V(y) + \beta^5 n^2 (2\pi m)^{-3}(2\pi)^3 \tag{4.49}$$
$$\times \left\{\left[2\left(p+\frac{k^1}{2}\right)^2 + 4k^2 - 4k\left(p+\frac{k^1}{2}\right)\right]V(y) - \frac{\partial^2 V}{\partial^2 y}\right\}.$$

Only the β^5 term gives a nonzero contribution to $D(p\,t)$. This can be shown to be

$$D(p\,t) = \frac{-16\beta^5 n^2}{m}(2\pi)^6\left(\frac{mk}{4t}\right)^4 (2\pi m)^{-3} V^2(k) \underset{k=0}{|}. \tag{4.50}$$

Thus, in the approximation, the initial correlations decay at least as t^{-4}. This was first shown by Lee, Fujita and Wu (Lee *et al.*, 1970). However, the Born series diverges. This suggests that this is true also for $D(p,t)$. This can be seen simply. The bound state contribution to the total Green's function is $G_{cm}\mathfrak{S}_\beta(t)$, where

$$\mathfrak{S}_\beta(t) = \sum_n |n\rangle\langle n| \exp(-iE_n t) \tag{4.51}$$
$$G_{cm} = \exp(-iH_{cm}t),$$

and for homogeneous systems, we have

$$D(p,t) = 16i\,\mathrm{Im}\int dk dk' dy dq \tag{4.52}$$
$$\sum_{mm^1}\Big[\exp(-iyk')\exp(-i(E_n - E_{n'})t)$$
$$\langle p-q|V|n\rangle\left\langle n \mid k+\frac{k'}{2}\right\rangle\left\langle k-\frac{k'}{2} \mid n'\right\rangle\langle n' \mid p-q\rangle\overline{G}(Ry, qt, q+k)\Big].$$

This obviously does *not* decay because of the oscillatory contributions. Initial bound state correlations do not decay in time.

There is an interesting special case of pure states, $\rho^2 = \rho$:

$$\Omega(t)\rho^2\Omega^\dagger(t) = [\Omega(t)\rho]\left[\rho\Omega^\dagger\right].$$

A theorem states that $A(t)B(t) \to AB$ weakly if $A(t) \Rightarrow A,\quad B(t) \Rightarrow B$ (strongly). By weakly convergent, remember $(\psi, A(t)\phi) \to (\psi, A\phi)$ in a Hilbert space ψ, so that the convergence $\rho\Omega^\dagger(t) \Rightarrow \rho\Omega^\dagger$ is required. $\rho\Omega^\dagger(t)$ converges strongly to $\Omega^\dagger$ only if no bound states contribute. Thus, if ρ is a projection operator on scattering states, we have $\rho\Omega^\dagger(t) \Rightarrow \rho\Omega^\dagger$. We may expect if the subspace of interest is the scattering states only, then an asymptotic evolution may be proven. This idea has not been carried through with any rigor. Some aspects have been considered by Snider and Sanctuary (1971).

McLennan (1989) carried this formal argument forward in the hierarchy. To follow this argument, we define initially

$$\rho_2(12, t_0) = \mathfrak{P}\rho_1(1, t_0)\,\rho_1(2, t_0) + D(1, 2, t_0). \tag{4.53}$$

$\mathfrak{P}$ is the projection operator of ρ on the scattering states, $\mathfrak{P}\rho = (1 - \lambda)\rho(1 - \lambda) = \Omega\Omega^{\dagger}\rho\Omega\Omega^{\dagger}$. The additional term D is to be determined. Now $\Omega\pm = \lim_{t\to\pm\infty} H_2(t)\,H_0(-t)$ (Taylor, 1972). Taking $t_0 \to -\infty$, as in the classical argument (McLennan, 1989), we have

$$\rho_2(12) = \Omega\rho_1(1)\,\rho_1(2)\,\Omega^{\dagger} + D(12). \tag{4.54}$$

If there are no bound states *initially*, then $D = 0$. This is true both for attractive and repulsive potentials. The limit $t_0 \to -\infty$ is expected to exist for scattering states. This is true of the first term. But what about D? We will now consider this. The first term of the hierarchy is now

$$\dot{\rho}_1(t) + \frac{1}{i}\left[\rho_1, H_1\right] = \frac{n}{i}\{\mathrm{Tr}_2\left[V(1,2), \Omega\rho_1(1)\,\rho_1(2)\,\Omega^{\dagger}\right] + D(12)\}. \tag{4.55}$$

To this order in the density,

$$\dot{D} + \frac{1}{i}[D, H_2] = 0. \tag{4.56}$$

Taking the Tr_2 and defining,

$$\rho_A(1) = \rho_1(1) - n\mathop{\mathrm{Tr}}_{(2)} D(12). \tag{4.57}$$

The subscript A here means atoms found in collision. We obtain, to lowest order in n_0,

$$\dot{\rho}_A(t) + \frac{1}{i}\left[\rho_A, H_1\right] = \frac{n_0}{i}\mathop{\mathrm{Tr}}_{(2)}\left[V(12), \Omega\rho_A(1)\,\rho_A(2)\,\Omega^{\dagger}\right]. \tag{4.58}$$

From Eq. (4.55) and Eq. (4.58), D may be obtained to this order in n_0. We may expect the asymptotic equation for the "atom" contribution to be well defined. The real question is the next order in n^0, and very little is known. However, see McLennan's book (1989). Here we have accomplished the goal of writing a kinetic equation with bound states by use of the scattering states only, where $\Omega\rho\Omega^{\dagger}$ is expected to be well defined.

This completes our overly long discussion of attractive forces and points out that much must yet be done on this interesting topic. Let us now turn briefly to the quantum Vlasov equation to complete the discussion of kinetic theory.

4.6 Quantum Vlasov equation

The operator Vlasov equation may be most quickly obtained by factoring $\rho_2 = \rho_1(1)\rho_1(2)$ in the first equation of the hierarchy. We have

$$i\dot{\rho}_1 = L_1^0\rho_1 + \mathrm{Tr}_2[V_{1,}\Omega\rho_1(1)\,\rho_1(2)\,\Omega^{\dagger}]. \tag{4.59}$$

To make the spacial dependence explicit, we must, as before for the Boltzmann equation, introduce the Wigner function. This we have implicitly done already. We take $\Omega = \Omega^{\dagger} = 1$ in Eq. (4.27). After doing the δ function integration, we have

$$\begin{aligned}\dot{w} + m^{-1}\mathbf{p}\cdot\nabla w = 16\,(2\pi)^{-3}\ \mathrm{Im}\int dq dq_1 dq_2 dx dy \\ \times\left\{\exp(-i\,(xq_1 + yq_2))V\left(\frac{q_1 - q_2}{2}\right)\right. \\ \times\, w\left(R + \frac{x+y}{2}, p - \frac{q_1 - q_2}{2}\right) \\ \left.\times\, w\left(R + \frac{x-y}{2}, p - 2q + \frac{q_1 - q_2}{2}\right)\right\}.\end{aligned}$$

Changing variables to $\frac{q_1 - q_2}{2} = k,\ q_1 + q_2 = k^1$. The k^1 integration leads to $(2\pi)^3\,\delta\left(\frac{x+y}{2}\right)$. Then doing the y integration and introducing the Fourier transform,

$$V(k) = (2\pi)^{-3}\int dr V(r)\exp(iky),$$

we have the quantum Vlasov equation (Balescu, 1963, 1975):

$$\begin{aligned}\dot{w} + m^{-1}\mathbf{p}\cdot\nabla w = -i\,(2\pi)^{-3}\int d\rho dl dp' dp''\{\exp il\,(p - p') \\ \times\left[V\left(R - \rho - \frac{l}{2}\right) - V\left(R - \rho + \frac{l}{2}\right)w\,(Rp')\,V\,(\rho, p'')\right].\end{aligned} \tag{4.60}$$

Let us consider the collisional transfer approximation to this Vlasov equation. Let

$$w\,(R - r, p'') = w\,(Rp'') - \mathbf{r}\cdot\nabla w\,(Rp'').$$

The first term does not contribute. Upon carrying out an r and p integration, the latter using a δ function, we have

$$\dot{w}(R, p) + m^{-1}\mathbf{p}\cdot\nabla w = \frac{\partial w\,(R, p)\,(R, p)}{\partial p}\cdot\nabla\iint dx dp'' V\,(x)\,w\,(R, p)\,(R, p''). \tag{4.61}$$

This is the exact form of the classical counterpart to this order. The quantum effects appear to higher order in the gradient expansion. If we localize the quantum Vlasov to first order in the spacial gradient expansion, we obtain exactly the classical Vlasov equation, which is still nonlocal. We see that the right side is of order $V(r)$ in the interaction, in contrast to the Boltzmann equation discussed earlier in this chapter.

This completes our discussion of kinetic equations. The profound results of these equations and the master equations discussed in the previous chapter will be considered in Chapters 5 and 6.

Appendix 4A: phase space distribution functions

Phase space distribution functions are introduced to transform quantum mechanics into a form similar to probability distributions in classical statistical mechanics. They are similar but not equivalent. In this chapter, we have utilized exclusively the Wigner function, whose properties we have discussed. In this appendix, we will look at a more general representation, after the work of Cohen (1966). The main point is that the phase space distributions are not unique, and we will see how they are determined. Phase space distributions functions are also utilized in quantum optics (Schleich, 2001) and in kinetic theory (see our previous discussion).

We map $\rho \to F$ by the relation

$$\left\langle \hat{O} \right\rangle = \mathrm{Tr}\left[\hat{O}\rho\right] = \int dRdpO\,(Rp)\,F\,(Rp)\,. \tag{4A.1}$$

Here $\hat{O}$ is the quantum observable, ρ the density operator, $O\,(Rp)$ the *classical* phase space operator, and $F\,(Rp)$ the appropriate phase space distribution. The relation of $O\,(Rp)$ to $\hat{O}$ is crucial. We assume

$$F\,(Rp) = (2\pi)^{-6} \int dydkdk' \\ \times \exp\left(ik'R\right)\exp\left(-iy\,(p-k)\right) g\left(k'y\right)\left\langle k + \frac{k'}{2}\right|\rho\left|k - \frac{k'}{2}\right\rangle. \tag{4A.2}$$

$g\,(ky)$ is the generating function for this mapping. Three reasonable conditions are assumed:

1. $F\,(Rp)$ is real.
2. There is a marginal momentum distribution,

$$\phi\,(p) = \int dRF\,(Rp)\,. \tag{4A.3}$$

3. There is also a marginal position distribution, a number density $n\,(R) = \int dpF\,(Rp)\,.$

The first condition requires

$$g(k, y) = g^*(-k, -y), \tag{4A.4}$$

and the second and third conditions imply

$$\langle p|\rho|p\rangle = \phi(p) \qquad g(0, y) = 1 \tag{4A.5}$$

$$\langle R|\rho|R\rangle = n(R) \qquad g(k, 0) = 1. \tag{4A.6}$$

It is easily seen that the dispersion in p and R space *separately* are independent of further properties of $F(R, p)$. Note that we may write

$$F(R, p) = \int drdkG(Rk)\, w(R - r, p - k), \tag{4A.7}$$

where w is the Wigner function, $g = 1$, in Eq. 4A.2, and

$$G(Rk) = (2\pi)^{-6} \int dk'dy \exp(ir \cdot k) \exp\left(-iy \cdot k'\right) g\left(k'y\right), \tag{4A.8}$$

which is the Fourier transform of $g(ky)$. Thus, $F(Rp)$ need not be positive, since w is not positive, as we discussed earlier.

We now turn to the role of correspondence rules. Given a classical observable, what is the correspondence to that observable in quantum mechanics (Cohen, 1966; Margenau and Cohen, 1967)? Some choices are:

1. symmetrization rule:

$$q^n p^n \to \frac{1}{2}\left(\hat{q}^n \hat{p}^m + \hat{p}^m \hat{q}^n\right),$$

2. Born–Jordan rule:

$$q^n p^m \to (m+1)^{-1} \sum_{l=0}^{m} \hat{p}^{m-l} \hat{q}^n \hat{p}^l,$$

3. Weyl rule:

$$q^n p^m \to 2^{-n} \sum_{l=0}^{n} \binom{n}{l} \hat{q}^{n-l} \hat{p}^m \hat{q}^l.$$

There is also a Dirac rule:

$$\{\} \to [,]$$

$$\text{classical Poisson bracket} \to \text{quantum commutator.}$$

In the rules listed, the "hat" (as a common notation) indicates a quantum operator. This notation is used only when necessary. In the context of the earlier discussion, one of the possibilities listed (and any other) will put further conditions on g. Let us discuss this now, following Cohen.

Let $\gamma\left(k', k\right)$ be the momentum space Fourier transform of $O\left(R, p\right)$. Using the phase shift

$$\exp\left(i\hat{r}\cdot\frac{k'}{2}\right)|k\rangle = \left|k - \frac{k'}{2}\right\rangle,$$

we have, using Eq. (4A.1) and Eq. (4A.2),

$$\int dRdpF\left(Rp\right)O\left(Rp\right) = \langle O\rangle = \int dkdk'dx[g\left(k', x\right)\gamma\left(k'x\right)$$
$$\times\langle k|\exp i\left(\hat{r}\cdot\frac{k'}{2}\right)\exp\left(i\hat{p}\cdot x\right)\exp\left(i\hat{r}\frac{k'}{2}\right)\rho\,|k\rangle.$$

Now $\exp\left(i\hat{r}\cdot\frac{k'}{2}\right)\exp\left(i\hat{p}x\right)\exp\left(i\hat{r}\frac{k'}{2}\right) = \exp\left(i\left(\hat{p}\cdot x + \hat{r}k'\right)\right)$. See Louisell (1973) in Section 3.3. We have then

$$\mathrm{Tr}\hat{O}\rho = \mathrm{Tr}\left[\int\int dk'dxg\left(k'x\right)\gamma\left(k'x\right)\exp i\left(\hat{p}\cdot x + \hat{r}\cdot k'\right)\right]\rho. \qquad (4A.9)$$

We identify that

$$\hat{O} = \int\int dk'dxg\left(k'x\right)\gamma\left(k'x\right)\times\exp\left(i\left(\hat{p}x + \hat{r}k'\right)\right). \qquad (4A.10)$$

This is a general correspondence rule, given $\gamma\left(k'x\right)$ and g. The latter generates both the correspondence rule and the phase space distribution, $F\left(Rp\right)$. To proceed further by means of a power series expansion, we write Eq. (4A.6) in normal ordered form, that is, $\hat{\mathbf{r}}$ preceding $\hat{\mathbf{p}}$. Then we may replace the operators by c numbers $\mathbf{r}$, $\mathbf{p}$, and carrying out the integration, we obtain

$$\hat{O}\left(\hat{r}\hat{p}\right) = g\left(-i\frac{\partial}{\partial\mathbf{r}}, i\frac{\partial}{\partial\mathbf{p}}\right)\exp\left(-i\frac{1}{2}\frac{\partial}{\partial r\partial p}\right)O\left(r, p\right)|_{p\to\hat{p}}^{r\to\hat{r}}. \qquad (4A.11)$$

Thus, we compute (4A.11), normal order, and let $r \to \hat{r}$, $p \to \hat{p}$. This gives $\hat{O}\left(\hat{r}\hat{p}\right)$. We may show that $g = 1$ yields the Weyl rule and the Wigner function. $g\left(k, x\right) = \cos\left(k - \frac{x}{2}\right)$ gives the symmetrization rule, and $g\left(kx\right) = \frac{\sin\left(k - \frac{x}{x}\right)}{k - \frac{x}{2}}$ yields the Born–Jordan prescription.

There is an infinity of rules. What is the "correct" unique rule? There appears to be none. One can only adopt a rule.

Then Eq. (4.A11) gives $\hat{O}\left(R, p\right)$. With the choice of g, Eq. (4A.11) may be inverted (in principle) to obtain $\gamma\left(k'x\right)$ and thus the appropriate distribution function. All this ambiguity is due to the fact that $F\left(R, p\right)$ is defined by "weak" conditions, Eqs. (4A.1), (4A.4) and (4A.5). We should emphasize that $F\left(R, p\right) \not> 0$ are not probabilities but rather calculational aids.

A final remark concerns the uncertainty relation. This is definitely true for the three rules mentioned in this appendix. Can we make more general comments? That is a problem. Even if so, the quantum averages would be properly calculated by $\mathrm{Tr}\hat{O}\rho$, with $\hat{O}$ given by Eq. (4A.11).

References

Balescu, R. (1963). *Statistical Mechanics of Charged Particles* (New York, Interscience).

Balescu, R. (1975). *Equilibrium and Non-equilibrium Statistical Mechanics* (New York, Wiley), rev. 1999 as *Matter out of Equilibrium* (London, Imperial College Press).

Bhatnager, D., Gross, E. and Krook, M. (1954). *Phys. Rev.* **94**, 511.

Bocchieri, P. and Loingier, A. (1957). *Phys. Rev.* **107**, 337.

Bogoliubov, N. N. (1962). Problems of a dynamical theory in statistical physics. In *Studies in Statistical Mechanics* **I**, ed. J. de Boer and G. E. Uhlenbeck (Amsterdam, North-Holland).

Boltzmann, L. (1872). Further studies on the thermal equilibrium of gas molecules. In *Sitzungsberichte der Kaiserlichen Akademie* in *Wissenschaften, Vienna II*, **66**, 275–370, trans. 1966 by S. Brush in *Kinetic Theory* **2** (New York, Pergamon).

Born, M. and Green, H. S. (1949). *A General Kinetic Theory of Liquids* (London, Cambridge University Press).

Chapman, S. and Cowling, T. G. (1960). *The Mathematical Theory of Non-uniform Gases* (London, Cambridge University Press).

Cohen, L. (1966). *J. Math. Phys.* **7**, 781–786.

Farrar, J., Schafer, T. and Lee, Y. (1973). Intermolecular potentials of symmetric rare gas pairs from elastic differential cross section measurements. In *Transport Phenomena*, ed. J. Kestin (American Institute of Physics).

Gallavotti, G., Lanford, O. and Lebowitz, J. (1970). *J. Math. Phys.* **11**, 2898.

Green, H. S. (1961). *J. Math. Phys.* **2**, 344.

Huang, K. (1987). *Statistical Mechanics,* 2nd edn. (New York, Wiley).

Hudson, R. (1974). *Rep. Math. Phys.* **6**, 249–250.

Irving, J. and Zwanzig, R. (1951). *J. Chem. Phys.* **19,** 1173.

Lanford, O. E. (1975). *Lecture Notes in Physics*, no. 38 (New York, Springer).

Lee, D., Fujita, S. and Wu, F. (1970). *Phys. Rev. A* **2**, 854.

Louisell, W. H. (1973). *Quantum Statistical Properties of Radiation* (New York, Wiley).

Margenau, H. and Cohen, L. (1967). *Quantum Theory and Reality*, 71–90, ed. M. Bunge (New York, Springer).

McLennan, J. A. (1989). *Introduction to Non-equilibrium Statistical Mechanics* (Englewood Cliffs, Prentice Hall).

Mori, H., Oppenheim, I. and Ross, J. (1962). *Studies in Statistical Mechanics* **1**, ed. J. de Boer and G. E. Uhlenbeck (Amsterdam, North-Holland).

Petrosky, T. and Schieve, W. C. (1982). *J. Stat. Phys.* **28,** no. 4, 711.

Piquet, C. R. (1974). *Acad. Sci. Paris* **279A**, 107–109.

Schleich, W. (2001). *Quantum Optics in Phase Space* (Berlin, Wiley–VCH).

Snider, R. S. and Sanctuary, B. C. (1971). *J. Chem. Phys.* **55**, 1055.

Taylor, J. R. (1972). *Scattering Theory* (New York, Wiley).

Thomas, M. W. and Snider, R. F. (1970). *J. Stat. Phys.* **2**, 61.

Tolman, R. D. (1938). *The Principles of Statistical Mechanics* (New York, Oxford University Press), reissued by Dover.

Uehling, E. A. and Uhlenbeck, G. (1933). *Phys. Rev.* **43**, 552.
Van Hove, L. (1955). *Physica* **21**, 517.
Van Hove, L. (1957). *Physica* **23**, 441.
Van Hove, L. (1962). *Fundamental Problems in Statistical Mechanics,* ed. E. G. D. Cohen (Amsterdam, North-Holland).
Vlassov, A. (1938). *Zh. Eksp. Terr. Fiz.* **8**, 291.
Wigner, E. P. (1932). *Phys. Rev.* **40**, 749–759.
Yvon, J. (1935). *La Theorie Statistique des Fluides et l'Equation d'Etat* (Paris, Hermann).

5
Quantum irreversibility

Here we will examine dynamic irreversibility of the master equations discussed in Chapter 3 and of the kinetic equations of the previous chapter. Irreversibility is one of their important properties. First, what is irreversibility, quantum and classical (Tolman, 1938; Farquahar, 1964)?

5.1 Quantum reversibility

Let us first consider classical reversibility and then its generalization to quantum mechanics. The Hamiltonian equations are

$$\begin{aligned} \dot{q}_i &= \frac{\partial H}{\partial p_i} \\ \dot{p}_i &= -\frac{\partial H}{\partial q_i} \\ i &= 1 \ldots N \\ -\infty \leq\ & t \leq \infty. \end{aligned} \tag{5.1}$$

We take $H\,(p_i q_i)$ to be time independent and to be even in p_i. Thus,

$$H\,(p_i q_i) = H\,(-p_i, q_i) \equiv H^T. \tag{5.2}$$

The time reversal transformation, or dynamic reversal T is

$$\begin{aligned} Tp &\to -p^T \\ TH &= H^T \equiv H \\ Tq_i &\equiv q_i^T. \end{aligned} \tag{5.3}$$

The T transformation on Eq. (5.1) gives

$$\dot{q}_i^T = \frac{-\partial H^T}{\partial p_i^T}$$
$$-\dot{p}_i^T = -\frac{\partial H^T}{\partial q_i^T},$$

where $\bullet \equiv \frac{d}{dt}$. So Tq_i, Tp_i obey Eq. (5.1) for $-t$. Thus,

$$q_i^T = q_i(-t) \qquad -\infty < t < +\infty \tag{5.4}$$
$$p_i^T = -p_i(-t).$$

For every solution of Eq. (5.1), there is a dynamic reversed solution for $-\infty \leq t \leq \infty$. This is time reversal invariance. Note that an external magnetic field will force us to modify the statement of this invariance. (See the appendix to this chapter.)

Of course, the Liouville equation also has this property. After a T operation on the Liouville equation, we have

$$\frac{T\partial f}{\partial t} = \frac{\partial f^T}{\partial t} = -\sum_{i=1}^{N}\left(\frac{\partial f^T}{\partial q_i^T}\frac{\partial H^T}{\partial p_i^T} - \frac{\partial H^T}{\partial q_i^T}\frac{\partial f^T}{\partial p_i^T}\right). \tag{5.5}$$

So, by comparison,

$$f^T = f(q_i, -p_i, -t); \qquad -\infty \leq t \leq +\infty.$$

These classical symmetries are termed *reversibility.* Here, from a solution $f(qpt)$ to the Liouville equation, we may construct *by the same equation* and *at all time* the reversed solution f^T.

Now, how must this be generalized to quantum mechanics? A hint is in the Schrödinger representation $\hat{p} \to -i\hbar\frac{\partial}{\partial q}$. Since q and $\hbar$ are unchanged, $\hat{p} \to -p$ is complex conjugation. So T is the operation of complex conjugation (see the chapter appendix). Since H is real,

$$T\hat{H}T^{-1} = \hat{H}^* = \hat{H}^T \qquad \text{(time independent)} \tag{5.6}$$
$$T\hat{p}T^{-1} = -\hat{p}.$$

We also take $T\hat{J}T^{-1} = -\hat{J}$, since we wish to preserve $\left[\hat{J}_i, \hat{J}_j\right] = i\varepsilon_{ijk}\hat{J}_k$. Now $\frac{d\hat{\rho}}{dt} = i\left[\hat{H}\hat{\rho} - \hat{\rho}\hat{H}\right]$ for $-\infty \leq t \leq \infty$. Dropping the "hat" and again setting $\hbar = 1$, we have

$$T\dot{\rho}T^{-1} = \dot{\rho}^T = -i\left[H^*\rho^T - \rho^T H^*\right]. \tag{5.7}$$

Thus, $\rho^T(-t)$ obeys the same von Neumann equation. More is said of the operator properties of the time reversal transformation in the appendix.

5.2 Master equations and irreversibility

The Pauli master equation, Eq. (3.25), is

$$\frac{dP}{dt}(\alpha,t)=\sum_{\alpha'}W_{\alpha a'}\left[P\left(\alpha',t\right)-P\left(\alpha,t\right)\right]\quad t>0, \tag{5.8}$$

where

$$W_{\alpha a'}=2\pi\lambda^2\delta\left(E^0_{\alpha'}-E^0_{\alpha}\right)|<\alpha|\,V\,|\alpha'>|^2.$$

We have emphasized here that the original repeated random phase argument of Pauli is for $t\geq t_0=0$. It has already been mentioned that Pauli's argument carried backward for $t<0$ gives this equation a sign change and thus the inconsistency mentioned in Chapter 3, pointed out by Van Hove. This difficulty was overcome by later arguments and is discussed in Chapter 3.

Let us consider the dynamic reversal issue. If we operate with T on Eq. (5.3) the equation is invariant, since $W_{\alpha a'}$ and $P(\alpha,t)$ are real. Thus, $T P(\alpha,t)\,T^{-1}=P(\alpha,t)$ for $t>0$. There is no dynamic time inversion possible with such an equation. For the full unitary group governing the solution to the von Neumann equation $\rho(t)$, $-\infty\leq t\leq\infty$, we have for the Schrödinger solution

$$|\psi(t)\rangle=U^{\dagger}(t)\,|\psi(0)\rangle$$

and

$$\rho(t)=U^{\dagger}(t)\,\rho(0)\,U(t)\equiv S_t\rho(0)\quad t\geq 0,$$

where, of course, $U^{\dagger}(t)=\exp(-iHt)$. Being a group, $U^{\dagger}(\tau)=U^{-1}(\tau)=U(-\tau)$, and $S_{-t}S_t=I$. Also, $S_{-t}S_t=S_tS_{-t}$ for $-\infty\leq t\leq\infty$. This last property does not hold for solutions to Eq. (5.8) or, as we now note, most equations to be discussed in this section. They are irreversible. For an evolution $S_{-t}S_t$ $(t\geq 0)$, there is no reversed solution. Of course the evolution is not governed by S_t either, but rather a different linear semigroup operator. $P(\alpha,t)$ are real, and there is a profound change in this reduced evolution which concerns only the diagonal elements. The appearance of the semigroup is a manifestation of irreversibility called *non-invertible* in the mathematics literature (Mackey, 1992).

5.3 Time irreversibility of the generalized master and Pauli equations

As a first step in obtaining the Pauli master equation for open systems by means of the projection operator P, the generalized master equation was obtained in Eq. (3.13), Eq. (3.14) and Eq. (3.15). "Is it irreversible?" is a common question. The answer is, certainly! A formal exact solution to Eq. (3.14) is obtained by Eq. (3.15) with the initial value of $(1-P)\,\rho(0)$ and $P\rho(t=\tau)=P\rho(0)$ at time $t=0$. It is valid for $0 \leq t \leq \infty$. Of course, it is possible by means of another initial value to find a solution to another equation for negative time. We shall not write it here. All the subsequent derivation is in terms of a non-Markovian evolution for $0 \leq t \leq \infty$. The Pauli master equation subsequently obtained by these more exact methods, Eq. (3.25), is a Markovian semigroup equation.

Balescu (1963) has discussed this clearly in his treatment of the classical Liouville equation. Let $\mathfrak{L}$ be the classical Liouville operator (Poisson bracket with H). He considers the Green's equation,

$$\mathfrak{L}G\left(xvt \mid x'v't'\right) = \delta\left(x - x'\right) d\left(v - v'\right) \delta\left(t - t'\right),$$

and introduces the semigroup causality condition,

$$G\left(xvt \mid x'v't'\right) = 0; \quad t \leq t'.$$

The appropriate solution is for

$$\mathfrak{L} f_N\left(xv, t\right) = \delta\left(xvt\right) \quad t \geq 0,$$

such that

$$f_N\left(xv, 0\right) = q\left(xv\right).$$

The solution is for homogeneous time,

$$G\left(xvt \mid x'v't'\right) = \theta\left(t - t'\right) G\left(xv \mid x'v', t - t'\right),$$

where the well-known Heaviside function is

$$\theta\left(x\right) = \begin{cases} 1 & x > 0 \\ 0 & x < 0 \end{cases}.$$

The Heaviside function leads to a *one-sided* (semigroup) Laplace transform,

$$R\left(xv \mid x'v', z\right) = \int_0^\infty d\tau \exp\left(iz\tau\right) G\left(xv \mid x'v', \tau\right).$$

All this is very clear but still is a cause of confusion. It holds in terms of quantum operators also. The discussion of open systems does not change matters at all. The appropriate reservoir projection operators Eq. (3.32) and Eq. (3.33) give us Eq. (3.34).

The more difficult task of understanding entropy and the approach to equilibrium for the Pauli equation and its generalizations will come in the next chapter. It is important to separate the intertwining of irreversibility from what we may term dissipative behavior.

Now let us turn to the Boltzmann equation and the B.B.G.Y.K. hierarchy.

5.4 Irreversibility of the quantum operator Boltzmann equation

Clearly the operator B.B.G.Y.K. hierarchy of Chapter 4, Eq. (4.9a, b, c), is reversible, since it is derived from the original von Neumann equation by $\mathrm{Tr}_{(2\ldots N)}\rho_N$ etc. operations. It is, in fact, a reduced vector representation of $\rho_N(t)$, as emphasized by Balescu (1975).

The Boltzmann equation (classical or quantum) is irreversible. This apparent paradox was first pointed out to Boltzmann by Loschmidt (1876). What is the source of irreversibility in the derivation? This can be seen in Eq. (4.10) and the following, where the equation for $g_{12}(t)$ is solved formally for $0 \le t \le \infty$. This equation is then used to obtain a non-Markovian and irreversible equation for $\rho_{12}(t)$. This is used to obtain the time asymptotic equation Eq. (4.14) for the one-body operator $\rho_1(t)$. We have called this equation the Boltzmann operator equation.

The Boltzmann operator equation is irreversible because $0 \le t \le \infty$, just as with the master equation already discussed. The source of the irreversible behavior is in the initial causality assumption.

The subsequent form of the asymptotic operator equation, Eq. (4.19),

$$i\dot{\rho}_1 = L_1^0\rho_1 + n_0 tr\left[V_{12,}\Omega_{12}\rho_1(t)\,\rho_2(t)\,\Omega_{12}^\dagger\right] \qquad t \ge 0$$

with

$$\Omega_{12} = \lim_{t\to\infty} \mathfrak{G}_{12}(t)\, G_{12}^\dagger(t),$$

makes the irreversibility all very transparent. Ω_{12} is the asymptotic Möller wave operator. (See the appendix to this chapter.) Usually this is called Ω_-, but we have not needed Ω_+. Recall that the time reverse of Ω_- is Ω_+.

The introduction of phase space Wigner function $\omega(\mathbf{R},\mathbf{p})$ in Section 4.5 does not at all change the irreversibility of the quantum phase space Boltzmann equation. As emphasized there, Eq. (4.31) is exactly the same form as the irreversible nonlinear Boltzmann equation. It contains the quantum differential cross section

$$\left|T_{qk}\right|^2 = \left[m^2(2\pi)^4\right]^{-1}\frac{d\sigma\,(k\to q)}{d\Omega}.$$

The consistency of irreversibility is even more apparent in the scattering from the "in" state to the "out" state q. We have not written this equation in its simplest form. For spacially local scattering, from Eq. (4.31), we have

$$\partial_t w + m^{-1}\mathbf{p}\cdot\nabla w = \int\int dv_2' d\Omega' \left\{ d\sigma \frac{(v_1 v_2 \to v_1' v_2')}{d\Omega} |v_1 - v_2| \quad t > 0 \right.$$
$$\left. \times \left[w(v_1')\,w(v_2') - w(v_1)\,w(v_2)\right] \right\} \qquad (5.9)$$

with

$$|v_1 - v_2| = |v_1' - v_2'| \; ; \; v_1^2 + v_2^2 = v_1'^2 + v_2'^2.$$

We have not yet discussed the gain–loss structure of this equation. It is pertinent to do that now. Let

$$R\left(v_1 v_2 \to v_1' v_2'\right) = \frac{d\sigma\left(v_1 v_2 \to v_1' v_2'\right)}{d\Omega} |v_1 - v_2| \, .$$

Time reversal invariance (discussed in the appendix) implies

$$R\left(v_1 v_2 \to v_1' v_2'\right) = R\left(-v_1' - v_2' \to -v_1 - v_2\right),$$

and parity invariance is

$$R\left(-v \to -v'\right) = R\left(v \to v'\right),$$

so the PT invariance suggested in the appendix gives

$$R\left(v_1 v_2 \to v_1' v_2'\right) = R\left(v_1' v_2' \to -v_1 - v_2\right).$$

This has been called inverse collision symmetry in the sense of the classical Boltzmann equation (Huang, 1987). We may then write in a reverse order to the usual derivation

$$\begin{aligned} \partial_t w + m^{-1} \mathbf{p} \cdot \nabla w = \int dv_2' d\Omega' \{ & R\left(v_1' v_2' \to v_1 v_2\right) w\left(v_1'\right) w\left(w_2'\right) \\ & - R\left(v_1 v_2 \to v_1' v_2'\right) w\left(v_1\right) w\left(v_2\right)\} \qquad t > 0. \end{aligned} \tag{5.10}$$

The first term of Eq. (5.0) is the gain in the correlations in $v_{1,} v_2$. The second term is the loss of correlations in $v_1 v_2$. Here, in the Boltzmann equation, the correlations are factored into a product of one body $w_1(v)$, as is seen in the derivation of Chapter 4. We recall the two-body scattering picture, which gives the well-known depiction of the binary scattering growth and loss of these correlations. See Balescu (1975) and Huang (1987). Here we have the picture of the reversible instantaneous gain and loss of correlations to cause a temporal change in w. The process is irreversible ($t > 0$). Markovian master equations usually have this form, also. See the discussion of the Kolmogorov equation in the appendix to Chapter 3.

5.5 Reversibility of the quantum Vlasov equation

Let us recall briefly the derivation of the Vlasov equation, in Chapter 4. The derivation began with the reversible hierarchy, as with the Boltzmann equation discussed in Section 5.4. By a simple instantaneous factorization of the first equation of the B.B.G.Y.K. hierarchy, letting

$$\rho_2(1, 2, t) = \rho_1(1, t)\, \rho_1(2, t),$$

we obtained Eq. (4.62),

$$i\dot{\rho}_1(t) = L_1^0 \rho_1(t) + \mathrm{Tr}_{(2)}\left[V_{12,}\rho_1(1,t)\,\rho_1(2,t)\right] \qquad (-\infty \leq t \leq \infty).$$

This is valid for the time range $-\infty < t < +\infty$. No formal causal solution of $\rho_2|1,2,t|$ has been used. This equation is reversible. It is highly nonlocal in space, as is apparent in the derivation of the subsequent equation, Eq. (4.31), for the Wigner function $w(Rp,t)$. However no irreversibility has been used there, either.

The above equation is similar to the von Neumann equation itself, and since V_{12} is real, it is invariant under the time reversal transformation of the appendix. The time reversed equation is

$$-i\frac{dT\rho_1 T^{-1}}{dt} = L_1 T\rho_1 T^{-1} + \mathrm{Tr}_2\left[V_{12}, T\rho_1\rho_2 T^{-1}\right].$$

Hence, as in the general discussion, $T\rho(-t)\,T^{-1}$ obeys the same Vlasov equation as the original solution. This fact of reversibility is fundamental in discussions of plasma physics, as given by Balescu (1963). We refer the reader to these applications. However, there is a form of damping in the solutions not connected to irreversibility. There are transient oscillations set up by initial perturbations. The damping is due to destructive interference produced by the distribution of initial velocities.

The early time solution to the Vlasoff equation may be expressed as a Fourier transform:

$$w(k,p,t) = \int dk \exp(-ikt)\, a_k w(kp,0).$$

The details of the solution do not concern us here (Balescu, 1963). We obtain

$$w(k,p,t) = a_k \exp(-ikp_0 t)\exp\left(-k\mu_0 t\right)$$

for a sufficiently broad initial $w(kp,0)$ characterized by μ_0 and a group momentum p_0. The damped oscillatory motion depends on the *initial value*. It may cause $w(R,t)$ to decay in long time as a power law due to the Riemann–Lebesgue theorem. This is the source of Landau damping (Landau, 1946). Such a reversible damping has been called phase mixing (Balescu, 1975) to distinguish it from irreversibility defined in this section. The difference should be apparent. Phase mixing is possible in free particle motion due to initial values, in states which have a continuum of values, as here.

5.6 Completely positive dynamical semigroup: a model

As discussed in Chapters 2 and 3, the Lindblad–Kossakowski equation,

$$\dot{\rho}(t) = -i[L, \rho] + \sum_{\alpha} [L_{\alpha}^{\dagger}, \rho] L_a + L_{\alpha}^{\dagger} [\rho_1 L_{\alpha}] \qquad t > 0$$

(Gorini *et al.*, 1976; Lindblad, 1977), represents a general mathematical class of quantum, non-Markovian irreversible master equations, termed completely positive. It is a dynamic map

$$\rho^1 = \int d\alpha \varepsilon(\alpha) B(\alpha) \rho B^{\dagger}(\alpha), \qquad \int d\alpha \varepsilon(\alpha) B^{\dagger}(\alpha) B(\alpha) = 1,$$

such that $B^{\dagger} = B$ and particularly $\varepsilon(\alpha) = 1$ *for all* α. (See the simple derivation of the Lindblad–Kossakowski equation in Section 3.5.) It represents the "if and only if" condition for complete positivity. This equation is irreversible by construction. Gorini has also shown that it is equivalent to the Pauli equation for a class of singular reservoir interactions (Gorini *et al.*, 1976). We discussed this in Chapter 3 and use it in Chapter 7.

A simple example is a harmonic oscillator in interaction with an equilibrium electromagnetic field as the reservoir R. This has been extensively discussed by Agarwal (1973). Another example is the Milburn–Walls model in Chapter 2. The interaction of the harmonic oscillator of frequency ω with the field frequency ω_k is

$$H = \omega a^{\dagger} a + \sum_k \omega_k a_k^{\dagger} a_k + \lambda V,$$

where $V = \sum_k V_k \left(a^{\dagger} a_k + h.c.\right)$. Assuming weak coupling and an equilibrium reservoir (the field), we may obtain a Pauli-type equation for the oscillator system S, as discussed extensively in Chapter 2. This equation may easily be put into a Lindblad–Kossakowski form, as proved by Gorini *et al.* We obtain

$$i\frac{d\rho}{dt} = \omega\left[a^{\dagger} a, \rho\right] - \frac{i}{2} \sum_{j=1}^{2} \left[L_j^{\dagger} L_j \rho + \rho L_j^{\dagger} L_j - L_j^{\dagger} \rho L_j\right],$$

where

$$L_1 = \left[\sum_k \gamma_k \left(\langle n_k \rangle + 1\right)\right]^{\frac{1}{2}} a$$

$$L_2 = \left[\sum_k \gamma_k \langle nk \rangle\right]^{\frac{1}{2}} a^{\dagger}$$

and

$$\gamma_k = 2\pi\lambda^2 |V_k|^2 \delta(\omega_k - \omega)$$
$$\gamma = \sum_k \gamma_k.$$

As previously discussed in Chapter 2, a simple and soluble form of the reservoir is called phase damping (Walls and Milburn, 1985; Gardiner, 1991), where the reservoir harmonic oscillator interaction may be written $\lambda V = a^\dagger a\Gamma$, where Γ is the simple reservoir damping contribution. Remember that in the number representation, the matrix elements $\langle n|\dot{\rho}|m\rangle$ are very simple:

$$\langle n|\dot{\rho}|m\rangle = -i\omega(n-m)\langle n|\rho|m\rangle + iK\left(2\langle n_k\rangle + 1\right)(n-m)^2\langle n|\rho|m\rangle,$$

where we write

$$\eta = \sqrt{km}$$
$$K = \pi\kappa\frac{\omega^2}{2\eta}.$$

The solution is

$$\langle n|\rho|m\rangle = \exp\left(-i\omega(n-m)t\right) \exp\left(-\left(2\langle n_k\rangle + 1\right)K(n-m)^2\frac{t}{2}\right).$$

The off-diagonal elements decay to zero as $(2\langle n_k\rangle + 1)$, K being the damping constant. The decay is proportional to $(n-m)$, the "distance" between the off-diagonal states. At long time, the remaining contributions are $\langle n|\rho|n\rangle = \langle n|\rho(0)|n\rangle$, which plays the role of a time-unchanging equilibrium state. Again, these results illustrate *decoherence*, which will be discussed later.

Let us make a final comment here on completely positive dynamics. Recently, in a discussion of quantum damping, Monroe and Gardiner (1996) have shown that a master equation more general than that of the Lindblad and Kossakowski form is valid when the rotating wave approximation of quantum optics does not hold. This leads to unphysical short-time transients. Consequently, the most general form of the quantum Brownian motion is not fully understood. Anil Shaji, in his 2005 University of Texas Ph.D. dissertation, "Dynamics of Initially Entangled Open Quantum," has found that in a simple dynamic map, complete positivity does not hold (Shaji, 2005).

Appendix 5A: the quantum time reversal operator

Let us examine in more detail the structure of T introduced in this chapter (Wigner, 1932; Mathews and Venkatesan, 1975; Bohm, 1993). Now define T in a Hilbert space, not that of superspace, although we use the same notation:

$$\begin{aligned} TQ_iT^{-1} &= Q_i \\ TP_iT^{-1} &= -P_i \\ TLT^{-1} &= -L. \end{aligned} \tag{5A.1}$$

Consider

$$T\,[Q, P]T^{-1} = -[Q, P].$$

We have $T\,(i\hbar)\,T^{-1} = -i\hbar$. Thus, we must require that T is not linear but rather *antilinear* (Wigner, 1932):

$$T\,(c_1\sigma_1 + c_2\sigma_2) = c_1^*T\sigma_1 + c_2^*T\sigma_2. \tag{5A.2}$$

Let us mention some properties of antilinear operators. An adjoint operator is antilinear:

$$(Af, g) = \left(A^\dagger g, f\right) = \left(f, A^\dagger g\right)^*.$$

Now we let $TH = HT$ and operate on the Schrödinger equation using the antilinearity of T. We obtain

$$-i\frac{\delta\sigma'}{\delta t}(xt) = H\sigma'(xt), \tag{5A.3}$$

taking

$$TH = HT$$

for

$$\sigma'(xt) = T\sigma(x, t).$$

There is, thus, another solution to the Schrödinger equation for $t, \sigma(t)$. It is $\sigma' = T\sigma$ for $-t$, the time reversal solution. We note that the time t has the range $-\infty \le t \le +\infty$. Now we introduce $T = UK$, where K is the complex conjugation operator and U is any linear operator. If $U = I$, then $TH = HT$ implies that H must be real, and Eq. (A5.3) is just the complex conjugate Schrödinger equation for $t' \ge 0$. T. Jordan has proved a stronger property for H than simply reality. He proved the theorem, "If the negative part of the spectrum of H has a lower bound and the positive part is unbounded then P, the parity operator, is linear and T, the time reversal operator, is antilinear" (Jordan, 1969). Further, using $U^{\dagger}U = 1$ and $T = UK$,

$$(T\phi, T\sigma) = (\phi, \sigma)^* = (\sigma, \phi).$$

For any operator A,

$$(\phi, A\sigma) = (T\phi, TA\sigma)^* = \left(T\phi, A^{'}T\sigma\right)^*,$$

where $A' = TAT^{-1}$.

In the case of particles with angular momentum, such as spin, the third relation in Eq. (5A.1) requires $T\mathbf{S}T^{-1} = -\mathbf{S}$. We may fulfill this by choosing $TG = \exp\left(i\pi S_y\right)K$. Since $K^2 = 1$, then $T^2 = \exp 2\pi i S_y = (-1)^S$. Hence,

$$T^2 = \begin{matrix} +1 \text{ integer spin} \\ -1 \text{half odd integer spin} \end{matrix}.$$

Now let us consider time reversal in quantum scattering (Taylor, 1972). This has much to do with our discussion of the Boltzmann equation in the previous chapter. For scattering of particles without spin,

$$T\exp(iHt) = \exp(-iHt).$$

Thus, the Möller wave operators

$$T\Omega_{\pm} = T\left[\lim_{t\to\mp\infty} \exp(iHt)\exp\left(-iH^0t\right)\right] = \Omega_{\mp}T,$$

since $T^{\dagger}T = 1$. Hence,

$$\Omega_{\pm} = T^{\dagger}\Omega_{\mp}T. \tag{5A.4}$$

T interchanges Ω_+ and Ω_-. Only the latter has been used in our previous discussion in Chapter 4. From this, it follows that for the S matrix, $S \equiv \Omega_-\Omega_+$, $T^{-1}S^{\dagger}T = S$. Forming matrix elements of S between asymptotic "in" state ϕ and "out" state χ, we have

$$\langle\chi\,|S|\,\phi\rangle = \left\langle\chi, T^{\dagger}S^{\dagger}T\phi\right\rangle = \left\langle T\chi, S^{\dagger}T\phi\right\rangle^* = \left\langle\chi_T, S^{\dagger}\phi_T\right\rangle^*. \tag{5A.5}$$

Thus,

$$\langle\chi\,|S|\,\phi\rangle = \left\langle\phi_T\,|S|\,\chi_T\right\rangle. \tag{5A.6}$$

We see that the S-matrix elements between "in" and "out" states are T invariant. We conclude that the scattering transition probability $W\,(\chi \leftarrow \phi)$ is the same for $W\left(\phi_T \leftarrow \chi_T\right)$. This is a form of microreversibility. The case of particles with spin is discussed in detail in the book by Taylor.

Finally, let us comment on the *TCP* invariance theorem of quantum field theories (Luders, 1957; Schweber, 1962). This is discussed all too often in the context of irreversibility or its failure. We are treating the fields classically here, and the only effect is to reverse **A**, the vector potential under T. Consider the current source of **A** (or **B**). T reverses the currents and thus **A** (Wigner, 1932). The natural choice in this classical limit is to choose $C \Rightarrow I$, so the invariance becomes PT invariance. We will take this to be the case. Hence, $[H, T\mathbf{P}] = 0$. This is consistent with the previous theorem of Jordan and its requirements on the spectrum. The main point is that if T and P invariance are not separately found true, then PT invariance must hold. We note that PT is antilinear, since P is linear and T antilinear.

The charge-classical field interaction Hamiltonian is

$$H = \frac{1}{2m}\left(\mathbf{P} - \frac{e}{c}\mathbf{A}\right)^2 + e\Phi$$

and is invariant under PT if Φ is a central potential, and $T\mathbf{A} = -\mathbf{A}$ (Wigner, 1932), since the sources of the external field corresponding to **A** are currents. The canonical $\mathbf{P} = m\dot{\boldsymbol{q}} + \frac{e}{c}\mathbf{A}$. Thus, $\mathbf{P} \to -\mathbf{P}$ under more this, as must be.

References

Agarwal, G. S. (1973). Master equation methods in quantum optics. In *Progress in Optics* **10**, 1 ed. E. Wolf (New York, North-Holland).
Balescu, R. (1963). *Statistical Mechanics of Charged Particles* (New York, Interscience).
Balescu, R. (1975). *Equilibrium and Nonequilibrium Statistical Mechanics* (New York, Wiley), rev. 1999 as *Matter out of Equilibrium* (London, Imperial College Press).
Bohm, A. (1993). *Quantum Mechanics: Foundations and Applications* (Berlin, Springer).
Farquahar, I. E. (1964). *Ergodic Theory in Statistical Mechanics* (London, Interscience).
Gardiner, C. W. (1991). *Quantum Noise* (New York, Springer).
Gorini, V., Kossakowski, A. and Sudarshan, E. C. G. (1976). *J. Math. Phys.* **17**, 821.
Huang, K. (1987). *Statistical Mechanics,* 2nd edn. (New York, Wiley).
Jordan, T. (1969). *Linear Operators for Quantum Mechanics* (Duluth, Minn., Thomas Jordan).
Landau, L. D. (1946). *J. Phys. USSR* **10**, 25.
Lindblad, G. (1977). *Commun. Math. Phys.* **48**, 119.
Loschmidt, J. (1876). *Sitzber. Akad. Wiss. Wien* **73**, 139.
Loschmidt, J. (1877). *Sitzber. Akad. Wiss. Wien* **75**, 287.
Louisell, W. (1973). *Quantum Statistical Properties of Radiation* (New York, Wiley).
Luders, G. (1957). *An. Phys. N.Y.* **2**, 1.
Mackey, M. C. (1992). *Time's Arrow: The Origin of Thermodynamic Behaviour* (Berlin, Springer).
Mathews, P. M. and Venkatesan, K. (1975). *A Textbook of Quantum Mechanics* (New Delhi, Tata McGraw-Hill).
Monroe, W. and Gardiner, C. (1996). *Phys. Rev.* **53**, 2633.
Schweber, S. (1962). *An Introduction to Relativistic Quantum Field Theory* (New York, Harper & Row).
Shaji, A. (2005). Dynamics of initially entangled open quantum, *Phys. Rev. Lett.*, to appear.
Taylor, J. R. (1972). *Scattering Theory* (New York, Wiley).

Tolman, R. C. (1938). *Principles of Statistical Mechanics* (London, Oxford University Press).
Vlassov, A. (1938). *Zh. Eksp. Terr. Fiz.* **8**, 291.
Walls, D. and Milburn, G. (1985). *Phys. Rev.* **31**, 2403.
Wigner, E. P. (1932). *Gottinger, Nachr.* **31**, 546.

6

Entropy and dissipation: the microscopic theory

6.1 Introduction

The microscopic theory of dissipation in open quantum systems will be discussed in this chapter. The central issue is the approach of an open quantum system to a local or global equilibrium, thermodynamic equilibrium. Of course, this was begun by Boltzmann in statistical mechanics, classically, in his famous work (Boltzmann, 1872; Balescu, 1975; Huang, 1987; McLennan, 1989).

The irreversible equations of the previous chapter, the quantum master equations and the quantum Boltzmann equation, will be utilized to follow the wonderful path set out in Boltzmann's work. To some extent we will have success, yet not entirely, since the trail is not at its end.

Central to the issue of dissipation is the entropy production theorem for an inhomogeneous or homogeneous system. We will now turn to so-called non-equilibrium thermodynamics to outline the macroscopic picture of what needs to be achieved from the microscopic theory.

6.2 Macroscopic non-equilibrium thermodynamics

Macroscopic non-equilibrium thermodynamics will be outlined for a fluid system. This thermodynamics is, of course, far more general than this system. This particular example is used in order to make a connection with the microscopic quantum Boltzmann equation of Chapter 4 (de Groot and Mazur, 1962; Prigogine, 1967; Callen, 1985; McLennan, 1989).

The macroscopic conservation laws for a fluid, for instance, are written in a laboratory inertial frame as

$$\frac{\partial \rho}{\partial t} + \nabla \cdot \mathbf{J}_m = 0 \tag{6.1}$$

$$\frac{\partial \varepsilon}{\partial t} + \nabla \cdot \mathbf{S} = 0 \tag{6.2}$$

$$\frac{\partial g_i}{\partial t} + \frac{\partial t_{ji}}{\partial x_j} = 0. \tag{6.3}$$

Here ρ is the mass density, $\rho(x, t)$, ε the energy density, and g the momentum density. All these are functions of $\mathbf{x}$,t, which we shall not explicitly indicate. t_{ij} is the pressure tensor, $\mathbf{J}_m$ the mass flux, and $\mathbf{S}$ the energy flux. Of course, repeated indices are summed one to three. These equations can be derived from the Boltzmann equation and the B.B.G.Y.K. hierarchy, as discussed in Chapter 4. Here, however, they are phenomenological equations. For a non-isolated system, we have

$$\frac{\partial g_i}{\partial t} + \frac{\partial t_{ji}}{\partial x_j} = F_i \tag{6.4}$$

and

$$\frac{\partial \varepsilon}{\partial t} + \nabla \cdot \mathbf{S} = W. \tag{6.5}$$

F_i is the external force per unit volume, and W the rate of doing work. If $\boldsymbol{\mu}$ is the fluid velocity, $\mathbf{g} = \rho\boldsymbol{\mu}$, and we may write Eq. (6.3) as

$$\rho\left(\frac{\partial \mu_i}{\partial t} + \boldsymbol{\mu}\cdot\nabla\mu_i\right) = \frac{\partial \sigma_{ji}}{\partial x_j}, \tag{6.6}$$

where $\sigma_{ji} = \rho\mu_i\mu_j - t_{ji}$ is the stress tensor.

The continuity equations must be equally true in all inertial frames. They are not form invariant. Let us make a Galilean transformation of a fluid element moving with velocity $\boldsymbol{\mu}$ at t.

The transformation to the new inertial frame is $\mathbf{x}' = \mathbf{x} - \mathbf{w}t$. ρ is invariant, and the fluid velocity in the moving coordinate is $\boldsymbol{\mu}'\left(\mathbf{x}'t\right) = \boldsymbol{\mu}\left(\mathbf{x}, t\right) - \mathbf{w}$. We obtain, since ρ is invariant and $\Delta' = \Delta$,

$$\left(\frac{\partial \rho}{\partial t} + \nabla\cdot\mathbf{g}\right)' = \frac{\partial \rho}{\partial t} + \nabla\cdot\mathbf{g}.$$

We leave it as a problem to show that the transformations are also

$$t'_{ij} = t_{ij} - w_i g_j - w_j g_i - \rho w_i w_j$$

$$\varepsilon' = \varepsilon - \mathbf{g}\cdot\mathbf{w} + \frac{1}{2}\rho w^2 \tag{6.7}$$

and

$$s'_i = s_i - \left(\varepsilon - \mathbf{w}\cdot\mathbf{g} + \frac{1}{2}\rho w^2\right) w_i - t_{ij}w_i + \frac{1}{2}w^2 g_i.$$

With this in mind, let us consider the thermodynamics of a local moving frame with velocity $\mu\left(x, t\right)$ in the fluid. w is a function of a particle position and time, that is, $\mathbf{w} = \mu(\mathbf{x}, t)$. There is a succession of rest frames for each $\mathbf{x}, t$ of the fluid. We consider the thermodynamics in these various frames. This is why the term "non-equilibrium thermodynamics" is adopted. Let $\rho_0 = \rho$ indicate a local rest frame, at $\mathbf{x}, t$. We have

$$\begin{aligned}
\rho_0 &= \rho \\
\mathbf{g}_0 &= 0 \\
\varepsilon_0 &= \varepsilon - \frac{1}{2}\rho\mu^2 \\
t_{0,ij} &= t_{ij} - \rho\mu_i\mu_j \\
s_{0,i} &= s_i - \left(\varepsilon_0 + \frac{1}{2}\rho\mu^2\right)\mu_i - t_{0ij}\mu_j.
\end{aligned} \tag{6.8}$$

For simplicity we ignore internal variables. The local *intensive* (additive) thermodynamic variables are ρ and ε_0. Hence it is reasonable to assume that the entropy per unit mass, $s = s(\rho, \varepsilon_0)$, and its derivative, $\frac{\partial s}{\partial \varepsilon_0}\,|_\rho \equiv \frac{1}{T}$, are functions of x, t. T here is the local thermodynamic Kelvin temperature. s is not to be confused with the vector $\mathbf{s}$, the energy flux. The pressure and chemical potential may be similarly defined locally (Callen, 1985). The total (global) entropy is

$$S = \int d_3 x \rho s. \tag{6.9}$$

This is a result of the general property of additivity of the entropy. We have then the first law,

$$\begin{aligned}
T ds &= d\left(\frac{\varepsilon_0}{\rho}\right) + pd\left(\frac{1}{\rho}\right) \\
dp &= \rho\,(d\mu + s dT),
\end{aligned} \tag{6.10}$$

where the chemical potential μ is given by

$$\mu\rho = \varepsilon_0 + p - T\rho s.$$

Let us now turn to *dissipative* fluxes and entropy production. If the local fluid element is in equilibrium $t_{0,ij} = p\delta_{ij}$ and the energy flux $\mathbf{s}_0 = 0$, then Eq. (6.8) becomes

$$\begin{aligned}
t_{ij} &= p\delta_{ij} + \rho\mu_i\mu_j \\
s_i &= \left(\varepsilon_0 + p + \frac{1}{2}\rho\mu^2\right)\mu_i.
\end{aligned} \tag{6.11}$$

However, if this is not so, additional dissipative terms are added:

$$
\begin{aligned}
t_{ij} &= p\delta_{ij} + \rho\mu_i\mu_j + t^*_{ij} \\
s_i &= \left(\varepsilon_0\mu_i + p\mu_i + \frac{1}{2}\rho\mu^2\mu_i + t^*_{ij}\mu_j + s^*_i\right).
\end{aligned}
\tag{6.12}
$$

In the local rest frame, $t_{0ij} = p\delta_{ij} + t^*_{ij}$, $\mathbf{s}_0 = \mathbf{s}^*$. The terms with * are the dissipative (also called irreversible!) parts. By means of Eq. (6.12), the conservation laws may be rewritten as follows:

$$
\begin{aligned}
&\rho\left(\frac{\partial}{\partial t} + \boldsymbol{\mu}\cdot\nabla\right)\boldsymbol{\mu} = \nabla p - \nabla\cdot\mathbf{t}^* \\
&\left(\frac{\partial}{\partial t} + \boldsymbol{\mu}\cdot\nabla\right)\varepsilon_0 = -\left(\varepsilon_0 + p\right)\nabla\cdot\boldsymbol{\mu} - t^*_{ji}\frac{\partial\mu_i}{\partial x_j} - \nabla\cdot\mathbf{s}^*
\end{aligned}
\tag{6.13}
$$

and

$$
\left(\frac{\partial}{\partial t} + \boldsymbol{\mu}\cdot\nabla\right)\rho = -\rho\nabla\cdot\boldsymbol{\mu}.
$$

Let us introduce the substantial derivative,

$$
\frac{D}{Dt} \equiv \frac{\partial}{\partial t} + \boldsymbol{\mu}\cdot\mathbf{D}.
$$

Form

$$
\begin{aligned}
\frac{DT}{Dt} &= \left(\frac{\partial T}{\partial\rho}\right)_{\varepsilon_0}\frac{D\rho}{Dt} + \left(\frac{\partial T}{\partial\varepsilon_0}\right)_\rho\frac{D\varepsilon_0}{Dt} \\
&= -\left[\rho\left(\frac{\partial T}{\partial\rho}\right)_{\varepsilon_0} + h\left(\frac{\partial T}{\partial\varepsilon_0}\right)_\rho\right]\nabla\cdot\boldsymbol{\mu} - \left(\frac{\partial T}{\partial\varepsilon_0}\right)_\rho\left[t^*_{ji}\frac{\partial\mu_i}{\partial x_j} + \nabla\cdot s^*\right].
\end{aligned}
$$

The enthalpy $h = \varepsilon_0 + p$ may be rewritten using the $Td\rho s$ equation as

$$
h = T\left(\frac{\partial p}{\partial T}\right)_\rho + \rho\left(\frac{\partial\varepsilon_0}{\partial\rho}\right)_T.
$$

Introducing the specific heat, we have

$$
\frac{DT}{Dt} = -T\left(\frac{\partial p}{\partial\varepsilon_0}\right)_\rho\nabla\cdot\boldsymbol{\mu} - \left(\rho c_v\right)^{-1}\left(t^*_{ji}\frac{\partial\mu_i}{\partial x_j} + \nabla s^*\right).
$$

Now consider

$$
\frac{Ds}{Dt} = \left(\frac{\partial s}{\partial T}\right)_\rho\frac{DT}{Dt} + \left(\frac{\partial s}{\partial\rho}\right)_T\frac{D\rho}{Dt}.
$$

With this, we obtain the important relationship

$$\left(\frac{\partial}{\partial t}+\nabla\cdot\boldsymbol{\mu}\right)\rho s=-\left(\rho T\right)^{-1}\left[t_{ji}^{*}\frac{\partial\mu_i}{\partial x_j}+\nabla\cdot\mathbf{s}^{*}\right]. \tag{6.14}$$

This shows that the entropy continuity depends only on the dissipative quantities s^*, t_{ij}^*. This is its importance.

From Eq. (6.14) we may write the time change of the global entropy,

$$\frac{dS}{dt}=\int d_3x\left[\mathbf{s}^{*}\nabla T^{-1}-T^{-1}t_{ji}^{*}\frac{\partial\mu_i}{\partial x_j}\right]-\int d\mathbf{A}\cdot\left(\rho s\boldsymbol{\mu}+T^{-1}s^{*}\right). \tag{6.15}$$

Here the entropy flow appears as a flux across the area $d\mathbf{A}$, and the dissipative *entropy production* is given by

$$\sigma=\mathbf{s}^{*}\cdot\nabla T^{-1}-T^{-1}t_{ji}^{*}\frac{\partial\mu_i}{\partial x_j}. \tag{6.16}$$

This separation of entropy change into a flow and spontaneous production is the principal point of this section on dissipative thermodynamics. If there is no flow, then we expect $\sigma\geq 0$. This will be examined from the microscopic theory with the irreversible equations we have obtained. In the case of zero flux, then,

$$\frac{dS}{dt}\geq 0, \tag{6.17}$$

the second law of thermodynamics.

We shall not, at this point, extend this discussion to the linear transport laws and Onsager's reciprocal relations. Comments were made on transport laws and the Boltzmann equation in the previous chapter. The main focus has been to introduce the entropy production due to dissipation.

In the chemical literature, see the book by Kondipudi and Prigogine (1998). The introduction of a local entropy production and the dissipative quantities to a local thermodynamics has been termed an extended irreversible thermodynamics. Particularly, see the work of Jou (1993, 1996). We prefer to utilize the title "dissipative" to distinguish it from irreversible, the reasons now being clear. The approach here taken is due to McLennan (1989).

Let us now consider the transport relations. t^* and s^* are functions of $\mathbf{x}$, t through their dependence on $T(x,t)$, $\rho(x,t)$ and $\boldsymbol{\mu}(x,t)$. These relations may be nonlinear and have all order spacial derivatives. Considering Taylor expansions, the simplest form is a linear one. For a fluid or gas with spherical symmetry, the so-called linear transport laws are uniquely

$$\mathbf{s}^{*}=-\lambda\nabla T \tag{6.18}$$
$$t_{ij}^{*}=-2\eta\mathbf{D}_{ij}-\eta' D_{ii}.$$

Here

$$D_{ij} = \frac{1}{2}\left(\frac{\partial \mu_i}{\partial x_j} + \frac{\partial \mu_j}{\partial x_i}\right) \tag{6.19}$$
$$D = D_{ii} = \nabla \cdot \boldsymbol{\mu}$$
$$\mathbf{D} = D_{ij} - \frac{1}{3}\delta_{ij} D.$$

Here λ is the thermal conductivity, η the shear, and η' the bulk viscosity. It has been emphasized that from a microscopic point of view, these relations are derived in the process of obtaining the steady solution to the Boltzmann equation.

By means of these, an expression for the entropy production discussed above may be found:

$$\sigma = \lambda T^{-2} (\Delta T)^2 + T^{-1} \left[2\eta \mathbf{D}_{ij} \mathbf{D}_{ij} + \eta' D^2\right]. \tag{6.20}$$

λ, η, η' are by hypothesis positive and thus also σ. Equation (6.20) is a special form of a postulate of steady non-equilibrium thermodynamics (Callen, 1985; Kondipudi and Prigogine, 1998).

The transport laws, Eq. (6.18), are written generally as

$$J_k = \sum_j^{\prime} L_{jk} \mathfrak{F}_j, \tag{6.21}$$

where the fluxes J_k are linearly dependent on the generalized force $\mathfrak{F}_j$, also called affinity. L_{jk} are the linear transport coefficients, generally tensors. The entropy production is (Callen, 1985)

$$\dot{S} = \sum_k \nabla \mathfrak{F}_k \mathbf{J}_k, \tag{6.22}$$

with $\mathfrak{F}_k = F_k - F_k^0$. F_k^0 is the equilibrium value. Eq. (6.20) is Eq. (6.22) for the system being discussed.

Onsager (1931) proved that

$$L_{jk} = L_{kj}. \tag{6.23}$$

This symmetry is the major content of steady non-equilibrium thermodynamics and has been verified extensively (Callen, 1985; Kondipudi and Prigogine, 1998). For instance, in the case of the thermoelectric effect, the coefficient of the Thomson effect is related to the derivative of the thermoelectric power (Callen, 1985).

For the flux of the system being discussed here, we may write, generally,

$$\begin{aligned} s_i^* &= \lambda_{ij}\frac{\partial T}{\partial x_j} - a_{ijk}\frac{\partial \mu_k}{\partial x_j} \\ t_{ij}^* &= -b_{ijk}\frac{\partial T}{\partial x_k} - \eta_{ijkl}\frac{\partial \mu_k}{\partial x_i}. \end{aligned} \tag{6.24}$$

The Onsager theorem gives $a_{ijk} = -Tb_{jki}$. These relations may be microscopically derived by Green–Kubo formulas (Green, 1951; Kubo, 1957). We shall not go into this approach now but will treat it later in Chapter 15. However, see the detailed discussion in McLennan (1989). Onsager brilliantly argued these results by the consideration of the average *equilibrium* correlation function fluctuation, $\langle \delta X_j, \delta X_k(\tau)\rangle$ of the extensive parameters X_j, X_k. This is a delayed correlation moment between $\tau = 0$ and τ. He assumed that there should be a time-*reversible* microscopic symmetry,

$$\langle \delta X_j \delta X_k(\tau)\rangle = \langle \delta X_j \delta X_k(-\tau)\rangle. \tag{6.25}$$

From this we may obtain, at $\tau = 0$,

$$\langle \delta X_j \delta \dot{X}_k\rangle = \langle \delta \dot{X}_j \delta X_k\rangle, \tag{6.26}$$

and using the *macroscopic* law,

$$\delta \dot{X}_k = \sum_i L_{ik}\delta\mathfrak{F}_i. \tag{6.27}$$

From Eq. (6.26) and Eq. (6.27) there follows

$$\sum_i L_{ik}\langle \delta X_j \delta\mathfrak{F}_i\rangle = \sum_i L_{ij}\langle \delta\mathfrak{F}_i \delta X_k\rangle.$$

We then obtain Eq. (6.23) also. This is indeed puzzling, in the light of the discussion in the previous chapter. Eq. (6.25) is a reversible equation, whereas Eq. (6.26) is irreversible and dissipative, containing transport relations.

The answer to this conundrum is that Eq. (6.25) is an equilibrium relationship. It is due to microscopic reversibility in the equilibrium solution. This will be seen in detail in Section 6.5, when we consider the derivation of the Onsager symmetry from the point of view of the open system Pauli equation.

6.3 Dissipation and the quantum Boltzmann equation

In Chapter 4 we wrote the Wigner function form of the quantum Boltzmann equation, Eq. (4.31), as

$$\partial_t w(\mathbf{R}, \mathbf{V}) + m^{-1}\mathbf{P} \cdot \nabla_R w(\mathbf{R}, \mathbf{V}) = \int\int d\mathbf{V}_1 d\Omega \left\{ \begin{array}{c} \frac{d\sigma\left(\mathbf{V}_1, \mathbf{V}_2 \to \mathbf{V}_1', \mathbf{V}_2'\right)}{d\Omega} |\mathbf{V}_1 - \mathbf{V}_2| \\ \times \left[w\left(\mathbf{R}, \mathbf{V}_1'\right) w\left(\mathbf{R}, \mathbf{V}_2'\right) - w\left(\mathbf{R}, \mathbf{V}_1\right) w\left(\mathbf{R}, \mathbf{V}\right)\right] \end{array} \right\}. \quad t \geq 0 \tag{6.28}$$

This quantum Boltzmann equation has exactly the classical form except that $\frac{d\sigma}{d\Omega}$ is the quantum differential cross section and w is the Wigner function. This makes for significant differences, because $w \not> 0$.

Already in Eq. (4.32) and following, we have derived the continuity equations, Eq. (4.33), corresponding to Eq. (6.18) and Eq. (6.19). We have followed the Chapman–Enskog work (Chapman and Cowling, 1970; McLennan, 1989), and we obtained the formulas for the dissipative transport coefficients, λ and η, the thermal conductivity and shear viscosity. The Chapman–Enskog expansion was based upon the *assumption* of

$$w' \equiv f^0 (1 + \Phi),$$

where we interpreted

$$f^0 (RV) = n \left(\frac{m}{2\pi kT}\right)^{\frac{3}{2}} \exp\left[-\left(\frac{m}{2kT}\right)(V - \mu)^2\right] \tag{6.29}$$

as the local equilibrium solution to the Boltzmann equation, Eq. (6.28). This is simply proved, classically (Balescu, 1975; McLennan, 1989).

But it is more subtle in the quantum case. (At this point we drop the explicit vector notation for R, V.) Now $f^0 (R, V)$ must obey

$$\left[f^0 \left(R, V'\right) f^0 \left(R, V_2'\right) - f^0 (R, V_1) f (R, V_2)\right] = 0$$

and

$$f^0 (R, V) \geq 0.$$

R. L. Hudson (1974) has proved that the necessary and sufficient condition for the Wigner function to be positive is that it correspond to a wave function, which is quadratically positive,

$$\psi (x) = \exp\left(-\frac{2}{w}\left(ax^2 + 2bx + c\right)\right),$$

i.e. a coherent state (see Chapter 2). Thus, the only density matrix satisfying the condition of Eq. (6.31) is Eq. (6.29). Here we make a similar argument, as in the classical case (McLennan, 1989), to assign the constants.

f^0 is now positive. We also have the microscopic conservation laws for the invariants $m\mathbf{V}$, $m\frac{V^2}{2}$ and 1, of the two-body elastic scattering. We then have

$$\left[\ln f^0\left(V_1'\right) + \ln f^0\left(V_2'\right) - \ln f^0\left(V_1\right) - \ln f^0\left(V_2\right)\right] = 0,$$

thus satisfying the condition of Eq. (6.30). In local equilibrium f^0 is a positive Gaussian in V and parameterized by $u\,(R)$ and $T\,(R)$, and is *uniquely* positive.

The use of the ln is not possible out of equilibrium, assuming that $w\,(R, V, t)$ is positive. The fact that local equilibrium is the only positive solution to Eq. (6.28) would imply that all time-dependent solutions are negative. This is a serious pitfall in interpreting Eq. (6.28) as closely analogous to the classical case. How close? This is as yet an unanswered problem suggested by the Hudson theorem.

One way to continue, of course, is to reduce the spacially dependent to the spacially independent case and obtain an equation for the marginal distribution function

$$\varphi\,(V, t) = \int dRw\,(R, V, t)\,.$$

It is

$$\partial_t \phi\,(V, t) = \int\int d\mathbf{V}_1 d\Omega \left\{ \begin{array}{c} \frac{d\sigma\left(V_1 V_2 \to V_1' V_2'\right)}{d\Omega} \left|V_1 - V_2\right| \\ \times \left[\phi\left(V_1', t\right)\phi\left(V_2', t\right) - \phi\left(V_{1,}t\right)\phi\left(V_2, t\right)\right] \end{array} \right\} \equiv J\,(\phi) \tag{6.30}$$

and has a form exactly like the classical case. Since $\phi(V, t) \geq 0$, the same analysis may now be made (Balescu, 1975; Huang, 1987). For equilibrium,

$$\left[\phi_0\left(V_1'\right)\phi_0\left(V_2'\right) - \phi_0\left(V_1\right)\phi_0\left(V_2\right)\right] = 0,$$

or

$$\ln\phi_0\left(V_1'\right) + \ln\phi_0\left(V_2'\right) = \ln\phi_0\left(V_1\right) + \ln\phi_0\left(V_2\right).$$

The collisional constants of the two-body scattering then give

$$\phi_0\,(V) = n\left(\frac{m}{2\pi kT}\right)^{\frac{3}{2}} \exp\left[-\left(\frac{m}{2kT}\right)(V - \mu)^2\right], \tag{6.31}$$

the global Maxwellian (McLennan, 1989). Here $\boldsymbol{\mu}$, T are not spacially dependent but equilibrium thermodynamic properties of the entire homogeneous system. We must note that Eq. (6.30), except for the nonlinearity, is of a form of the Pauli equation discussed in the previous chapter.

The famous Boltzmann $\mathfrak{H}$ theorem (for a homogeneous system) may now be obtained quantum mechanically. We define

$$\mathfrak{H} = \int d_3 V \phi \ln \phi, \tag{6.32}$$

and from the symmetries of the two-body scattering, we may write the fundamental relation

$$\begin{aligned} \int d_3 V h J(\phi) = \frac{1}{4} \int & d_3 V d_3 V_1 d\Omega \, |V_1 - V| \, \sigma \\ & \times \left[h + h_1 - h' - h_1'\right] \left[\phi(V') \phi(V_1') - \phi(V) \phi(V_1)\right], \end{aligned} \tag{6.33}$$

where h is any function of $\mathbf{V}$ (Balescu, 1975; Huang, 1987; McLennan, 1989). Using this, we write the entropy production:

$$\begin{aligned} \frac{d\mathfrak{H}}{dt} &= \int d_3 V \, (1 + \ln \phi) \, J \\ &= \frac{1}{4} \int d_3 V d_3 V_1 d\Omega \left|V_1 - V_1'\right| \sigma \left(\phi' \phi_1' - \phi \phi_1\right) \ln \frac{\phi \phi_1}{\phi' \phi_1'}. \end{aligned} \tag{6.34}$$

And since

$$(y - x) \ln \frac{x}{y} < 0, \qquad y > 0, \ x > 0, \tag{6.35}$$

therefore,

$$\frac{d\mathfrak{H}}{dt} \leq 0. \tag{6.36}$$

For equilibrium we choose

$$\mathfrak{H}_0 = \int d_3 V \phi_0 \ln \phi_0, \tag{6.37}$$

and we have

$$\begin{aligned} \int d_3 V \phi \ln \phi_0 &= \int d_3 V \phi_0 \ln \phi_0 \\ \mathfrak{H} - \mathfrak{H}_0 &= \int d_3 V \phi \ln \frac{\phi}{\phi_0} \\ &= \int d_3 V \phi_0 \left\{1 + \frac{\phi}{\phi_0} \left[\ln \frac{\phi}{\phi_0} - 1\right]\right\}. \end{aligned} \tag{6.38}$$

Thus,

$$\mathfrak{H} \geq \mathfrak{H}_0. \tag{6.39}$$

Here $\mathfrak{H}_0$ is the lower bound of $\mathfrak{H}$. $\mathfrak{H} > 0$ and is monotonically decreasing as $t \to \infty$. Thus, the equilibrium value is obtained as $t \to \infty$ and $\dot{\mathfrak{H}} = 0$. It is a Liapunov function (Liapunov, 1949; Lasalle and Lefschitz, 1961) showing that asymptotically as $t \to \infty$, $\phi(p,t) \to \phi_0(p,t)$, the equilibrium global Maxwellian.

Choose $S = k\mathfrak{H}$. S is the microscopic representation of entropy, and $\dot{S}$ the entropy production. k is Boltzmann's constant, $k = 1.380658 \times 10^{-23}\frac{J}{K}$, introduced for historical reasons. Boltzmann proved Eq. (6.39) classically for $f(\mathbf{R}, \mathbf{V}, \mathbf{t}) \ln f(\mathbf{R}, \mathbf{V}, t)$. We remark that we have here assumed, using $\phi(\mathbf{p}, t)$, that the system is initially homogeneous and evolves homogeneously to the global Maxwellian, ϕ_0. There is no strong proof classically that an initially inhomogeneous system governed by Eq. (6.28) evolves to a homogeneous state governed by ϕ_0. In fact, it is probably not so.

The $\mathfrak{H}$ theorem led to the famous discussion of Boltzmann with Zermelo, the recurrence paradox or the *Wiederkehreinwand.* We invite the student to read this in the marvelous compilation of Brush (1966). Zermelo first argued by the Poincaré recurrence theorem (Poincaré, 1890) that in an isolated classical system, any initial phase state must recur with near precision in a finite time. This being so, how can the monotonic decreasing $\mathfrak{H}$ function be correct, having been derived by his dynamic equation (albeit approximate)? We must note that Zermelo formulated the recurrence theorem in a new way based upon the conservation dynamically of phase. This proof is repeated in the book of Huang (1987) and in the article on stochastic processes by Chandresekhar (1943). The proof that a similar result holds also in quantum mechanics is given by Bocchieri and Loinger (1957). We give this proof in the appendix to this chapter.

How did Boltzmann answer? He agreed that the Poincaré theorem is valid but then introduced the new element of interpreting his equation in a probabilistic sense, as is done universally today. Poincaré recurrences are thus fluctuations from the average, which at some long time may, infrequently, be very large. This picture Boltzmann sketched in his final paper of the series. He said, "I've also emphasized that the second law of thermodynamics is from the molecular point of view merely a statistical law" (Boltzmann, 1896). In addition, in an appendix to his paper, he estimated a macroscopic recurrence time, a task repeated in subsequent years by others. He wrote, for his estimate,

$$n = \text{even} \qquad \frac{N}{b} = \frac{2\,(2\pi)^{\frac{(3n-4)}{2}}\,a^{3(n-1)}}{3 \bullet 5 \bullet\bullet\bullet 3(n-1)}\ \text{sec.}$$

n is the number of molecules, $a = 5 \times 10^{11}$ m/sec, and $b = 2 \times 10^{27}$ collisions per sec. The point is made that a large complete macroscopic recurrence is super-astronomical in time.

In his second paper, Zermelo does not accept the probabilistic view and insists that the entropy principle has not yet been obtained from pure mechanical arguments as *he* desires. He also raises the question of the special role of initial states, saying, "But as long as one cannot make comprehensible the *physical* origin of the initial state one must merely assume what one wants to prove" (Zermelo, 1896). He then turns to the choice of irreversible conditions and says, "Not only is it impossible to explain the general *principle* of irreversibility it is also impossible to explain individual irreversible processes themselves without introducing new physical assumptions, at least as far as the time direction is concerned" (Zermelo, 1896).

Faced with this, Boltzmann, in his second rejoinder, turns easily to the justification of the use of probability in what we might call the law of large numbers. He simply asserts that his theory is designed to be applied to a large system in which n (his) is large, almost macroscopic.

The second question concerning the choice of improbable initial conditions is more difficult. He suggests two possible assumptions: (1) the universe is in an improbable state, and the system chosen from it and isolated from it at some time is also in an improbable state, and the entropy must increase; (2) the universe is in equilibrium. A subsystem fluctuates from equilibrium, and in it, the direction of time is chosen for there always to be an increase in entropy.

The discussion stops, although Zermelo implies he will turn to it again with a purely mechanical answer. At any rate, the great debate has begun which will be taken up with enthusiasm in the next hundred years or longer. I cannot possibly do a complete bibliography here, but only mention the more recent articles of Prigogine (1973) and of Lebowitz (1993).

Now, in order to proceed further with an inhomogeneous case in obtaining the $\mathfrak{H}$ theorem, let us linearize the Boltzmann equation by the Chapman–Enskog procedure. This has been done earlier in Eq. (4.36).

We expand

$$w\left(Rp,t\right)=f_1^0\left(1+\Phi^1+\Phi^2+\cdots\right), \tag{6.40}$$

where the local equilibrium is given by Eq. (6.29). Now, as earlier,

$$\int d_3p\psi\left(w-f_1^0\right)=0$$

for the summational invariants ψ. Thus,

$$\int d_3p\psi\phi f_1^0=\int d_3p\psi\phi^i=0. \qquad i=1,2,\ldots$$

ϕ^1 is linear in the first-order spacial gradients. With this the Boltzmann equation reduces to the *linear* form

$$\left(\frac{\partial}{\partial t}+\mathbf{v}\cdot\boldsymbol{\nabla}\right)\ln f^0=-nI\Phi^1,$$

where

$$nI\Phi=\int d_3v_1 d\Omega\,|\mathbf{g}|\,\sigma f_1^0\left[\Phi+\Phi_1-\Phi'-\Phi_1'\right]. \tag{6.41}$$

Here, as customary, we changed variables from $\mathbf{p}$ to $\mathbf{v}$, $\mathbf{g}=\left|\mathbf{v}-\mathbf{v}'\right|$, and now dropped the superscript 1 in Φ. Again, as discussed in Chapter 3, we may use the hydrodynamic equations to write the right side of Eq. (6.41) as

$$I\Phi=\frac{-1}{nkT^2}\left(\frac{1}{2}mv^2-\frac{5}{2}kT\right)\mathbf{v}\cdot\boldsymbol{\nabla}T-\frac{1}{nkT}m\left(v_iv_j-\frac{1}{3}\delta_{ij}v^2\right)\frac{\partial\mu_i}{\partial x_j}. \tag{6.42}$$

The solution to this linear integral equation may be written as

$$\Phi=-\frac{1}{nkT^2}\mathfrak{S}_i\frac{\partial T}{\partial x_i}-\frac{1}{nkT}\mathfrak{T}_{ij}\frac{\partial\mu_i}{\partial x_j}, \tag{6.43}$$

where the two integral equations are now

$$I\mathfrak{S}_i=\left(\frac{1}{2}mv^2-\frac{5}{2}kT\right)v_i\equiv S_i \tag{6.44}$$

$$I\mathfrak{T}_{ij}=m\left(v_iv_j-\frac{1}{3}\delta_{ij}v^2\right)\equiv T_{ij}.$$

With Eq. (6.43) we may obtain the transport laws for λ, the thermal conductivity, and η, the viscosity. We have outlined this in Chapter 4.

Consider the scalar product,

$$(k,Ih)\equiv n^2\int d_3v d_3v_1 d\Omega\,(\mathbf{g})\,\sigma f^0 k\left[h+h_1-h'-h_1'\right]. \tag{6.45}$$

We may show, similarly to the proof of the fundamental lemma in Eq. (6.33), that

$$(k,Ih)=(Ik,h)\,. \tag{6.46}$$

Thus, for the linearized Boltzmann equation,

$$(h,Ih)\geq 0. \tag{6.47}$$

This is the important result. The equality holds if h is a summational invariant. Using the integral equations given as Eq. (6.44) and the general expressions for λ and η,

$$\lambda = \frac{1}{3kT^2}\left(S_i, \mathfrak{S}_i\right)$$
$$\eta = \frac{1}{kT}\left(T_{xy}, \mathfrak{T}_{xy}\right). \tag{6.48}$$

We recognize

$$n^2 k\left(\Phi, I\Phi\right) = -T^{-2}\mathbf{S}^{*}\cdot\nabla T - T^{-1}t^{*}_{ij}\frac{\partial \mu}{\partial x_j}. \tag{6.49}$$

k is here Boltzmann's constant. This is the expression for the irreversible thermodynamic entropy production. See Eq. (6.16). We have for the entropy production

$$\sigma = n^2 k\left(\Phi, I\Phi\right) \geq 0. \tag{6.50}$$

We have obtained the Boltzmann entropy production theorem for inhomogeneous systems by utilizing the linearized Boltzmann equation and the associated Chapman–Enskog procedure, in order to arrive at the transport coefficients with Eq. (6.48) and Eq. (6.49) (McLennan, 1989). The central points are the inequality, Eq. (6.47), the use of the general thermodynamic definition of entropy production, Eq. (6.16), and the microscopic relation Eq. (6.46). For the inhomogeneous case, we can do no more. Eq. (6.50) may be used as a basis for a variational solution to the linearized Boltzmann equation. One can verify that Eq. (6.42) has solutions, providing

$$\left(\psi, S_i\right) = \left(\psi, T_{ij}\right) = 0.$$

This can be shown to be the case.

6.4 Negative probability and the quantum $\mathfrak{H}$ theorem

In an essay dedicated to David Bohm, Feynman (1988) argued the possibility of negative probabilities in classical and quantum mechanics. There he gave a number of interesting simple examples, from a roulette wheel to a two-level spin system. We invite the reader to enjoy this. He pointed out the Wigner function as a quantum manifestation of negative probability, arguing that such a concept is a helpful and useful approach which need not or should not be an observable quantity.

We will carry this idea further to treat the Wigner function seriously as a negative probability in the derivation of the quantum Boltzmann $\mathfrak{H}$ theorem, directly alleviating some of the difficulties discussed in the previous section.

Consider the Wigner function Boltzmann equation, Eq. (6.28). Define again

$$\mathfrak{H} = \int d\mathbf{R}d\mathbf{v}w\left(\mathbf{R},\mathbf{v},t\right)\ln w\left(\mathbf{R},\mathbf{v},t\right). \tag{6.51}$$

By means of Eq. (6.28) we may form an equation for $\dot{\mathfrak{H}}$ as before. We operate there with

$$\int d\mathbf{R}d\mathbf{v}\left(1+\ln w\right)$$

and obtain

$$\frac{\partial}{\partial t}\int d\mathbf{R}d\mathbf{v}w\ln w = \frac{d\mathfrak{H}}{dt}. \tag{6.52}$$

The external force has been neglected, and so have surface flows at large volume. Here the collisional functional is

$$I\left(\phi\right) = \int d\mathbf{v}d\mathbf{v}_1 d\Omega\left|v-v_1\right|\frac{d\sigma}{d\Omega}\times\phi\left[w_1w-w_1'w'\right]. \tag{6.53}$$

By scattering symmetries we write

$$4I\left(1+\ln w\right) = I\left(1+\ln w\right)+I\left(1+\ln w_1\right)-I\left(1+\ln w_1'\right)-I\left(1+\ln w\right). \tag{6.54}$$

Now consider the difference of the quantum from the classical in the meaning of $\ln w$. Since w may be negative, we must consider the complex representation of $\ln z$:

$$\log z = \log r + i\left(\theta \pm 2n\pi\right); \qquad n = 0, 1, \ldots \tag{6.55}$$

We will choose the principal branch ($n = 0$), and then

$$\log z = \log r + i\theta; \qquad \left(-\pi < \theta \leq \pi\right). \tag{6.56}$$

Maintaining the principal branch, we have the properties

$$\log z_1 + \log z_2 = \log\left(z_1 z_2\right) \tag{6.57}$$

$$\log z_1 - \log z_2 = \log\left(\frac{z_1}{z_2}\right). \tag{6.58}$$

It is the Eq. (6.57) relation that we wish to maintain as a thermodynamic additive property. For the special case of negative probabilities, $z = |w|\exp\left(+i\pi\right)$, and

$$\begin{aligned}\log w_1 + \log w_2 &= \log|w_1| + \log|w_2| + 2i\pi\\ &= \log|w_1| + \log|w_2|\end{aligned} \tag{6.59}$$

on the principal branch. And

$$\log|w_1| - \log|w_2| = \log\frac{|w_1|}{|w_2|}. \tag{6.60}$$

Now, for negative probabilities, we may use Eq. (6.59) and Eq. (6.60) in Eq. (6.52) and obtain from Eq. (6.54) exactly the classical result:

$$4I(\Phi) = -\int dv \int dv_1 \, |v - v_1| \, \sigma d\Omega L(x, y), \tag{6.61}$$

where

$$L(x, y) = (x - y)\ln\left(\frac{x}{y}\right) \tag{6.62}$$

and

$$x = |w'| \, |w'_1| > 0$$
$$y = |w| \, |w_1| > 0.$$

All this is as in the classical case, and $L(xy) \geq 0$, since x, y are positive. Hence,

$$\frac{d\mathfrak{H}}{dt} \leq 0, \tag{6.63}$$

or $S = -k\mathfrak{H}$.

$$\frac{dS}{dt} \geq 0. \tag{6.64}$$

This is true for positive and negative w. The equilibrium value is $\frac{dS}{dt} = 0$, which requires $|w'_1| \, |w'_2| = |w_1| \, |w_2|$. By the familiar argument, already made at Eq. (6.31), we obtain a positive Gaussian distribution. Both the positive and the negative approach the Gaussian distribution. This is consistent with the Hudson theorem previously discussed.

Now out of equilibrium S is complex:

$$S = -k \int d\mathbf{v} d\mathbf{R} w \ln w = k \int d\mathbf{v} d\mathbf{R} \, |w| \left[\ln|w| + \pi i\right]. \tag{6.65}$$

This precludes the "counting" interpretation of entropy. A physical interpretation is not apparent. A few more remarks on this will be made in the section on equilibrium statistical thermodynamics.

6.5 Entropy and master equations

Quantum master equations were discussed and derived in Chapter 3 for open systems. In the previous chapter the elements of irreversibility were derived for these equations.

Here we will turn to dissipation and entropy for such equations and the physics and chemistry they describe. This is parallel to the preceding discussions of kinetic theory. We will see that the discussion is technically quite different and the limits are different from those previously discussed.

First we consider the Pauli equation for *closed* systems (Pauli, 1928). Eq. (3.25) was

$$\frac{d}{dt}P(\alpha,t)=\sum_{\beta}\left[W_{\beta a}P(\beta,t)-W_{a\beta}P(\alpha,t)\right];\qquad t\geq 0. \tag{6.66}$$

$P(\alpha,t)=\rho_{\alpha\alpha}(t)$, the probability of state $|\alpha\rangle$. The interaction potential V is Hermitian, and

$$\begin{aligned} W_{\alpha\beta}&=2\pi\left|V_{\alpha\beta}\right|^{2}\delta\left(E(\alpha)-E(\beta)\right)\\ W_{\alpha\beta}&=W_{\beta\alpha}. \end{aligned} \tag{6.67}$$

We also note that $P(\alpha,t)$ are positive. It is quite simple to obtain an $\mathfrak{H}$ theorem from this equation. We define

$$\mathfrak{H}=\sum_{\alpha}P(\alpha,t)\ln P(\alpha,t). \tag{6.68}$$

It has the additive property for independent systems. We operate on Eq. (6.51) with $\sum_{\gamma}(1+\ln P(\gamma,t))$. We have

$$\frac{d\mathfrak{H}}{dt}=\sum_{\gamma\beta}(1+\ln P(\gamma,t))\left(P(\beta,t)W_{\beta\gamma}-P(\gamma,t)W_{\gamma\beta}\right).$$

Since

$$\sum_{g\beta}P(\beta)W_{\beta\gamma}=\sum_{\gamma\beta}P(\gamma)W_{\gamma\beta},$$

we have

$$\frac{d\mathfrak{H}}{dt}=\sum_{\beta\gamma}\ln P(\gamma,t)\left(P(\beta,t)W_{\beta\gamma}-P(\gamma,t)W_{\gamma\beta}\right).$$

This may be written

$$\frac{d\mathfrak{H}}{dt}=\sum_{\beta\gamma}P(\beta,t)W_{\beta\gamma}\ln\left(\frac{P(\gamma,t)}{P(\beta,t)}\right).$$

By changing indices and using $W_{\gamma\beta}=W_{\beta\gamma}$, we obtain

$$\frac{d\mathfrak{H}}{dt}=\frac{1}{2}\sum_{\beta\gamma}W_{\gamma\beta}\left(P(\beta,t)-P(\gamma,t)\right)\ln\frac{P(\gamma,t)}{P(\beta,t)}. \tag{6.69}$$

We use the same inequality as in the Boltzmann equation, Eq. (6.35). Since $P \geq 0$,

$$(P(\beta, t) - P(\gamma, t)) \ln \frac{P(\gamma, t)}{P(\beta, t)} \leq 0, \tag{6.70}$$

and since $W_{\gamma\beta} > 0$, we obtain the Pauli $\mathfrak{H}$ theorem,

$$\frac{d\mathfrak{H}}{dt} \leq 0; \qquad t \geq 0. \tag{6.71}$$

Not surprisingly, it is the same result as with the spacially independent Boltzmann equation. $\mathfrak{H}$ is again a Liapunov function (Liapunov, 1949), assuring the time-asymptotic stability of the $t \to \infty$ solution. We may again define the thermodynamic entropy, S:

$$S = -k\mathfrak{H}, \tag{6.72}$$

k being Boltzmann's constant. At equilibrium, S is a maximum as desired, and because of stability,

$$\frac{dS}{dt} = 0. \tag{6.73}$$

From Eq. (6.53) we see that the equilibrium solution is

$$\begin{aligned} P(\beta, \infty) &= P(\gamma, \infty) \text{ for } E(\alpha) = E(\gamma) \\ &= 0 \text{ otherwise,} \end{aligned} \tag{6.74}$$

where $|\beta\rangle, |\gamma\rangle$ are states of the unperturbed energy $H^0 |\alpha\rangle = E(\alpha) |\alpha\rangle$, the unperturbed energy shell.

The equilibrium solution is microcanonical. Further, $S_{eq} = +k \sum_\alpha P_{micro}(\alpha) \ln P_{micro}(\alpha)$. We should note that this is microcanonical equilibrium on the unperturbed energy states and not on the state of $H = H^0 + V$, which would be the beginning of a discussion of an equilibrium thermodynamics. This will be discussed in Chapter 7.

Let us now turn to the case of open systems, which is our focus. Here the situation is far more difficult and, as we shall see, less finished. We will begin with the Lindblad–Kossakowski equation (Kossakowski, 1972; Lindblad, 1976). As discussed and derived in Chapter 4, the solutions of this master equation are the necessary and sufficient conditions for $\langle \phi, \rho\phi \rangle$ being positive for any $|\phi\rangle$. Its importance is that it represents a mathematically rigorous quantum, though limited Brownian motion description, concerning which there is much recent interest. However, the physical content is limited, as has been recently discussed by Monro and Gardiner (1996). We consider the Lindblad–Kossakowski equation in a special form (Gorini *et al.,* 1978):

$$\dot{\rho} = L\rho \qquad t \geq 0 \tag{6.75}$$

$$L\rho = \frac{1}{2}\sum_{\alpha,\beta=1}^{N^2-1} C_{\alpha\beta}\left\{\left[V_\alpha^S \rho(t), V_\beta^{\dagger S}\right] + \left[V_\alpha^S, \rho(t) V_\beta^{\dagger S}\right]\right\} \tag{6.76}$$

for an N-level system where

$$H^{SR} = \sum_{\alpha=1}^{N^2} V_\alpha^S \bigotimes V_\alpha^R,\ V^\dagger = V,\ \text{and } C_{\alpha\beta} > 0 \text{ and symmetric.}$$

This results from the Born approximation (weak coupling) with a free Bose or Fermi reservoir. Another point of view which gives this result is the singular limit discussed in Chapter 4. In this case (see Gorini *et al.,* 1978),

$$h_{\alpha\beta}(t) \to C_{\alpha\beta}\delta(t). \tag{6.77}$$

We may overlook the mathematical details but simply be concerned with this restricted form of the Lindblad equation. This will serve our purposes here. In terms of system states $|i\rangle$,

$$H^S = \sum_i E_i |i\rangle\langle i|.$$

We may write the commutators in Eq. (6.76) as

$$\begin{aligned}&[V_\alpha |j\rangle\langle j|\rho|k\rangle\langle k| V_\beta - V_\beta |j\rangle\langle j| V_\alpha |k\rangle\langle k|\rho]\\ &\quad - [V_\alpha |j\rangle\langle j|\rho|k\rangle\langle k| V_\beta - \rho|j\rangle\langle j| V_\beta |k\rangle\langle k| V_\alpha].\end{aligned}$$

Now, in the above, by an argument of Spohn and Lebowitz (1978), we consider diagonal elements of ρ *only,* taking always

$$\dot{\rho} = \sum_i |i\rangle\langle i|\dot{p}_i; \qquad \dot{p}_i \geq 0. \tag{6.78}$$

This is justified by

$$\exp(Lt)(\exp(-iH^S)\rho\exp(iH^S t)) = \exp(-iH^S t)\exp(Lt)\rho\exp(iH^S t). \tag{6.79}$$

The diagonal elements are an invariant space, and we obtain

$$\dot{p}_i = \sum_j\sum_{\alpha\beta} C_{\alpha\beta}\left[(V_\alpha)_{dj}\left(V_\beta^\dagger\right)_{dj} p_j - (V_\beta)_{dj}\left(V_\alpha^\dagger\right)_{dj} p_i\right]; \qquad t \geq 0. \tag{6.80}$$

This is a Pauli equation for the probabilities $p_j(t) \geq 0$. Now

$$W_{ij} = \sum_{\alpha\beta} C_{\alpha\beta} (V_\alpha)_{ij} \left(V_\beta\right)_{ji}. \tag{6.81}$$

We note $W_{ij} \neq W_{ji}$. However, $W_{ij} > 0$, since $V_\beta^\dagger = V_\beta$.

To define the equilibrium state, we assume the KMS (Kubo–Martin–Schwinger) boundary conditions (Huang, 1987) for the density matrix satisfying

$$\mathrm{Tr}\left[\rho A B(t)\right] = \mathrm{Tr}\left[\rho B(t - i\beta) A\right] \tag{6.82}$$

for all observables A, B. From this the reservoir correlation function is

$$h_{\alpha\gamma}(\omega) = \int_{-\infty}^{+\infty} dt \exp(-i\omega t) \,\mathrm{Tr}_R\left(\omega^R V_\gamma V_\alpha(t)\right).$$

This has the time invariance of ω^R, the equilibrium density matrix for the reservoirs. For a simple system, $\omega^R = \exp(-\beta H) Z^{-1}$. We have

$$h_{\alpha\gamma}(\omega) = h_{\gamma\alpha}(-\omega) \exp \beta\omega, \tag{6.83}$$

and thus from the Pauli equation at equilibrium where $\dot{p}_j = 0$,

$$W_{jk} \exp\left(-\beta E_k^S\right) = W_{kj} \exp\left(-\beta E_j^S\right) \geq 0. \tag{6.84}$$

We will define the conditional entropy

$$S_C\left(\frac{f}{g}\right) = -k\mathrm{Tr} f \log \frac{f}{g}. \tag{6.85}$$

f and g are positive probabilities. We will now use a theorem due to Voigt (1981). We may view Eq. (6.80) with Eq. (6.81) as a Markov equation for p_i. We write Eq. (6.80) as

$$\dot{p}_i(t) = P_t p_i(t); \qquad t \geq 0. \tag{6.86}$$

It has the property $p_i(t) \geq 0$. The Voigt theorem is from Mackey's book, with slight rewording (Mackey, 1992).

Let P_t be a Markov operator; then

$$S_C\left(\frac{P_t f}{P_t g}\right) \geq S_C\left(\frac{f}{g}\right) \tag{6.87}$$

for $f \geq 0$ and for all probabilities g.

Now let $g = p_i^*$, the equilibrium solution $p_i^* = P_t p_i^*$, and $f = p_i(t)$. We have

$$S_C\left(\frac{P_t p_i}{p_i^*}\right) \geq S_C\left(\frac{p_i(t)}{p_i^*}\right); \qquad t > 0. \tag{6.88}$$

The conditional entropy is an increasing nondecreasing function which has as a maximum $S_C\left(p_i^* \mid p_i^*\right) = 0$. Note that the Markov process is irreversible, since the Pauli equation, Eq. (6.80), also is irreversible. The main limitation for a more general argument based on the Lindblad equation, Eq. (6.75), is the following: although ρ is of a completely positive form, it cannot, in general, be construed to be of the Pauli form, and thus Voigt's theorem cannot be assured. However, we could have used this argument earlier for the closed system Pauli equation, Eq. (6.76). There the system has a microcanonical equilibrium, and $S\left(P_t p_i\right) \geq S\left(p_i\right)$ for all p_i.

Let us consider a different interpretation of this result by returning to entropy production and flows previously considered in this chapter (Spohn and Lebowitz, 1978; Mackey, 1992). We begin with the entropy continuity law,

$$\frac{\partial S}{\partial t} = -\mathrm{div}\mathbf{J}_s = \sigma, \tag{6.89}$$

σ being the local entropy production and $\mathbf{J}_s$ the flow. We integrate on the system coordinates and obtain

$$\frac{dS_{\text{total}}}{dt} = \sigma_{\text{total}} - J_{\text{total}}. \tag{6.90}$$

J_{total} is the total entropy flow into the system due to the energy loss in the reservoir:

$$J_{\text{total}} = -\frac{\beta dQ}{dt} = \frac{d\mathrm{Tr}\rho\left(t\right)}{dt} \log \rho_\beta. \tag{6.91}$$

Here $\rho_\beta = Z^{-1} \exp\left(-\beta H^R\right)$; we assume a single reservoir in canonical equilibrium as discussed in this section. A steady state may not be possible, nor is it necessary for the discussion now being presented. Now we integrate over time. We have

$$S\left(t\right) + \mathrm{Tr}\rho\left(t\right) \log \rho_\beta - S\left(0\right) - \mathrm{Tr}\rho\left(0\right) \log \rho_\beta = \int_0^t \sigma_{l\text{total}}\left(t\right) dt.$$

We introduce the conditional entropy, $S_C\left(\frac{f}{g}\right)$, and obtain

$$\left[S_C\left(\rho\left(t\right) \mid \rho^*\right) - S_C \rho\left(0\right) \mid \rho^*)\right] = \int_0^t \sigma_{\text{total}}\left(t\right) dt = \bar{\sigma}\left(t\right). \tag{6.92}$$

Thus, the Voigt theorem gives

$$\int_0^t \sigma_{\text{total}}\left(t\right) dt \geq 0; \qquad t \geq 0. \tag{6.93}$$

The time average of the total entropy production over any infinitesimal time is positive. The change in the conditional entropy is related to the time average entropy production.

Let us now return and comment on the fundamental result, which assures that $S_C\left(P_t p_i \mid p_i^*\right)$ is ever increasing to its maximum, $S_i\left(p_i^* \mid p_i^*\right) = S_C(1) = 0$, where

$$P_t p_i^* = p_i^* \tag{6.94}$$

is the equilibrium state. We have assumed that it exists and is unique. We must then examine the zero eigenvalues of the Pauli equation, Eq. (6.80), which we will not do. A slightly more general derivation of these results has been given by Spohn and Lebowitz (1978). It is also more complicated.

Spohn and Lebowitz have argued that the results may be generalized to more than one independent reservoir, $L = L_1 + \bullet\bullet + L_r,\ \beta_1 \bullet\bullet \beta_r$ and

$$\bar{\sigma}_{\text{total}} = \sum_{k=1}^{r} \bar{\sigma}_{k\ \text{total}}. \tag{6.95}$$

From here we will drop the “total” in the notation, leaving it to be understood.

Now we assume the steady state thermodynamic postulate

$$\sigma = \sum_{r=1}^{k} X_k J_k, \tag{6.96}$$

and for the heat flow case we take

$$X_k = \left(\beta - \beta_k\right). \tag{6.97}$$

Further, we introduce the linear transport coefficient assumption

$$J_k = \sum_{j=1}^{r} L_{kj} X_j. \tag{6.98}$$

Thus, as designed, we have

$$\sigma = \sum_{k,j=1}^{r} L_{kj}(\beta) X_k X_j. \tag{6.99}$$

The entropy production is quadratic in X_k, and since σ is positive and real, so are $L_{kj}(\beta)$.

The symmetry will now be examined. This is the Onsager result (Onsager, 1931). Let us now use the Green–Kubo formula for the thermal conductivity matrix (Lebowitz and Shimony, 1962; McLennan, 1989). This will be discussed in Chapter 15. Here

$$L_{kj}(\beta) = \int_0^\infty dt\ \mathrm{Tr}\left[J_{k\beta}(t)\, J_{j\beta}\rho_\beta\right], \tag{6.100}$$

where $J_{k\beta}(t) = \exp(L^* t) J_{k\beta}$, L^* being again the Markovian weak coupling Heisenberg operator.

The steady state is $L_S(\rho_S) = 0$. We may now write the detail balance condition (Kossakowski *et al.*, 1977), since $V^* = V$, and $L^* = L$ in the model considered, Eq. (6.76). For steady state $L_S \rho_S = 0$,

$$\exp(L_S t)\left(L_{kj}\rho_\beta\right) = \exp\left(L_S^* t\right)(L_{kj}\rho_\beta). \tag{6.101}$$

This is difficult to discuss in more general cases (Gorini *et al.*, 1978; Spohn and Lebowitz, 1978) and is a major obstacle to general proofs of Onsager's result. To continue, and using the condition $\left[J_{kj}, \rho_\beta\right] = 0$, we have

$$\begin{aligned}\operatorname{Tr}\left[\exp\left(L_S^* t\right) J_{k\beta} J_{j\beta}\rho_\beta\right] &= \operatorname{Tr}\left[J_{k\beta}\exp(L_S t) J_{j\beta}\rho_\beta\right] = \\ \operatorname{Tr}\left[J_{k\beta}\exp\left(L_S^* t\right) J_{j\beta}\rho_\beta\right] &= \operatorname{Tr}\left[\exp\left(L^* t\right) J_{j\beta} J_{k\beta}\rho_\beta\right],\end{aligned} \tag{6.102}$$

leading to the Onsager symmetry

$$L_{kj} = L_{jk}. \tag{6.103}$$

Appendix 6A: quantum recurrence

The proof of quantum recurrence (Bocchieri and Loinger, 1957) is quite direct. It says that, given a discrete energy Schrödinger state $\psi(t)$, having its value $\psi(t_0)$ at $t = t_0$, there exists a time T for which $\|\psi(T) - \psi(t_0)\| < \varepsilon$ for an arbitrary small ε.

The formal proof is to consider the solution

$$\psi(t) = \sum_{n=0}^{\infty} r_n \exp(iE_n t)\, u_n,$$

where $Hu_n = E_n u_n$. r_n are real and positive. Thus,

$$\|\psi(T) - \psi(t_0)\| = 2\sum_{n=0}^{\infty} r_n^2 \left(1 - \cos E_n (T - t_0)\right).$$

We may choose N such that

$$\sum_{n=N}^{\infty} \left(1 - \cos E_n (T - t_0)\right) < \varepsilon.$$

Then we prove by the property of almost periodic functions that there exists a $T - t_0$ such that

$$\sum_{n=0}^{N-1} (1 - \cos E_n (T - t_0)) < \varepsilon.$$

The theorem fails for a continuous spectrum.

A considerable discussion is given in the book of Schleich (2001) of estimates of recurrent times, particularly in quantum optic models. Experimental results are also discussed extensively. The theorem was worked out for density matrices by Percival (1961). He placed conditions on the interaction potentials for quantum recurrences to occur in the entropy. No estimates were made of T.

References

Balescu, R. (1975). *Equilibrium and Non-equilibrium Statistical Mechanics* (New York, Wiley), rev. 1999 as *Matter out of Equilibrium* (London, Imperial College Press).

Bocchieri, P. and Loinger, A. (1957). *Phys. Rev.* **107**, 337.

Boltzmann, L. (1872). Further studies on the thermal equilibrium of gas molecules. Trans. S. Brush (1966) in *Kinetic Theory* **2** (London, Pergamon).

Boltzmann, L. (1896). Reply to Zermello's remarks on the theory of heat. Trans. S. Brush (1966) in *Kinetic Theory* **2** (London, Pergamon).

Brush, S. (1966). *Kinetic Theory* **2** (Oxford, Pergamon).

Callen, H. (1985). *Thermodynamics*, 2nd edn. (New York, Wiley).

Chandresekhar, S. (1943). *Rev. Mod. Phys.* **15**, 1.

Chapman, S. and Cowling, T. G. (1970). *The Mathematical Theory of Non-uniform Gases*, 3rd edn. (London, Cambridge University Press).

de Groot, S. R. and Mazur, P. (1962). *Non-equilibrium Thermodynamics* (Amsterdam, North-Holland).

Feynman, R. P. (1988). *Quantum Implications in Honor of David Bohm*, ed. B. J. Hiley and F. D. Peat (London, Routledge).

Gorini, V., Frigerio, A., Kossakowski, A. and Sudarshan, E. C. G. (1978). *Rep. Math. Phys.* **13**, 149.

Green, H. S. (1951). *J. Math. Phys.* **2**, 344.

Huang, K. (1987). *Statistical Mechanics*, 2nd edn. (New York, Wiley).

Hudson, R. (1974). *Rep. Math. Phys.* **6**, 249.

Jou, D. (1993, 1996). *Extended Irreversible Thermodynamics* (New York, Berlin, Springer).

Kondipudi, D. and Prigogine, I. (1998). *Modern Thermodynamics* (New York, Berlin, Wiley).

Kossakowski, A. (1972). *Rep. Math. Phys.* **3**, 247.

Kossakowski, A., Frigerio, A., Gorini, V. and Verri, M. (1977). *Commun. Math. Phys.* **57**.

Kubo, R. (1957). *J. Phys. Soc. Jpn.* **12**, 570.

Lasalle, J. P. and Lefschitz, S. (1961). *Stability by Liapunov's Direct Method with Applications* (New York, Academic Press).

Lebowitz, J. (1993). *Phys. Today,* **Sept.**, 32.

Lebowitz, J. and Shimony, A. (1962). *Phys. Rev.* **128**, 391.

Liapunov, A. M. (1949). Problème général de la stabilité du movement. *Ann. Math. Studies* **17** (Princeton, Princeton University Press).

Lindblad, G. L. (1976). *Commun. Math. Phys.* **48**, 119.

Mackey, M. G. (1992). *Time's Arrow: The Origins of Thermodynamic Behaviour* (New York, Springer).
McLennan, J. A. (1989). *Introduction to Non-equilibrium Statistical Mechanics* (Englewood Cliffs, N.J.: Prentice Hall).
Munro, W. J. and Gardiner, C. W. (1996). *Phys. Rev. A* **53**, 2633.
Onsager, L. (1931). *Phys. Rev.* **37**, 405 and **38**, 2265.
Pauli, W. (1928). *Festschrift zum 60 Geburtstag A. Sommerfeld.* (Leipzig, Hirzl).
Percival, I. (1961). *J. Math. Phys.* **2**, 235.
Poincaré, H. (1890). "Sur le probléme des trois corps et les équations de la dynamique," *Acta Math.,* 13, 1.
Prigogine, I. (1967). *Introduction to Thermodynamics of Irreversible Processes* (New York, Interscience).
Prigogine, I. (1973). Statistical interpretation of non-equilibrium entropy. In *Acta Physica Austriaca,* **Suppl. 10**, *The Boltzmann Equation,* ed. G. D. Cohen and W. Thirring (New York, Springer).
Schleich, W. P. (2001). *Quantum Optics in Phase Space* (Berlin, Wiley–VCH).
Spohn, H. and Lebowitz, J. L. (1978). *Adv. Chem. Phys.* **38**, 109, ed. S. A. Rice and I. Prigogine. (New York, Wiley).
Voigt, J. (1981). *Commun. Math. Phys.* **81**, 31.
Zermelo, E. (1896). On the mechanical explanation of irreversible processes. Trans. S. Brush (1966) in *Kinetic Theory,* **2** (London, Pergamon).

7

Global equilibrium: thermostatics and the microcanonical ensemble

We shall assume here that the total system $H = H_S + H_R + V \equiv E$ is isolated. Thus,

$$\left[H_{S,}\rho_{eq}\right] = 0, \tag{7.1}$$

and

$$\frac{d\rho_{eq}}{dt} = 0; \qquad \rho_{eq}(t) = \rho_{eq}(0).$$

There has been no proof of this state, in the preceding chapters, for the total H. There were arguments that the system in interaction with the reservoir, in some approximation (basically weak coupling), approaches a state for which $\left[H_S^0, \rho\right] = 0$. Eq. (7.1) is a fundamental assumption whose justification is based on empirical results. It further carries with it an additional ansatz that the constant of the motion H is unique for systems with many degrees of freedom.

In classical dynamics a sufficient number of other constants may lead to the integrability of the dynamic equations (Farquahar, 1964; Balescu, 1975; Lichtenberg and Lieberman, 1991; Scheck, 1999). Then the motion may be defined on a subset of the "energy surface." It must be conjectured that most systems of a large number of degrees of freedom have only the energy as a constant. This is born out in the proof of J. G. Sinai (1963) that a system of N hard spheres in a box has no integrals other than the energy.

These matters may have only indirect effect in quantum mechanics, where the question of the number of simultaneously commuting observables plays a similar role. We will assume that, for a large system, only the total energy may be observed. Thus, in the equilibrium state, ρ_{eq}, we may choose

$$\rho_{mn} = a_m \delta_{mn}, \tag{7.2}$$

where $H\,|m\rangle = E_m\,|m\rangle$. We now drop the explicit notation "eq." Since we know nothing concerning a fine structure on the surface of constant energy, we make the equal a-priori hypothesis of Tolman (1938) and choose

$$a_m = 1 \quad E \leq E_m \leq E + \Delta E \tag{7.3}$$
$$a_m = 0 \quad \text{otherwise.}$$

This is the microcanonical ensemble and very much a classical distribution. It is a mixture, as can be seen by writing

$$\rho = \Omega^{-1} \sum_{E \leq E_m \leq E_m + \Delta E} |m\rangle \langle m| \, . \tag{7.4}$$

Now normalization gives

$$\text{Tr}\rho = 1; \qquad \Omega = \sum_{E \leq E_m \leq E_m + \Delta E} \, . \tag{7.5}$$

Ω is a sum of the states in $E \leq E_m \leq E + \Delta E$ and is a function of E, N and V, the last being the macroscopic number of particles and volume.

7.1 Boltzmann's thermostatic entropy

Carved on the tombstone of Boltzmann in the Zentral Friedhof in Vienna is the formula

$$S = k \log W. \qquad (W \text{ is here } \Omega.) \tag{7.6}$$

This is the remarkable connection of a macroscopic quantity, the thermostatic entropy, to probability and the number of microstates. However, in his famous paper of 1877 (Boltzmann, 1877), he introduced entropy in the $f \ln f$ form, which served the same purpose for him. The formula itself, in the form of Eq. (7.6), is apparently due to Planck (1923). k is Boltzmann's constant, as mentioned in Chapter 4.

How do we understand this? The thermostatic entropy for a homogeneous isolated system must be a function $f(\Omega)$ of the number of microstates leading to the macroscopic S. Ω is termed the thermodynamic probability. Now two independent systems, ϕ_1 and ϕ_2 in Hilbert space, form a resulting state $\phi_1 \bigotimes \phi_2$, and consequently

$$\Omega = \Omega_1 \Omega_2. \tag{7.7}$$

Thus, by the law of independent classical probabilities,

$$f(\Omega_1 \Omega_2) = f(\Omega_1) + f(\Omega_2) \, .$$

In the light of Chapter 1, this is a reasonable assumption, since the assumption of equal a-priori probabilities leading to Ω is classical. The only way for this to be true is if $S = k \ln \Omega$, k being a constant entering for dimensional reasons $1.38 \times 10^{-23} J K^{-1}$. The really important point is Boltzmann's connection of S to microscopic probabilities. This, of course, is also true in the modern interpretations of the Boltzmann equation and its consequences, already discussed extensively in Chapters 4 and 6.

7.2 Thermostatics

Once we have the equation for entropy,

$$S(E, V) = k \ln \Omega(E, V), \tag{7.8}$$

we are in a position to obtain thermostatics from the microcanonical distribution (Callen, 1985). From Eq. (7.8) we solve for $S(E, V)$, knowing $\Omega(E, V)$. Then

$$dS(E, V) = \left(\frac{\partial S}{\partial E}\right)_V dE + \left(\frac{\partial S}{\partial V}\right)_E dV, \tag{7.9}$$

and we now define the absolute temperature and pressure as

$$T^{-1} = \frac{\partial S}{\partial E} |_V \tag{7.10}$$

$$P = -\frac{\partial E}{\partial V} |_S . \tag{7.11}$$

Now we may write

$$TdS = dE - đW, \tag{7.12}$$

where the quasistatic work is

$$đW = -PdV.$$

Also, we may identify heat flux as

$$đQ \equiv TdS. \tag{7.13}$$

For systems with fixed particle number, being considered here, we have the first law of thermodynamics and the definition of S. We, with Callen, further assume that $T > 0$. Further results of this are done by Callen extensively in his book. With these results, simply from micro statistical ensembles, we have derived the macroscopic thermostatic (thermodynamic!) laws. It is all based on Boltzmann's assertion in Eq. (7.1).

The Einstein model of a lattice is a nice illustrative model of the microscopic view. There are $3N$ vibrational modes. E is quantized with $\frac{E}{\hbar\omega_0}$ quanta. This is the problem of placing $\frac{U}{\hbar\omega_0}$ (integer) indistinguishable balls in $3N$ distinguishable states (boxes). The result is

$$\Omega = \frac{\left(3N + \frac{E}{\hbar\omega_0}\right)!}{(3N!)\left(\frac{E}{\hbar\omega_0}\right)!}. \tag{7.14}$$

Using Stirling's formula for $N!$,

$$\ln(N!) = N\ln - N,$$

we obtain the molar entropy

$$s \equiv \frac{S}{N_A} = 3R\ln\left(1 + \frac{\mu}{\omega_0}\right) + 3R\frac{\mu}{\omega_0}\ln\left(1 + \frac{\omega_0}{\mu}\right). \tag{7.15}$$

N_A is Avogadro's number where the equation of state is

$$\mu \equiv 3N_A\hbar\omega_0. \tag{7.16}$$

It is characteristic of this model that $T\frac{\partial S}{\partial V} = 0$. We have

$$\frac{1}{T} = \frac{\partial S}{\partial E} = \frac{k}{\hbar\omega_0}\ln\left(1 + \frac{3N}{E}N_A\hbar\omega_0\right).$$

7.3 Canonical and grand canonical distribution of Gibbs

We will take the entire isolated universe, system plus reservoir, to be in microcanonical equilibrium and from this obtain the system statistical state, which will be canonically characterized by a parameter, β. For the moment, no particle interchange is possible with the reservoir. Only energy may change in the system. Let P_j be the probability that system S is in state E_j. We have

$$P_j = \frac{\Omega_R\left(E_T - E_j\right)}{\Omega_T\left(E_T\right)}. \tag{7.17}$$

The reservoir is assumed to be so large that it is microcanonically distributed. The numerator is the number of reservoir states which are in $E_T - E_j$. These are *a priori* uniformly distributed. Thus Eq. (7.17) is the functional number of reservoir states for which $E_T - E_j$, thus reflecting indirectly thc system probability P_j, a remarkable result indeed! Using Boltzmann's formula we have

$$P_j = \frac{\exp\left(k^{-1}S_R\left(E_T - E_j\right)\right)}{\exp\left(k^{-1}S_{S+R}\left(E_T\right)\right)}. \tag{7.18}$$

Let U be the average energy of S. We may expand

$$S_R\left(E_T - E_j\right) = S_R\left(E_T - U\right) - \frac{\left(E_j - U\right)}{T}$$

and obtain

$$P_j = \exp \beta F \exp\left(-\beta E_j\right). \tag{7.19}$$

This is Gibbs's canonical distribution (Gibbs, 1961); see also Tolman, 1938; Balescu, 1975; Callen, 1985). Here $\beta = 1/kT$, and we identify

$$F = U - TS \tag{7.20}$$

as the Helmholtz free energy. Define the partition function, $Z = \exp(-\beta F)$. Then normalization of Eq. (7.19) gives

$$Z = \sum_j \exp\left(-\beta E_j\right). \tag{7.21}$$

This is the cornerstone of equilibrium calculations. We obtain

$$-\beta F = \ln Z \tag{7.22}$$

and

$$U = -\frac{\partial(\ln Z)}{\partial \beta}$$

$$P = \beta^{-1}\frac{\partial(\ln Z)}{\partial V} \tag{7.23}$$

$$S = k \ln Z + k\beta U. \tag{7.24}$$

From Eq. (7.24) using Eq. (7.19), we find that

$$S = -k \sum_j P_j \ln P_j \tag{7.25}$$

in terms of the canonical distribution of the system, S.

Comments should be made here concerning the equilibrium entropy. We note that Eq. (7.11) is for the system in interaction with the reservoir, in a sense a "reduced" entropy. The Boltzmann entropy, Eq. (7.1) upon which it is based, is the entropy of the universe (system plus surroundings). Both are in time-independent equilibrium. The entropy production form, Eq. (7.25), has been achieved previously for the time-dependent Pauli equation and Boltzmann equation in special cases, as in the asymptotic time limit $t \to \infty$ (see Chapter 6). This does not at all justify Eq. (7.11) as used here. We further note that $S = -k\mathfrak{H}$, $\mathfrak{H}$ being the function $\mathfrak{H}$ of previous chapters. From general statistical considerations, Shannon has related this to statistical uncertainty or disorder (Shannon and Weaver, 1949). This has important engineering applications.

Let us now generalize the results to allow exchange of particles between the reservoir and the system. For the system plus reservoir, we use the simultaneous eigenstate

$$\hat{H}_{S+R} \left|\mu\left(E_j, N\right)\right\rangle = E_j N \left|\mu\left(E_j, N\right)\right\rangle$$
$$\hat{N} \left|\mu\left(E_j, N\right)\right\rangle = N_j \left|\mu\left(E_j, N\right)\right\rangle.$$

Here $\hat{N}$ is the number operator which commutes with $\hat{H}_{S+R}$. It is assumed that $N = \sum_j N_j$. Now we write for the universe $\Omega(E, N)$. Using this and the same argument as in the canonical case,

$$P\left(E_j, N_j\right) = \frac{\Omega_R\left(E_T - E_j, N_T - N_j\right)}{\Omega_T\left(E_T N_T\right)}. \tag{7.26}$$

As before, $E_T \equiv E$ and $N_T \equiv N$ emphasize the conserved quantities of the system plus reservoir. We now obtain

$$P\left(E_j, N_j\right) = \exp\left[\frac{1}{k} S_R\left(E_T - E_j, N_T - N_j\right) - \frac{1}{k} S_{S+R}\left(E_T, N_T\right)\right] \tag{7.27}$$

and consequently, on using a Taylor series expansion,

$$P\left(E_j, N_j\right) = \exp\left(\beta\psi\right)\exp\left(-\beta\left(E_j - \mu N_j\right)\right). \tag{7.28}$$

This is the grand canonical distribution where ψ is the grand canonical potential

$$\psi = U - TS - \mu N. \tag{7.29}$$

Normalization of Eq. (7.28) gives

$$\exp(-\beta\psi) = Z_G = \sum_j \exp\left(-\beta\left(E_j - \mu N_j\right)\right). \tag{7.30}$$

This is the grand partition function. It is a function of β and μ, which must be obtained by an additional condition on the number of particles. It is the chemical potential or the Gibbs potential. We may show that

$$d\mu = T ds - P dv. \tag{7.31}$$

Introducing $U = \langle N\rangle u,\; S = \langle N\rangle s$ and $V = \langle N\rangle v$, we obtain

$$dU = T ds - P dV + (u - Ts + Pv)\, d\langle N\rangle \tag{7.32}$$

and

$$g = (u - Ts + Pv) = \mu, \tag{7.33}$$

the Gibbs free energy per mole. Finally, we may collect all the statistical mechanics connections to thermodynamics for this ensemble:

$$\begin{aligned} U &= \langle E \rangle \\ S &= k\left[\ln Z_G + \beta \langle E \rangle - \beta\mu \langle N \rangle\right] \\ g &= \mu \\ \langle E \rangle &= -\frac{\partial}{\partial \beta}(\ln Z_G) + \mu kT \frac{\partial}{\partial \mu}(\ln Z_G) \end{aligned} \tag{7.34a}$$

$$\begin{aligned} P &= kT \frac{\partial}{\partial V}(\ln Z_G) \\ \langle N \rangle &= kT \frac{\partial}{\partial \mu}(\ln Z_G) \end{aligned} \tag{7.34b}$$

We have tabulated these separately to emphasize that Eqs. (7.34b) represent simultaneous relations determining the equation of state *and* the Gibbs potential. Examples given later will emphasize this. Note that there is a variety of notations for Z_G, the grand partition function. To finish, we will calculate the equilibrium fluctuations about $\langle E \rangle$ and $\langle N \rangle$ with this ensemble (Tolman, 1938; Callen, 1985).

7.4 Equilibrium fluctuations

When the system is in interaction with the reservoir in equilibrium, we may expect that there are fluctuations in the thermodynamic variables. Fundamental references are Einstein (1910); Landau and Lifschitz (1967); and Callen (1985).

Consider first the canonical ensemble. We have

$$\begin{aligned} \langle E \rangle \equiv \bar{E} &= Z^{-1} \sum_k E_k \exp(-\beta E_k) \\ &= -Z^{-1} \frac{\partial Z}{\partial \beta}. \end{aligned} \tag{7.35}$$

Now

$$\Delta^2 E \equiv \bar{E^2} - (\bar{E})^2 \tag{7.36}$$

is the variance of E. We find

$$\bar{E^2} = Z^{-1} \frac{\partial^2 Z}{\partial \beta^2}.$$

Thus

$$\Delta^2 E = \frac{-\partial U}{\partial T}\frac{\partial T}{\partial \beta} = kT^2 C_V \tag{7.37}$$

in the case of PdV work, C_V being the heat capacity and positive. This is, in fact, a macroscopic stability criterion. If we expand $P(E)$ about equilibrium, the positive C_V of the resulting Gaussian behavior assures this. The magnitude of these fluctuations may be estimated as

$$\frac{\Delta(E)}{\bar{E}} = \frac{1}{\left(\frac{3\bar{N}}{2}\right)^{\frac{1}{2}}}$$

and is unimportant for the system in equilibrium with a large number of particles. Similarly from the grand ensemble we may find that

$$\beta\bar{N} = \frac{\partial}{\partial\mu}(\ln Z_G)$$

and

$$\Delta^2(N) = \beta^{-1}\left(\frac{\partial\bar{N}}{\partial\mu}\right)_{V,T}. \tag{7.38}$$

We expect $\frac{\partial\bar{N}}{\partial\mu} \approx \bar{N}$, so $\frac{\Delta N}{\bar{N}} \approx \bar{N}^{-\frac{1}{2}}$ with the same conclusion as earlier.

There is another useful thermodynamic relation. The isothermal compressibility is

$$K_T = \frac{-1}{V}\left(\frac{\partial V}{\partial P}\right)_{N,T}. \tag{7.39}$$

We may obtain $d\mu = +vdP - sdT$ and show $\left(\frac{\partial\mu}{\partial v}\right)_T = v\left(\frac{\partial P}{\partial v}\right)_T$ and thus

$$\frac{-N^2}{V}\left(\frac{\partial\mu}{\partial N}\right)_{V,T} = V\left(\frac{\partial P}{\partial V}\right)_{N,T}. \tag{7.40}$$

Consequently

$$\frac{(\Delta N)^2}{\bar{N}} = \frac{K_T}{\beta v}. \tag{7.41}$$

Hence K_T must also be positive. A main point of these results is that the thermostatic laws become *exact* in the limit $N \to \infty$. T. G. Kurtz (1972) has proved a similar result for chemical kinetics. The chemical reaction equations are exact as $\bar{N} \to \infty$, and the stochastic effects play no role.

7.5 Negative probability in equilibrium

Let us return to the possibility of negative probabilities in the light of the Wigner representation. This discussion was begun in Chapter 6. In the canonical ensemble formula, Eq. (7.19), we assume that w may have negative values for P_j. Consequently,

$$S = -k \sum_j |P_j| \ln |P_j| - k\pi i \sum_j |P_j| . \tag{7.42}$$

After Feynman (1988), we normalize the "probability" as

$$\sum_j P_j = 1 \tag{7.43}$$

and write

$$S = S_r + S_i, \tag{7.44}$$

where

$$S_r = -k \sum_i |P_i| \ln |P_i| \tag{7.45a}$$

$$S_i = -ki\phi, \tag{7.45b}$$

where the "angle"

$$\phi = \pi \sum_j |P_j| < \pi . \tag{7.46}$$

The entropy is complex, having a phase ϕ. This phase is obtained from Eq. (7.46). We may view the entropy as a function $S(E, V, \phi)$, ϕ being a *macroscopic* thermodynamic phase variable. Hence

$$dS = \frac{\partial S}{\partial E} |_{V,\phi}\, dE + \frac{\partial S}{\partial V} |_{E\phi}\, dV + \frac{\partial S}{\partial \phi} |_{V,E}\, d\phi, \tag{7.47}$$

where, as before,

$$\frac{\partial S}{\partial E} |_{V\phi} = \frac{1}{T}$$

$$\frac{\partial S}{\partial V} |_{E\phi} = \frac{P}{T}$$

$$\frac{\partial S}{\partial \phi} |_{V,E} = -ki \, ,$$

$-ki$ being a small imaginary constant. The real part of the entropy obeys

$$TdS_r = dE + PdV, \tag{7.48a}$$

assuming the heat flow and work are real. (Need this be true?)

$$dS_i = -kid\phi \tag{7.48b}$$

is governed by the distribution of the "positivity" of the probability distribution

$$\pi \sum_i P_i - \pi \sum_i |P_i| = d\phi. \tag{7.49}$$

This is determined by the reservoir.

What is the mechanism for this macroscopic quantum effect? If $d\phi = 0$, we would then insist on positive probabilities, P_i. Of course, we may attempt to maintain Eq. (7.49) on an equilibrium thermodynamic scale. Feynman obtains such results for a microscopic model based on the assumption of the possibility of negative P_i.

7.6 Non-interacting fermions and bosons

Let us consider the important examples of systems of *non-interacting* Fermi and Bose particles. Assume the energy of a single-particle quantum state to be E_k $(k = 1\ldots)$. The total energy of the non-interacting system of identical particles is $E_{\{n\}} = \sum_k n_k E_k$, where $\{n\} = (n_1, n_2, \ldots, n_k, \ldots)$, n_k being the number of particles in single particle state k. This, of course, is the occupation number representation which may be systematically developed by the methods of second quantization (see Balescu, 1975; Plischke and Bergersen, 1989).

The total number of particles is $N = \sum_k n_k$. We write the grand ensemble partition function as

$$Z_G = \sum_{N=0}^{\infty} \exp(\beta N \mu) \sum_{\{n\}}{}' \exp\left(-\beta \sum_k n_k E_k\right). \tag{7.50}$$

The restrictive prime on the summation in Eq. (7.50), $N = \sum_k n_k$, is removed by the first summation on all N. Thus,

$$Z_G = \sum_{n_1} \sum_{n_2} \sum_{n_3} \ldots \exp\left(\beta \sum_k (\mu - E_k)\right) n_k = \Pi_k \left\{\sum_{n_k} \exp \beta (\mu - E_k) n_k\right\}.$$

There are two occupation number possibilities brought out in the symmetries of the many independent particle wave functions. The Fermi states are Slater determinants, and the bosons are so-called permanents. For fermions, $n_k = 0, +1$ only, as a result of the Pauli exclusion principle, whereas the boson states allow $n_k = 0, 1, 2, \ldots$ This is because of the fundamental commutation laws of the latter and anticommutation in the case of the former.

For Fermi–Dirac we obtain

$$\sum_{n_i=0,1} \exp\left(\beta\left(\mu - E_k\right) n_i\right) = 1 + \exp\left(\beta\left(\mu - E_k\right)\right),$$

and for Bose

$$\sum_{n=0}^{\infty} \exp\left(\beta\left(\mu - E_k\right)\right) = \left(1 - \exp\beta\left(\mu - E_k\right)\right)^{-1}.$$

We may write concisely

$$Z_G = \Pi_k \left\{1 \mp \exp\left(\beta\left(\mu - E_k\right)\right)\right\}^{\mp 1}. \tag{7.51}$$

(−1) Bose

(+1) Fermi–Dirac

Now we may show that $P(n_k)$, the probability that n_k particles occupy state E_k, is

$$P(n_k) = \frac{\exp\beta\left(\mu - E_k\right) n_k}{\sum_{n_k} \exp\beta\left(\mu - E_k\right) n_k}.$$

Hence,

$$\langle n_k \rangle = \frac{\sum_{n_k} n_k \exp\beta\left(\mu - E_{,}\right) n_k}{\sum_{n_k} \exp\beta\left(\mu - E_k\right) n_k}. \tag{7.52}$$

Thus,

$$\langle n_k \rangle = \frac{\exp\beta\left(\mu - E_k\right)}{1 \mp \exp\beta\left(\mu E_k\right)}, \tag{7.53}$$

which are the Fermi–Dirac (+) and Bose (−) distributions. Now we may show that

$$PV = kT \ln Z_G \tag{7.54}$$

and obtain

$$PV = \mp kT \sum_{n_k} \ln\left(1 + \langle n_k \rangle\right). \tag{7.55}$$

The thermodynamic quantities are

$$N = \sum_k \langle n_k \rangle \text{ and } E = \sum_k \langle n_k \rangle E_k$$

$$N = \sum_k \left(\exp\beta\left(E_k - \mu\right) \mp 1\right)^{-1}$$

$$E = \sum_k E_k\left(\exp\beta\left(E_k - \mu\right) \mp 1\right).$$

With the above we can calculate S from Eq. (7.46):

$$S = k \sum_k \left[\langle n_k \rangle \log \langle n_k \rangle \pm (1 \pm \langle n_k \rangle) \log (1 \pm \langle n_k \rangle) \right]. \tag{7.56}$$

Quite generally, we have the thermodynamics of ideal quantum non-interacting gas. Now let $z = \exp \mu$, the fugacity. We then have a pair of equations which implicitly are the equation of state:

$$\frac{PV}{kT} = \ln Z_G = \mp \sum_k \log (1 \mp z \exp (-\beta \varepsilon_k)) \tag{7.57}$$

$$N = z \frac{\partial}{\partial z} \ln Z_G = \sum_k \frac{z \exp (-\beta \varepsilon_k)}{1 \mp z \exp (-\beta \varepsilon_k)}. \tag{7.58}$$

For the purpose of physical parameterization, let us adopt a continuum state model:

$$H_i = \frac{p_i^2}{2m} \qquad p_i = \hbar k_i \qquad \mathbf{k}_i = \frac{\mathbf{m}\pi}{L} \qquad \mathbf{m} = 0, 1, 2, \ldots \tag{7.59}$$

Then as $L \to \infty$,

$$\sum_k \to \int_{-\infty}^{+\infty} d_3 p \to \int_0^{\infty} g(\varepsilon)\, d\varepsilon, \tag{7.60}$$

where the energy density of states is

$$g(\varepsilon)\, d\varepsilon = 4\pi \left(\frac{2m}{h^2} \right) V \sqrt{\varepsilon} d\varepsilon. \tag{7.61}$$

We then obtain

$$\begin{aligned} \frac{P}{kT} &= \frac{1}{\lambda^3} f_{\frac{5}{2}}(z) \\ \frac{N}{V} &= \frac{1}{v} = \frac{1}{\lambda^3} f_{\frac{3}{2}}(z) \end{aligned} \tag{7.62}$$

for fermions. For bosons,

$$\begin{aligned} \frac{P}{kT} &= \frac{1}{\lambda^3} g_{\frac{5}{2}}(z) \\ \frac{1}{v} &= \frac{1}{\lambda^3} g_{\frac{3}{2}}(z). \end{aligned} \tag{7.63}$$

Here

$$\lambda = \sqrt{\frac{2\pi \hbar^2}{mkT}}, \tag{7.64}$$

which is a measure of quantum wave properties, the thermal de Broglie wavelength. Here

$$f_{\frac{5}{2}}(z) = \frac{4}{\sqrt{\pi}} \int_0^\infty dx x^2 \log\left(1 + z\exp\left(-z^2\right)\right) = \sum_{l=1}^{\infty} \frac{(-1)^{l+1} z^l}{l^{\frac{5}{2}}} \tag{7.65}$$

$$f_{\frac{3}{2}}(z) = z\frac{\partial}{\partial z} f_{\frac{5}{2}}(z) = \sum_{l=1}^{\infty} \frac{(-1)^{l+1} z^l}{l^{\frac{3}{2}}},$$

and similarly

$$g_{\frac{5}{2}}(z) = \frac{-4}{\sqrt{\pi}} \int_0^\infty dx x^2 \log\left(1 - z\exp\left(-x^2\right)\right) = \sum_{l=1}^{\infty} \frac{z^l}{l^{\frac{3}{2}}} \tag{7.66}$$

and

$$g_{\frac{3}{2}}(z) = z\frac{\partial}{\partial z} g_{\frac{5}{2}}(z) = \sum_{l=1}^{\infty} \frac{z^l}{l^{\frac{3}{2}}}.$$

For the preceding expansions of the integrals to be valid, $z < 1$. In the Bose case, this continuum limit has not treated properly the near-ground states. Much more will be said in the next chapter about this.

We may compactly express the small z approximation

$$P = kT\lambda^{-3} z\left(1 + \theta 2^{-\frac{5}{2}} z + 3^{\frac{5}{2}} z^2 + \cdots\right) \tag{7.67}$$

$$\frac{N}{V} = \frac{1}{v} = c\lambda^{-3}\left(z + \theta 2^{-\frac{3}{2}} z^2 + 3^{-\frac{3}{2}} z^3 + \cdots\right) \tag{7.68}$$

$\theta = +1$ for bosons and -1 for fermions.

By iteration, for low density, Eq. (7.67) is solved for $z = \exp \mu$:

$$z = \lambda^3 c[1 - \theta 2^{-\frac{3}{2}} \lambda^3 c + \cdots] \tag{7.69}$$

This is used in Eq. (7.67) to obtain

$$P = ckT\left[1 - \theta 2^{-\frac{5}{2}} \lambda^3 c \ldots\right] \tag{7.70}$$

$$U = \frac{3}{2} PV = \frac{3}{2} NkT\left[1 - \theta 2^{-\frac{5}{2}} \lambda^3 c + \cdots\right]. \tag{7.71}$$

The expansion parameter is now apparent:

$$\lambda^3 c = \left(\frac{h^2}{2\pi mkT}\right)^{\frac{3}{2}} c \ll 1.$$

The first term is, of course, the classical Boltzmann result. For hydrogen at standard conditions, $T = 300\,\mathrm{K}$, $c\lambda^3 = 10^{-4}$ and at $10^{-2}\,\mathrm{K}$, $c\lambda^3 = 10^{-2}$. The quantum effects are negligible for such gases.

7.7 Equilibrium limit theorems

A very important question is the existence of the partition function (microcanonical, canonical, grand canonical) for reasonable potentials in the thermodynamic limit. Considerable work has been done in this regard, mostly classical. The first work was that of Van Hove (1949). Ruelle, in his book of rigorous results, treats uniquely full quantum systems (Ruelle, 1969). It is beyond the scope of the remarks to be made here, but a mathematically mature reader is encouraged to look at this book. Concerning the canonical partition function, it is outlined in detail by Munster (1969). Balescu (1975), in a very readable fashion, outlines the Van Hove work.

We will follow Munster's discussion of the microcanonical case, since it is the simplest and contains the weakest assumptions. This is due to Van der Linden (Van der Linden, 1966; Van der Linden and Mazur, 1967). Our principal purpose will be to state the resulting theorem and the physical conditions for the proofs.

We write the entropy per particle as $s\,(e, v)$,

$$N s\,(e, v) = \ln \Omega\,(E, N, V),$$

where the quasi-quantum phase volume (Balescu, 1975) is

$$\Omega\,(E) = \frac{1}{h^n \prod_i N_i!} \int^E d\Omega, \tag{7.72}$$

which may be written after momentum integration as

$$\begin{aligned} \Omega\,(EV, N) = {} & \frac{(2\pi m)^{\frac{3}{2}N}}{h^{3N} N!\Gamma\left(\frac{3}{2}N + 1\right)} \\ & \times \int_V dq_N \left[E - U^N\left(q^N\right)\right]^{\frac{3}{2}N} \theta\left[E - U^N\left(q^N\right)\right]. \end{aligned} \tag{7.73}$$

The Heaviside function, θ, contains the N-body interaction potential $U^N\left(q^N\right) \equiv U^N\,(q_1 \ldots q_N)$, and the $N!$ is because we are assuming particle identity. The factors multiplying the integral contain free particle de Broglie wavelengths. In this sense, this is quasi-quantum. To discuss the thermodynamic limit, we assume that V is a cylinder of constant cross section parallel to the z-axis. The upper and lower surfaces at $z = h$ and h' have walls of constant thickness $\frac{1}{2}R_0$. The so-called free volume is $(h-h')A$. In the thermodynamic limit $(h-h')N$ and A are held constant.

The initial assumptions concerning the *classical* many-body potential are more general than with the B.B.G.Y.K. hierarchy discussion (see Chapter 4). Assume that

$$U^N(q_1 \dots q_N) = \sum_i u^{(1)}(q_1) + \sum_{N \geq i > j \geq 1} u^{(2)}(q_i q_j) + \sum_{N \geq i > j > k \geq 1} u^{(3)}(q_i q_j q_k) + \cdots \quad (7.74)$$

Here the cluster decomposition is evident:

$$u^{(1)}(q_1) = U^1(q_1)$$
$$u^{(2)}(q_1 q_2) = U^2(q_1 q_2) - U^1(q_1) - U^1(q_2) \text{ etc.}$$

The basic assumptions beginning the proof are:

1. $U^N(q_2 \dots q_N)$ is symmetric in N.
2. $U^N(q_1 \dots q_N)$ is translationally invariant.
3. $U^N(q_N)$ is piece-wise continuous in $U^N(q^N) < E$.
4. Stability condition: $U^N(q_1 \dots q_N) \geq -N\mu_A$ for all $q_1 \dots q_N$ and all N. We now also assume the so-called tempering condition. Here it is called strong tempering, making the proof weak.
5. Strong tempering: $U^{(N_1 N_2)}\left(q_1 \dots q_{N_1};\, q_1' \dots q_{N_2}'\right) \leq 0$ for $\left|q_i - q_j'\right| \geq R_0$ for all q_i, q_j.

Let us examine these conditions. As earlier, the condition (1) is classical particle identity. Condition (2) excludes external fields, and thus transport phenomena, as discussed elsewhere. Condition (3) implies that $U^N(q_1 \dots q_N)$ is bounded from below and may allow Lebesque integrals. The stability criterion (4) appears to be due to Onsager (1939). This may be examined for pair potentials (see Ruelle, 1969). The violation of stability is termed catastrophic potentials. An example is an attractive square well with *no* hard core. Here the bound is $\mu_A = 0$. Ruelle has stated the proposition that the pair potential of the form

$$U(x) \geq \phi_1(|x|) \qquad |x| \leq a_1$$
$$U(x) \geq -\phi_2(|x|) \qquad |x| \geq a_2$$

is stable. Lennard–Jones potentials are of this type with

$$\phi_1(|x|) = \phi_2(|x|) = |x|^{-\lambda};\ \lambda > 0.$$

The tempering condition (5) is more difficult. It may be shown to hold for pair interactions if and only if

$$U(x) \leq A\,|x|^{-\lambda} \qquad |x| \geq R,$$

particularly if A corresponds to Van der Walls and finite long-range pair potentials. More generally, the mutual interaction energy of two separated groups of particles, N_1 and N_2, is

$$U\left(q_1 \ldots q_{N_1};\ q'_1 \ldots q'_{N_2}\right) - U\left(q_1 \ldots q_{N_1}\right) - U\left(q'_1 \ldots q'_{N_2}\right).$$

Here there are no particles in the distinct groups at a distance less than d, and the net interaction is purely attractive. The distance between the distinct groups is R. There are no long-range repulsions which would cause the groups $N_1,\ N_2$ to explode. The positive part of the interaction is small at large distances.

Let us now state the important theorem in detail:

Theorem *If conditions 1, 2, 3, 4, 5 are satisfied and the thermodynamic limit is carried out* ($E \to \infty,\ N \to \infty,\ V \to \infty,\ e =$ constant, $v =$ constant) *with a sequence of cylinders of constant cross section* A, *then the function* $s(e, v, N_k)$, *the entropy density, converges uniformly to* $s^{\infty}(e, v)$ *for* $e_{\min} < \infty$, *and* $v_{\min} \leq v \leq v_1 < \infty$, $s^{\infty}(e, v)$ *has the desirable properties:*

1. $s^{\infty}(ev)$ *is continuous and convex in* e *and* v.
2. $s^{\infty}(e, v)$ *is a nondecreasing function of* e *for constant* v *and also a nondecreasing function of* v *for constant* e.
3. *The derivatives with respect to* e, v *exist almost everywhere and are nonnegative. And the derivative with constant* v, *with respect to* e, *is a nonincreasing function of* e. *Also, at constant* e, *the derivative with respect to* v *is a nonincreasing function of* v.
4. *The* $\frac{\partial^2 s^{\infty}}{\partial^2 e}$ *and* $\frac{\partial^2 s^{\infty}}{\partial^2 v}$ *exist almost everywhere and are nonpositive.*

Two lemmas lead to the theorem.

Lemma 1 *Lemma 1 follows from the stability condition property 4. It is the inequality*

$$s(e, v, N) \leq Ln\left[\frac{4\pi m}{3h^2}\left(e + \mu_A\right)\right]^{\frac{3}{2}} + \ln v + \frac{5}{2}. \tag{7.75}$$

Lemma 2 follows from strong tempering condition 5. We state it here:

Lemma 2 *If volume* V *is divided into two subsets* V_1, V_2 *in such a way that for* $V_1, -h - \frac{1}{2}R_0 \leq z \leq h''$ *and for* $V_2,\ \ h'' \leq z < h' + \frac{1}{2}R_0$, *and in* V_1 *there are* N_1 *particles and in* V_2 *there are* $N_2 = N - N_1$ *particles, then*

$$Ns\,(evN) \geq N_1 s\,(evN_1) + N_2 s\,(e, vN_2) \tag{7.76}$$

for all N and $N_1 \leq N$.

Functions obeying the inequality, Eq. (7.76), are such that $s\,(e, v, N)$ are sub-additive in N. Using these two limits, Van der Linden proved, using the sub-additive property, that

$$\lim_{N_k \to \infty} s\,(e, v, N_k) = \sup_{N_k \to \infty} s\,(e, v, N_k) = s^{\infty}\,(e, v)\,. \tag{7.77}$$

A similar argument also shows that for $e\,(v, N_k)$,

$$\lim_{N_k \to \infty} e_{\min}\,(v_1 N_k) = \inf_{N_k \to \infty} e_{\min}\,(v N_k) \equiv e_{\min}\,(v)\,.$$

In addition, from sub-additivity, following conditions 4 and 5, we may show that $s^{\infty}\,(e, v)$ is convex:

$$s^{\infty}\left[\frac{1}{2}\,(e_1 + e_2)\,, \frac{1}{2}\,(v_1 + v_2)\right] \geq \frac{1}{2} s^{\infty}\,(e_1 v_1) + \frac{1}{2} s^{\infty}\,(e_2 v_2)\,. \tag{7.78}$$

From this, continuity in e, v follows. It also follows that the remaining results in this theorem hold. The details are outlined in Munster's discussion. The point here is to give the reader an idea of how the physical conditions lead to the theorem. This, and such theorems for the other ensembles, are the foundations of equilibrium statistical mechanics as the basis of macroscopic thermodynamics.

References

Balescu, R. (1975). *Equilibrium and Non-equilibrium Statistical Mechanics* (New York, Wiley), revised 1999 as *Matter out of Equilibrium* (London, Imperial College Press).

Boltzmann, L. (1877). Uber die Bezichungen Zwischen dem II Hauptsatz der Mechanischen Wärmtheorie und Wahrscheinlichkeitsrechnung. *Wien. Ber.* **77**, 373.

Callen, H. (1985). *Thermodynamics,* 2nd edn. (New York, Wiley).

Einstein, A. (1910). *Ann. Phys.* **33**, 1275.

Farquahar, I. E. (1964). *Ergodic Theory in Statistical Mechanics* (London, Interscience).

Feynman, R. P. (1988). *Quantum Implications in Honor of David Bohm,* ed. B. J. Hiley and F. D. Peat (London, Routledge).

Gibbs, J. W. (1961). *The Scientific Papers of J. Willard Gibbs* (New York, Dover).

Kurtz, T. G. (1972). *J. Chem. Phys.* **57**, 29–76.

Landau, L. and Lifshitz, E. (1967). *Physique Statistics* (Moscow, Mir).

Lichtenberg, A. J. and Lieberman, M. A. (1991). *Regular and Chaotic Motion,* 2nd edn. (New York, Springer).

Munster, A. (1969). *Statistical Thermodynamics* (Berlin, Springer).

Onsager, L. (1939). *J. Phys. Chem.* **43**, 189.

Planck, M. (1923). *Vorlesungen uber der Theorie der Wärmstrahlung* (Leipzig, Barth), 119.

Plischke, M. and Bergersen, B. (1989). *Equilibrium Statistical Mechanics* (New York, Prentice Hall).

Ruelle, D. (1969). *Statistical Mechanics* (New York, W. A. Benjamin).
Scheck, F. (1999). *Mechanics,* 3rd edn. (New York, Springer).
Shannon, C. E. and Weaver, W. (1949). *The Mathematical Theory of Communication* (Chicago, University of Illinois Press).
Sinai, J. G. (1963). *Duklady Akad. Nauk S.S.S.R.* **153**, 1261.
Tolman, R. D. (1938). *The Principles of Statistical Mechanics* (New York, Oxford University Press), reissued by Dover.
Van der Linden, J. (1966). *Physica* **32**, 642.
Van der Linden, J. and Mazur, P. (1967). *Physica* **36**, 491.
Van Hove, L. (1949). *Physica* **15**, 951.

8

Bose–Einstein ideal gas condensation

8.1 Introduction

Let us turn to the unusual and exciting quantum effects first suggested by Einstein (1924a,b). After translating the paper by Bose (1924) for the *Zeitschrift Physik*, Einstein generalized it and noted, because of the particle identity, that there would be a statistical tendency for the particles to "condense" into their ground state, the state of momentum zero. Further, he stated that the condensation would begin at a critical temperature. For a three-dimensional box, volume V, with N particles,

$$\frac{N}{V}\left(\frac{h^2}{2\pi mkT_c}\right)^{\frac{3}{2}} = \sum_{j=1}^{\infty} j^{-\frac{3}{2}} = 2.612. \tag{8.1}$$

We recognize this as $\lambda_c^3 = 2.612$, much beyond the limits of the expansion discussed at the end of the previous chapter, Eq. (7.70).

Fritz London was one of the few to note Einstein's suggestion and in the continuum approximation gave detailed calculation of the thermodynamic properties of the condensate state for a box in the thermodynamic limit $N \to \infty$, $V \to \infty$, c = constant, Eq. (7.60) and Eq. (7.61). He boldly associated the resulting transition (phase transition) at $T_C = 3.1$ K. with that for the super fluid transition in ^{4}He at $T_\gamma = 2.17$ K. We will discuss this further and go through the London calculation in detail in the next section.

The London continuum approximation was examined in detail by de Groot (de Groot *et al.*, 1950) in a heroic calculation of the grand ensemble for a variety of trapping potentials. He examined in detail the apparent transition for finite N.

Much later the technical development of supercooled dilute atomic traps in ^{87}Rb (Anderson *et al.*, 1995) and ^{23}Na (Davis *et al.*, 1995) led to the creation of dilute condensates for finite numbers of particles in these systems of trapped condensates, no longer spacially homogeneous (see also Pethick and Smith, 2002; Pitaevskii and Stringari, 2003).

This led to a renewed interest in finite N ideal Bose–Einstein condensation (Bagnato *et al.*, 1987; Grossman and Holthaus, 1995; Ketterle and Van Druten, 1996). These Bose–Einstein condensates are marvelous examples of spacially inhomogeneous systems showing "exotic" quantum hydrodynamic properties. It is possible to use the early theories of Bogoliubov and others because of the dilute nature of the system. We will not have space or time to go into the hydrodynamic inhomogeneous properties of these Bose–Einstein condensates but refer the reader to the recent book of Pethick and Smith (2002) and that of Pitaevskii and Stringari (2003).

In Section 8.5 we will examine fluctuations in the ground state. There we will show, after Ziff (Ziff *et al.*, 1977), that the grand canonical approach cannot be relied upon to estimate fluctuations in the ground state. In Section 8.6 we will return to the master equation methods of Chapter 3 and consider the recent master equation for boson condensation of Scully (1998) and Kocharovsky (Scully, 1996; Kocharovsky *et al.*, 2000). Finally, in the chapter appendix, we outline the theory of finite trap condensation of de Groot (de Groot *et al.*, 1950).

8.2 Continuum box model of condensation

Again, for the grand ensemble Bose–Einstein continuum model, Eq. (7.63) and Eq. (7.66),

$$\frac{P}{kT} = \frac{1}{\lambda^3} g_{\frac{5}{2}}(z) = \frac{1}{\lambda^3} \sum_{1}^{\infty} \frac{z^l}{l^{\frac{5}{2}}} \tag{8.2}$$

$$\frac{N}{V} = \frac{1}{v} = \frac{1}{\lambda^3} g_{\frac{3}{2}}(z) = \frac{1}{\lambda^3} \sum_{1}^{\infty} \frac{z^l}{l^{\frac{3}{2}}}. \tag{8.3}$$

The difficulty begins to appear at $z = 1$, $\mu = 0$. Note Eq. (8.3) in the limit $0 < z \leq 1$ (or $\mu < 0$). We define the critical particle number, $N = N_c$:

$$N_c = \frac{V}{\lambda^3} \cdot 2.612.$$

There can be no larger particle number even though we have taken $N \to \infty$. What is the origin of this unphysical limitation? As London (1938) pointed out, this is the result of the unphysical neglect of the discrete ground and adjacent states. We may also define a corresponding T_c,

$$kT_c = \frac{2\pi\hbar^2}{m} \left[\frac{N}{V} \frac{1}{2.612} \right]^{\frac{2}{3}}, \tag{8.4}$$

which Einstein mentioned. For fixed density, no lower temperature can be achieved.

Let us break the discussion here and treat the single ground state in Eq. (7.57) and Eq. (7.58) separately. Let it be $\mathbf{p} = 0$. The remaining states will be treated in the continuum approximation. Thus we have a better approximation to Eq. (7.57) and Eq. (7.58):

$$\frac{PV}{kT} = \ln Z_G = -\ln(1-z) + \frac{1}{\lambda^3} V g_{\frac{5}{2}}(z), \tag{8.5}$$

and

$$N = \frac{z}{1+z} + \frac{1}{\lambda^3} g_{\frac{3}{2}}(z). \tag{8.6}$$

We may now extend this to $z > 1$ and approximate, in this regime, $g_{\frac{3}{2}}(z)$ by $g_{\frac{3}{2}}(1)$, defining again the critical temperature with $\lambda_c^3 \frac{N}{V} = 2.612$, obtaining

$$N = N_0 + N \frac{\lambda_c^3}{\lambda^3}. \tag{8.7}$$

Therefore, the ground state occupation density is

$$N_0 = N \left[1 - \left(\frac{T}{T_c} \right)^{\frac{3}{2}} \right]. \tag{8.8}$$

We call 3/2 the critical index.

Below T_c the ground state rapidly accumulates to N particles. Above T_c there is no ground state occupation. Does this have a physical effect? By the same argument we examine PV/kT, Eq. (7.57):

$$\text{For } T < T_c \qquad \frac{P}{kT} = \frac{1}{V} \ln(1-z) + \frac{1}{\lambda^3} V g_{\frac{5}{2}} \ (z=1) \tag{8.9}$$

$$T > T_c \qquad \frac{P}{kT} = \frac{1}{\lambda^3} g_{\frac{5}{2}} \ (z < 1). \tag{8.10}$$

We must examine the first term in Eq. (8.9), $(1-z) = 0\left(\frac{1}{V}\right)$ as $T < T_c$. Hence $\frac{1}{V} \ln(1-z) \to 0$ as $V \to \infty$. Thus P is independent of V for $T < T_c$, whereas for $z < 1$ Eq. (8.2) holds, and for $z \ll 1$, as discussed, $P = ckT$. The ground state has no contribution to the pressure, which is natural, since this is the zero-momentum state. In the zero-momentum state the system is spacially homogeneous, so there is no spacial evidence of this condensation below T_c.

Now, from the exact formula for ideal quantum gas in the continuum limit, $PV = \frac{2}{3}E$. We have, from the above,

$$E = \frac{3}{2}\frac{VkT}{\lambda^3}g_{\frac{5}{2}}(1) \qquad T < T_c \tag{8.11}$$

$$E = \frac{3}{2}RT\frac{g_{\frac{5}{2}}(z)}{g_{\frac{3}{2}}(z)} \qquad T > T_c. \tag{8.12}$$

We may eliminate z from Eq. (8.12). We need $g_{\frac{5}{2}}(z)$ in terms of $g_{\frac{3}{2}}(z)$. This was done by London in the appendix to his book *Superfluids*. The result is

$$E = \frac{3}{2}RT\left[1 - 0.4618\left(\frac{T_c}{T}\right)^{\frac{3}{2}} - 0.0226\left(\frac{T_c}{T}\right)^3 - \dots\right]; \qquad T > T_c. \tag{8.13}$$

With these results we obtain

$$C_V = 1.926R\left(\frac{T}{T_c}\right)^{\frac{3}{2}} \qquad T < T_c$$

$$C_V = \frac{3}{2}R\left[1 + 0.231\left(\frac{T_c}{T}\right)^{\frac{3}{2}} + 0.045\left(\frac{T_c}{T}\right)^3 \dots\right] \qquad T > T_c. \tag{8.14}$$

As pointed out by London, a more careful analysis must be done at $T \approx T_c$. He examined

$$C_V^+ = \lim_{(T-T_c)\to 0^+} C_V = \frac{3}{2}R\left(\frac{5}{2}\frac{g_{\frac{5}{2}}(z)}{g_{\frac{3}{2}}(z)} - \frac{3}{2}\frac{g_{\frac{3}{2}}(z)}{g_{\frac{1}{2}}(z)}\right). \tag{8.15}$$

By the inversion, eliminating z in the limit, London showed the second term vanishes and thus C_V is continuous at $T = T_c$. A similar analysis by London was used to examine the discontinuity in $\partial C_V/\partial T$ at $T = T_c$. We find

$$\Delta\left(\frac{\partial C_V}{\partial T}\right) = 3.66\frac{R}{T_c}. \tag{8.16}$$

London then associated this "phase transition" with the experiments on ^{4}He of Keesom (Keesom and Clusius, 1932). These wonderful experiments exhibit an extremely sharp derivative discontinuity at $T_c = 2.12\,\text{K}$. The formula (Eq. 8.3) for this experiment gives $T_c = 3.1\,\text{K}$, as already mentioned. We will come back to this association in the last section of this chapter.

In the preceding analysis the important limits $N \to \infty$, $V \to \infty$, $N/V = c$ have been implicitly used. This is termed the thermodynamic limit. The entropy is given by $S = -\left(\frac{\partial F}{\partial T}\right)|_{\mu,v} = \frac{\partial(PV)}{\partial T}|_{\mu,v}$. Using this and Eq. (8.9) and Eq. (8.10),

$$S = \frac{5}{2} k V \frac{1}{\lambda^3} g_{\frac{5}{2}}(1) = \frac{5}{2} n_n V \frac{g_{\frac{5}{2}(1)}}{g_{\frac{3}{2}(1)}}, \qquad T < T_c, \tag{8.17}$$

where we introduced $n_m(T) = \frac{1}{\lambda^3} g_{\frac{3}{2}}(1)$ as the density of particles *not* in the ground state. As expected, we see that the entropy below T_c decreases with normal component density. The ground state, of course, has zero entropy. The latent heat is proportional to this entropy.

Let us compare these approximate results with the *exact* results (without continuum approximation) of de Groot (de Groot *et al.*, 1950) for the box. They find, in the finite N, V limit, a continuous curve for $z(T)$, which means that $z(T)$, $E(T)$ and all their derivatives are continuous. Then, in the thermodynamic limit, they show $z = 1$ for $T < T_c$ given by Eq. (8.1). E is given exactly by Eq. (8.8) and Eq. (8.9). In addition, they obtain C_+ and C_- to be continuous, and the derivative discontinuous. Also, the formula for the ground state density is Eq. (8.6). De Groot did not use London's periodic boundary conditions. The same equation of state was found qualitatively:

$$PV = 0.5133 \left(\frac{T}{T_c}\right)^{\frac{3}{2}}, \qquad T < T_c,$$

compared with 1.342. In the appendix we will illustrate their calculation by considering their theorems.

8.3 Harmonic oscillator trap and condensation

For the ideal Bose–Einstein gas in a harmonic oscillator container, there is no natural thermodynamic limit. Here

$$E_{n_1 n_2 n_3} = \frac{1}{2}(\omega_1 n_1 + \omega_2 n_2 + \omega_3 n_3) + E_0 \tag{8.18}$$

$$n_1 n_2 n_3 \in 0, 1, 2, \ldots$$

$$N = \sum_{n_1 n_2 n_3} \left[\exp\left(\beta\left(\omega_1 n_1 + \omega_2 n_2 + \omega_3 n_3\right)\right) + \beta\left(E_0 - \mu\right) - 1\right]^{-1}.$$

De Groot and coauthors discussed three possibilities:

1. $N = n(a)^3$, where n is a mean density and a is the radius of a sphere containing the particles in their ground state. Thus, take $N \to \infty$, $a \to \infty$, and $n =$ constant. However, this is unphysical, since $E_{n_1 n_2 n_3} \to 0$.
2. Let $v = \frac{n}{a^3} = \frac{N}{a^6} =$ constant as $N \to \infty$, $a \to \infty$. This choice suffers from the same criticism as the first.
3. A finite $N \to \infty$. $z = 1$ for all T. In this case it may be seen that

$$N - N_0 = \lim \sum_{E_i}{}' \frac{\exp\left(\frac{-E_i}{kT}\right)}{1 - \exp\left(\frac{-E_i}{kT}\right)}.$$

$N - N_0$ is finite except at $T \to \infty$, where $N \to \infty,\ N_0 \to \infty$. The lowest state is excluded.

Despite these difficulties, Bagnato (Bagnato *et al.*, 1987; Grossman and Holthaus, 1995; Kirsten and Toms, 1997) has introduced a continuum approximation to discuss the harmonic oscillator traps, motivated by the experiments in progress. Since T is of the order of a few micro Kelvin and ω_i $100H_z$, $\beta\omega_i \ll 1$, and $\beta\omega_i$ is closely spaced, a continuum (though not exact) is expected to be a good approximation. It is straightforward to construct a continuum approximation for the symmetric harmonic trap $\omega = \omega_1 = \omega_2 = \omega_3$ in three dimensions.

The number of lattice points is neglecting the zero energy:

$$\int_{(\omega n_1 + \omega n_2 + \omega n_3 \leq E)} dn_1 dn_2 dn_3 = \nu(E).$$

The latter integral is

$$\nu(E) = \frac{1}{\omega^3} \int_0^E d\varepsilon_1 \int_0^{E-\varepsilon_1} d\varepsilon_2 \int_0^{E-\varepsilon_1-\varepsilon_2} d\varepsilon_3 = \frac{1}{6} \frac{E^3}{\omega^3}. \tag{8.19}$$

The answer for an asymmetric harmonic trap is

$$\nu(E) = \frac{1}{6} \frac{E^3}{\Omega^3}, \tag{8.20}$$

where $\Omega = (\omega_1 \omega_2 \omega_3)^{\frac{1}{3}}$. De Groot and coauthors have obtained the density for a general trap in w dimensions:

$$E_{s_1 \ldots s_w} = \frac{Mk}{d^2} \sum_{\nu=1}^{w} s_\nu^\alpha \qquad (s_1 \ldots s_w = 0, 1, 2, \ldots). \tag{8.21}$$

They show

$$\nu(E) = E^w I = E^w \int_{s_1^\alpha + \cdots + s_w^\alpha = 1} ds_1 \cdots ds_w = \frac{\left\{(\alpha)^{-1}!\right\}^w}{q!} E^w. \tag{8.22}$$

Here $1 \leq q \leq 2$. This is the same as above for the harmonic trap $\alpha = 1$. Bagnato (Bagnato *et al.*, 1987) also obtained such results.

Using such continuous density of states, the argument of London may be carried out, separating out the ground state contribution. Now, as before,

$$N = N_0 + \frac{1}{2}\frac{1}{(\Omega)^3}\int_0^\infty \frac{E^2 dE}{\exp\left[\beta\left(E+E_0-\mu\right)\right]-1} \tag{8.23}$$

$$= N_0 + \left(\frac{kT}{\Omega}\right)^2 g_3\left(z\right), \tag{8.24}$$

where

$$N_0 = \frac{z}{z-1};\quad g_3\left(z\right) = \sum_{l=1}^{\infty}\frac{z^l}{l!},$$

and the limiting value $z = 1$ gives $g\left(1\right) = 1.202$. Thus,

$$N_0 = N - \left(\frac{kT}{\Omega}\right)^3 g_3\left(1\right). \tag{8.25}$$

Defining, for the trap, a temperature associated with the onset of condensation into the ground state,

$$kT_c = \Omega\left(\frac{N}{1.202}\right)^{\frac{1}{3}}. \tag{8.26}$$

We may find for the trap for *finite* N

$$\frac{N_0}{N} = 1 - \left(T_c\right)^3. \tag{8.27}$$

Since no thermodynamic limit may be taken, we do not expect a sharp change at T_c. Note the difference in temperature dependence from the box. Values N ranging from 10^4 to 10^7 have been achieved so that $T_C \cong 10^2 nK$. A sudden transition is seen in the Ensher experiments (Ensher *et al.,* 1995). There the formula Eq. (8.27) is obeyed very well. The condensate fraction $\frac{N_0}{N}$ approaches 1 as $T \to 0$. We must reiterate that this is not strictly speaking a phase transition, since no true discontinuity is found in the derivative of the specific heat.

Grossman and Holthaus (1995) have suggested an additional small correction due to the $\frac{(n+1)(n+2)}{2}$ degeneracy of state $|n\rangle$. They write

$$g\left(E\right) = \frac{1}{2}\frac{E^2}{(\Omega)^3} + \frac{3}{2}\frac{E}{\Omega^2}.$$

This leads to a small shift in the critical temperature,

$$\frac{\Delta T}{T_c} = -\frac{3}{2}N^{-\frac{1}{3}}.$$

This is verified by numerical computation. Kirsten and Toms (1997) wrote an interesting formula in general for such effects.

Let us finish this section by remarking on the heat capacity of the harmonic trap (de Groot *et al.,* 1950; Grossman and Holthaus, 1995). If the method of London is used to make the continuous approximation,

$$\ln Z_G = \left(\frac{kT}{\Omega}\right)^3 g_4(z) - \ln(1-z). \tag{8.28}$$

$g_n(z)$ is, as in Chapter 7, $g_n(z) = \sum_{l=1}^{\infty} \frac{z^l}{l^n}$. With this we find

$$E = E_0 + 3kT\left(\frac{kT}{\Omega}\right)^3 g_3(z) \qquad T \geq T_c \tag{8.29}$$

$$E = E_0 + 3kT\left(\frac{kT}{\Omega}\right)^3 g_3(1) \qquad T < T_c \tag{8.30}$$

$$E = 3kT\left(\frac{kT}{\Omega}\right)^3 g_4(z) \qquad T > T_c.$$

The heat capacity is in the N very large limit,

$$C_- = 12Nk\frac{g_4(1)}{g_3(1)}\left(\frac{T}{T_C}\right)^3 \qquad T < T_c \tag{8.31}$$

$$C_+ = 3\left[4\frac{g+(z)}{g_3(z)} - \frac{g_3(z)}{g_2(z)}\right] \qquad T > T_c$$

$$z < 1,$$

a formula similar to Eq. (8.15). An examination reveals that there is a discontinuity in C itself at $T = T_c$. The magnitude is $6.6Nk$. De Groot *et al.* (1950) had previously recognized this in their exact treatment of the harmonic oscillator trap. Grossman and Holthaus further calculated numerically the behavior of $C_\pm$ for small N. It looked like a "rounded" version of the λ transition for $T = T_c$, the discontinuity being less in evidence.

8.4 ^{4}He: the λ transition

London's purpose in discussing the Bose–Einstein condensation was to explain the experiment of Keesom showing the λ transition in the heat capacity of ^{4}He. This phase transition occurs between $\mathrm{He_{II}}$, superfluid liquid phase, and $\mathrm{He_I}$, liquid phase at $T = 2.17$ K for pressure zero. See the nice discussion of Pitaeveskii and Stringari (2003). London's calculation gives $T = 3.1$ K, so he felt strongly that this was the Einstein-suggested condensation. The reasons for thinking this could be true, in this dense system where the average interaction distance is only a few angstroms, were mentioned by London (1954) and Munster (1964). The most compelling reason was that the experiments of Osborne (unpublished) found no superfluidity at low temperature in ^{3}He, a Fermi liquid.

In the modern sense, He_{II} is a superfluid. This will be discussed later. No matter what the pressure is, there is no liquid–solid phase transition, and there are a number of "exotic" hydrodynamic effects such as zero sound velocity, quantized vortices, zero entropy and free capillary flow. This leads to the two-fluid model of Tisza (1940) and Landau (1941). The agreement between modern many-particle calculations (Ceperley, 1995) and the λ transition experiment is now good. The temperature T_c is correct. Since ^{4}He is a fluid, the atomic interactions are important. A relatively simple estimate of their importance was made by Penrose and Onsager (1956). We shall follow the brief discussion of Munster in his book.

The wave function in the ground state is $\psi_0 (q_1 \ldots q_N)$. It is real. For $T = 0$, the two-point density matrix for one particle is

$$\rho_1 (q' - q) = N \int \cdots \int \psi_0 (q \ldots q^N) \psi_0 (q' \ldots q^N) \, dq_2 \ldots dq_N. \qquad (8.32)$$

Writing this in the momentum representation, we may prove

$$\lim r \to \infty \qquad \rho_1 (\mathbf{r}) = \frac{\bar{N}_0}{V} \qquad (8.33)$$

because of the rapid oscillations of $\exp(-i2\pi \mathbf{p} \cdot \mathbf{r})$. We consider the pair correlation and write

$$\psi_0 (q_1 \ldots q_N) = \left[Q^{(N)}\right]^{\frac{-1}{2}} F^{(N)} (q_1 \ldots q_N),$$

where

$$F^{(n)} (q_1 \ldots q_N) = 0; \qquad \left|q_i - q_j\right| \leq \sigma. \qquad (8.34)$$

This is the quantum hard sphere gas. $Q^{(N)}$ is the classical normalizing factor and is the configuration integral of the classical partition function. With this,

$$\rho_1 \left(q - q'\right) = \frac{N}{Q^{(N)}} \int \cdots \int F^{(N+1)} \left(qq', q_2 \ldots q_N\right) dq_2 \ldots dq_N. \qquad (8.35)$$

Now the classical pair distribution hard sphere value is

$$\rho_{\text{pair}} \left(q - q'\right) = \frac{(N+1)\, N}{Q^{(N)}} \int \cdots \int F^{(N+1)} \left(qq'q_2 \ldots q_N\right) dq_2 \ldots dq_N. \quad (8.36)$$

We find

$$\rho_1 \left(q - q'\right) = \frac{1}{N} \frac{Q^{(N+1)}}{Q^{(N)}} \rho_{\text{pair}} (\mathbf{r}). \qquad (8.37)$$

Making use of the approximation to the classical pair correlation function,

$$\lim r \to \infty \qquad \rho_{\text{pair}}(\mathbf{r}) = \left(\frac{N}{V}\right)^2 = c^2. \tag{8.38}$$

Thus, Eq. (8.33) and Eq. (8.38) give the ratio of the ground state occupation to the total particle number

$$\frac{\bar{N}_0}{N} = \frac{1}{V}\frac{Q^{(N+1)}}{Q^{(N)}}. \tag{8.39}$$

This is simply estimated for ^{4}He by Munster after Penrose and Onsager. Taking $\sigma = 2.56$ angstroms, they find 0.08, which is much less than unity, the ideal gas Bose–Einstein condensation answer.

The experimental ratio was obtained by neutron scattering after some numerical calculations (Sokol, 1995). The result was 10%. This is in remarkable agreement with the simple Penrose and Onsager estimate. These results show that the London calculations of the Bose–Einstein condensation properties are indeed too simple, as was expected. However, the ^{4}He transition may still be considered an Einstein condensation with interactions.

8.5 Fluctuations: comparison of the grand canonical and canonical ensemble

Ziff, Uhlenbeck and Kac (1977), in a comprehensive article, showed that in the thermodynamic limit for the ideal Bose gas, the grand canonical and canonical ensembles give the same result for the intensive bulk thermodynamic quantities p, u, s. The results may again be written, for completeness, as

$$s = \frac{S}{V} = \frac{5}{2}\frac{k}{\lambda^3} g_{\frac{5}{2}}(z) + k\rho\mu \qquad \rho < \rho_c \tag{8.40}$$

$$= \frac{5}{2}\frac{k}{\lambda^3} g_{\frac{5}{2}}(1) \qquad \rho > \rho_c$$

$$\mu \equiv g = -kTu \qquad \rho < \rho_c \tag{8.41}$$

$$= 0 \qquad \rho > \rho_c$$

$$\frac{u}{\rho} = \frac{3}{2}\frac{g_{\frac{5}{2}}(z)}{g_{\frac{3}{2}}(z)} \qquad T > T_c \tag{8.42}$$

$$= \frac{3}{2}kT\left(\frac{T}{T_c}\right)^{\frac{3}{2}} \frac{g_{\frac{5}{2}}(1)}{g_{\frac{3}{2}}(1)} \qquad T < T_c$$

and

$$\frac{C_V}{k} = \frac{15}{4}\frac{g_{\frac{5}{2}}(z)}{g_{\frac{3}{2}}(z)} - \frac{9}{4}\frac{g_{\frac{3}{2}}(z)}{g_{\frac{1}{2}}(z)} \qquad T > T_c \tag{8.43}$$
$$= \frac{15}{4}\left(\frac{T}{T_c}\right)^{\frac{3}{2}} \frac{g_{\frac{5}{2}}(1)}{g_{\frac{3}{2}}(1)} \qquad T < T_c.$$

Here $\rho = \frac{N}{V}$ and $\rho_c = \frac{g_{\frac{3}{2}}(1)}{\lambda^3}$. Now it may be shown, for the canonical ensemble, that

$$\lim_{V\to\infty} \frac{\langle n_k^2\rangle - \langle n_k\rangle^2}{V^2} = 0 \text{ for all } n_k \text{ and } \rho, \tag{8.44}$$

including the ground state. This is done utilizing

$$\langle n_i\rangle = \frac{1}{Z}\sum_{\{n_k\}}^{N} n_i \exp\left(-\beta \sum_k n_k \varepsilon_k\right) \tag{8.45}$$

and

$$\langle n_i^2\rangle = \frac{1}{Z}\sum_{\{n_k\}}^{N} n_i^2 \exp\left(-\beta \sum_k n_k \varepsilon_k\right). \tag{8.46}$$

N is fixed. Since N is fixed and finite, Eq. (8.44) is reasonable. Now, utilizing the well-known general relation for the grand canonical case, $\langle n_k^2\rangle - (\langle n_k\rangle)^2 = (\langle n_k\rangle)(1 + \langle n_k\rangle)$ for all k. In particular, we find that for the ground state,

$$\lim_{V\to\infty} \frac{\langle n_0\rangle}{V} = 0 \qquad \rho < \rho_c \tag{8.47}$$
$$= \rho - \rho_c \qquad \rho > \rho_c$$

$$\lim_{V\to\infty} \frac{\langle n_0^2\rangle}{V^2} = 0 \qquad \rho < \rho_c \tag{8.48}$$
$$= 2\left(\rho - \rho_c\right)^2 \qquad \rho > \rho_c$$

and thus the anomaly

$$\lim_{V\to\infty} \frac{\langle n_0^2\rangle - (\langle n_0\rangle)^2}{V^2} = \left(\rho - \rho_c\right)^2. \tag{8.49}$$

This exhibits large uncontrolled fluctuations in densities for $\rho > \rho_c$. For the excited state

$$\lim_{V\to\infty} \frac{\langle n_k\rangle}{V} = \lim_{V\to\infty} \frac{\langle n_k^2\rangle}{V^2} = 0. \tag{8.50}$$

Eq. (8.50) is not in agreement with the canonical result, since in this case $N \to \infty$ and is *not* fixed but is consistent, since we may show for $\langle N \rangle = N$

$$\lim_{V\to\infty} \frac{\langle (\Delta N)^2 \rangle}{V^2} = 0 \qquad \rho < \rho_c \tag{8.51}$$
$$= (\rho - \rho_c)^2 \qquad \rho > \rho_c.$$

For the condensed phase the fluctuations are those of the entire system. Ziff *et al.* (1977) have explained this by considering a system λ' containing a smaller subsystem λ having a boundary $\lambda' - \lambda$. The region λ is determined by the canonical ensemble being an open system in the limit V, $V' \to \infty$. λ' must be determined by the canonical ensemble in contradiction to the results in the condensed phase. We note that this is a difficulty with the average ground state number density and its moments.

8.6 A master equation view of Bose condensation

A recent suggestion of Willis Lamb induced M. Scully and his colleagues (Kocharovsky *et al.,* 2000) to reconsider the laser transition analogy to a phase transition (Degiorgio and Scully, 1970). They utilized the density matrix master equation of Scully and Lamb (Scully and Zubairy, 1999) to possibly describe the Bose–Einstein phase transition which we are considering in this chapter. Since it deals with the master equation description of an open system (see Chapter 3), it is pertinent to consider this here. The original quantum optics application will be looked at extensively in Chapter 9.

Let us look at the ideal gas Einstein condensate from a master equation and possibly non-equilibrium point of view. The reservoir is taken to be a system of harmonic oscillators with $b_j^\dagger$ creation operators and $a_k^\dagger$ for the condensing Bose atoms in state k, $\hbar\nu_k$ being the energy of the particular trap, not yet specified.

The interaction is

$$V = \sum_j \sum_{k>l} g_{j,kl} b_j^\dagger a_k a_l^\dagger \times \exp\left(-i\left(\omega_j - \nu_k + \nu_l\right)t\right) + Hc. \tag{8.52}$$

In the Markov approximation, with basically the assumptions of Chapter 3, assuming $\rho(0) = \rho_S(0) \otimes \rho_R(0)$, the infinite reservoir is taken to be in equilibrium. The von Neumann equation for the Bose system is (see Kocharovsky *et al.,* 2000; Scully and Zubairy, 1999)

$$\dot{\rho}_S = \frac{-\kappa}{2} \sum_{k>l} (n_{kl}+1)\left[a_k^\dagger a_l a_l^\dagger \rho_S - 2a_l^\dagger a_k \rho_S a_k^\dagger a_l + \rho_S a_k^\dagger a_l a_l^\dagger a_k\right] \tag{8.53}$$
$$-\frac{\kappa}{2}\sum_{k>l} n_{kl}\left[a_k a_l^\dagger a_l a_k^\dagger \rho_S - 2\left(a_l a_k^\dagger \rho_S a_k a_l^\dagger + \rho_S a_k a_l^\dagger a_l a_k^\dagger\right)\right].$$

The reservoir energies have been taken as continuous with densities $D(\omega_{kl})$ and assumed to be constant. Also, the off diagonal contributions $\omega_{kl} \neq \omega_{k'l'}$ are neglected. Then $\kappa = \frac{2\pi D g^2}{\hbar^2}$. Also, $n_{kl} = \left[\exp\left(\frac{\hbar\omega_{kl}}{T}\right) - 1\right]^{-1}$, being the equilibrium occupation of the infinite heat baths.

For the condensate system we will obtain a further reduced equation for the conditional diagonal probability:

$$P_{n0} = \sum_{\{nk\}'} P_{n0,\{nk\}'}. \tag{8.54}$$

The prime means $n_0 + \sum_{k\neq 0} n_k = N$. Here N is fixed, and $\{n_k\}$ is summed over all n_k but not over n_0. This is consistent with the soon to be obtained canonical equilibrium ensemble.

Further, assume that the excited states $\{n_k\}'$ are in thermal equilibrium at temperature T, the bath temperature. This factors the intermediate nonlinear equation for P_{n0}. It results in

$$\langle n_k\rangle = \sum_{\{n_k\}'} n_{k'} \frac{P_{n0,\{n_k\}'}}{P_{n0}}, \qquad k \neq 0,$$

characteristic of a conditional probability. The excited states are now in equilibrium at T and subject to $\sum_{k>0} n_k = N - n_0$, given n_0 and N.

Finally, the simple linear "working" equation is obtained for P_{n0}:

$$\begin{aligned}\frac{dP_{n0}}{dt} = -\kappa \{&K_{n0}(n_0+1)P_{n0} - K_{n0-1}n_0P_{n0-1} + H_{n0}n_0P_{n0}\\ &-H_{n0+1}(n_0+1)P_{n0+1}\},\end{aligned} \tag{8.55}$$

where

$$\begin{aligned}K_{n0} &= \sum_{k'>0} (n_{k'}+1)\langle n_{k'}\rangle \\ H_{n0} &= \sum_{k'>0} n_{k'}(\langle n_{k'}\rangle + 1).\end{aligned} \tag{8.56}$$

The averages in Eq. (8.56) are conditional. This is a linear irreversible birth–death equation for the probabilities P_{n0} in the ground state. The occupation probability P_{n0} is changed from the states $n_0 \pm 1$, both increased by $K_{n0-1}n_0P_{n0-1}$ and $H_{n0+1}(n_0+1)P_{n0+1}$ and decreased by $K_{n0+1}(n_0+1)P_{n0}$ and $H_{n0}n_0P_{n0}$. It is a master equation in the sense of Pauli but not necessarily weak coupling in the reservoir condensate coupling. It is better understood to be in the Van Hove limit, as in Chapter 3 ($\lambda^2 t$ finite, $\lambda \to 0, t \to \infty$).

For such an equation as Eq. (8.55), the steady "equilibrium" solution may be readily obtained, as is the object of this discussion. The authors (Kocharovsky *et al.*, 2000) have done this for a number of traps and condensate numbers. By standard procedure, using detailed balance (Gardiner, 1985), one may obtain the time-independent (steady) solution:

$$P_{n_0} = Z_{N-1}\Pi_{i=n_0+1}^{N}\frac{H_i}{K_{i-1}}, \tag{8.57}$$

and with this and normalization we write

$$Z_N = \sum_{n_0=0}^{N} \tau\Pi_{i=n_0+1}^{N}\left(\frac{K_{i-1}}{H_i}\right)^{-1}, \tag{8.58}$$

the canonical partition function.

Scully and his colleagues made the approximation for low temperature $n_{k'}+1 \approx 1$ and also made a constant coefficient approximation

$$K_{n_0} = N - n_0. \tag{8.59}$$

$N - n_0$ is the number of noncondensed atoms. In this case,

$$H_{n_0} = \sum_{k>1} n_k, \tag{8.60}$$

and

$$P_{n_0} = \frac{1}{Z_N}\frac{\left(H_{n_0}\right)^{N-n_0}}{(N-n_0)!}. \tag{8.61}$$

Normalizing to obtain Z_N, there results, for the noncondensed probability,

$$P_{N-n} = \frac{\exp\left(-H_{i_0}\right)N!}{\Gamma\left(N+1, H_{n_0}\right)} \times \frac{\left(H_{n_0}\right)^n}{n!}. \tag{8.62}$$

Also, immediately,

$$\langle n_0\rangle = N - H_{n_0} + \frac{\left(H_{n_0}\right)^{N+1}}{Z_N N!} \tag{8.63}$$

$$\left\langle n_0^2\right\rangle - \langle n_0\rangle^2 = \left(H_{n_0}\right)\left(1 - (\langle n_0\rangle + 1)\frac{H_{n_0}^N}{Z_N N!}\right).$$

They show that these approximations are valid in the weak trapping limit, $T >> \varepsilon_1$, ε_1 being the energy difference of the ground and excited state. They appear to be qualitatively true, in general, for the harmonic oscillator trap, which we will turn to now.

Consider again the three-dimensional (3-D) harmonic oscillator trap where

$$\varepsilon_{\mathbf{k}} = \hbar\left(k_1\omega_1 + k_2\omega_2 + k_3\omega_3\right).$$

$$\text{Now} \quad H_{n_0} = \sum_{\mathbf{k}>0} (\exp\beta\hbar\omega - 1)$$

$$\text{and} \quad \eta H_{n_0} = \sum_{k>0} (\exp\beta\hbar k - 1)^2 .$$

As we have already done in Section 8.2, Eq. (8.3) and following, we approximate the sums by integrals and also define the critical temperature as

$$kT_c = \Omega\left(\frac{N}{1.202}\right)^{\frac{1}{3}},$$

where $\Omega = (\omega_1\omega_2\omega_3)^{\frac{1}{3}}$. We have

$$H_{n_0} = \left(\frac{T}{T_c}\right)^3 N \tag{8.64}$$

and

$$\eta H_{n_0} = \left(\frac{T}{T_c}\right)^3 N\left(\frac{g_2(1) - g_3(1)}{g_3(1)}\right).$$

The number of particles in the condensate is thus approximately $N - H_{n_0}$, where $n_0 = N\left(1 - \left(\frac{T}{T_c}\right)^3\right)$, as we found in Section 8.3. The formula, Eq. (8.61), has been numerically evaluated in the harmonic oscillator case and found to agree well with the numerical results of Wilkens, Grossmann and Holthaus (Grossman and Holthaus, 1995) for finite N.

Appendix 8A: exact treatment of condensate traps

We will outline here the exact evaluation of rather general traps by the summation of the Bose–Einstein grand ensemble of de Groot (de Groot *et al.*, 1950). Particular attention will be paid to the box and harmonic traps. We suggest the reader look into this impressive and rather complete work.

They choose generally the energy

$$\beta\left(\varepsilon_{s_1}\ldots\varepsilon_w\right) = \frac{M}{T}\sum_{\nu=1}^{w}\frac{s_\nu^\alpha}{a_\nu^2} \qquad (s_1\ldots s_w = 1, 2, \ldots). \tag{8A.1}$$

Here $a_1 \ldots a_w$ have dimension of length in w dimensions, λ is a parameter ranging $1 \leq \lambda \leq 2$. $\lambda = 1$ is the harmonic oscillator (dimension possibly $w = 3$), and $\lambda = 2$ is the spectrum of the particle in a box. M is a constant. Then one has

$$N = \sum_{j=1}^{\infty} z^j \Pi_{\nu=1}^{w} G_\alpha (x_\nu, j) \tag{8A.2}$$

and

$$\bar{\varepsilon} = \frac{kT}{N} \sum_j z^j \frac{d}{dj} \Pi_{\nu=1}^{w} G_\alpha (x_\nu, j). \tag{8A.3}$$

Here

$$x_\nu = \frac{M}{a_\nu^2 T}$$

$$G_\alpha (x) = \exp x \sum_{s=1}^{\infty} \exp (-x s^\alpha). \tag{8A.4}$$

$G_\alpha (x)$ (for $x > 0$) and its first and second derivatives are continuous monotonically decreasing. But $G_\alpha (0) = \infty$, and $G_\alpha (\infty) = 1$. However, $\lim_{x \to 0} x^{\frac{1}{2}} G_\alpha (x) = (\alpha^{-1})!$.

Now consider z and the series

$$R_q (z) \equiv \sum_{j=1}^{\infty} z^j j^{-q}. \tag{8A.5}$$

As we already know for $0 \le z < 1$, Eq. (8A.5) converges and reaches the g_q functions at $g_q (z = 1)$. The derivative

$$\frac{dR_q}{dz} = \frac{R_{q-1} (z)}{z}$$

is zero for $z = 0$. But for $q < 1$, $z \to 1$ and $R_q (z) \to \infty$ such that

$$(1 - z)^{-q+2} \frac{dR_q}{dz} \to (-q + 1)!.$$

Let us consider the inversion $z_q (R)$. We may obtain three regimes:

$$1 < q < \frac{3}{2} \quad \frac{d^2 z}{dR^2} = 0 \quad \left(R = R_q (1)\right) \tag{8A.6}$$

$$q = \frac{3}{2} \quad \frac{d^2 z}{dR^2} = \frac{-1}{2\pi} \quad \left(R = R_q (1)\right) \tag{8A.7}$$

$$\frac{3}{2} < q \le 2 \quad \frac{d^2 z}{dR^2} \to \infty \quad \left(R = R_q (1)\right), \tag{8A.8}$$

and $q > 2$ $\frac{dz}{dR} > 0$ for $0 \le R \le R_q (1)$. De Groot and colleagues proved the central theorem for $a_\nu \to \infty$, $N \to \infty$ and ν =constant where

$$\nu = \frac{N}{\Pi_{\nu=1}^{w} a_{\nu}^{\frac{2}{\alpha}}} \qquad \text{with } q = \frac{w}{\alpha}. \tag{8A.9}$$

There are two cases:

(1) $q = 1$ for all T. $z(R)$ is defined by

$$u(T) \equiv \nu \left\{ \left(\alpha^{-1} \right)! \right\}^{-w} \left(\frac{M}{T} \right)^{q} = R_q(z). \tag{8A.10}$$

(2) $q > 1$, then Eq. (8A.10) is valid only for $T > T_c$ where T_c is determined by the limiting value

$$\nu \left\{ \left(\alpha^{-1} \right)! \right\}^{-w} \left(\frac{M}{T_c} \right)^{q} = R_q\,(z = 1) \quad \text{and for } T < T_c. \tag{8A.11}$$

$$z = 1 \tag{8A.12}$$

These two cases distinguish the behavior of $z(T)$ and its derivatives. For case 1, Eq. (8A.10) defines z as a continuous function of T decreasing monotonically from $z = 1$ at $T = 0$ to $z = 0$ at $T = \infty$. Case 2 is more interesting. $R_q(z) \le R_q(1)$, and the functions meet at $T = T_c$. T_c is determined by Eq. (8A.11), and now from Eqs. (8A.7), (8A.8) and (8A.9) we see the character of this transition in terms of $\frac{dz}{dT}$ and $\frac{d^2z}{dT^2}$. We find:

For $1 \le q < \frac{3}{2}$: E, $\frac{dE}{dT}$ and $\frac{d^2E}{dT^2}$ are continuous at $T = T_c$.

For $q = \frac{3}{2}$ (the box, $w = 3$, $\alpha = 2$) : $\frac{d^2E}{dT^2}$ shows a finite discontinuity.

For $\frac{3}{2} < q \le 2$: $\frac{d^2E}{dT^2}$ has an infinite discontinuity.

For $q > 2$: $\frac{dE}{dT}$ has finite discontinuity and thus shows a λ transition at T_c.

Only a finite transition T_c appears if $N \to \infty$ for ν finite. Now, for the box, $a = 2$, and ν is the mean gas density when $N \to \infty$. If $\nu \to 0$ or ∞, $T_c \to 0$ or ∞, respectively.

Generally, de Groot *et al.* prove that where $N \to \infty$, finite ν, for the case $q > 1$ and $T < T_C$

$$\frac{N_0}{N} = \left(1 - \left(\frac{T}{T_c} \right)^{q} \right). \tag{8A.13}$$

This is London's result with $q = \frac{3}{2}$. It was later proved by Grossman and Holthaus, also.

References

Anderson, M. H., Ensher, T. R., Mathews, M. R., Wieman, G. E. and Cornell, E. A. (1995). *Science* **269**, 198.
Bagnato, V., Pritcherd, D. E. and Kleppner, D. (1987). *Phys. Rev. A* **35**, 4354.
Bose, S. M. (1924). *Z. Phys.* **26,** 178.
Ceperley, D. M. (1995). *Rev. Mod. Phys.* **67**, 279.
Davis, K. B., Mewes, M. O., Andrews, M. R., Druten, N. J., Durfee, D. S., Kurn, P. M. and Ketterle, W. (1995). *Phys. Rev. Lett.* **75**, 3969.
Degiorgio, V. and Scully, M. (1970). *Phys. Rev. A* **2**, 1170.
Einstein, A. (1924a). *Preus. Akad. Wiss.* **22,** 261.
Einstein, A. (1924b). *Preus. Akad. Wiss.* **23**, 3.
Ensher, J. R., Din, D. S., Mathews, M. R., Williams, C. E. and Cornell, E. A. (1995). *Phys. Rev. Lett.* **77**, 4984.
Gardiner, G. W. (1985). *Handbook of Stochastic Methods* (Berlin, Springer).
Grossman, S. and Holthaus, M. (1995). *Z. Naturforsch. A* **50**, 921.
de Groot, S. R., Hooyman, G. J., and ten Seldam, C. A. (1950). *Proc. R. Soc. Lond. A* **203**, 266.
Keesom, W. H. and Clusius, K. (1932). *Proc. R. Acad. Amstr.* **35**, 307.
Keesom, W. H. and Keesom, A. P. (1932). *Proc. R. Acad. Amstr.* **39**, 736.
Ketterle, W. and van Druten, N. J. (1996). *Phys. Rev.* A54, 656
Kirsten, K. and Toms, D. J. (1997). *Phys. Lett. A* **222**, 148.
Kocharovsky, V. V., Scully, M. O., Zhu, S. and Zubairy, M. S. (2000). *Phys. Rev.* **61**, 023609-1.
Landau, L. (1941). *Zurn. Eksp. Teor. Fiz.* **5**, 71.
London, F. (1938). *Phys. Rev.* **54**, 947.
London, F. (1954). *Superfluids,* vol. 2 (New York, Wiley).
Munster, A. (1964). *Statistical Thermodynamics* (New York, Springer).
Penrose, O. and Onsager, L. (1956). *Phys. Rev.* **104**, 576.
Pethick, C. J. and Smith, H. (2002). *Bose–Einstein Condensation in Dilute Bose Gases* (London, Cambridge University Press).
Pitaeveskii, L. and Stringari, S. (2003). *Bose–Einstein Condensation* (Oxford, Clarendon).
Scully, M. O. (1996). *Phys. Rev. Lett.* **17**, 3927.
Scully, M. O. and Zubairy, M. S. (1999). *Quantum Optics* (Cambridge, Cambridge University Press).
Sokol, P. (1995). *Bose–Einstein Condensation,* ed. A. Griffin, D. W. Snoke and S. Stringari, 51 (London, Cambridge University Press).
Tisza, L. (1940). *J. Phys. Radium* **1**, 164, 350.
Ziff, R. M., Uhlenbeck, G. E. and Kac, M. (1977). *Phys. Rep.* **32**, 164.

9

Scaling, renormalization and the Ising model

9.1 Introduction

The Heisenberg ferromagnetic Hamiltonian is

$$\hat{H} = -\sum_{\mathbf{ij}} J_{\mathbf{ij}} \hat{\mathbf{s}}_{\mathbf{i}} \cdot \hat{\mathbf{s}}_{\mathbf{j}} - h_z \sum_{\mathbf{i}} \hat{s}_{\mathbf{i}z}. \tag{9.1}$$

Here $\hat{\mathbf{s}}_{\mathbf{i}}$ is the electron spin operator at lattice site **i**. **i** is an arbitrary lattice vector in d dimensions. $J_{\mathbf{ij}}$ is the spin coupling constant, the details of which do not interest us except to say that it is strong, near unity. The details of the lattice are also not important, and we will choose a square array and simply write J_{ij} without the explicit vector notation. The second term is the paramagnetic effect of the external dimensionless field **h** in the z direction. As already discussed in Chapter 2, we may write

$$\hat{\mathbf{s}} = \frac{1}{2}\hbar\hat{\boldsymbol{\sigma}}, \tag{9.2}$$

where $\hat{\boldsymbol{\sigma}}$ are the Pauli spin matrices having the properties

$$\hat{\sigma}_x^2 = \hat{\sigma}_y^2 = \hat{\sigma}_z^2 = 1 \qquad \hat{\boldsymbol{\sigma}}^2 = 3 \tag{9.3}$$

$$\begin{aligned} \hat{\sigma}_x\hat{\sigma}_y &= -\hat{\sigma}_y\hat{\sigma}_x = i\hat{\sigma}_z \\ \hat{\sigma}_y\hat{\sigma}_z &= -\hat{\sigma}_z\hat{\sigma}_y = i\hat{\sigma}_x \\ \hat{\sigma}_z\hat{\sigma}_x &= -\hat{\sigma}_x\hat{\sigma}_z = i\hat{\sigma}_y \end{aligned} \tag{9.4}$$

for a given spin site **i**. Consequently Eq. (9.1) cannot be diagonalized in the spin $\frac{1}{2}$ basis set.

It is thought that the partition function written with Eq. (9.1) is the basis for the description of the ferromagnetic phase transition, although this has never been exactly demonstrated. A simplification of this was introduced by Ising (1925) in a model of a spin chain. This will be the basis of the discussion here, as elsewhere in many good books (Huang, 1987; Chandler, 1987; Plischke and Bergersen, 1989; Kadanoff, 2000).

The Ising model is a severe replacement of Eq. (9.1) by

$$H_I = -\sum_{ij} J_{ij} \hat{s}_{iz} \hat{s}_{jz} - h_z \sum_i \hat{s}_{iz}. \tag{9.5}$$

Now the partition function may be evaluated in the basis which diagonalizes $\hat{\sigma}_{iz}$. We have the diagonal form

$$\langle \sigma_1 \ldots \sigma_N | H_I | \sigma_1 \ldots \sigma_N \rangle = -J \sum_{ij}{}' \sigma_i \sigma_j - h_z \sum_i \sigma_i \equiv E(\sigma_1 \ldots \sigma_N), \tag{9.6}$$

where $\sigma_1 \ldots \sigma_N = \pm 1$. Here appropriate factors of $\hbar$ and the number of nearest neighbors have been incorporated in J and h.

In his famous work, Onsager (1944) obtained the exact specific heat for the 2–D Ising model showing a phase transition. We shall not repeat this here, since good descriptions of the transfer matrix technique (Schultz *et al.*, 1964) already exist. See the books of Huang (1987) and Plischke and Bergersen (1989). We shall discuss this later by an approximation.

Let us begin with Landau (1941) mean field theory to introduce critical indices. We will then turn to the phenomenological Widom scaling (Widom, 1965) and then Kadanoff's block spin scaling (Kadanoff *et al.*, 1967). These are a prelude to the consideration of the renormalization theory of Wilson (1971). Here we will consider the use of these methods and results for the 2–D Ising model. This will follow the nice introduction of Maris and Kadanoff (1978).

9.2 Mean field theory and critical indices

Recall the thermodynamic properties of magnetic systems. The work done by the system with constant applied field is $\eth W = -\mathbf{H} \cdot d\mathbf{M}$, where $\mathbf{H}$ is the applied magnetic field and $\mathbf{M}$ the total magnetization (Callen, 1985).

The Helmholtz free energy is given by

$$dF \equiv dA = \mathbf{H} \cdot d\mathbf{M} - SdT \tag{9.7}$$

and for the canonical ensemble, as earlier,

$$A = (kT) \ln Z(T, H), \tag{9.8}$$

Z again being the partition function

$$Z = Tr \exp(-\beta H). \tag{9.9}$$

The Gibbs free energy, which will play an important role, is

$$dG = -SdT - MdH. \tag{9.10}$$

We drop the vector notation on M and H. It may be shown for the grand canonical ensemble that

$$G(H, T, N) = \mu N,$$

so the g per particle is μ. Thus we have

$$S = -\left(\frac{\partial A}{\partial T}\right)_M \tag{9.11}$$

and

$$H = \left(\frac{\partial A}{\partial M}\right)_T. \tag{9.12}$$

The specific heats are

$$C_M = -T\left(\frac{\partial^2 A}{\partial T^2}\right)_M \tag{9.13}$$

and

$$C_H = -T\left(\frac{\partial^2 G}{\partial T^2}\right)_H. \tag{9.14}$$

The susceptibilities are

$$\chi_S = \left(\frac{\partial M}{\partial H}\right)_S \tag{9.15}$$

and

$$\chi_T = \left(\frac{\partial M}{\partial H}\right)_T. \tag{9.16}$$

Two coexisting phases at equilibrium lead to the Gibbs phase rule. We have for two phases, "1" and "2,"

$$\begin{aligned} T_1 &= T_2 \\ H_1 &= H_2 \\ \mu_1 &= \mu_2 \end{aligned} \tag{9.17}$$

and consequently

$$g_1(T, H) = g_2(T, H). \tag{9.18}$$

From this, along the phase coexistence curve, the Clausius–Clapyron equation follows:

$$\left(\frac{d\mu}{dT}\right)_{\text{coex}} = \frac{s_1 - s_2}{v_1 - v_2}. \tag{9.19}$$

Critical indices or exponents became important experimentally and theoretically (Fisher, 1974). Near the critical point, it is assumed that the singularities are power laws:

$$\begin{aligned} C_+ &= A_+ (T - T_c)^{-\alpha} \qquad && T > T_c \\ C_- &= A_- (T_c - T)^{-\alpha'} && T < T_c \end{aligned} \tag{9.20}$$

$$\begin{aligned} \chi_+ &= D_+ (T - T_c)^{-\gamma} \qquad && T > T_c \\ \chi_- &= D_- (T_c - T)^{-\gamma'} && T < T_c \end{aligned} \tag{9.21}$$

and

$$M = B (T_c - T)^{\beta} \qquad T < T_c \tag{9.22}$$

as well as the equation of state,

$$M = H^{\frac{1}{\delta}}. \tag{9.23}$$

The spin correlation length is

$$\langle s_i - \langle s \rangle \rangle \langle s_j - \langle s \rangle \rangle \approx \exp\left(\frac{-r_{ij}}{\xi}\right), \tag{9.24}$$

where

$$\xi \approx (T - T_c)^{\nu} \qquad T > T_c$$

and

$$\approx (T_c - T)^{-\nu'} \qquad T < T_c.$$

The parameters (apparently disparate) $\alpha,\ \alpha',\ \gamma,\ \gamma',\ \beta,\ \delta,\ \nu,\ \nu'$ will be the central focus of the subsequent discussion.

The mean field approximation, used many places in physics, replaces σ_j in Eq. (9.6) by the average value $\langle \sigma_j \rangle$ in the first term. $\langle \sigma_j \rangle = m$, the magnetization per atomic site, is independent of j. Thus we have an effective $E(\sigma_1 \ldots \sigma_N)$:

$$E_{\text{eff}}(\sigma_1 \ldots \sigma_N) = -Jm \sum_i \sigma_i - h_z \sum_i \sigma_i. \tag{9.25}$$

The idea is that the effect of fluctuations is small in the interaction and that the nearest neighbor sites effect a spin through their average value. This, of course, depends nonlinearly on all adjacent spins in a self-consistent fashion. The validity of the mean field approach depends on the smallness of these fluctuations and may or may not be true, as we will see.

We may immediately easily calculate m in this approximation:

$$m = \frac{\mathrm{Tr}\sigma_0 \exp\left(-\beta E_{\mathrm{eff}}\right)}{\mathrm{Tr}\exp\left(-\beta E_{\mathrm{eff}}\right)} = \tanh\left[\beta\left(Jm + h\right)\right]. \tag{9.26}$$

We obtain a transcendental equation of state for m. Here we have incorporated the number of nearest neighbors into J.

We may solve Eq. (9.26) approximately or numerically. For small βJ and $h = 0$,

$$m_0 = \beta J m_0 - \frac{1}{3}\left(\beta J\right)^3 m_0^3.$$

This has solutions

$$m_0 = 0 \tag{9.27}$$

and

$$1 = \beta J - \frac{1}{3}\left(\beta J\right)^3 m_0^2. \tag{9.28}$$

Eq. (9.28) is used to define the critical temperature, $\beta_c J = 1$ and $m_0 \neq 0$. Using this, we find

$$m_0 = \pm 3^{\frac{1}{2}}\left(\frac{T}{T_c}\right)^{\frac{3}{2}}\left(\frac{T_c}{T} - 1\right)^{\frac{1}{2}}. \tag{9.29}$$

Which solution applies? The Gibbs free energy is for small m

$$\beta\left[G\left(m_0 T\right) - G\left(0, T\right)\right] = \frac{m_0^2}{2}\left(1 - \beta J\right) + \frac{1}{12} m_0^4 - \ln 2. \tag{9.30}$$

The solution to Eq. (9.29) has the lowest value of G for $T < T_c$, $\frac{J}{kT_c} = 1$. It is the stable phase. Thus, Eq. (9.29) represents the spontaneous magnetization in the temperature range. For $T > T_c$, $m_0 = 0$ is the stable phase. $\beta = \frac{1}{2}$ is the mean field critical index already mentioned.

Consider now the susceptibility per spin site

$$\chi\left(h, T\right) = \frac{\partial m}{\partial h}\,|_T\,. \tag{9.31}$$

From the expression for m,

$$\chi\left(0, T\right) = \frac{\beta \sec h^2\left(\beta J m\right)}{1 - \beta J \sec h^2\left(\beta J m\right)}, \tag{9.32}$$

as $T \to T_c^+$, we obtain

$$\chi^+(0,T) = \frac{1}{kT_c\left(1 - \frac{T_c}{T}\right)}; \qquad T > T_c.$$

By writing

$$m = \frac{T}{T_c} \tanh^{-1} m \tag{9.33}$$

we expand and obtain

$$\chi^-(0,T) = \frac{1}{kT\left(1 - \frac{T_c}{T}\right)}; \qquad T < T_c.$$

The susceptibility mean field critical exponent is $\gamma = \gamma' = 1$.

The specific heat is obtained by writing

$$\langle H \rangle = -J \sum_{ij} \langle \sigma_i \rangle \langle \sigma_j \rangle = -\frac{1}{2} J N m^2. \tag{9.34}$$

$C_h^+ = 0$, $T > T_c$ and $C_h^+ = \frac{3}{2} Nk$ for $T < T_c$. Thus, the mean field does *not* have conventional critical indices for the specific heat. This failure is what led Onsager to examine the 2–D Ising model exactly, which led to the famous $\ln\left(1 - \frac{T}{T_c}\right)$ result. Also, we should note from Eq. (9.26) how, taking $h \neq 0$,

$$m = m + \beta h - \frac{1}{3}(m + \beta h)^2. \tag{9.35}$$

Near $h = 0$,

$$h = m^3, \tag{9.36}$$

and the mean field critical index is here $\delta = 3$.

Eq. (9.30) is a special case of the general Landau approach to mean field theory. Near the critical point, it is assumed that

$$G(m,T) = G(0,T) + \frac{1}{2} b(T) m^2 + \frac{1}{4} c(T) m^4 + \frac{1}{6} d(T) m^6 \ \ldots \tag{9.37}$$

$b(T)$, $c(T)$, $d(T)$ are unspecified macroscopic coefficients. As with the Ising model, it is assumed $G(m,T) = G(-m,T)$, and only even powers in the expansion appear. Assume $C(T), d(T) > 0$ and $b(T) = b_0 (T - T_c)$. Then $G(m,T) - G(0,T)$ may have the double symmetric well curve for $T < T_c$, which disappears for $T > T_c$. $m = 0$ is now a local maximum with symmetric minima on either side.

We write the free energy extremum condition

$$\frac{\partial G}{\partial m} \mid_T = 0 \tag{9.38}$$

$$bm + cm^3 + dm^5 + \cdots = 0$$

and obtain to m^3 order

$$m = \pm \left[\frac{b_0}{c\,(T_c)}\right]^{\frac{1}{2}} (T_c - T)^{\frac{1}{2}} \qquad T < T_c, \tag{9.39}$$

having the critical index already obtained. Consider $S = -\frac{\partial G}{\partial T}$, and obtain the specific heat

$$\begin{aligned} C &= T\left(\frac{\partial S}{\partial T}\right) \\ &= -T\frac{d^2a}{dT^2} + \frac{1}{2}\left(\frac{Tb_0^2}{c}\right) \qquad T < T_c \\ &= -T\frac{d^2a}{dT^2} \qquad T > T_c. \end{aligned} \tag{9.40}$$

We have not yet obtained the critical index for ξ, the correlation length. To do this, we will treat the spacial dependence $\mathbf{r}$ by the Landau–Ginsburg macroscopic fluctuation theory (Ginsburg and Landau, 1950). The total magnetization is

$$M = \int d_3 r m\,(\mathbf{r}), \tag{9.41}$$

and the Gibbs free energy is

$$G\,(h\,(\mathbf{r}), T, m\,(\mathbf{r})) = A - \int d_3 r h\,(\mathbf{r})\, m\,(\mathbf{r}). \tag{9.42}$$

Expand the Helmholtz free energy as a function of $m\,(\mathbf{r})$:

$$A\,(\{m\,(r)\}, T) = \int d_3 r \left\{a\,(T) + \frac{b}{2}m^2\,(\mathbf{r}) + \frac{c}{4}m^4\,(r)\cdots + \frac{f}{2}\,[\Delta m\,(r)]^2\right\}. \tag{9.43}$$

This is a spacially dependent generalization of the previous expansion, Eq. (9.37). Assume f positive, and this guarantees the fluctuation term increases the Helmholtz free energy. This term is the simplest assumption which is invariant under $m \to -m$, which also determines the form of the other terms. Consider the functional derivative

$$h\,(r) = \frac{\delta A}{\delta m\,(r)}. \tag{9.44}$$

Taking the variation of Eq. (9.43), we may write the Ginsburg–Landau equation

$$h(\mathbf{r}) = bm(\mathbf{r}) + cm^3(\mathbf{r}) - f\Delta^2 m(r), \tag{9.45}$$

where we have integrated by parts and assumed $\delta m(\mathbf{r}) = 0$ on the boundaries.

Assume an expansion around weak fluctuations, $\phi(\mathbf{r})$,

$$m(\mathbf{r}) = m_0(T) + \phi(\mathbf{r}) \tag{9.46}$$

and the inhomogeneity

$$h(r) = \frac{h_0}{f}\delta(r) \quad \text{locally;} \tag{9.47}$$

$$\begin{aligned} m_0(r) &= 0 \qquad && T > T_c \\ m_0^2 &= -\frac{b}{c} \qquad && T < T_c \end{aligned} \tag{9.48}$$

as in the deterministic theory discussed earlier.

The linearized equation for ϕ is then

$$\begin{aligned} \Delta^2\phi - \frac{b}{f}\phi &= -\frac{h_0}{f}\delta(\mathbf{r}) \qquad && T > T_c \\ \Delta^2\phi + \frac{2b}{f}\phi &= -\frac{h_0}{c}\delta(r) \qquad && T < T_c. \end{aligned} \tag{9.49}$$

The solution is

$$\phi = \frac{h_0 f}{4\pi f r}\exp\left(-\frac{r}{\xi}\right). \tag{9.50}$$

The spherical correlation length becomes

$$\begin{aligned} \xi &= \left[\frac{f}{b(T)}\right]^{\frac{1}{2}} \qquad && T > T_c \\ \xi &= \left[\frac{1f}{2b(T)}\right]^{\frac{1}{2}} \qquad && T < T_c. \end{aligned} \tag{9.51}$$

Since $b(T) = b_0(T - T_c)$, $\xi(T)$ has the critical index $\nu = \nu' = \frac{1}{2}$, being symmetric around T_c. $\phi(\mathbf{r})$ is indeed the correlation function, since

$$\frac{\delta m(\mathbf{r})}{\delta h(0)} = \frac{\phi(\mathbf{r})}{h_0} = \beta\left(\langle m(r)\, m(0)\rangle - \langle m(r)\rangle\langle m(0)\rangle\right). \tag{9.52}$$

From the foregoing mean field considerations, a criterion may be obtained for the self-consistency of the mean field approach known as the Ginsburg criterion, which we write down as

$$d > 2 + \frac{2\beta}{\nu}, \tag{9.53}$$

d being the dimension. For Landau and mean field theories, $\beta = \frac{1}{2}$ and $\nu = \frac{1}{2}$ and hence are valid for $d > 4$ and not $d = 3$. This is consistent with the fact that the exact 2–D Ising values are $\beta = \frac{1}{8}$, $\nu = 1$.

9.3 Scaling

The failure of the mean field theory and its expression in Eq. (9.53) led to macroscopic scaling, due to Widom (1965) and Kadanoff (Kadanoff *et al.,* 1967), which we shall now examine. For the magnetic case we generalize

$$h = \frac{\partial A}{\partial m} = m\chi\left(t, m^{\frac{1}{\beta}}\right). \tag{9.54}$$

We no longer use the Landau expansion. Let $t = \frac{T-T_c}{T_c}$, and β is the critical index already introduced. Following Widom's brilliant suggestion, take χ to be a homogeneous function of two variables:

$$\chi\left(\lambda^{\frac{1}{\gamma}} t, \lambda^{\frac{1}{\gamma}} m^{\frac{1}{\beta}}\right) = \lambda\chi\left(t, m^{\frac{1}{\beta}}\right), \tag{9.55}$$

or equivalently, assume the Gibbs free energy singular part

$$G(t, h) = \lambda G\left(\lambda^{s} t, \lambda^{r} h\right). \tag{9.56}$$

The parameters r, s will be determined. λ is the scale parameter. Using $m = \frac{-\partial G}{\partial h}$ and $\chi = \frac{\partial m}{\partial h}\,|_t$, we may obtain

$$m(t, h) = \lambda^{r+1} m\left(\lambda^{s} t, \lambda^{r} h\right) \tag{9.57a}$$

$$\chi(t, h) = \lambda^{2r+1} \chi\left(\lambda^{s} t, \lambda^{r} h\right). \tag{9.57b}$$

Also, $C_h = -T\left(\frac{\partial^2 G}{\partial t^2}\right)|_h$, so

$$C_h(t, h) = \lambda^{2s+1} C_h\left(\lambda^{s} t, \lambda^{r} h\right). \tag{9.57c}$$

Now we examine this near both sides of the critical point, $t = 0$, for t small, positive and negative. First we take $h = 0$ and assume $\lambda = -t^{\frac{-1}{s}}$. We have from Eq. (9.57a) and Eq. (9.57b)

$$m(t, 0) = (-t)^{-\frac{(1+r)}{s}} m(-1, 0) \tag{9.58a}$$

$$\chi(t, 0) = (-t)^{-\frac{(2r+1)}{s}} \chi(-1, 0) \tag{9.58b}$$

$$C_h(t, 0) = (-t)^{-\frac{(2s+1)}{s}} C_h(-1, 0). \tag{9.58c}$$

Now we choose $t = 0$ and the scale λ as $\lambda = +h^{\frac{-1}{r}}$ for small h. Eq. (9.57a) becomes

$$m(0, h) = h^{\frac{-r+1}{s}} m(0, +1). \tag{9.58d}$$

We may obtain similar equations for t positive and also h negative.

Now, comparing with Eq. (9.21), Eq. (9.22) etc., we find

$$\gamma = \gamma' = \frac{2r+1}{s} \tag{9.59a}$$

$$\alpha = \alpha' = \frac{2s+1}{s}$$

and from Eq. (9.58d)

$$\delta^{-1} = \frac{-(r+1)}{r}. \tag{9.59b}$$

Also, Eq. (9.58a) gives

$$\beta = \frac{-(r+1)}{s}. \tag{9.59c}$$

These relations may be rewritten compactly, as follows:

$$\alpha + 2\beta + \gamma = 2 \tag{9.60}$$

$$\beta(\delta - 1) = \gamma.$$

We have the remarkable result that there are *only* two independent critical indices and these relations. The first was deduced as an inequality from thermodynamics by Rushbrook (Rushbrook, 1963). The scaling laws are thought to be exact and valid even when mean field theory holds.

There is another relation called hyperscaling,

$$d\nu = 2 - \alpha, \tag{9.61}$$

in which ν is the correlation length index and d the dimension. The physical content of the scaling assumption, Eq. (9.60), is not clear. Kadanoff took an important step by introducing the notion of a block spin Hamiltonian. We will see that the scaling relations may be obtained from this. If we are near the critical point, $t = 0$, we may expect that neighbor aggregates of micro spins may be statistically correlated, such that $\langle \sigma_i, \sigma_j \rangle = 1$ for $|\mathbf{i} - j| \leq \mathbf{R}$ and $|k| >> a_0$. We may form blocks of these spins. These R^d volume blocks of d dimensions may then form a new lattice with an effective (macro?) spin $\sigma_{\mathbf{R}}$ on this lattice:

$$\sigma_{\mathbf{R}} = \sum_{\mathbf{jk} \text{ in } R} \sigma_{\mathbf{jk}}. \tag{9.62}$$

We assume that this new set of $\sigma_{\mathbf{R}}$ are governed by an effective Ising Hamiltonian with h' and t' and also $K' = \beta' J'$. The number of block spins is $N' = |R|^{-d} N$. The spins are "thinned," to use a future terminology. We further assume $h' \to -h'$, $t' \to t'$ when $h = -h$. We then take

$$h' = hR^x \qquad (9.63)$$
$$t' = tR^y,$$

$x,\ y$ being as yet unspecified parameters and positive.

The rescaled Gibbs free energy per site is now

$$g(t, h) = R^{\ d} g\left(R^x t, R^y h\right), \qquad (9.64)$$

and the correlation length is

$$\xi(t, h) = R\xi\left(R^x t, R^y h\right). \qquad (9.65)$$

Thus, scaling of the Gibbs free energy in R appears naturally from the block picture. The λ of Widom is R, and s is dy, r is dx in general.

Eq. (9.65) is a new result of the block scaling. Assume $R = t^{\frac{-1}{y}}$ and for $h = 0$,

$$\xi(t, 0) = t^{\frac{-1}{y}} \xi(\pm 1, 0). \qquad (9.66)$$

We obtain $\nu = 1/y$ and show

$$\frac{d}{y} = 2 - \alpha = d\nu. \qquad (9.67)$$

We have used $d/y = 1/s = 2 - \alpha$. ν is the correlation length. Eq. (9.67) is the hyper scaling relation depending on dimension. All scaling laws, including hyper scaling relation Eq. (9.67), hold for the 2–D Ising model exactly. Eq. (9.67) is not true for mean field theory except for $d = 4$.

9.4 Renormalization

From the previous section we do not see the reason for two independent critical indices, nor do we have a method of calculating these indices. The fundamental method and deeper understanding of how to do this is due to Wilson, and we call it renormalization theory.

To see the elements of this, we will consider the one-dimensional Ising model, as Wilson did in his first paper (Wilson, 1971). The methods are far more general, as nicely discussed in the book of Kadanoff (2000) and also in Wilson and Kogut (1974).

Consider again the 1 – D nearest neighbor, Ising Hamiltonian energy

$$H = -K \sum_{i=1}^{N} \sigma_i \sigma_{i+1} - h \sum_{i=1}^{N} \sigma_i, \tag{9.68}$$

where the coupling constants are $K = \beta J$, $h' = \beta h$. Now we will drop the prime. The partition function is

$$Z(N, Kh) \equiv \sum_{\{\sigma_i\}=\pm 1} \exp\left(\sum_{i=1}^{N} K \sigma_i \sigma_{i+1} + \frac{1}{2} h (\sigma_i + \sigma_{i+1}) \right). \tag{9.69}$$

We wish to write this in a block spin representation with new coupling constants K', h'. We assert that there is a transformation (mapping) from $N \to N' = N/2$, $K \to K'$, $h \to h'$. Clearly it is possible to reduce the number of sites and introduce blocks by summation (integration!), but is it then of the Ising form with simply K', h' ? In fact, it is of a more general form:

$$\begin{aligned} Z\left(\frac{1}{2} N K' h'\right) &= Z(N, K, h) \exp -N g(Kh) \\ &= \left[f(K)\right]^{-N} Z(Kh). \end{aligned} \tag{9.70}$$

This is the Kadanoff transformation. The factor $f(K)$ is proportional to the free energy, and $g(K)$ is independent of the system size. In the 1 – D example for $h = 0$, we sum on all even spin sites and introduce K':

$$f^{-1}(K)\left[\exp K\left(\sigma + \sigma'\right) + \exp\left(-K\left(\sigma + \sigma'\right)\right)\right] = \exp\left(K' \sigma \sigma'\right), \tag{9.71}$$

or, since $\sigma, \sigma' = \pm 1$,

$$\begin{aligned} K' &= \frac{1}{2} \ln \cosh(2K) \\ f(K) &= 2 \cosh^{\frac{1}{2}}(2K). \end{aligned} \tag{9.72a}$$

Using

$$g(K) = \frac{1}{2} \ln f(K) + \frac{1}{2} g\left(K'\right),$$

we have

$$g\left(K'\right) = 2g(K) - \ln\left[2\sqrt{\cosh 2K}\right]. \tag{9.72b}$$

Eq. (9.72a) and Eq. (9.72b) represent the transformation of $Z(NK)$ when the number of sites is reduced by 1/2, forming blocks of effective spins. They are termed the renormalization group equations. More generally,

$$K' = \frac{1}{4}\ln\frac{\cosh(2K+h)\cosh(2K-h)}{\cosh^2 h} \tag{9.73}$$

$$h' = h + \frac{1}{2}\ln\frac{\cosh(2K+h)}{\cosh(2K-h)}.$$

$g = \frac{1}{8}\ln\left[16\cosh(2K+h)\cosh(2K-h)\cosh^2 h\right]$, which we leave as a problem for the student.

This procedure may be repeated from $N' \to \frac{N}{2}$, $N'' = \frac{N'}{2} = \frac{N}{4}$ etc., introducing repeatedly larger blocks with effective Ising constants $K,\ K',\ K'',\ h,\ h',\ h'',\ldots,$ which change the scale of description. Wilson emphasized that this may be viewed as a continuous transformation in block size and not necessarily discrete. We have not made this explicit but have maintained a block picture for simplicity, which is not strictly valid, but physical (Wilson, 1971). This is the essence of the renormalization maps, the iteration of Eq. (9.73). We may write

$$-\beta G(NKh) = Ng(Kh) + \ln Z\left(\frac{1}{2}N,\ K'h'\right) \tag{9.74}$$

$$= \sum_{j=0}^{\infty}\left(\frac{1}{2}\right)^j g(K_j, h_j).$$

Higher dimension is more complicated, but the procedure is similar. There are now $\{K\} = (K_1 \ldots K_n)$ coupling constants in d dimensions and b^d degrees of freedom, one of which is h. The Hamiltonian energy is written

$$H = N\sum_{\alpha=1}^{h} K_\alpha \psi_\alpha(\sigma_i). \tag{9.75}$$

The renormalization transformation gives new constants,

$$K'_\alpha = R_\alpha(K_1 \ldots K_n), \tag{9.76}$$

and the Kadanoff transformation is

$$\mathrm{Tr}_{\{\sigma_i\}}\exp H = \exp Ng(K) \times \mathrm{Tr}_{\{\sigma'_i\}}\exp H'(K'), \tag{9.77}$$

where $\mathrm{Tr}\exp H' - \exp\left(\frac{Nf(K')}{b^d}\right)$. Therefore,

$$f(\{K\}) = g(\{K\}) + b^{-d} f(K'). \tag{9.78}$$

Eq. (9.76) and Eq. (9.78) are the renormalization group transformations. $b^{-d}f(K')$ is identified with the singular part of the free energy (Niemeijer and van Leeuwen, 1976). In general, Eq. (9.76) is a continuous scale transformation and is *analytic* at the fixed point. Wilson wrote down these *differential* equations explicitly. Eq. (9.76) represents the solutions.

9.5 Renormalization and scaling

This change of scale is represented by the sequence of coupling constant evolutions. In a dynamic sense, there is a flow governed by Wilson's equations in the coupling constant space (Wilson, 1971). By construction, $Z\left(\mathbf{K}', N'\right) = Z\left(\mathbf{K}N\right)$. $\mathbf{K}$ is a vector made of the components (K_α). Hence we may conclude that the singular part of the free energy $f\left(\mathbf{K}'\right) = b^{-d} f\left(\mathbf{K}\right)$ (Niemeijer and van Leeuwen, 1976), which is Widon's scaling.

In the flow in $\mathbf{K}$ space under the renormalization, the fixed points play a special role. This can be seen in the 1–D Ising example. We have for $h = 0$

$$K' = \frac{1}{2} \ln \cosh 2K \leq K.$$

$K = K'$ for $K = 0$ and $K = \infty$. $K = 0$ corresponds to $J = 0$ and no spontaneous magnetization, or to finite J for weak coupling. From the map the $K' = \infty$ fixed point is unstable, and the flow is toward the no-interaction fixed point. This, of course, corresponds to the fact that in the 1–D Ising model there is no spontaneous magnetization.

In higher dimensions the fixed points are

$$\mathbf{K}^* = R\left(\mathbf{K}^*\right). \tag{9.79}$$

We may also argue physically, from the block picture, that the spin correlation length obeys

$$\xi\left(\mathbf{K}'\right) = b^{-d} \xi\left(K\right). \tag{9.80}$$

We see that at the fixed point $\xi\left(K^*\right) = b^{-d} \xi\left(K^*\right)$, which has two solutions, $\xi\left(K^*\right) = 0$ or ∞ for finite b. One of these we have already met in zero magnetization. The other is the critical point.

Let us examine this further in two dimensions with K_1, K_2. We parameterize the critical point $(K_{1c}, K_{2c}) = \left(\frac{J_1}{kT_c}, \frac{J_2}{kT_c}\right)$ by fixing $\frac{J_1}{J_2}$. Criticality is now determined by T_c. The flow induced by the renormalization transformations is for $T > T_c$ toward $K_1 = K_2 = 0$ and away for $T < T_c$ toward a zero-temperature ground state. For $\xi = \infty$ it is along a line of invariant criticality. This may be a saddle in the K_1, K_2 space. The flow must be away from $\xi = \infty$, since the renormalization increases the block size. Consequently, $\xi\left(K^*\right) = \infty$ is an unstable fixed point, since ξ decreases on repeated mapping.

Now let the generally nonlinear map be

$$\begin{aligned} K_1^* &= R_1\left(K_1 K_2\right) \\ K_2^* &= R_2\left(K_1 K_2\right). \end{aligned} \tag{9.81}$$

We examine the solution by standard linear stability analysis. Let

$$\delta K_1 = K_1 - K_1^* \tag{9.82}$$

and

$$\delta K_2 = K_2 - K_2^*$$

be small. To first order we have the linearized map, by Taylor expansion,

$$\begin{aligned} \delta K_1' &= M_{11}\delta K_1 + M_{21}\delta K_2 \\ \delta K_2^1 &= M_{21}\delta K_1 + M_{22}\delta K_2, \end{aligned}$$

where $M_{ij} = \frac{\partial R_i}{\partial K_j} |_{K_i^* K_j^*} \neq M_{ji}$ is, in general, not Hermitian. We may diagonalize this by obtaining the left eigenvalue and eigenfunction of

$$\sum_{ij} \phi_{\alpha i} M_{ij} = \lambda_\alpha \phi_{\alpha j}. \tag{9.83}$$

Since $\lambda_\alpha(b)\,\lambda_\alpha(b) = \lambda_\alpha\left(b^2\right)$, we may write, by the group property,

$$\lambda_\alpha = b^{y_\alpha}, \tag{9.84}$$

introducing the parameter y_α. Using the $\phi_{\alpha i}$ coordinates we write generally

$$U_\alpha = \delta K_1 \phi_{\alpha 1} + \delta K_2 \phi_{\alpha 2}$$

and

$$U_\alpha^1 = \delta K_1' \phi_{\alpha 1} + \delta K_2' \phi_{\alpha 2}.$$

Hence $U_\alpha' = \lambda_\alpha U_\alpha$. The U_α scale under the linear transformation by λ_α. Thus, $U_\alpha' = b^{y_\alpha} U_\alpha$.

Generally, we have the free energy recursion under the map

$$f(\{\mathbf{K}\}) = g(\{\mathbf{K}\}) + b^{-d} f\left(\{\mathbf{K}'\}\right). \tag{9.85}$$

The singular part of the free energy is the second term. Near the critical point we may express $\mathbf{K}$ in terms of $\mathbf{U}$. Thus, for the singular part,

$$f(U_1, U_2) = b^{-d} f\left(b^{y_1} U, b^{y_2} U_2\right), \tag{9.86}$$

which is the Kadanoff scaling form of the free energy. y_1 and y_2 may be related to critical indices, as we have already done. Since the fixed point is hyperbolic, y_1, y_2 must have opposite sign. We choose y_1 positive. The U_α for $y_\alpha > 0$ are called relevant scaling fields. Wilson has examined in detail the structure of a two-dimensional renormalization set of equations and their solution. For $h = 0$ having the gradient form, the saddle fixed point is possible with the well-known properties. We may show

$$\frac{d}{y_1} = 2 - \alpha \tag{9.87}$$

with

$$y_1 = \frac{\ln \lambda_1}{\ln b}$$
$$y_1' = \nu^{-1}.$$

The important point here is that we have a method of *calculating* λ by solving the linearized eigenvalue problem at the unstable critical point. In addition, all the scaling relations hold. We only need $M_{ij} = \frac{\partial R_i}{\partial K_{ij}} |_{K_1^* K_2^*}$ and the solution to Wilson renormalization equations.

9.6 Two-dimensional Ising model renormalization

We will return to the two-dimensional Ising model which, in the light of the famous Onsager result, exhibits a phase transition in the specific heat. We will use the renormalization approach to illustrate the technique of obtaining an approximate solution. We follow closely the simple paper of Maris and Kadanoff (1978). See also the book of David Chandler (1987). The thinning or block size mapping will be carried out directly on the Ising 2–D partition function. The procedure was already begun earlier in 2–D when discussing the block scaling of Kadanoff. Another approach would be to solve the renormalization group equations of Wilson. Other approximation methods are discussed in the book by Plischke and Bergersen (1989).

We again consider the 2–D nearest neighbor Ising partition function for the square lattice and write the partition function for $h = 0$:

$$Z = \sum_{\{\sigma\}} \ldots \exp\left[K\sigma_5\left(\sigma_1 + \sigma_2 + \sigma_3 + \sigma_4\right)\right] \times \exp\left[K\sigma_6\left(\sigma_2 + \sigma_3 + \sigma_7 + \sigma_8\right)\right] \ldots \tag{9.88}$$

The site "5" has neighbors 1, 2, 3, 4, and site "6" has 2, 3, 7, 8, etc. $K = J/KT$. We reduce the degrees of freedom, as in the 1–D case, by summing over 1/2 the spins that have "5" and "6" and also other nearest neighbors to 1, 2, 3, 4, 7, 8. At this point unlabeled sites in the original block also remain unsummed. We obtain, for the two relevant summations,

$$Z = \sum_{\{\sigma\}'} \left\{ \begin{array}{c} \exp\left[K\left(\sigma_1+\sigma_2+\sigma_3+\sigma_4\right)\right] \\ +\exp\left[-K\left(\sigma_1+\sigma_2+\sigma_3+\sigma_4\right)\right] \end{array} \right\}$$

$$\times \left\{ \begin{array}{c} +\exp\left[K\left(\sigma_2+\sigma_3+\sigma_7+\sigma_8\right)\right] \\ +\exp\left[-K\left(\sigma_2+\sigma_3+\sigma_7+\sigma_8\right)\right] \end{array} \right\}.$$

$\{\sigma\}'$ means the remaining sums.

Now, is this of the Kadanoff transformation form? This would assume that the summed partition function is effective 2–D Ising. For this special case it would read

$$\begin{aligned} I\left(K\sigma\right) &= \exp\left[K\left(\sigma_1+\sigma_2+\sigma_3+\sigma_4\right)\right]+\exp\left[-K\left(\sigma_1+\sigma_2+\sigma_3+\sigma_4\right)\right] \\ &= f\left(K\right)\exp\left[K'\left(\sigma_1\sigma_2+\sigma_1\sigma_4+\sigma_2\sigma_3+\sigma_3\sigma_4\right)\right]. \end{aligned} \tag{9.89}$$

There are two parameters, K' and $f(K)$, and four $\sigma_i = \pm 1$. This cannot hold. The Kadanoff transformation must be modified. Unlike 1–D, we cannot in 2–D obtain a renormalized *exact* Ising partition function for nearest neighbor blocks. A possibility is to enlarge the block interaction and introduce new constants, K_2 and K_3, such that

$$\begin{aligned} I\left(K\sigma\right) = f\left(K\right)\exp\Bigg[\left(\frac{1}{2}K_1\right)\left(\sigma_1\sigma_2+\sigma_2\sigma_3+\sigma_3\sigma_4+\sigma_4\sigma_1\right) \\ + K_2\left(\sigma_1\sigma_3+\sigma_2\sigma_4\right)+K_3\left(\sigma_1\sigma_2\sigma_3\sigma_4\right)\Bigg]. \end{aligned} \tag{9.90}$$

We obtain

$$\begin{aligned} K_1 &= \frac{1}{4}\ln\cosh\left(4K\right) \\ K_2 &= \frac{1}{8}\ln\cosh\left(4K\right) \\ K_3 &= \frac{1}{8}\ln\cosh\left(4K\right)-\frac{1}{2}\ln\cosh\left(2K\right). \end{aligned} \tag{9.91}$$

To obtain an Ising block partition function, K_2 and K_3 must be approximately zero. Setting $K_2 = K_3 = 0$, however, reduces the problem to 1–D where there is no phase transition. Another approximation is essential.

Let us, after Maris and Kadanoff, at least keep approximately K_2, letting $K_3 = 0$. Assume that the K_2 and K_3 terms in Eq. (9.90) may be written $K'\left(K_1K_2\right)\sum_{ij}\sigma_i\sigma_j$, an effective nearest neighbor interaction. This gives the Ising-like expression

$$Z\left(K,N\right) = f\left(K\right)^{\frac{1}{2}} Z\left(K'\left(K_1K_2\right), \frac{N}{2}\right) \tag{9.92}$$

and

$$g(K) = \frac{1}{2} \ln f(K) + \frac{1}{2} g(K') \tag{9.93}$$

or

$$g(K') = 2g(K) - \ln Z(K_1, K_2), \tag{9.94}$$

with K_1, K_2 given by Eq. (9.91). This is the approximate renormalization transformation. For the 2–D cubic lattice of $N/2$ spins, there are N nearest neighbors and N next nearest. We may approximate $K' = K_1 + K_2$. Thus, the renormalization transformation solution, Eq. (9.79), is

$$K' = \frac{3}{8} \ln \cosh 4K. \tag{9.95}$$

The fixed points to this are

$$K_c = \frac{3}{8} \ln \cosh 4K_c,$$

which are $K_c = 0, \infty$ and 0.50698. The latter is unstable. The exact Onsager answer is

$$\frac{J}{kT_c} = \frac{1}{2} \sinh^{-1}(1) = 0.44069. \tag{9.96}$$

Now we follow the Wilson procedure discussed in the previous section. We expand around the fixed point. Assume a nonanalytic part of g which contributes to the scaling $(K - K_c)^{2-\alpha}$. Thus, using near K_c,

$$K' = K_c + (K - K_c) \frac{dK'}{dK} |_{K=K_c},$$

which is the equation for M_{ij} discussed in the previous section. We have, from Eq. (9.95),

$$\alpha = 2 - \frac{\ln 2}{\ln \left(\frac{dK'}{dK}\right)_{k=K_c}} = 0.131, \tag{9.97}$$

giving $\alpha = 0.131$ compared with the Onsager answer of zero where the singularity is logarithmic. The formula for the specific heat index α may be obtained by an expansion around K_c of the singular part of the free energy. We leave this as a problem.

The main point of the renormalization theory is that it provides a tool for the application of approximation methods. They are more systematic than what has been done in this simple model. See Plischke and Bergersen (1989) for an introduction. For instance, the position space cumulant approach (Niemeijer and van Leeuwen, 1976) gives to first order $\alpha = -0.267$ but in the next systematic approximation gives $\alpha = 0.081$. It must be emphasized that these methods are applied to a much wider and realistic group of problems than the 2– D Ising model. However, it shows that the Onsager solution is a touchstone for examining a multitude of approaches.

References

Callen, H. (1985). *Thermodynamics and an Introduction to Thermostatics*, 2nd edn. (New York, Wiley).

Chandler, D. (1987). *Introduction to Modern Statistical Mechanics* (New York, Oxford University Press).

Fischer, M. E. (1967). Rep. Progr. Phys. 30, 615.

Ginsburg, V. L. and Landau, L. (1950). *Zh. Eksp. Teor. Fiz.* **26**, 1064.

Huang, K. (1987). *Statistical Mechanics*, 2nd edn. (New York, Wiley).

Ising, E. (1925). *Z. Phys.* **1**, 253.

Kadanoff, L. P. (2000). *Statistical Physics* (Hackensack, NJ, World Scientific).

Kadanoff, L. P., Götze, W., Hamblen, D., Hecht, R., Lewis, E. A. S., Palciaukas, V. V., Rayl, M., Swift, J., Aspnes, D. and Kane, J. (1967). *Rev. Mod. Phys.* **39**, 395.

Landau, L. (1941). *J. Phys. USSR* **5**, 71.

Maris, M. J. and Kadanoff, C. J. (1978). *Am. J. Phys.* **46**, 652.

Niemeijer, T. and van Leeuwen, J. M. J. (1976). *Phase Transitions and Critical Phenomena,* vol. 6, ed. C. Domb and M. S. Green (New York, Academic).

Onsager, L. (1944). *Phys. Rev.* **65**, 117.

Plischke, M. and Bergersen, B. (1989). *Equilibrium Statistical Mechanics* (Englewood Cliffs, NJ., Prentice Hall).

Rushbrook, G. S. (1963). *J. Chem. Phys.* **39**, 842.

Schultz, T., Mattis, D. and Lieb, E. (1964). *Rev. Mod. Phys.* **36**, 856.

Widom, B. (1965). *J. Chem. Phys.* **43**, 3898.

Wilson, K. G. (1971). *Phys. Rev. B* **4**, 3174, 3184.

Wilson, K. G. and Kogut, J. (1974). *Phys. Rep.* **12**, 75.

10

Relativistic covariant statistical mechanics of many particles

10.1 Introduction

We will focus here principally on quantum relativistic kinetic theory in a *covariant* form. Much of the work that has been done on classical relativistic kinetic theory is summarized in the fine book by de Groot, van Leeuwen and van Weert (de Groot *et al.,* 1980). A short review of this noncovariant point of view is in the book of Liboff (1998). Pauli, in his classical review of special relativity (Pauli, 1958), touches on the early work of Jüttner (1911). Ehlers (1974) has reviewed the kinetic theory in the context of classical general relativity, but we shall limit ourselves to a discussion of special theory. This may come as a surprise to the reader. However, it must be remembered that even the two-body classical and quantum Schrödinger equation solutions have not been obtained exactly (Bethe and Saltpeter, 1957).

The noncovariant point of view starts with a Hamiltonian

$$\mathfrak{H} = \sum_i H_i + \frac{1}{8\pi} \int \left(E^2 + H^2\right) d^3x, \tag{10.1}$$

where $i = 1 \ldots N$ particles, H_i, with the fields being H and E and

$$H_i = \sqrt{\left[\mathbf{p}_i - \mathbf{e}\mathbf{A}_i\left(\mathbf{x}_i, t\right)\right]^2 + m_i^2} \; + V\left(x, t\right); \qquad c = 1. \tag{10.2}$$

$\mathbf{p}_i, \mathbf{x}_i$ are three vectors, and the time t associated with the dynamics is the lab frame time. This is the basis of the work of Balescu, Hakim and Kandrup (Balescu, 1964; Havas, 1965; Balescu and Kotera, 1967; Hakim, 1967; Kandrup, 1984) among others. These theories are said to be "on mass shell," since for each particle

$$E_i^2 = c^2 p_i^2 + m_i^2 c^4, \tag{10.3}$$

m_i being the particle rest mass.

The formulation of a truly relativistic theory of *many particles* as distinct from field theories has only recently been achieved. Here we will discuss the statistical mechanics of this approach with emphasis first on non-equilibrium and the relativistic quantum Boltzmann equation of *events* and then turn generally to the Gibbs-equilibrium ensembles. Comments will be made on new properties in this theory of quantum equilibrium ensembles.

There is a misunderstanding (Goldstein, 1980) that because we may write a covariant Lorentz–Dirac equation for a single particle in interaction with the electromagnetic field, as

$$\begin{aligned} \frac{d\mu^{\mu}}{ds} &= \frac{e}{mc} F^{\mu\nu}\mu_{\nu} + R^{\mu} \\ F^{\mu\nu} &= \partial^{\mu} A^{\nu} - \partial^{\nu} A_{\mu} \,, \end{aligned} \tag{10.4}$$

that this may be easily generalized to many particles. s in this case is the *single* proper time of the accelerating particle. This is, in fact, difficult to accomplish. It appears to be true that the Lorentz–Einstein coordinate time, s, cannot be used as a dynamic time ($ds^2 = d\mathbf{x} \cdot d\mathbf{x} - dt^2$).

Among a number of possibilities, we will adopt what we might call the universal time formalism. Let us consider a succession of local clocks evolving with the particles time τ_i. These are, of course, for a particular observer the particle properties $d\tau_i^2 = d\mathbf{x}_i \cdot d\mathbf{x}_i - dt_i^2$. The evolution of $x_i^{\mu}(\tau_i)$ in time τ_i we term *events.* We will at first take the number of event times τ as discrete and equal to N. This should not be confused with the $8n$ degrees of freedom of the n particles. We will correlate these events by means of a global universal covariant parameter τ, where

$$\tau = \tau_1 = \tau_2 = \tau_3 \ldots = \tau_i \ldots = \tau_n \,. \tag{10.5}$$

This approach was begun by Stueckelberg (1941) and Feynman (1949) and later enlarged and completed by Horwitz and Piron, and independently by Cook (Cook, 1972; Horwitz and Piron, 1973; Trump and Schieve, 1999). With this it is assumed that there exists a total invariant energy K:

$$\frac{dK}{d\tau} = 0. \tag{10.6}$$

We define generalized coordinates and velocities of n particles:

$$\begin{aligned} \mathbf{x}_i^{\mu}(\tau) &= (x_i(\tau), t_i(\tau)) \\ v_i^{\mu}(\tau) &= \frac{dx_i^{\mu}(\tau)}{d\tau} \qquad i = 1 \ldots n. \end{aligned} \tag{10.7}$$

Then we define the invariant action at a distance interaction potential:

$$v_{ij} = v_{ij}\left(\rho_{ij}(\tau)\right),$$

where

$$\rho_{ij} = \left|\left(x_i(\tau) - x_j(\tau)\right)\right|.$$

The covariant momentum is defined as

$$p_i^{\mu}(\tau) = m_i \frac{dv_i^{\mu}(\tau)}{d\tau}. \tag{10.8}$$

m_i is a scalar particle constant. The $\left|p_i^{\mu}\right| \neq m_i$, and the dynamics are off particle energy shell. The Hamiltonian function is now

$$K = T + v = \frac{1}{2}\sum_{i=1}^{n} \frac{1}{m_i} p_i^{\mu} p_{\mu i} + \sum_{i>j} v\left(\rho_{ij}\right). \tag{10.9}$$

Thus we obtain a classically covariant many-body Hamiltonian set,

$$\begin{aligned} \frac{dp_i^{\mu}}{d\tau} &= -\frac{\partial K}{\partial x_{\mu i}} \\ \frac{dx_{\mu i}}{d\tau} &= \frac{\partial K}{\partial p_{\mu i}}. \end{aligned} \tag{10.10}$$

We have here an $8n$-dimensional phase space, p_i^{μ}, x_i^{μ}. The p_i^{μ}, x_i^{μ} transform by the Lorentz–Einstein transformations. The motion in the space, $p_i^{\mu}(\tau), x_i^{\mu}(\tau)$ is generated by the invariant Hamiltonian, K.

10.2 Quantum many-particle dynamics: the event picture

Utilizing these ideas, we generalize to quantum mechanics (Horwitz *et al.*, 1989), introducing a scalar many-body wave function $\psi\left(x_i^{\mu}, \tau\right)$ in the $4n$-dimensional space x_i^{μ} (not τ). The assumed Schrödinger equation for the events x_i^{μ}, ($\mu = 0, 1, 2, 3$; $i = 1$ to n) is

$$i\hbar \frac{\partial \psi\left(x_i^{\mu}, \tau\right)}{\partial \tau} = \bar{K}\psi\left(x_i^{\mu}, \tau\right). \tag{10.11}$$

This equation has been named the Stueckelberg equation (Stueckelberg, 1941; Fanchi, 1993). τ is an invariant, as is $\hat{K}$, so the Stueckelberg or Schrödinger equation for the scalar $\psi\left(x_i^{\mu}, \tau\right)$ is also invariant. For the case of a single free particle, $\hat{K}_0 = \frac{1}{2m}\left(\partial_t^2 - \bar{\nabla}^2\right)$, which gives the invariant Klein–Gordon equation in the steady state:

$$i\partial_{\tau}\psi\left(x^{\mu}, \tau\right) = 0.$$

This is not the equation of motion but an *event* eigenstate, the eigenmodes for a spin zero function.

Now we must assume, also, that

$$\int dx_i^\mu \left|\psi\left(x_i^\mu\right)\right|^2 < \infty \tag{10.12}$$

and invariant. $\psi\left(x_i^m\right)$ here is not a function of $\left|x_i^\mu\right|^2$, so it is not necessarily invariant.

Further, for a single free particle, the solution

$$\psi\left(x^\mu, \tau\right) = \frac{1}{(2\pi)^2} \int d_4 p \exp\left(-i\frac{p^2}{2m}\tau\right) \exp\left(i\mathbf{p} \cdot \frac{\mathbf{q}}{\hbar}\right) \psi_0(p) \tag{10.13}$$

gives a wave packet at center $x_c^\mu = \frac{p_c^\mu}{m}\tau$, moving along the classical world line. The events are distributed around this.

Let us introduce an event, ket $\left|x_i^\mu\right\rangle$, being not invariant. We may write the Schrödinger wave function as $\psi\left(x_i^\mu \tau\right) = \left\langle x_i^\mu \mid \psi(\tau)\right\rangle$ with the assumed scalar product. We introduce the pure state density matrix,

$$\left\langle x^{\mu'} |\rho| x^\mu \right\rangle = \psi\left(x^{\mu'}, \tau\right) \psi^*\left(x^\mu, \tau\right). \tag{10.14}$$

From the Schrödinger equation we obtain the von Neumann equation:

$$i\hbar \frac{d}{d\tau} \left\langle x^{\mu'} |\rho| x^\mu \right\rangle = [K, \rho]_{x^{\mu'}, x^\mu}. \tag{10.15}$$

We might proceed differently by assuming that Eq. (10.11) can be generalized to Heisenberg operator equations for the observables

$$\begin{aligned} p_i^\mu &\rightarrow \hat{p}_i^\mu \\ x_i^\mu &\rightarrow \hat{x}_i^\mu \\ K &\rightarrow \hat{K} \end{aligned} \tag{10.16}$$

$$\begin{aligned} \frac{d\hat{p}_i^\mu}{d\tau} &= -\frac{\partial \hat{K}}{\partial \hat{x}_i^\mu} \\ \frac{d\hat{x}_i^\mu}{d\tau} &= \frac{\partial \hat{K}}{\partial \hat{p}_i^\mu} \end{aligned} \tag{10.17}$$

with commutation laws

$$\left[\hat{x}_i^\mu, \hat{p}_i^\nu\right] = i\hbar g^{\mu\nu} \delta_{ij}. \tag{10.18}$$

As in the classical case, a quantum event, $\hat{x}_i^\mu, \hat{p}_i^\mu$ evolves by operator Heisenberg equations as $\hat{x}_i^\mu(\tau), \hat{p}_i^\mu(\tau)$, but abstractly, not in a Minkowski picture. In the Schrödinger picture we may view events as wave packets $\psi\left(x_i^\mu, \tau\right)$ on or near particle world lines in the $4n$ x_i^μ space. In this sense there are "particles" in this picture which may be localized in space-time $(\mathbf{x}, t)$. The entire history of a packet with τ *is the particle.* This is realized as a summation on all τ, a concatenation or (Latin) *vincula.* To introduce ρ we assume that any average of a Heisenberg operator at time τ is

$$\left\langle \hat{A}(\tau) \right\rangle = \mathrm{Tr}\hat{\rho}(0)\,\hat{A}(\tau)\,. \tag{10.19}$$

Then, using the cyclic trace property, we introduce

$$\hat{\rho}(\tau) = \exp\left(-i\hat{K}\tau\right)\rho(0)\exp(iK\tau)\,. \tag{10.20}$$

We obtain, by differentiation, the operator form of the von Neumann equation, Eq. 10.15, above.

To write a quasi distribution function for events, we will utilize the second quantization form of the wave function, assuming for the event $x^{\mu} \equiv q, q = \mathbf{q}, t$ with the operator

$$\begin{aligned}
\psi(q) &= \frac{1}{(2\pi\hbar)^2}\int d^4p\,\psi(p)\exp(ip\cdot q) \qquad (10.21)\\
\left[\psi(p), \psi(p')\right]_{\pm} &= 0\\
\left[\psi(p), \psi^{\dagger}(p')\right]_{\pm} &= \delta^4\left(P - P'\right)\\
p &\equiv \left(\mathbf{p}, \frac{E}{c}\right).
\end{aligned}$$

An operator in the space of N events may be written

$$\mathbf{A} = \sum_{i=1}^{N}\frac{1}{i!}\int d^4q_1\ldots d^4q_i\psi^{\dagger}(q_1)\ldots\psi^{\dagger}(q_i)\,\mathbf{A}_i\psi(q_1)\ldots\psi(q_i)\,. \tag{10.22}$$

$\mathbf{A}_i$ is an operator on the subset of the N-event space, a reduced operator. In the following we will fix N. The idea is that events leading to a realization of a particle (with positive energy) trajectory should not disappear in a finite space–time volume.

Important examples are the single-particle kinetic energy representation in terms of quantum fields,

$$K_0 = -\hbar^2\int d^4q\psi^{\dagger}(q)\frac{\partial_{\mu}\partial^{\mu}}{2m}\psi(q)\,,$$

and the two- "body" covariant interaction potential,

$$V = \frac{1}{2}\int d^4q'd^4q''\psi^{\dagger}\left(q'\right)\psi^{\dagger}\left(q''\right)V\left(\left|q' - q''\right|\right)\psi\left(q'\right)\psi\left(q''\right). \tag{10.23}$$

This should be called the two-event interaction. This interaction is weighted by the event distribution through $\psi(q)$, $\psi(q')$. Also, the potential is taken as covariant with

$$|q' - q''|^2 = c^2 \Delta t^2 - \Delta \mathbf{x}^2.$$

The potential is found phenomenologically or through field theories. It is here an action at a distance between events.

10.3 Two-event Boltzmann equation

In this section we will see that Boltzmann's profound ideas on microscopic statistical dynamics may be carried over to a covariant form which treats the binary statistical dynamics of quantum event interaction (Boltzmann, 1872; Horwitz *et al.*, 1989). The global time, covariant τ, plays the central role. Let us proceed quickly with the outline of this development in which you will see Boltzmann's ideas. We must add that this may be more rigorously done and with fewer assumptions. Some comments will be made on this later.

We will adopt, as the simplest quantum event distribution function, the Wigner function (Wigner, 1932). This is discussed in detail in Chapter 4. The one-event relativistic Wigner function is in the four-momentum representation

$$f_1(q, p) = \frac{1}{(2\pi)^4} \int d_4k \mathrm{Tr}\left[\rho\psi^\dagger\left(p - \frac{\hbar k}{2}\right)\psi\left(p + \frac{\hbar k}{2}\right)\right] \exp(i\mathbf{k}\cdot\mathbf{q}).$$

f_1 was called w previously. With this,

$$\langle A_1\rangle = \int d_4q d_4p A_1(q, p) f_1(q, p). \tag{10.24}$$

The two-event Wigner function Fourier transform is

$$f_2(k_1 p_1 k_2 p_2) = \mathrm{Tr}\rho\psi^\dagger\left(p_1 - \frac{\hbar k_1}{2}\right)\psi^\dagger\left(p_2 - \frac{\hbar k_2}{2}\right) \times \psi\left(p_2 + \frac{\hbar k_2}{2}\right)\psi\left(p_1 + \frac{\hbar k_1}{2}\right). \tag{10.25}$$

This may, of course, be extended to N events. From Eq. (10.24), $f_1(q, p)$ seems to play a role of a classical distribution function. However, remember that this is not true, since

$$f_s(q, p) \ngeq 0, \tag{10.26}$$

but the marginal distribution functions have the property

$$\int dq f_1(q_1 p_1) \geqslant 0$$
$$\int dp f_1(q_1 p_1) \geqslant 0.$$

Recall that a very important property of phase space distribution functions is that they are associated with correspondence rules. In Eq. (10.24), $A_1(q, p)$ is the classical operator associated with the quantum, operator $\hat{A}_1$ by the Weyl correspondence rule

$$q^n p^n \to 2^{-n} \sum_{l=0}^{n} \binom{n}{l} q^{n-l} p^m q^l. \tag{10.27}$$

We will further normalize $f_1(qp)$ in the eight-dimensional phase space

$$\int d_4 q d_4 p f_1(qp) = N, \tag{10.28}$$

N being the total number of events, which we assume to be fixed in τ. The reduced event Wigner distributions may be formed into a B.B.G.Y.K. hierarchy as in the nonrelativistic classical and quantum cases. (See Chapters 4 and 6.)

We can use this hierarchy to derive the Boltzmann event equation by the methods of Green and Bogoliubov for the quantum case, as shown in Chapter 4. However, we will not do this but rather, for simplicity, follow directly a Boltzmann-type argument, filling in important points of the more general approach. We will operate on the event von Neumann equation above and form an equation for $\frac{\partial f_1}{\partial \tau}$. It is in the eight-dimensional position momentum space:

$$\partial_\tau f(q_1 p_{1,} \tau) + \frac{1}{m}(q_1 \cdot p_1) f(q_1 p_{1,} \tau) = \tag{10.29}$$
$$\int d_4 p_2 d_4 q_2 \delta^4(q_2) L_{12} f_2 (q_1 p_1 q_2 p_{2,} \tau) \equiv J(q_1 p_1).$$

This is the first equation of the hierarchy in f_s already mentioned. It is not closed, since $f_2(q_1 p_1 q_2 p_{2,} \tau)$ appears on the right. The basic problem is to obtain f_2 from the second equation of the hierarchy or make an ad hoc approximation and evaluate the right side.

The *Stosszahlansatz* may be made at global time τ. We replace

$$f_2(q_1 p_1 q_2 p_2, \tau) \to f_1(q_1 p_1, \tau) f_1(q_2 p_2, \tau) \tag{10.30}$$

and write the right side as an event transition in a Chapman–Kolmogorov gain loss form. We drop the "1" now:

$$J(qp,\tau) = R^{+} f(qp,\tau) - R^{-} f(qp,\tau) \tag{10.31}$$

$$R^{+} f(qp) = \int d_4p_1 d_4p_1' d_4p' \dot{P}\left(p_1' p' \to p_1 p\right) f\left(qp'\right) f\left(qp_1'\right) \tag{10.32}$$

$$R^{-} f(qp) = \int d_4p_1 d_4p_1' d_4p' \dot{P}\left(pp_1 \to p_1' p'\right) f\left(qp_1\right) f\left(qp_1\right).$$

$\dot{P}$ is the event transition rate. One must not think too physically about the event transitions due to binary interaction. Here events are not particles, nor is f a measure of particle density. It is not a probability. All this is written by analogy.

We must estimate the transition rate from binary event interaction. We assume a dilute event density. The event density is a covariant idea; thus we assume no three-particle world line interactions at any time τ.

We will estimate $\dot{P}$ from binary event scattering (Horwitz and Lavie, 1982). We think of an event wave packet of an incoming event beam ψ_{in} having, by means of the scattering, an outgoing wave event packet. We have

$$\psi_{out}(p) = \int d_4p' \langle p\,|S|\,p'\rangle \psi_{in}\left(p'\right), \tag{10.33}$$

where

$$\langle p\,|S|\,p'\rangle = \delta^4\left(p - p'\right) - 2\pi i \delta\left(\frac{p^2}{2m} - \frac{p'^2}{2m}\right) T\left(p' \to p\right). \tag{10.34}$$

In center-of- "mass" coordinates, the event Möller operator is

$$\Omega_+ = \lim_{\tau\to-\infty} \exp(iK\tau)\exp(-iK_0\tau) \tag{10.35}$$

$$S = \Omega_-^{\dagger}\Omega_+ .$$

It has been shown that the wave operator exists and is asymptotically complete for a wide range of interactions. This follows from the well-known methods of formal scattering theory.

The differential event scattering cross section is

$$d\sigma\left(\psi_n' \to d_4p\right) = \frac{N_{sc}}{N_{inc}}(d_4p). \tag{10.36}$$

We may show that

$$d\sigma\left(\psi_{\text{in}} \to d_4p\right) = d_4p\,(2\pi)^4\, 2m^2 \int d_4p' \frac{1}{|p'|} \tag{10.37}$$

$$\times\, \delta\left(p^2 - p'^2\right) \left|T\left(p' \to p\right)\right|^2 \left|\psi_{\text{in}}\left(p'\right)\right|^2 .$$

With this, the rate of event scattering in relative event coordinates is

$$\dot{P}\left(p_r' \to p_r, P\right) = (2\pi)^3 m \left|T_{p_r' p_r}\right|^2 \delta\left(P^2 - p_r'^2\right), \tag{10.38}$$

and we may write the binary event Boltzmann equation as

$$\begin{aligned} &\partial_\tau f(qp,\tau) + \frac{p_\mu}{m}\frac{\partial}{\partial q_\mu} f(qp,\tau) \\ &= \int d_3 p_r d_3 p_r' d_3 p_r'' \frac{\left|p_r'\right|}{m} \frac{d\sigma\left(p_r' \to p_r, P\right)}{d_3 p_r} \\ &\times \left[f\left(qp_1',\tau\right) f\left(qp',\tau\right) - f(qp,\tau) f\left(qp_1,\tau\right)\right]. \end{aligned} \tag{10.39}$$

Not surprisingly, this has the same form as the quantum Wigner–Boltzmann equation neglecting exchange symmetries. The obvious difference is the increased dimensionality to the phase space; the cross section is now of dimension L^3, and τ is a dynamic parameter. The cross section may be reduced in dimension to an experimental comparison by an integration of dp_r^0 over an initial mass distribution. The gradient is obviously four-dimensional.

Let us make some remarks. The event potential for fixed τ is taken to be covariant $V(\rho)$, $\rho^2 = q^\mu q_\mu \equiv x^\mu x_\mu$, identically. For two-event scattering it may be shown that

$$\lim_{\tau \to +\infty} \left\| \exp\left(-iK_r\tau\right)\psi - \exp\left(-iK_{01}\tau\right)\phi_{\text{out}}\right\| = 0$$

for a dense set ϕ_{out}, and

$$\int_T^\infty d_{\hat{\upsilon}} \left\| V \exp\left(-iK_{0r}\right)\phi_{\text{out}}\right\| < \infty$$

if $V(\rho) = \left(\frac{1}{\rho^2}\right)^\alpha$ with $\alpha = \frac{1}{2}+\delta$, $\delta > 0$ (Horwitz and Lavie, 1982). We might note that in the case of simple central force scattering, asymptotic condition requires $V(r) = \frac{1}{r^{\frac{3}{2}}}\left[\int d_3x \left|V(x)\right|^2\right]^{\frac{1}{2}} < \infty$ (Taylor, 1972). The difficulties here start at $r^{\frac{-3}{2}}$.

The necessity for low-density events is not clear from the derivation, but it is from the B.B.G.Y.K. hierarchy. Here, three event correlations $f_3(q_1p_1, q_2p_2, q_3p_3)$ are neglected. Briefly, one may write

$$f_2(12) = f_1(1) f_1(2) + g_{12}(12) \tag{10.40}$$

and show that from the second-hierarchy equation

$$\begin{aligned} i\hbar\partial_\tau g_{12} &= L_{12} g_{12} + L_{12} f_1(1) f_1(2) \\ &\quad - i\hbar\partial_\tau f_1(1) f_1(2) + n_0 \mathrm{Tr}\left\{\left(L_{13}' + L_{23}'\right) f_3(123)\right\}. \end{aligned} \tag{10.41}$$

We treat f_s to zero order in $N_0 = N/V$ = constant as $N \to \infty$ and $V \to \infty$. N_0 is the density of *events* (K. Hawker, unpublished 1975 Ph.D. thesis, Contributions to Quantum Kinetic Theory, University of Texas, Austin). See Chapter 4. The form of L_{12} is not necessary here. Thus, to low "density," f_3 of Eq. (10.41) may be neglected, and we have

$$ik\partial_\tau g_{12} = L_{12}g_{12} + L'_{12}f_1(1)\, f_1(2)\,. \tag{10.42}$$

The *Stosszahlansatz* may be treated in a similar way. This latter assumption is probably the weakest point. To do better, we follow the method of Bogoliubov (1946). See also the work of McLennan (1989). Eq. (10.42) may be formally solved for $0 < \tau \leqslant \infty$. We obtain

$$f_{12}(\tau) = \exp\left(-iL_{12}\frac{\tau}{\hbar}\right) g_{12}(0) + \exp\left(-iL_{12}\frac{\tau}{h}\right)\exp\left(+iL^0_{12}\frac{\tau}{h}\right) \times f_1(1,\tau)\, f_1(2,\tau)\, \tau \geq 0. \tag{10.43}$$

Now we assume *initially* (not at all τ) that $g_{12}(0) = 0$ to obtain the *protokinetic* equation. We call it the operator Boltzmann equation:

$$i\hbar\partial_\tau f_1(1,\tau) = L^0_1 f_1(1,\tau) + n_0\mathrm{Tr}\{L'_{12}\left[\exp\left(-iL_{12}\frac{\tau}{\hbar}\right)\exp\left(+iL^0_{12}\frac{\tau}{\hbar}\right)\right] \times f_1(1,\tau)\, f_1(2,\tau)\}; \quad \tau \geq 0. \tag{10.44}$$

This equation, after much detailed calculation, leads to the Boltzmann equation, Eq. (10.39). Two very important points appear:

1. The factorization is *initial* only.
2. The equation is irreversible, since $\tau \geqslant 0$.

10.4 Some results of the quantum event Boltzmann equation

Time reversal is, in this case, defined by

$$\begin{aligned}\psi'_\tau(x,t) &= T\psi_\tau(x,t) \\ &= \psi^*_\tau(x,-t)\,.\end{aligned} \tag{10.45}$$

Since $f(xt)$ is real, the above Boltzmann equation, Eq. (10.44), is *not* time reversal invariant. This is not surprising, since it is derived for $\tau \geqslant 0$.

Let us now examine the equilibrium solution. The binary event collision has the following invariants:

$$\frac{1}{2m}\left(P_1^2 + P_2^2\right) = \frac{1}{2m}P^2 + \frac{p_{r^2}}{m}.$$

Thus p^2, p^μ are conserved. $M^{\mu\nu} = q^\mu p^\nu - q^\nu p^\mu$ is also invariant, but we will constrain the system so this does not play a role.

To achieve a positive $f_0(p, q)$, which causes the right side of the Boltzmann equation to vanish, we choose the Gaussian as discussed in Chapter 6. It is unique and positive because of the theorem of Hudson (1974):

$$f_0(qp) = c(q)\exp\left(-A(q)(p - p_c(q))^2\right) \qquad (10.46)$$
$$p = \left(\mathbf{p}, \frac{E}{c}\right).$$

In the $q = (\mathbf{x}, t)$ space, the events distribution is a time dependent wave packet parameterized by the functions $c(q)$, $A(q)$, $p_c(q)$. This is a local equilibrium solution very much like hydrodynamics or the notion of coherent states.

Some subtlety of this approach is the thought that this theory does *not* generally maintain the mass shell condition $\left|P_i^\mu\right| = m_i$ for particle momentum. In terms of dynamics, this would generally lead to a loss of n degrees of freedom in the phase space of n particles. In a Stueckelberg theoretical approach, $\left|P_i^\mu\right| = \pi_i(\tau) \neq m_i =$ constant. In a sense, $\left|P_i^\mu\right|$ is a dynamic mass. In K_0 m is a property of an event.

We now restrict P^2 to a small region of fixed m, i.e. $P^2 \cong -m^2$. Then, with some calculation, we identify $2Am_c = 1/kT$, which is the definition of equilibrium absolute temperature as suggested by Synge from a mass-shell theory (Synge, 1957). In this approximation

$$f_0(qp) = c(q)\exp\left\{A\left(m^2 + m_c^2\right)\right\} \times \exp(2Ap_\mu p_c^\mu)$$

when $\mathrm{p}_c^2 = -m_c^2$.

In the local rest energy frame where $\boldsymbol{\mu} = \frac{\mathbf{P}_r}{E}$, $E' = \frac{E_q - \mu\cdot\mathbf{p}}{\sqrt{1-\mu^2}}$ we have the interesting result $\left\langle E'\right\rangle = m\frac{K_2(2Am_c)}{K_1(2Am_c)}$ where K_i are Bessel functions of the third type (Horwitz *et al.*, 1989). We find that

$$\left\langle E'\right\rangle = \frac{3}{2}kT + m \qquad T \to 0 \qquad (10.47a)$$

and

$$\left\langle E'\right\rangle = 2kT \qquad T \to \infty. \qquad (10.47b)$$

The latter important result was obtained from the equilibrium Gibbs theory (Horwitz *et al.*, 1981). The first result agrees with Pauli in his famous article "Relativistats Theorie" (Pauli, 1958). In the $T \to \infty$ limit, 2 is replaced by 3 in Pauli's result. This remains one of the significant tests of the event time theory being discussed here, as yet *not* determined experimentally.

From the pressure tensor we may obtain, in the local rest frame in the previously stated limits,

$$P = \frac{N_0}{2Am_c} = N_0 kT, \qquad (10.48)$$

the ideal gas law. Here N_0 is the number of *particles* per unit *space* volume. $\langle N_0 \rangle_q = \langle J^0 (q) \rangle_q$ where by means of a *concatenation* over events, τ, we write the conserved four-particle current as

$$J^\mu (q) = \sum_i \int \frac{p_i^\mu}{m} \delta^4 (q - q_i (\tau)) \, d\tau, \tag{10.49}$$

a weighted event history.

Let us turn now to the local entropy production. We define

$$s (qp) = -k\mathfrak{H} (qp), \tag{10.50}$$

assume additivity

$$s_A + s_B = s_{AB}, \tag{10.51}$$

and take $s (q) = \int dp f_0 (qp) \ln f_0 (qp)$. In Eq. (10.51), $f_0 (qp) \geqslant 0$, so this is possible. We have, then,

$$s_0 (q) = c (q) A (q) \langle (p - p_c)^2 \rangle .$$

Then the entropy production is $\frac{ds_0}{dt} = \sigma$, and $\dot{c} (q) =$ constant in the steady state. From the conservation laws, as discussed in Chapter 6, we would obtain $\sigma \geqslant 0$, which is a general steady non-equilibrium thermodynamic result (McLennan, 1989). This has not been carried out in detail, but there is no doubt of the result.

We may consider another global quantity. Assume q independence, i.e. homogeneity in time as well as in space. We utilize the marginal Wigner function,

$$\phi (p) = \int dq f (qp) ; \quad \geqslant 0. \tag{10.52}$$

We may write a Boltzmann equation for $\phi (p)$ and obtain

$$\begin{aligned} \partial_\tau \phi (p, \tau) = 2 \int d_3 p_r d_3 p_r' dp_r^0 \frac{|p_r'|}{m} \frac{d\sigma}{d^3 p_r} \left(p_r' \to p_r ; P \right) \\ \times \left\{ \phi \left(p', \tau \right) \phi \left(p_1', \tau \right) \quad \phi (p, \tau) \phi (p_1, \tau) \right\} . \end{aligned} \tag{10.53}$$

Now we can define the global $\mathfrak{H}$, assuming it is bounded:

$$\mathfrak{H} (\tau) = \int dp \phi (p, \tau) \ln \phi (p, \tau) . \tag{10.54}$$

Forming $\frac{d\mathfrak{H}}{d\tau}$ and utilizing the well-known property of integral invariants,

$$4I (F) = I (F) + I (F_1) - I \left(F' \right) - I \left(F_1' \right), \tag{10.55}$$

where $I(F)$ is a function of the right side of the above Boltzmann equation, Eq. (10.53). With this, as in Chapter 6,

$$4I(1+\ln\phi) = \int dp \int dp_1 \dot{R}\left(\phi(p_1')\phi(p') - \phi(p_1)\phi(p)\right) \ln \frac{\phi(p_1')\phi(p')}{\phi(p_1)\phi(p)}. \tag{10.56}$$

It then follows that

$$4I(1+\ln\phi) \leqslant 0, \tag{10.57}$$

so

$$\frac{d\mathfrak{H}}{d\tau} \leqslant 0.$$

This is exactly the form of Boltzmann's $\mathfrak{H}$ theorem. $\mathfrak{H}$ is a Lyapunov function and guarantees that $\phi(p,\tau)$ approaches $\phi_0(p,\infty)$. Here $\phi_0(p,\infty)$ is a global Maxwellian given by $f_0(p)$ with $c(q)$, $A(q)$ independent of space and time. The initially inhomogeneous system approaches a spacial-temporal independent system characterized by a Gaussian in (p, E) and is physically characterized by an event density and temperature. This is not very surprising.

The situation with respect to $f(qp,\tau)$ is more problematic. If we assume

$$\mathfrak{H} = \int dqdpf(qp)\,lnf(qp) \tag{10.58}$$

near equilibrium, then the inhomogeneous Boltzmann event equation, by precisely the same argument, gives

$$\frac{d\mathfrak{H}}{d\tau} \leqslant 0; \quad q \text{ independent.}$$

This would seem to imply that an inhomogeneous event distribution approaches homogeneity. This is doubtful.

As shown in Chapter 5, by identifying the collisional invariants here, p^μ, p^2, and $M^{\mu\nu} = q^\mu p^\nu - q^\nu p^\mu$, we may obtain the macroscopic conservation laws from the Boltzmann equation (Horwitz *et al.*, 1989). An important point to be mentioned is that we must define the particle densities' currents as *vincula* (concatenation) of the historical events, such as

$$J^\mu(q) = \int_{-\infty}^{+\infty} \frac{n}{M}\langle p^\mu\rangle_q \, d\tau, \tag{10.59}$$

having the conservation property

$$\frac{\partial}{\partial q_\mu} J^\mu(q) = 0, \tag{10.60}$$

assuming the event density n vanishes at $\tau = \pm\infty$. Similar arguments are made to obtain the other local in $q = (\mathbf{x}, t)$ conservation laws.

An important program yet to be done would be to follow the well-known Chapman–Enskog procedure (see Chapter 5) and calculate the transport coefficients, and then to compare the results with the noncovariant calculation so completely described by de Groot, van Leeuwen and van Weert in their 1980 book.

10.5 Relativistic quantum equilibrium event ensembles

Let us now consider equilibrium and some thermodynamic consequences of the covariant event formulation (Horwitz *et al.,* 1981). We will consider the quantum aspects. The classical ensembles have also been treated in full detail in Horwitz *et al.* (1981).

Utilizing the covariant Hamiltonian K, Eq. (10.8), we construct in the usual fashion (see Chapter 8) the microcanonical ensemble, the event density operator

$$\rho = \sum_{kE\varepsilon\Delta, m_i\varepsilon\mu_i} \psi_{kE}\psi^*_{kE}. \tag{10.61}$$

$\psi_{k,E}$ is eigenfunction of the operator $\hat{K}$, having the four-dimensional eigenvalues $\mathbf{K}$, $E \equiv (kE)$. These are the *event invariant* eigenvalues for which, from Eq. (10.15),

$$\frac{d\rho}{d\tau} = 0, \tag{10.62}$$

the equilibrium state. In the classical $8N$ (N being the number of events) phase space, this is the event (τ) invariant *distribution function.* Also, we are *not* on mass shell, Eq. (10.3). Consequently, the particle parameters m_i may vary since E, $\mathbf{p}$, which are also independent of one another. We will confine the m_i to some small regions μ_i which are in the range of the particle-free mass, M_i. Thus μ_i (M_i). Free particle masses M_i are assumed on the mass shell.

We assume that the number of event states,

$$\Gamma(k, E) = \mathrm{Tr}\rho = \sum_{kE\varepsilon\Delta;\ \mu_i\varepsilon\mu_i(M_i)}, \tag{10.63}$$

is bounded. With Boltzmann's famous formula (see Chapter 7), we assume that the thermodynamic entropy is

$$S = k \ln \Gamma(k, E). \tag{10.64}$$

From this the other microcanonical thermodynamic quantities follow (Tolman, 1967). The classical microcanonical density is

$$\Gamma(k, E) = \int_{m_i \varepsilon \mu_i, q_i \varepsilon \sigma_i} dE_1 \ldots dE_n d_3p_1 \ldots d_3p_n d_4q_1 \ldots d_4q_N \delta(K - k). \tag{10.65}$$

Here $c = 1$, $E = \sum_i E_i$, and $q = (\mathbf{q}, t)$.

The interparticle forces are assumed weak, and hence $m_i = M_i\left(1 + 0\left(1/c^2\right)\right)$, M_i being the free-particle mass. For a free-particle gas, it has been shown that in the ultra relativistic limit where

$$\frac{dE_i}{c} \cong c^2 \frac{m_i dm_i}{|p_i|}$$

and with $K = -\frac{1}{2}Mc^2$, that

$$\Gamma(k, E) \cong (4\pi)^N V^N T^N c^{3N-1} \int_{m_i \varepsilon \mu_i} m_1 dm_1 \ldots dm_N p_1 dp_1 \ldots dp_N$$
$$\times \delta\left(\sum_i \frac{m_i^2}{M_i} - M\right) \delta\left(\sum_i E_i - E\right)$$

for a finite range of τ, T. From this it follows, with $p_i dp_i = (1/c^2) E_i dE_i$, that

$$\Gamma(k, t) \cong E^{2N}, \tag{10.66}$$

and hence

$$E = 2NkT,$$

as shown in the discussion of the Boltzmann equation in the earlier section. As stated earlier in this chapter, the classical Jüttner result (Jüttner, 1911) of 3 rather than 2 in Eq. (10.66) remains the principal test of the event covariant approach being outlined here. Experiments have not yet achieved the precision necessary for such a decision.

Now let us adopt a model to further investigate the free-particle *quantum* event microcanonical ensemble, which will further elucidate the difference between the results. Restrict the system to L, T $\left(V^{(4)} = L^3 T\right)$ with $\psi_0(xyz, t) = 0$ on the time and space limits. Here we take the parameter m_i to M for all particles. Then the variables separate in the eigenvalue solution. The event modes are obtained from

$$\hat{K}_0 \psi = K\psi.$$

$\hat{K}_0$ is the free-particle kinetic energy operator, and

$$\sum_{i=1}^{N} \frac{h^2}{2M}\left(\frac{\partial^2}{\partial^2 t_i} - \Delta_i^2\right) \psi(x_i y_i z_i t_i) \equiv \sum_{i=1}^{N} K_i \psi(x_i y_i z_i t_i). \tag{10.67}$$

Hence,

$$2MK_i = \hbar^2 \left(k_1^{i^2} + k_2^{i^2} + k_3^{i^2} - k_0^{i^2}\right). \tag{10.68}$$

For $d\rho/dt = 0$, the zero eigenmode gives $\dot{\psi} = 0$ and $K_i = 0 = \left(k_1^{i^2} + k_2^{i^2} + k_3^{i^2} - k_0^{i^2}\right)$. For the clock determining τ placed in the center of mass of the system, the modes for $T \to \infty$ are light-like, moving with velocity c on the forward light cone of the center of mass. For T finite, the modes are distributed *near* the light cone around $\mu\,(M_i)$. For $\mathbf{p} = \hbar\mathbf{k}$, $e = \hbar k_0$,

$$\mathbf{p} = \frac{2\pi\hbar}{L}\boldsymbol{\nu},\ \varepsilon = \frac{2\pi\hbar}{T}\nu_0, \tag{10.69}$$

with $\nu_0, \nu_j = 0, \pm 1, \ldots$ We must consider only $\nu_0 \geq 0$ to exclude the antiparticle modes.

Now the eigenvalue spectrum is four-dimensional for each independent mode. Let $n_{\mathbf{p},\varepsilon}$ be the number of event modes with energy momentum $\mathbf{p},\varepsilon$. There is a mass parameter constraint, $m\varepsilon\mu_i$. We further divide the eigenvalue space into these mass regions, labeling it with i. Further, we coarse-grain. Let g_i = the number of *mass and momentum* states in each cell, a mass degeneracy parameter. Also, $n_i = \sum_{\mathbf{p},\varepsilon\varepsilon i} n_{\mathbf{p},\varepsilon}$, the number of modes within the cell i.

The constraints are

$$E = \sum_i \bar{\varepsilon}_i n_i \tag{10.70}$$

$$K = \sum_i \bar{K}_i n_i\ ;$$

$N = \sum_i n_i$ =the total number of events. i now labels *cells*. Note that, in contrast to the usual three-dimensional space, there is an additional constraint on K. Because of the four-dimensional eigenspace, $\bar{K}_i$, $\bar{\varepsilon}_i$ are the average values in each cell, μ_i.

Now we distribute the $\{n_i\}$ events into the mass cells with equal a-priori probability subject to the foregoing constraints. This number of possibilities is $\Gamma\,(E, K)$ of the microcanonical ensemble. A good estimate is to maximize the entropy subject to the constraints. Boltzmann showed that we may maximize

$$S = k ln \Gamma\,(E, K)\,.$$

For each cell with n_j identical events and g_j energy mass levels, we have the statistical weight, assuming

$$W\left(\{n_j\}\right) = \frac{g_j!}{n_j!\,(g_j - n_j)!} \qquad \text{(Fermi–Dirac)} \tag{10.71}$$

$$W\left(\{n_j\}\right) = \frac{\left(n_j + g_j - 1\right)!}{n_j!\,(g_j - 1)!} \qquad \text{(Bose–Einstein)},$$

where at most one mode may occupy a state in the Fermi case and any number in the Bose–Einstein case. With this we have

$$\ln \Gamma_{FD} = \sum_i \left[g_i \ln g_i - n_i \ln n_i - (g_i - n_i) \ln (g_i - n_i) \right],$$

and

$$\ln \Gamma_{BE} = \sum_i \left[(n_i + g_i - 1) \ln (n_i + g_i - 1) - n_i \ln n_i - (g_i - 1) \ln (g_i - 1) \right].$$

Introducing Lagrange multipliers α, β, γ, we find the constrained maximization in the usual fashion

$$\ln \left(\frac{g_j}{n_j} - 1 \right) = -\alpha - \beta \bar{\varepsilon}_j - \gamma \bar{K}_j \qquad \text{(Fermi–Dirac)}$$

$$\text{and } \ln \left(\frac{n_j + g_j}{n_j} - 1 \right) = -\alpha - \beta \bar{e}_j - \gamma \bar{K}_j \qquad \text{(Bose–Einstein).}$$

Note here that $\bar{K}_i < 0$, since K is time-like. Defining $z = \exp(\alpha)$ and $\zeta = \exp \gamma$, we have

$$\begin{aligned} n_i^{\text{avg.}} &= \frac{g_i}{z^{-1} \zeta^{-K_i} \exp(\beta \bar{\varepsilon}) \mp 1} \qquad -1\ B.E.; \quad +1\ F.D. \qquad (10.72) \\ &= \text{average number of events in the cell } i. \end{aligned}$$

This is, of course, the usual form. We have assumed that all cells have the same z, ζ, β. In a sense they are in thermal equilibrium together. The new aspect to this relativistic event description of thermodynamic equilibrium is the parameter ζ, which is associated with the constraint due to K. We find a mass fugacity ζ, which determines the distribution of mass in the cell. Before saying more, we might mention that the same result has been achieved from the grand canonical ensemble for equilibrium events. This is also discussed in the paper of Horwitz, Schieve and Piron (Horwitz *et al.*, 1981). There the grand partition function is

$$Z_G \left(V^{(4)}, \zeta, z, \beta \right) = \sum_N z^N \hat{Q}_N \left(V^{(4)}, \zeta, \beta \right), \qquad (10.73)$$

which is, for the non-interacting relativistic quantum event gas,

$$\begin{aligned} Z_G \left(V^{(4)}, \zeta, z\beta \right) &= \Pi_{\mathbf{p}\varepsilon\mu(i)} \frac{1}{1 - z\zeta^K \exp(-\beta\varepsilon)} \qquad \text{(B.E.)} \qquad (10.74) \\ &= \Pi_{\mathbf{p}\varepsilon\mu(i)} \left(1 + z\zeta^K \exp(-\beta\varepsilon) \right) \qquad \text{(F.D.).} \end{aligned}$$

Here $V^{(4)} = VT$. From this, the total event normalization is

$$N = \sum_{\mathbf{p}\varepsilon\mu(i)} n_{\mathbf{p}}^{\text{avg}} = \sum_{\mathbf{p}\varepsilon\mu(i)} \frac{z\zeta^K \exp(-\beta\varepsilon)}{1 + z\zeta^K \exp(-\beta\varepsilon)}, \qquad (10.75)$$

in agreement with Eq. (10.72).

In a series of papers, Burakovski (Burakovski and Horwitz, 1993; Burakovski *et al.*, 1996a, 1996b; Burakovski and Horwitz, 1997) has investigated the consequences of the preceding formulation in detail. We shall consider a particular aspect here. Let us consider the relativistic Bose gas, Eq. (10.75) (without antiparticles). We have the event normalization

$$N = V^{(4)} \sum_{K\mu} \left(\exp\left(E - \mu - \mu_K \frac{m^2}{2M} \right) \frac{1}{T} - 1 \right)^{-1}.$$

We have taken $\hbar = c = k = 1$. Here $\zeta^K = \exp \beta\mu_K$, and $m^2 = -k^2 = -k^\mu k_\mu$. We wish to examine how μ_K may determine the form of the mass distribution and may consequently be termed a mass potential and ζ^K the mass fugacity.

Now $n_{k\mu}, k\varepsilon\,(\mathbf{p}\varepsilon)$ and $\mu = \mu\,(M_i)$ are necessarily positive. Thus,

$$M_i - \mu - \mu_K \frac{m^2}{2M} \geq 0. \tag{10.76}$$

Eq. (10.76) has the solution bounded by m_1 and m_2 given by

$$m_1 \equiv \frac{M}{\mu_K} \left(1 - \sqrt{1 - \frac{2\mu\mu_K}{M}} \right) \leq m \leq m_2, \tag{10.77}$$

where

$$m_2 \equiv \frac{M}{\mu_K} \left(1 + \sqrt{\frac{1 - 2\mu\mu_K}{M}} \right).$$

Thus, for small μ,

$$\mu \leq m \leq \frac{2M}{\mu_K}. \tag{10.78}$$

μ_K determines the upper bound to the mass spectrum, and μ the lower bound.

This may be carried further by making the continuum approximation to the sum on k:

$$n \equiv \frac{N}{V^{(4)}} = \int_{m_1}^{m_2} \int_{-\infty}^{+\infty} d\beta \frac{m^3 \sinh^2 \beta}{\exp\left(m \cosh \beta - 1 - \mu_K \frac{m^2}{2M} \right) \frac{1}{T} - 1}. \tag{10.79}$$

We have used four momentum hyperbolic coordinates (Horwitz *et al.*, 1989), where $0 \leq \pi, 0 \leq \phi < 2\pi$, and $-\infty \leq \beta < \infty$. At high temperature, $T >> \frac{M}{\mu_K}$, the $d\beta$ integral may be done, obtaining

$$n = \frac{T \exp\left(\frac{\mu}{T}\right)}{4\pi^3} \int_{m_1}^{m_2} dm m^2 K_1 \left(\frac{m}{T}\right) \exp\left(\mu_K \frac{m^2}{2MT} \right). \tag{10.80}$$

K_1 is a Bessel function with imaginary argument. We have $T >> \frac{M}{\mu_K}$. Thus, from Eq. (10.77),

$$\frac{\mu}{T} \leq \frac{m}{T} \leq \frac{2M}{T\mu_K} << 1. \tag{10.81}$$

Hence, using the formula

$$K_\nu(x) \cong \frac{1}{2}\pi(\nu)\left(\frac{x}{2}\right)^{-\nu} \qquad x << 1,$$

the average n becomes

$$n = \frac{T^2}{4\pi^3}\int_{m_1}^{m_2} dmm.$$

The density of mass states is as m. And

$$n = \frac{T^2}{2\pi^3}\left(\frac{M}{\mu_K}\right)^2\sqrt{1-\frac{2\mu\mu_K}{M}}. \tag{10.82}$$

The thermodynamic variables may be obtained. We find

$$\langle m\rangle = \frac{4}{3}\left(\frac{M}{\mu_K}\right)\left(1-\frac{\mu\mu_K}{2M}\right) \tag{10.83}$$
$$\left\langle m^2\right\rangle = \frac{3}{2}\left(\frac{M}{\mu_K}\right)\langle m\rangle .$$

The high-temperature quantum average is $\langle E\rangle = 2T$ and is in agreement with the earlier discussion. One also finds, generally,

$$N^0 = \left\langle J^0\right\rangle = \frac{T^3}{\pi^2}\frac{8-x^3K_3(x)}{x^2} \tag{10.84}$$
$$p = \frac{1}{3}\left(T_{ij}\right)g_{ij} = N_0T$$
$$\rho = \left\langle T^{00}\right\rangle = \frac{T^4}{\pi^2}\frac{40-x^4K_4(x)+x^3K_3(x)}{x^2},$$

where $x = \frac{2M}{T\mu_K}$. These *particle* quantities are defined by concatenation (vincula) as before:

$$J^\mu(q) = \sum_i\int d\tau\frac{p_{i\mu}}{\frac{M}{\mu_K}}\delta(q-q_i(\tau)),$$

and $\left\langle N^0\right\rangle = \left\langle J^0\right\rangle$. From this we again obtain the ideal gas law, $p = N_0T$.

The low-temperature limit may also be obtained using the formula

$$K_\nu(x) = \sqrt{\frac{\pi}{2x}}\exp\left(1+\frac{4\nu^2-1}{8x}\right); \qquad x >> 1.$$

We find

$$p = \frac{2T^6}{\pi^2 \left(\frac{2M}{\mu_K}\right)^2} \rho.$$

This is the same form as obtained from the phenomenological hadronic equation of state suggested by Shuryak (1988).

Antiparticles have also been considered by taking

$$N = V^{(4)} \sum_{K\mu} \left[\frac{1}{\exp\left(E - \mu - \mu_K \frac{m^2}{2\mathrm{M}} \frac{1}{T}\right) - 1} - \frac{1}{\exp\left(E + \mu - \mu_K \frac{m^2}{2M} \frac{1}{T}\right) - 1} \right],$$

and one finds

$$n = \frac{1}{\pi^2} \left(\frac{M}{\mu_K}\right)^2 \sqrt{1 - \frac{2\mu_K \mu T}{M}}.$$

Also, $p = 2p\,(|\mu|)$, and $\rho = 2\rho\,(|\mu|)$. p, ρ are the same as obtained earlier.

References

Balescu, R. (1964). *Cargese Summer School* (New York, Gordon and Breach).
Balescu, R. and Kotera, T. (1967). *Physica* **33**, 558.
Bethe, H. A. and Saltpeter, E. E. (1957). *The Quantum Mechanics of One and Two Electron Atoms* (New York, Academic Press).
Bogoliubov, N. N. (1946). *Problems Dynamical in Statistical Physics*, trans. E. K. Gora, in *Studies in Statistical Mechanics I*, ed. G. E. Uhlenbeck and J. de Boer (Amsterdam, North Holland).
Boltzmann, L. (1872). *Lectures on Gas Theory*, trans. S. G. Brush (Berkeley, University of California Press).
Burakovski, L. and Horwitz, L. P. (1993). *Physica A* **201**, 666.
Burakovski, L. and Horwitz, L. P. (1997). *Nucl. Phys.* **A614**, 373.
Burakovski, L., Horwitz, L. P. and Schieve, W. C. (1996a). *Phys. Rev. D* **54**, 4029.
Burakovski, L., Horwitz, L. P. and Schieve, W. C. (1996b). *Mass-Proper Time Uncertainty Relations in a Manifestly Covariant Relativistic Statistical Mechanics.* Preprint.
Cook, J. L. (1972). *Aust. J. Phys.* **25**, 117.
de Groot, S. R., van Leeuwen, C. K. and van Weert, G. (1980). *Relativistic Kinetic Theory* (New York, North-Holland).
Ehlers, J. (1974). In *Lectures in Statistical Physics* **28**, ed. W. C. Schieve and J. S. Turner (New York, Springer).
Fanchi, J. R. (1993). *Parametrized Relativistic Quantum Theory* (Dordrecht, Kluwer).
Feynman, R. P. (1949). *Phys. Rev.* **76**, 746.
Goldstein, H. (1980). *Classical Mechanics*, 2nd edn. (Reading, Mass., Addison-Wesley).
Hakim, R. (1967). *J. Math. Phys.* **8**, 1315, 1379.
Havas, P. (1965). *Statistical Mechanics of Equilibrium and Non-Equilibrium*, ed. Meixner (Amsterdam, North-Holland).
Horwitz, L. P. and Lavie, Y. (1982). *Phys. Rev. D* **26**, 819.
Horwitz, L. P. and Piron, C. (1973). *Helv. Physica Acta* **46**, 316.

Horwitz, L. P., Schieve, W. C. and Piron, C. (1981). *Ann. Phys. (N.Y.)* **137**, 306.
Horwitz, L. P., Shashoua, S. and Schieve, W. C. (1989). *Physica A* **161**, 300.
Hudson, R. (1974). *Rep. Math. Phys.* **6**, 249.
Jüttner, F. (1911). *Ann. Phys. (Leipzig)* **34**, 856.
Kandrup, H. (1984). *Ann. Phys. (N.Y.)* **153**, 44.
Liboff, R. L. (1998). *Kinetic Theory*, 3rd edn. (New York, Springer).
McLennan, J. A. (1989). *Introduction to Non-equilibrium Statistical Mechanics* (New York, Prentice Hall).
Pauli, W. (1958). *Theory of Relativity* (New York, Pergamon).
Shuryak, E. V. (1988). *The QCD Vacuum, Hadrons and Superdense Matter* (Singapore, World Scientific).
Stueckelberg, E. C. G. (1941). *Helv. Phys. Acta* **14**, 588.
Synge, J. L. (1957). *Relativistic Gas Theory* (Amsterdam, North-Holland).
Taylor, J. R. (1972). *Scattering Theory* (New York, Wiley).
Tolman, R. C. (1967). *Principles of Statistical Mechanics* (London, Oxford University Press), reprinted by Dover.
Trump, M. A. and Schieve, W. C. (1999). *Classical Relativistic Many-Body Dynamics* (Dordrecht, Kluwer).
Wigner, E. P. (1932). *Phys. Rev.* **40**, 2127.

11
Quantum optics and damping

11.1 Introduction

In this chapter we will turn to the arena of quantum optics for illustrative examples of the use of the master equation discussed in Chapters 3 and 6. In fact, quantum optics examples have already been utilized in Chapter 2 as introduction to the density matrix. The principal focus here will be quantum damping in these systems, that is, the damping effect on an atom in interaction with the electromagnetic field as a reservoir. Damping is discussed extensively in Chapter 17 in connection with decay-scattering systems. For this system general phase space distribution functions will be reexamined. To some degree, this has already been done in Chapter 2 with the introduction of the Glauber–Sudarshan $P(\alpha a^*)$ function. The micromaser will be discussed as a modern and interesting example of the dynamic interaction of an atom with an electromagnetic cavity not in equilibrium. For use of the student, an appendix to this chapter will briefly review the quantization of the free electromagnetic field and its atomic interaction.

There is no possibility of reviewing this extensive and growing field here. Our desire in this chapter is to connect the general topics of this book to this example. The books of Louisell (1973) and Scully and Zubairy (1997) are excellent. We are also indebted to the work of Nussenzweig, Schleich, and Mandel and Wolf (Nussenzweig, 1973; Mandel and Wolf, 1995; Schleich, 2001). We also recall the fine early introduction to this topic by Agarwal (1973).

11.2 Atomic damping: atomic master equation

In this section we shall consider the so-called quantum optics master equation for the reduced atomic density operator, $\rho_A(t)$, in the Born approximation. The elements of this derivation have already been discussed in Chapter 3 with the derivation of the Pauli equation for $\langle\alpha|\rho_A|\alpha\rangle$. Here the generalization to off-diagonal contributions only adds complication. Thus, we will outline the derivation and refer the reader to the work of Peier (1972), Louisell (1973) and Agarwall (1973, 1974) for more detail. We will use the resultant dynamic equation to discuss the important process of spontaneous emission first discussed by Einstein (1917) and later in detail by Weisskopf and Wigner (1930). This is also discussed in Chapter 17.

As in Chapter 3 we begin with the von Neumann operator for $\rho_{AR}(t)$:

$$i\dot{\rho}_{AR}(t) = L\rho_{AR}(t) \equiv \left[H, \rho_{AR}(t)\right]; \qquad -\infty \leq t \leq \infty. \tag{11.1}$$

We have not incorporated i in L here. The reservoir in the electromagnetic field, which is assumed *initially* to be the vacuum, is

$$\rho_R(0) = |\{0\}\rangle\langle\{0\}|,$$

and

$$\rho_{AR}(0) = \rho_A(0)\,\rho_R(0). \tag{11.2}$$

We introduce the projection operator again:

$$\begin{aligned} P\rho &= \rho_R(0)\,\mathrm{Tr}_R\rho_{AR} = \rho_R(0)\,\rho_A(t) \\ P^2 &= P. \end{aligned} \tag{11.3}$$

Assuming

$$H = H_A + H_R + H_{AR},$$

then

$$PL_A = L_A P, \quad PL_R = L_R P = 0. \tag{11.4}$$

We also assume

$$PL_{AR}P\ldots = 0.$$

We obtain the generalized master equation to lowest order in H_{AR} (the so-called Born approximation; see Chapter 3). We incorporate the free atom motion by going to the interaction picture, and obtain in that representation the *irreversible* equation

$$\partial_t P\rho_{AR}(t) = -\int_0^t d\tau\, PL_{AR}(t)\,L_{AR}(t-\tau)\,P\rho_{AR}(t-\tau) \qquad t \geq 0, \tag{11.5}$$

where

$$\rho(t) = \exp(-iH_A t)\,\rho_{\mathrm{int}}(t)\exp(+iH_A t). \tag{11.6}$$

In Eq. (11.5) we do not make the interaction picture explicit. The irreversibility of such equations was discussed in detail in Chapter 5. Eq. (11.5) is the same form as we met in Chapter 3.

Now we introduce a group of two-level atoms utilizing the Pauli spin representation already introduced in Chapter 2. The Hamiltonian of the two-level atoms in interaction with the quantized radiation field is

$$H = \omega \sum_i S_i^z + \sum_{l\sigma} \omega_{l\sigma} a_{l\sigma}^{\dagger} a_{l\sigma} + \sum_{il\sigma} \left\{ g_{il\sigma} a_{l\sigma} \left(S_i^+ + S_i^- \right) + h.c. \right\}. \tag{11.7}$$

(See the appendix to this chapter.) We recall that

$$\begin{aligned} S_i^+ &= |1_i\rangle \langle 2_i| , \\ S_i^- &= |2_i\rangle \langle 1_i| , \\ S_i^z &= \frac{1}{2} \{ |1_i\rangle \langle 1_i| - |2_i\rangle \langle 2_i| \} , \end{aligned} \tag{11.8}$$

$|1\rangle \equiv |+\rangle \equiv \alpha$ being the *excited* atomic state. The atomic dipole moment is $\mathbf{d}_j \equiv \mathbf{d}\left(S_j^+ + S_j^-\right)$, being off diagonal, and $\omega = E_1 - E_2$.

In Eq. (11.7) the $\mathbf{E} \cdot \mathbf{P}$ interaction discussed in the appendix gives

$$g_{ils} = -i \left(\frac{2\pi c e}{V} \right)^{\frac{1}{2}} (\mathbf{d} \cdot \boldsymbol{\varepsilon}_{ls}) \exp (i \mathbf{l} \cdot \mathbf{r}_i) . \tag{11.9}$$

It will be left as a problem to work this out in detail.

If we make the rotating wave approximation discussed in Chapter 2, we drop the $a_{l\sigma}^{\dagger} S_i^+$ term, and then

$$H' = \omega \sum_i S_i^z + \sum_{l\sigma} \omega_{l\sigma} a_{l\sigma}^{\dagger} a_{l\sigma} + \sum_{il\sigma} \left\{ g_{il\sigma} a_{l\sigma} S_i^+ + h.c. \right\}. \tag{11.10}$$

Now we make the Markov approximation to Eq. (11.5), letting $t \rightarrow \infty$. Just as in Chapter 3, we assume a collision time τ_c and take $\tau_B >> \tau_c$ or, in the special limit, $\tau_c \rightarrow 0$. For the case being considered, $\tau_c = \frac{|r_A^{\max}|}{c}$, and $\tau_B = \left(\gamma_{ij}\right)^{-1}$, γ_{ij} given by Eq. (11.13). This is discussed in detail in that chapter and will not be repeated. We also must take $V \rightarrow \infty$, so that the continuum approximation $\sum_{l\sigma} \rightarrow \frac{V}{(2\pi)^3} \int d_3 l \sum_{\sigma}$ may be made.

We make the rotating wave approximation in the course of this evaluation, ignoring the $S_i^+ S_j^+$ and $S_i^- S_j^-$ terms, to obtain

$$\begin{aligned} \frac{\partial \rho_A}{\partial t} &= -i \sum_{ij} \Omega_{ij} \left[S_i^+ S_j^-, \rho_A \right] \\ &\quad - \sum_{ij} \gamma_{ij} \left\{ S_i^+ S_j^- \rho_A - 2 S_i^- \rho_A S_i^+ + \rho_A S_i^+ S_j^- \right\}. \qquad (11.11) \\ t &\geq 0 \end{aligned}$$

In the Schrödinger picture this is

$$\begin{aligned} \frac{\partial \rho_A}{\partial t} &= -i \sum_i (\omega_0 + \Delta) \left[S_i^z, \rho_A \right] \qquad t \geq 0 \qquad (11.12) \\ &\quad - i \sum_{i \neq j} \Omega_{ij} \left[S_i^+ S_j^-, \rho_A \right] \\ &\quad - \sum_{ij} \gamma_{ij} \left(S_i^+ S_j^- \rho_A - 2 S_j^- \rho_A S_i^+ + \rho_A S_i^+ S_j^- \right). \end{aligned}$$

Here we may let $\omega_0 = \omega + \Omega_{ii}$, where

$$\gamma_{ij} = \pi \sum_{l\sigma} g_{l\sigma i} g_{l\sigma j}^* \delta \left(\omega - \omega_{l\sigma} \right) \qquad (11.13)$$

$$\text{and } \Delta \equiv \Omega_{ii} = - \sum_{l\sigma} |g_{l\sigma}|^2 \left\{ (\omega_{l\sigma} - \omega)^{-1} - (\omega_{l\sigma} + \omega)^{-1} \right\} \qquad (11.14)$$

$$\Omega_{ij} = - \sum_{l\sigma} g_{l\sigma i} g_{l\sigma j}^* \left\{ (\omega_{l\sigma} - \omega)^{-1} + (\omega_{l\sigma} + \omega)^{-1} \right\}. \qquad (11.15)$$

A detailed discussion of the rotating wave approximation is given by Agarwal (1973). He points out that the consistent use of this approximation is in the equation and not by use of H'.

Now we must note that Eq. (11.12) has the form of a Lindblad–Kossakowski equation, discussed first in Chapter 3. Thus the rotating-wave quantum master equation is of the completely positive form. $\rho_A(t)$ is assured, in this case, of being positive in the semi group time evolution. If we had taken the view that the dissipative evolution *should be* of the Lindblad form, we would have, for this open system of atoms and fields, chosen this equation. This result is consistent with similar comments made earlier concerning the Pauli equation. We have also mentioned that Monroe and Gardiner (1996) have discussed the failure of the Lindblad form in the more general nonrotating approximation case.

Before considering a single two-level atom in interaction with the field, we must evaluate Eq. (11.13) and Eq. (11.14). In the continuum limit,

$$\gamma_{ij} = 2\pi^2 c\,(2\pi)^{-3} \int d_3 k d\Omega \delta\,(\omega - kc)\,|d|^2 \left(1 - \cos\theta^2\right) \exp\left(i\mathbf{k}\cdot\mathbf{r}_{ij}\right).$$

For a single two-level atom, we take $\gamma_{ij} = 0$ for $i \neq j$ and evaluate the delta function in $d_3 k$. We obtain

$$\gamma_{ij} \to \gamma = \frac{2}{3}\,|d|^2\,\frac{\omega^3}{c^2}. \tag{11.16}$$

This is 1/2 the famous Einstein A coefficient (Einstein, 1917). The other terms are more involved. Agarwal (1974) obtained

$$\Delta \equiv \Omega_{ii} = -\left(\frac{\gamma}{\pi}\right) \ln\left\{\left|\left(\frac{\omega_c}{\omega} - 1\right)\left(\frac{\omega_c}{\omega} + 1\right)\right|\right\}, \tag{11.17}$$

having a logarithmic divergence necessitating the cutoff frequency ω_c. We will not consider further this frequency shift.

The master equation becomes simply

$$\frac{\partial \rho_A}{\partial t} = -i\omega\left[S^z, \rho\right] - \gamma\left(S^+ S^- \rho_A - 2S^- \rho_A S^+ + \rho_A S^+ S^-\right) \quad t \geq 0, \tag{11.18}$$

where $\omega = \omega_0 + \Omega_{ii}$. After renormalization there is a shift of both the ground and excited states.

Now we write an equation for $\rho_{12} \equiv \langle 1|\,\rho_A\,|2\rangle$ and $\rho_{11}(t)$, utilizing the properties of S^z, S^+ and S^-. Eq. (11.18) becomes

$$\begin{aligned} \frac{d\rho_{12}}{dt} &= -i\omega\rho_{12} - \gamma\rho_{12} \\ \frac{d\rho_{11}}{dt} &= -2\gamma\rho_{11}(t)\,. \end{aligned} \tag{11.19}$$

Thus, the solution is simply

$$\begin{aligned} \rho_{12}(t) &= \rho_{12}(0) \exp(-i\omega t) \exp(-\gamma t) \\ \rho_{11}(t) &= \rho_{11}(0) \exp(-2\gamma t) \qquad t \geq 0 \\ \rho_{11}(t) + \rho_{22}(t) &= 1. \end{aligned} \tag{11.20}$$

The probability decays by the interaction with the electromagnetic field reservoir in a time $\tau_B = 1/2\gamma$. The off-diagonal correlations decay spontaneously, slightly more rapidly due to the factor of 1/2. This is qualitatively similar to the Walls and Milborn example of Chapter 2 (Walls and Millborn, 1985). As already mentioned, this is an indication of the decoherence process extensively studied by Zurek (1991) and discussed in Chapter 12. Zurek argues that this is the source of classical behavior in quantum systems. The 2γ decay constant given by Eq. (11.11) was first obtained by Weisskopf and Wigner in their theory of spontaneous emission (Weisskopf and Wigner, 1930). We will discuss this extensively in Chapter 17. We must emphasize here that the source of the atomic decay resides in the dissipative open system equation, Eq. (11.18), treated in a consistent manner. This goes beyond the Schrödinger equation. We note that there is no induced emission or absorption in Eq. (11.9). This is due to the initial reservoir condition of Eq. (11.2).

If we assume the reservoir is an initial thermal state with

$$\rho_R(0) = \frac{\exp\left\{-\beta \sum_{ls} \omega_{ls} a_{ls}^{\dagger} a_{ls}\right\}}{Tr \exp\left(-\beta \sum_{ls} \omega_{ls} a_{ls}^{\dagger} a_{ls}\right)}, \tag{11.21}$$

then by similar arguments, as already made, the system master equation is

$$\begin{aligned} \frac{\partial \rho_A}{\partial t} &= i\omega \left[S^z, \rho_A\right] \\ &\quad - \gamma \left(1 + \langle n(\omega)\rangle\right) \left(S^+ S^- \rho_A - 2\rho S^- \rho_A S^+ + \rho_A S^+ S^-\right) \\ &\quad - \gamma \langle n(\omega)\rangle \left|S^- S^+ \rho_A - 2S^+ \rho_A S^- + \rho_A S^- S^+\right|, \end{aligned} \tag{11.22}$$

where $\langle n(\omega)\rangle = (\exp \beta\omega - 1)^{-1}$. It may easily be shown that this leads to the famous stochastic equation of Einstein with the induced, spontaneous and absorption terms. We leave it to the student to prove this.

In a number of respects, spontaneous emission is sometimes interpreted as due to the vacuum fluctuations of the field

$$\left\langle 0 \left| E^2 \right| 0 \right\rangle = \sum_l \hbar\omega_l. \tag{11.23}$$

This is a puzzle. Eqs. (11.19) indicate that the initial atomic state $|1\rangle$ is unstable and decays at least at long time when these equations are valid under the influence of the reservoir vacuum, Eq. (11.23). The details of this *initial* decay are not seen, since we have not written a short-time initial exact solution. As discussed in Chapter 3, it can easily be seen that at $t = 0$ for the diagonal part of $P\rho_{AR}$, $\Delta P\rho_{AR}$,

$$\frac{d\Delta P\rho_{AR}(t)}{dt} = 0; \qquad t = 0^+. \tag{11.24}$$

It is a fixed point in the nonlinear dynamics sense that asymptotically decays. But what is the cause? We shall turn to this now.

It is best to use the Heisenberg equations to discuss the short-time behavior, following Milloni in his *Physical Reports* review (Milloni, 1976). (He has done an extensive review of the literature. See also Senitzky, 1973.) The main theme of the discussion (and controversy) is whether vacuum fluctuations play the entire role or whether a quasi-classical picture of radiation reaction on the atom by the field is primary. (See any good text such as that of Panovsky and Phillips [1962] or the classic discussion of Becker [1964 edn.]). The simple derivation of the Lamb shift of Welton (Welton, 1948) would seem to support this latter view. Scully and Zubairy (1997) repeat this derivation. In doing this, they derive the following interesting formula:

$$\left\langle (\delta \mathbf{r}_{\text{vac}})^2 \right\rangle = \sum_k \left(\frac{e}{mc^2k^2} \right)^2 \langle 0 \mid \mathbf{E}_k \mid 0 \rangle^2 , \tag{11.25}$$

$(\delta r)^2$ being a fluctuation in atomic position.

Milloni, in his derivation of the Heisenberg equations of motion for the two-level atom, obtains in a self-consistent manner

$$\dot{S}_x = -\omega_0 S_y \qquad \omega = \omega + \Delta \tag{11.26a}$$

$$\dot{S}_y = \omega_0 S_x + \frac{2}{\hbar} \boldsymbol{\mu} \cdot \mathbf{E}^{\perp} (0, t)\, S_z \tag{11.26b}$$

$$\dot{S}_z = -\frac{2}{\hbar} \boldsymbol{\mu} \cdot \mathbf{E}^{\perp} (0, t)\, S_y, \tag{11.26c}$$

and

$$\nabla^2 \mathbf{E}^{\perp} (r, t) = \frac{1}{c^2} \ddot{\mathbf{E}}^{\perp} (\mathbf{r}, t) . \tag{11.27}$$

Now

$$\mathbf{E}^{\perp} (0, t) = E_0^{\perp} (0, t) + E_{RR}^{\perp} (t) , \tag{11.28a}$$

separating the particle source and homogeneous parts.

A number of approximations have been made. The foremost is the adiabatic approximation that the atom density matrix should follow the free evolution $\dot{\rho}_{ij} = -i\omega_{ij}\rho_{ij}$ in the *field part* of ρ_{AR}. This is equivalent to the Weisskopf–Wigner approximation. In addition, in the choice of the representation, a normal ordering is now assumed. Photon annihilation operators are put to the right of operator products. Milloni obtains the level shift and width and subsequently obtains the Bethe expression for the Lamb shift. The main point here is that Eqs. (11.26) are valid at short time. No time scaling has been used.

The atom radiation field is

$$\mathbf{E}_{RR}^{\perp} (t) = \left[\frac{2}{3c^3} \dddot{\sigma}_x - \frac{4K}{3\pi c^2} \ddot{\sigma}_x + \frac{4K^3}{9\pi} \sigma_x \right] \boldsymbol{\mu}, \tag{11.28b}$$

and $\mathbf{E}_0^{\perp} (0, t)$ is the homogeneous solution to Eq. (11.27), depending on the vacuum. $K = \frac{E_{\max}}{\hbar c}$ on introduction of the Bethe cutoff wave number, $\frac{mc}{\hbar}$. The third term in Eq. (11.28b) has no classical counterpart. We solve now in the adiabatic approximation. Taking the vacuum field expectation values, $\langle\rangle$, Eqs. (11.26) become

$$\begin{aligned}\langle\dot{\sigma}_{12}(t)\rangle &= -i(\omega_0 - \Delta - i\gamma)\langle\sigma_{12}(t)\rangle - i(\Delta - i\gamma)S^+(t) \\ \langle\dot{\sigma}_z\rangle &= -2\gamma(1 + \langle\sigma_z(t)\rangle),\end{aligned} \tag{11.29}$$

where again

$$\gamma = 2|\boldsymbol{\omega}_{12}|^2 \frac{\omega_0^3}{3\hbar c^3}. \tag{11.30}$$

The energy shift is now

$$\Delta = \frac{2|\mu_{12}|^2}{3\pi\hbar c^3}\omega_0^2 P\int d\omega\left(\frac{\omega}{\omega-\omega_0} - \frac{\omega}{\omega+\omega_0}\right). \tag{11.31}$$

Δ is apparently the effect of the vacuum fluctuations. However, it is not explicit, since the homogeneous solution to Eq. (11.27), $E_0^+(0)$, proportional to $a_{l\sigma}$, does not enter at all. It *appears* that Eq. (11.28b) plays the dominant role, which may be interpreted as a radiation reaction effect.

Is this physical conclusion independent of normal ordering? Senitzky and Milloni have redone the calculation with antinormal ordering (Senitzky, 1973; Milloni, 1976). Using the rotating wave approximation, it is found that $\mathbf{E}_0^+(0,t)$ plays an explicit role due to the new ordering. We have

$$\langle\dot{\sigma}_{12}(t)\rangle = -i(\omega_0 - \Delta - i\gamma)\langle\sigma_{12}(t)\rangle,$$

neglecting counterrotating terms. However, the physical interpretation differs, indicating that the original question is meaningless. We cannot say that spontaneous emission is due to radiation reaction *or* vacuum fluctuations. They are all one, as implied by Senitzky (1973).

11.3 Cavity damping: the micromaser: detection

Let us now turn to the master equation and damping in the field of the cavity. We will focus on the micromaser of Walther, Rempe and Klein (Rempe *et al.,* 1987), the Munich micromaser. See the review by Raithal (Raithal *et al.,* 1994). In a cavity of very high Q and very low temperature, a few atoms are sequentially injected and excite the field of the cavity. We will derive by physical arguments the birth–death equation for the density matrix of the field. The density matrix is off-diagonal. The transiting atoms are later observed, and it is these atoms which measure the field properties indirectly. The detection process will be included in the master equation.

The two atomic levels are rubidium $63P_{\frac{3}{2}}$ and $61D_{\frac{5}{2}}$ with frequency 21.5 MHz. The spontaneous decay time for the upper state is 488 μs, and the average number of thermal photons due to the environment is 0.054; the cavity quality factor is 3×10^{10}, with $T = 0.3$ K.

The theory of such a system without detection was first done by Filipowicz, Javanainen and Meystre (1986). A density matrix formulation was given early by Krause (Krause *et al.,* 1986). This is a form of the basic Scully–Lamb laser theory (Scully and Lamb, 1967, 1969). It is to be compared with the previous section; the system is the electromagnetic field in interaction with the injected atoms as the "reservoir." We will not approach this from the point of view of the generalized master equation. Rather, we shall simply give an argument similar to the Pauli and also the Scully–Lamb birth–death approach.

11.4 Detection master equation for the cavity field

For the two-level atom, we take $|A\rangle$ and $|B\rangle$ to be the upper and lower states. The macroscopic detector registers after the atomic passage through the cavity, $|+1\rangle$, $|-1\rangle$ and $|0\rangle$ for the atom in the upper state, lower state or *no register.* Superpositions are possible in the detector registration. Ionizing field channeltrons were used as detectors in the experiments. A postselection of phase may be made at the cavity exiting port. We will take the incoming atoms in the $|A\rangle$ state. In this we will adopt the simple "collapse" approach to measurement. This is discussed in some detail in Chapter 13, and references are there. Now consider the work of McGowan and Schieve (1997).

The atom, field and detector state before measurement is

$$\begin{aligned} \left|\psi_{afd}\right\rangle = {} & c_1 \left|\psi_f\right\rangle |A\rangle \, |+\rangle + c_2 \left|\psi_f\right\rangle |A\rangle \, |0\rangle + c_3 \left|\psi_f\right\rangle |A\rangle \, |-\rangle \\ & + c_4 \left|\psi_f\right\rangle |B\rangle \, |+\rangle + c_5 \left|\psi_f\right\rangle |B\rangle \, |0\rangle + c_6 \left|\psi_f\right\rangle |B\rangle \, |-\rangle \, . \end{aligned} \tag{11.32}$$

We assume no detector errors. We define then, on measurement,

$$|c_1|^2 = p_A \quad \text{(state } A \text{ detected atom in } |A\rangle\text{)} \tag{11.33a}$$

$$|c_6|^2 = p_B \quad \text{(state } B \text{ detected atom in } |B\rangle\text{)} \tag{11.33b}$$

$$|c_3|^2 = 0 \quad \text{(state } B \text{ detected atom in } |A\rangle\text{)} \tag{11.33c}$$

$$|c_4|^2 = 0 \quad \text{(state } A \text{ detected atom in } |B\rangle\text{)} \tag{11.33d}$$

$$\begin{aligned} |c_2|^2 &= 1 - p_A \quad \text{(no detection)} \\ |c_5|^2 &= 1 - p_B \quad \text{(no detection)} \end{aligned} \tag{11.33e}$$

and

$$c_1 c_4^* = c_1^* c_4 = c_3 c_6^* = c_3^* c_6 = 0. \tag{11.33f}$$

The only cross terms are $c_2 c_5^*$ and $c_2^* c_5$, which is a mixed state, "no check," of the detectors. It must be emphasized that the detectors are *macroscopic*, and thus the detector states are diagonal.

The resulting density matrix of the entire system after an atom passage of the cavity is

$$\begin{aligned} \rho_{afd} = & \; p_A \left|A\right\rangle \left|+\right\rangle \rho_{AA} \left\langle +\right| \left\langle A\right| \\ & + (1 - p_A) \left|A\right\rangle \left|0\right\rangle \rho_{AA} \left\langle 0\right| \left\langle A\right| \\ & + (1 - p_B) \left|B\right\rangle \left|0\right\rangle \rho_{BB} \left\langle 0\right| \left\langle B\right| \\ & + p_B \left|B\right\rangle \left|-\right\rangle \rho_{BB} \left\langle -\right| \left\langle B\right| \\ & + c_2 c_5^* \left|A\right\rangle \left|0\right\rangle \rho_{AB} \left\langle 0\right| \left\langle B\right| + h.c. \end{aligned} \tag{11.34}$$

We will now obtain a master equation for the field due to undetected atoms. We take

$$\mathrm{Tr}_A \left[\left\langle 0\right| \left|0\right\rangle\right] \tag{11.35}$$

and find that the field changes after the atoms' undetected passage leads to the *state reduction* of the field density matrix,

$$\rho_f(t) \to (1 - p_A) A \rho_f(t) + (1 - p_B) B \rho_f(t). \tag{11.36}$$

The operators A, B depend on the form of measurement. If A detector clicks when the atom is in the upper Ryberg state and B observes the lower state, then from the simple Jaynes–Cummings model (Jaynes and Cummings, 1963), one obtains the evolution in the phase-insensitive case:

$$\begin{aligned} A\rho_f &= S_A \rho_f S_A^\dagger \\ B\rho_f &= S_B \rho_f S_B^\dagger, \end{aligned} \tag{11.37}$$

where

$$\begin{aligned} S_A &= \cos\left(g\tau\sqrt{aa^\dagger}\right) \\ S_B &= \frac{a^\dagger \sin\left(g\tau\sqrt{aa^\dagger}\right)}{\sqrt{aa^\dagger}}. \end{aligned} \tag{11.38}$$

These super-operators describe this field change due to a single atom passage. The atom is in interaction for a period τ, which is a parameter. It may be statistically distributed, but we assume here that it is determined by a precise injection rate, r, and cavity length. This is not experimentally so. The operator coefficients $A(\tau)$, $B(\tau)$ are the principal differences between the micromaser and laser. (See the 1997 book of Scully and Zubairy.) There A, B are constants independent of this parameter.

A commonly used parameter is $\theta = \sqrt{N_{ex} g\tau}$, a pumping parameter having values (commonly) 1 to 10. This is left to the student as a problem. If we normalize Eq. (11.36) and calculate the change in $\rho_{f,} \Delta\rho_f(t)$, we find

$$\Delta\rho_f(t) = \frac{(1-p_A)\, A\rho_f(t) + (1-p_B)\, B\rho_f(t)}{1 - p_A \mathrm{Tr}\{A\rho_f(t)\} - p_B \mathrm{Tr}\{B\rho_f(t)\}} - \rho_f(t). \tag{11.39}$$

This is the product of the probability that there is an undetected atom in the cavity rdt and the probability that the atom is undetected. We obtain

$$\begin{aligned}\frac{\partial\rho_f(t)}{\partial t} &= r\left[A(\tau) + B(\tau) - 1\right]\rho_f(t) - r\left[p_A A(\tau) + p_B B(\tau)\right]\rho_f(t) \qquad (11.40)\\ &+ L\rho_f(t) + r\left[p_A \mathrm{Tr}\left[A(\tau)\rho_f(t)\right] + p_B \mathrm{Tr}\left[B(\tau)\rho_f(t)\right]\rho_f(t)\right].\end{aligned}$$

Such an equation was first obtained by Briegel (Briegel *et al.*, 1994). It is nonlinear, containing the inefficient $p_{A,} p_B$ detector coefficients. Here

$$L\rho(t) = \frac{-1}{N_{ex}}\begin{bmatrix}(n_b+1)\left(a^\dagger a\rho(t) - 2a\rho(t)a^\dagger + \rho(t)a^\dagger a\right)\\ + n_b\left(aa^\dagger\rho(t) - 2a^\dagger\rho(t)a + \rho(t)aa^\dagger\right)\end{bmatrix} \tag{11.41}$$

describes the *field* damping in the cavity. This has been discussed in Chapter 2. It may be obtained from the density matrix for the driven-damped single harmonic oscillator (Scully and Zubairy, 1997). Here n_b is the mean number of thermal photons, < 1 for the micromaser, $N_{ex} = r/\gamma$. r is the rate of atomic injection, and γ the mean photon decay rate. The equation without the Tr terms is the master equation of the isolated laser. References were given earlier (see also Lugiato *et al.*, 1987). We should remark that Johnson and Schieve (2001) have discussed how the nonnormalized form, Eq. (11.36), may be used in numerical calculations. This obviates the use of the nonlinear operations.

In the phase-sensitive case, we may form an *entanglement* detection scheme for states $\frac{1}{\sqrt{2}}(|A\rangle - |B\rangle)$ and $\frac{1}{\sqrt{2}}(|A\rangle + |B\rangle)$. (Entanglements are discussed in some detail in Chapter 12.) If the atoms are injected in the upper state $|A\rangle$, then, with a $\pi/2$ pulse before the detectors for postphase selection, the operators become for entanglement detection

$$\begin{aligned}\begin{Bmatrix}A\rho\\ B\rho\end{Bmatrix} &= \frac{1}{2}\left(S_A\rho S_A^\dagger + S_B\rho S_B^\dagger\right)\\ &\mp \frac{1}{2}\left(S_A\rho S_B^\dagger + S_B\rho S_A^\dagger\right),\end{aligned}$$

and Eq. (11.40) becomes appropriately modified. This is the Ramsey detection method (see Scully and Zubairy, 1997). It was also done by Herzog (2000).

Let us consider the *analytic* solution to Eq. (11.40) on totally inefficient detection, $p_A = p_B = 0$. (We now drop the f subscript on the field.) The equation then is

$$\partial_t \rho(t) = r[A + B - 1]\rho(t) + L\rho(t),$$

which we write in the number representation $|n\rangle$. We assume the injection atoms arrive in state $|A\rangle$. Then

$$\begin{aligned}\partial_t \rho_n^{(k)} = & -\gamma(n_b+1)\left[\left(n+\frac{k}{2}\right)\rho_n^{(k)}(t) - \sqrt{(n+1)(n+k+1)}\rho_{n+1}^{(k)}\right] \\ & -\gamma n_b\left[\left(n+1+\frac{k}{2}\right)\rho_n^{(k)}(t) - \sqrt{n(n+k)}\rho_{n-1}^{(k)}(t)\right] \\ & - r\rho_n^{(k)} \\ & + r\left[\cos\left(g\tau\sqrt{n+1}\right)\cos\left(g\tau\sqrt{n+k+1}\right)\right]\rho_n^{(k)}(t) \\ & + \sin\left(g\tau\sqrt{n}\right)\sin\left(g\tau\sqrt{n+k}\right)\rho_{n-1}^{(k)}(t).\end{aligned} \tag{11.42}$$

Here $\rho_n^{(k)} \equiv \rho_{n,n+k}$ is off-diagonal. n_b is the number of thermal photons present, γ the cavity decay constant, and r the rate of injection.

The solution to Eq. (11.42) was discussed in some detail by McGowan and Schieve (1997), using a method due to Scully (Scully and Lamb, 1967). Assuming $\gamma = 0$ and $n_b = 0$, we have, from Eq. (11.42), the recursion relations

$$\begin{aligned}&\left[\cos\left(g\tau\sqrt{n+1}\right)\cos\left(g\tau\sqrt{n+k+1}\right)\right]\rho_n^{(k)} \\ &\quad = -\sin\left(g\tau\sqrt{n}\right)\sin\left(g\tau\sqrt{n+k}\right)\rho_{n-1}^{(k)}.\end{aligned} \tag{11.43}$$

From this, as with birth–death equations generally, we obtain for $k = 0$,

$$\rho_{n,n}^S = \rho_{nn}(0)\,\Pi_{j=1}^n \frac{N_{ex}}{j}\sin^2\left(g\tau\sqrt{j}\right). \tag{11.44}$$

The recursion relation is interesting (Filipowicz *et al.*, 1986); the recursion truncates for n values both upward and downward for $k = 0$. For $n(q)$

$$\sin\left(g\tau\sqrt{n(q)+1}\right) = 0 \quad g\tau\sqrt{n(q)+1} = q\pi \quad q \text{ odd} \tag{11.45}$$

and $n(p)$

$$\sin(g\tau)\sqrt{n(p)} = 0 \quad g\tau\sqrt{n(p)} = p\pi \qquad p \text{ odd}. \tag{11.46}$$

These truncation points are called *trapping states* of the cavity field, $n\,(q)$ (up) and $n\,(p)$ (down). For $n\,(q)$,

$$\begin{aligned}\rho^{s}_{n(q),n(q)} &= \text{Eq. (11.44)} \qquad & n < n\,(q) \\ &= 0 & n > n\,(q)\,,\end{aligned} \tag{11.47}$$

and for $n\,(p)$,

$$\begin{aligned}\rho^{s}_{n(p),n(p)} &= 0 & n < n\,(p) \\ &= \text{Eq. (11.44)} \qquad & n > n\,(p)\,.\end{aligned} \tag{11.48}$$

We recognize from Eq. (11.43) that

$$\sum_n \dot{\rho}_n^{(k)} = \sum_n X_n^{(k)} \rho_n^{(k)}, \tag{11.49}$$

which suggests the solution

$$\rho_n^{(k)}\,(t) = \rho_n^{(k)}\,(0) \exp\left(X_n^{(k)} t\right), \tag{11.50}$$

where

$$\begin{aligned}X_n^{(k)} &= -r - \gamma\,(n_b + 1)\left(n + \frac{k}{2} - \sqrt{n\,(n+k)}\right) \\ &\quad - \gamma n_b\left[n + 1 + \frac{k}{2} - \sqrt{(n+1)\,(n+k+1)}\right] \\ &\quad + r\left[\begin{array}{l}\cos\left(g\tau\sqrt{n+1}\right)\cos\left(g\tau\sqrt{n+k+1}\right) \\ +\sin\left(g\tau\sqrt{n+1}\right)\sin\left(g\tau\sqrt{n+k+1}\right)\end{array}\right].\end{aligned} \tag{11.51}$$

This solution was utilized by McGowan and Schieve to obtain an approximate solution to the cavity *with* measurement. The γ-dependent terms here are, of course, the cavity decay due to the various photon loss mechanisms. The trigonometric terms are the new and interesting features in the cavity. These field points block diagonalize the Fock space and are one of the main features of the one-atom micromaser. The physical interpretation is that the injected atom undergoes integer Rabi oscillations, thus returning to its initial $|A\rangle$ state, leaving the field unchanged.

Trapping states have been observed in the Munich micromaser (Weidinger *et al.*, 1989). Dips in the inversion agree well with the preceding formulas. Johnson and Schieve have done an extensive comparison of the theory based upon the Jaynes–Cummings model (Jaynes and Cummings, 1963) with these experiments, as outlined here. They find the positions of the trapping states in excellent agreement with Eq. (11.45), Eq. (11.46), Eq. (11.47) and Eq. (11.48). However, the qualitative behavior elsewhere in the inversion—for instance, as a function of τ for this experimental condition—does not agree with the theory. A number of modifications of the theory were made, including atomic decay, velocity averaging and two successive atom events. Little improvement in the agreement was obtained (D. Johnson, unpublished Ph.D. thesis, University of Texas, 2003).

Let us return to the question of trapping states including detection. The steady-state condition is

$$r\left[(1-p_A)A+(1-p_B)B-1\right]\rho^s+r\left[p_A+rA\rho^s\right]+p_B\mathrm{Tr}\left[B\rho^s\right]\rho^s=0. \tag{11.52}$$

We again assume $n_b=0, \gamma=0$ and take $p_A << 1$. This becomes, for $k=0$ in $\rho_{n,n+k}\equiv\rho_n^s$,

$$\begin{aligned}&\left[\cos^2\left(g\tau\sqrt{n+1}\right)-1\right]\rho_n^s+\sin^2\left(g\tau\sqrt{n}\right)\rho_{n-1}^s \\ &\quad+\left[p_A\sum_{n'}\cos^2\left(g\tau\sqrt{n'+1}\right)\rho_{n'}^s+p_B\sum_{n'}\sin^2\left(g\tau\sqrt{n'}\right)\rho_{n'-1}^s\right]\rho_n^s=0.\end{aligned} \tag{11.53}$$

The steady state is a solution to this nonlinear equation. Because of the nonlinearity, no simple recursive form may be found. Suppose now, for instance, that the first term is approximately zero for the case $\sin\left(g\tau\sqrt{n(q)+1}\right)=0$, there being no states present for $n>n(q)$. This might be true for very small p_A, p_B, weak detection. The right side vanishes then. For $n<n'(q)$, we then have Eq. (11.53) having $\sum_{n'<n(q)}$ and $\rho_{n'}^s$ given on the right by Eq. (11.49). Thus we see that in weak detection we may expect, by this iterative argument, the conditions of Eq. (11.48) to be maintained. The distribution is altered by the p_A, p_B terms in the nonlinearity. We have argued from the detection theory that the trapping conditions can be observed. They are! Even for large p_A, p_B, this argument seems to be true. Monte Carlo simulations with 100% detector efficiencies, thus taking into account the measurement, agree well with the experimental trapping values.

Let us turn to time-dependent solutions, in the phase-insensitive case in Eq. (11.40), by a perturbation assuming p_A, p_B again small. We may use the $p_A=p_B=0$ solution and iterate with it in the Tr terms. The first-order equation is

$$\begin{aligned}\frac{d}{dt}p_n^k(t)&=X_n^k\rho_n^k(t)+\rho_n^k(0)\exp\left(X_n^kt\right) \\ &\quad\times\left[E_n^k+F_{n+1}^k+\sum_m\left\{G_m\rho_m^0(0)+H_m\rho_{m-1}^0(0)\right\}\right],\end{aligned} \tag{11.54}$$

where

$$E_n^k = -rp_A \cos\left(g\tau\sqrt{n+1}\right)\cos\left(g\tau\sqrt{n+k+1}\right) \tag{11.55}$$
$$F_n^k = -rp_B \sin\left(g\tau\sqrt{n}\right)\sin\left(g\tau\sqrt{n+k}\right)$$
$$G_m = rp_A \cos^2\left(g\tau\sqrt{m+1}\right)$$
$$H_m = rp_B \sin^2\left(g\tau\sqrt{m}\right).$$

The solution is to first order

$$\rho_n^k(t) = \rho_n^k(0)\exp\left(X_n^k t\right)\left[1 + tE_n^k + tF_{n+1}^k + t\sum_m\left\{G_m\rho_m^0(0) + H_m\rho_{m-1}^0(0)\right\}\right]. \tag{11.56}$$

With nonzero γ, the solution decays with the parameter X_n^k given by Eq. (11.51). Without damping, it diverges with t, meaning that it is then a short-time solution.

Now we will briefly mention the micromaser spectrum (Lu, 1993; McGowan and Schieve, 1997). The spectrum is defined as

$$S(\omega - \omega_c) = \text{Re}\int_0^\infty K(t)\exp\left(-i(\omega-\omega_c)t\right)dt, \tag{11.57}$$

where

$$K(t) = \left\langle a^\dagger(t)\, a(0)\right\rangle \tag{11.58}$$

is the two-time correlation function. This may be related to the Green's function,

$$K(t) = \sum_{nm} G_{n,n+1}^m(t)\sqrt{(n+1)(m+1)}\,P_{m+1},$$

where

$$P_{m+1} = \left\langle m+1\left|\rho_f^s\right|m+1\right\rangle, \tag{11.59}$$

and

$$G^{(m)}(t) = \text{Tr}_r\left[U(t)\,|m\rangle\langle m+1|\,\rho_r^s U^\dagger(t)\right]. \tag{11.60}$$

We may show in this approximation, using the initial conditions

$$G_{n,n+1}^{(m)}(0) = \delta_{n,m}$$
$$G_{n,n+k}^{(m)}(0) = 0 \quad k \neq 1,$$

that the analytic answer is

$$G_{n,n+1}^{(m)} = \delta_{n,m}\exp\left(X_n^{(1)}t\right).$$

As a result of this, we obtain, for the micromaser with measurement,

$$S(\omega) = \sum_m (m+1) P_{m+1} \left[\frac{\left|X_m^1\right|}{\left|X_m^1\right|^2 + \omega^2} + \left\{E_m^1 + F_{m+1}^1\right\} \frac{\left|X_n^1\right|^2 - \omega^2}{\left(\left|X_n^1\right|^2 + \omega^2\right)^2} \right]. \tag{11.61}$$

The second term contains the new additional measurement terms depending on p_A, p_B.

The micromaser line width was calculated from this by McGowan and Schieve (1997) in one case of 10% detector efficiency and also for no detection as a function of $\theta = g\tau\sqrt{N}$. Peaks appear in both cases. These were associated with trapping states by Lu (1993) by the formulas of Eq. (11.48). The result here was a significant increase in peak height with detection. No significant qualitative change in the curve structure was seen. The undetected case agrees well with Lu's more exact calculation.

This ends our brief discussion of quantum master equations in quantum optics. The topics chosen were obviously personal but should illustrate the master equation applications both to atoms and fields. The student should read the recent good texts cited for other applications.

Appendix 11A: the field quantization and interaction

The source free Maxwell equations are

$$\begin{aligned} \nabla \cdot \mathbf{B} &= 0 \\ \nabla \times \mathbf{E} &= \frac{-\partial \mathbf{B}}{\partial t} \\ \nabla \cdot \mathbf{E} &= 0 \\ \nabla \times \mathbf{H} &= \frac{\partial \mathbf{D}}{\partial t} \end{aligned} \tag{11A.1}$$

with

$$\begin{aligned} \mathbf{B} &= \mu_0 \mathbf{H} \\ \mathbf{D} &= \varepsilon_0 \mathbf{E} \\ \mu_0 \varepsilon_0 &= c^{-2}. \end{aligned} \tag{11A.2}$$

We introduce the vector and scalar potentials $\mathbf{A}, U$. We are dealing with the nonrelativistic electrodynamics. We choose the Coulomb gauge,

$$\nabla \cdot \mathbf{A} = 0; \qquad V = 0. \tag{11A.3}$$

From Eq. (11A.1) we have the wave equation for $\mathbf{A}(\mathbf{r}, t)$:

$$\nabla^2 \mathbf{A} = \frac{1}{c^2} \frac{\partial_2 \mathbf{A}}{\partial t^2}.$$

The Hamiltonian is

$$H = \frac{1}{2} \int d_3 r \left(\varepsilon_0 E^2 + \mu_0 H^2 \right) = \frac{1}{2} \int d_3 r \left[\varepsilon_0 \left(\frac{\partial \mathbf{A}}{\partial t} \right)^2 + \frac{1}{\mu_0} \left(\nabla \times \mathbf{A} \right)^2 \right]. \tag{11A.4}$$

We separate the variables $\mathbf{r}$, t by assuming $\mathbf{A}(\mathbf{r}, t) \approx \sum_l q_l(t) u_l(\mathbf{r})$. The structure of the mode l is determined by the boundary conditions of the cavity. We obtain

$$\nabla^2 \mathbf{u}_l(\mathbf{r}) + \frac{\omega_l^2}{c^2} \mathbf{u}_l(\mathbf{r}) = 0, \tag{11A.5}$$

and

$$\frac{d^2 q_l}{dt^2} + \omega_l^2 q_l = 0. \tag{11A.6}$$

Choosing standing wave boundary conditions at the wall, $\mathbf{u}_l \mid_{\parallel} = 0$, and $\nabla \times \mathbf{u}_l \mid_{\perp} = 0$. We find in this case that $\mathbf{u}_l(\mathbf{r})$ are $\sin \mathbf{k}_l \cdot \mathbf{r}$ and $\cos \mathbf{k}_l \cdot \mathbf{r}$. A plane wave representation is more convenient. Then we assume

$$A(\mathbf{r}, t) = \sum_l \sum_{\sigma=1}^{2} \sqrt{\frac{\hbar}{2\omega_l \varepsilon_0 V}} e_{l\sigma} \left[a_{l\sigma} \exp i \left(\mathbf{k}_{l\sigma} \cdot \mathbf{r} - \omega_l t \right) + a_{l\sigma}^{\dagger} \exp -i \left(\mathbf{k}_{l\sigma} \cdot \mathbf{r} - \omega_l t \right) \right]. \tag{11A.7}$$

The $a_{l\sigma}$ obey Eq. (11A.6) for harmonic oscillators. Here $\hbar$ and an $a\dagger$ in the complex conjugate part are prematurely introduced. The polarization vectors obey

$$\mathbf{e}_{l1} \cdot \mathbf{e}_{l2} = 0 \tag{11A.8}$$

and

$$\mathbf{e}_{l\sigma} \cdot \mathbf{k}_l = 0,$$

since $\nabla \cdot \mathbf{A} = 0$. To obtain a solution to the wave equation, we must have the dispersion relation

$$k_l^2 = \frac{\omega_l^2}{c^2}. \tag{11A.9}$$

For periodic boundary conditions,

$$\mathbf{k}_l = \frac{2\pi}{L} \left(l_1 \hat{\imath} + l_2 \hat{\jmath} + l_3 \hat{k} \right), \tag{11A.10}$$

where l_1, l_2, l_3 are infinite countable sets of integers from $-\infty$ to $+\infty$. With this we have a new harmonic oscillator (countably infinite) representation of the electromagnetic field in the cavity of volume $V = L^3$. We leave it as a problem to show, after obtaining **E** and **H** in this representation, that the energy is

$$H = \sum_{l,\sigma} \hbar\omega_l \left(a_{l\sigma} a_{l\sigma}^{\dagger} + a_{l\sigma}^{\dagger} a_{l\sigma} \right). \tag{11A.11}$$

$\hbar$ appears because $\hat{a}_{l\sigma} = \frac{1}{\sqrt{2\hbar\omega_l}} \left(\omega_l \hat{q}_{l\sigma} + i\hat{p}_{l\sigma} \right)$ and $\hat{q}_{l\sigma}, \hat{p}_{l\sigma}$ are the position and momentum and will obey canonical quantization commutation rules. We assume

$$\left[\hat{q}_{l\sigma}, \hat{p}_{l'\sigma'} \right] = i\hbar\delta_{l\sigma,l'\sigma'}.$$

$l \equiv (l_1 l_2 l_3)$, and now care has been taken in the ordering of $a_{l\sigma}$ and $a_{l\sigma}^{\dagger}$ in the derivation.

We quantize the electromagnetic field after that of harmonic oscillators, assuming

$$\hat{H} = \sum_{l\sigma} \hbar\omega_l \left(\hat{a}_{l\sigma} \hat{a}_{l\sigma}^{\dagger} + \hat{a}_{l\sigma}^{\dagger} \hat{a}_{l\sigma} \right), \tag{11A.12}$$

with

$$\left[\hat{a}_{l\sigma}, \hat{a}_{l'\sigma'}^{\dagger} \right] = \delta_{l\sigma,l'\sigma'} \tag{11A.13}$$

$$\left[\hat{a}_{l\sigma}, \hat{a}_{l'\sigma'} \right] = \left[\hat{a}_{l'\sigma'}^{\dagger}, \hat{a}_{l\sigma}^{\dagger} \right] = 0.$$

This quantization procedure is due to Dirac (1958).

Just as with the 1–D harmonic oscillator in Chapter 2, we may construct the number states utilizing

$$\left[\hat{a}_{l\sigma}, \hat{a}_{l\sigma}^{\dagger} \hat{a}_{l\sigma} \right] = \hat{a}_{l\sigma} \tag{11A.14}$$

$$\left[\hat{a}_{l\sigma}^{\dagger}, \hat{a}_{l\sigma}^{\dagger} \hat{a}_{l\sigma} \right] = -\hat{a}_{l\sigma}^{\dagger}.$$

We will now drop the operator "hat" notation. As in Section 11.2, we introduce the number operator for each mode $l\sigma$,

$$N_{l\sigma} = a_{l\sigma}^{\dagger} a_{l\sigma} = N_{l\sigma}^{\dagger}, \tag{11A.15}$$

and form the number states $|n_{l\sigma}\rangle$,

$$N_{l\sigma} \left| n_{l\sigma} \right\rangle = n_{l\sigma} \left| n_{l\sigma} \right\rangle,$$

by

$$N_{l\sigma} a_{l\sigma} \left| n_{l\sigma} \right\rangle = (n_{l\sigma} - 1) \left\langle a_{l\sigma} \mid n_{l\sigma} \right\rangle \tag{11A.16}$$

$$N_{l\sigma} a_{l\sigma}^{\dagger} \left| n_{l\sigma} \right\rangle = (n_{l\sigma} + 1) \left\langle a_{l\sigma}^{\dagger} \mid n_{l\sigma} \right\rangle.$$

The orthonormal states are

$$|n_{l\sigma}\rangle = \frac{a_{l\sigma}^{\dagger n}}{\sqrt{n!}}\,|0\rangle \tag{11A.17}$$

$$\left\langle n'_{l\sigma} \mid n_{l\sigma}\right\rangle = \delta_{n_{l\sigma}n'_{l\sigma}}$$

with

$$\sum_{n=0}^{\infty} |n_{l\sigma}\rangle\,\langle n_{l\sigma}| = I.$$

With this we have the many-particle harmonic oscilator Hamiltonian:

$$H = \frac{1}{2}\sum_{l\sigma}\left(p_{l\sigma}^2 + \omega_l^2 q_{l\sigma}^2\right) = \frac{1}{2}\sum_{l\sigma}\hbar\omega_l\left(a_{l\sigma}^{\dagger}a_{l\sigma} + a_{l\sigma}a_{l\sigma}^{\dagger}\right). \tag{11A.18}$$

The independent-mode pure field states are

$$|n_1\rangle\,|n_2\rangle\ldots|n_{l\sigma}\rangle\ldots \equiv |n_1 n_2\ldots n_{l\sigma}\ldots\rangle \tag{11A.19}$$

and obey the relations of Eq. (11A.15) and Eq. (11A.16). Thus,

$$\langle n_1\ldots n_{l\sigma}|\,H\,|n_1\ldots n_{l\sigma}\rangle = \sum_{l\sigma}\hbar\omega_l n_{l\sigma} + \frac{\hbar\omega_l}{2} \tag{11A.20}$$

$$n_{l\sigma} = 0, 1, \ldots$$

The ground state energy has been shifted to zero. These multimode states are bosons, and they obey boson commutation laws. See Schweber (1962) for a detailed discussion of this.

These modes are often thought of as "particles," i.e. photons. However, they cannot be localized, as seen from Eq. (11A.7), where both positive and negative frequencies appear. Some discussion of this is made in the book of Scully and Zubairy (1997). Here "photon" will simply mean the mode of the field as described above.

From the commutation laws for $a_{l\sigma}, a_{l\sigma}^{\dagger}$, we may obtain the field commutation laws for **E** and **H**. This is left as a problem. In the continuum limit,

$$\sum_{\mathbf{k}} \rightarrow 2\left(\frac{L}{2\pi}\right)^3 \int d_3k. \tag{11A.21}$$

They are for equal time

$$\left[E_i\left(\mathbf{r},t\right), H_j\left(\mathbf{r}',t\right)\right] = 0 \tag{11A.22}$$

$$\left[E_j\left(\mathbf{r},t\right), H_k\left(\mathbf{r}',t\right)\right] = -ic^2\frac{\partial}{\partial r_l}\delta^{(3)}\left(\mathbf{r}-\mathbf{r}'\right),$$

where i, j, l are 1, 2, 3 and cyclic. Parallel components of **E**, **H** may be measured simultaneously, and perpendicular components may not. In going to the continuum limit, we may introduce the density of modes in $d\omega$, $g(\omega) = \frac{2\omega^2}{(2\pi c)^3}$, since $\omega^2 = c^2k^2$. The vacuum state $n_{l\sigma} = 0$ is important:

$$\langle 0| H |0\rangle = \sum_l \frac{h\omega_l}{2} l. \tag{11A.23}$$

The average field in any state is

$$\langle n_1 \ldots n_{l\sigma} | \mathbf{E} \left| n_{1 \ldots n_{l\sigma}} \right\rangle = 0,$$

since **E** is linear in $a_{l\sigma}$, and $a_{l\sigma}^\dagger$. However, the field fluctuation is nonzero, and for $n_{l\sigma} = 0$ it is given by $\langle 0| \mathbf{E}^2 |0\rangle = \sum_{l\sigma} \frac{\hbar\omega_l}{2}$. The zero point field fluctuations give rise to possibly spontaneous emission and the famous Lamb shift of the $2P_{\frac{1}{2}} - 2S_{\frac{1}{2}}$ energy levels of hydrogen (Lamb and Retherford, 1947). See Schweber (1962) and the book of Scully and Zubairy (1997) for a good, brief introduction, and also see Section 11.2 of this chapter.

The interaction Hamiltonian of the field interacting with a charge e is an addition to the radiation field previously discussed:

$$H'_{fa} = \frac{1}{2m} \left[\mathbf{p} - e\mathbf{A}(\mathbf{r}, t)\right]^2 + cV(\mathbf{r}), \qquad c = 1. \tag{11A.24}$$

This is nonrelativistic, and electron spin is not included. The total Hamiltonian may be written in the Coulomb gauge as ($\mathbf{p} \cdot \mathbf{A} = \mathbf{A} \cdot \mathbf{p}$):

$$H = H_a + H_r + H',$$

where

$$H_a = \frac{\mathbf{p}^2}{2m} + eV(\mathbf{r}) \tag{11A.25}$$

$$H_r = \sum_{l\sigma} \hbar\omega_l \left(a_{l\sigma}^\dagger a_{l\sigma} + \frac{1}{2}\right) \tag{11A.26}$$

$$H' = -\mathbf{A} \cdot \mathbf{p} + \frac{e^2}{2m} \mathbf{A}^2. \tag{11A.27}$$

The second term in H' will be neglected as normally small. In some places the interaction is written

$$H'' = -e\mathbf{r} \cdot \mathbf{E}(\mathbf{r}_0, t). \tag{11A.28}$$

They are not equivalent, as discussed nicely by Scully and Zubairy (1997). It may be estimated that

$$\left| \frac{\langle f \left| H' \right| i \rangle}{\langle f \left| H'' \right| i \rangle} \right| = \frac{\omega}{\nu}.$$

$\omega = E_f - E_i$, the transition frequency, and ν is the field frequency. In our discussions in this book, we adopt Eq. (11A.28) for simplicity.

References

Agarwal, G. S. (1973). Master equation methods in quantum optics. In *Progress in Optics* **11,** ed. E. Wolf (Amsterdam, North Holland).
Agarwal, G. S. (1974). Quantum statistical theories of spontaneous emission and their relation to other approaches. In *Springer Tracts in Modern Physics* **70** (New York, Springer).
Becker, R. (1964 edn.) *Electromagnetic Fields and Interactions* (New York, Dover).
Briegel, H. J., Englert, B. C., Sterpi, N. and Walther, H. (1994). *Phys. Rev. A* **49,** 2962.
Dirac, P. A. M. (1958). *Principles of Mishionaries* (Oxford, Clarendon Press).
Einstein, A. (1917). *Phys. Z.* **18**, 121.
Filipowicz, P., Javanainen, J. and Meystre, P. (1986). *Phys. Rev. A* **34**, 3077.
Herzog, U. (2000). Phys. Rev. A61, 47801.
Jaynes, E. T. and Cummings, F. W. (1963). *Proc. IEEE* **51**, 89.
Johnson, D. and Schieve, W. C. (2001). *Phys. Rev. A* **63**, 1050.
Johnson and Schieve (2004) unpublished.
Krause, J., Scully, M. O. and Walther, H. (1986). *Phys. Rev. A* **34**, 2032.
Lamb, W. E. Jr. and Retherford, R. C. (1947). *Phys. Rev.* **72**, 241.
Louisell, W. H. (1973). *Quantum Statistical Properties of Radiation* (New York, Wiley).
Lu, N. (1993). *Opt. Comm.* **103**, 315.
Lugiato, L. A., Scully, M. O. and Walther, H. (1987). *Phys. Rev. A* **36**, 740
Mandel, L. and Wolf, E. (1995). *Optical Coherence and Quantum Optics* (New York, Cambridge University Press).
McGowan, R. R. and Schieve, W. C. (1997). *Phys. Rev. A* **56**, 56.
Milloni, P. W. (1976). *Phys. Rep.* **25** no. 1, 1.
Monroe, W. J. and Gardiner, C. W. (1996). *Phys. Rev. A* **53**, 2633.
Nussenzweig, H. M. (1973). *Introduction to Quantum Optics* (London, Gordon and Breach).
Panovsky, W. K. H. and Phillips, M. (1962). *Classical Electricity and Magnetism* (Reading, Addison-Wesley).
Peier, W. (1972). *Physica* **57,** 565 and **58**, 229.
Raithal, G., Wagner, C., Walther, H., Narducci, L. M. and Scully, M. O. (1994). *Adv. At. Mol. Opt. Phys.* (suppl. 2), **57**.
Rempe, G., Walther, H. and Klein, H. (1987). *Phys. Rev. Lett.* **58**, 353.
Schleich, W. (2001). *Quantum Optics in Phase Space* (Berlin, Wiley–VCH).
Schweber, S. S. (1962). *An Introduction to Relativistic Quantum Field Theory* (New York, Harper and Row).
Scully, M. O. and Lamb, W. E. Jr. (1967). *Phys. Rev.* **159**, 208.
Scully, M. O. and Lamb, W. E. Jr. (1969). *Phys. Rev.* **179**, 368.

Scully, M. O. and Zubairy, M. S. (1997). *Quantum Optics* (New York, Cambridge, Cambridge University Press).
Senitzky, I. R. (1973). *Phys. Rev. Lett.* **31**, 155.
Walls, D. F. and Milborn, G. J. (1985). *Phys. Rev. A* **31**, 2403.
Weidinger, M., Varcoc, B. T.H., Heerlein, R. and Walther, H. (1989). *Phys. Rev. Lett.* **63**, 2001.
Weisskopf, V. and Wigner, E. (1930). *Z. Phys.* **63**, 54.
Welton, T. A. (1948). *Phys. Rev. Lett.* **31**, 155.
Zurek, W. (1991). *Physics Today* **44**, 36.

12
Entanglements

12.1 Introduction

We will now turn to quantum entanglements and their contemporary, possibly practical, interest. Entanglements, first discussed in the E.P.R. paradox (Einstein, Podolsky, Rosen, 1935) are a perplexing nonlocal feature of quantum mechanics. This was immediately and succinctly discussed by Schrödinger (1935). The long history of this apparent paradox is outlined in the wonderful book of Jammer (1974). We will not focus on the central issue of "hidden variables" and their resolution by the Bell inequalities (Bell, 1964) and the test, nor the failure, of these in experiment (see Fry, 1998).

The distinctive quantum nature of entanglements has led to two quantum effects which we will discuss: quantum information teleportation (Zeilenger, 1998) and quantum computation by means of entangled states. A nice, recent, elementary introduction to the latter is in the Los Alamos reports of James and Kwiat (2002). The quantum correlations or entanglements are sensitive to environmental destruction. This was pointed out early by Zurek, who termed this "decoherence." The loss of coherence may occur on a short time scale. Recent discussions have been given by Zurek (2002, 2003). We have already given a theoretical example early in Chapter 2. It is pertinent to discuss this here, since it is a property of open-system quantum master equations that is a central part of our study in this book. In a sense, the decoherence of correlations has turned out to be a "practical" application of these theoretical notions. Are there remedies for unwanted decoherence? Quantum error correction is a possibility. This will be mentioned also.

12.2 Entanglements: foundations

Following the reading of the E.P.R. paper, Schrödinger (1935b) quickly repeated the argument from quantum theory, but more generally. He introduced the term "entanglement" to describe what he seemed to agree was a curious, if not unacceptable, property.

Let us follow his point of view. Given a state $\psi(x, y)$ of a composite system of particles x and y formed in their mutual interaction. $\psi(x, y)$ is not a product state. At the asymptotic separation, a complete measurement of "y" determines a set of normalized eigenstates, $f_n(y)$. The variables measured commute, and by the rules of quantum mechanics, we may write for the entangled state

$$\psi(x, y) = \sum_n c_n g_n(x) f_n(y), \tag{12.1}$$

and

$$c_k g_k(x) = \int f_k^*(y) \psi(x, y) \, dy. \tag{12.2}$$

The latter determine $g_k(x)$. The $c_k(x)$ are introduced for normalization. $|c_k|^2$ is the probability of "x" being in state $g_k(x)$, given that we *measured* $f_k(y)$. This is the point right here. A measurement of the recently separated system in $f_k(y)$ determines $g_k(x)$. The nonlocality is now apparent. The entangled state $\psi(x, y)$ produces this nonclassical (bizarre?) behavior *after* the interaction. Schrödinger proves further the conditions on $f_k(y)$ for which there is a unique orthogonal expansion of $\psi(x, y)$. It is that

$$\int dx \psi^*(x, y') \psi(x, y) < \infty \tag{12.3}$$

(Courant and Hilbert, 1966). Then we have the condition

$$|c_k|^2 f_k(y) = \int dx \int dy' f_k(y') \psi^*(x, y') \psi(x, y). \tag{12.4}$$

This is an integral equation for what Schrödinger called the relevant eigenfunctions $f_k(y)$. Knowing these, one knows *all* the $g_k(x)$. This is a program of measurement of these $f_k(y)$, with probability $|c_k|^2$, that determines the other "systems" $g_k(x)$.

Einstein, Podolsky and Rosen, in their paper (Einstein *et al.*, 1935), had considered the special model of two-particle scattering. After the interaction at a time $t > T$, they wrote the entanglement as

$$\psi(x_1 x_2) = \int dp \exp\left[\frac{-2\pi i (x_2 - x_0) p}{h}\right] \exp\left(\frac{2\pi i x_1 p}{h}\right), \tag{12.5}$$

the term $\exp\left(\frac{2\pi i x_1 p}{h}\right)$ being the eigenfunction $u_p(x_1)$ corresponding to eigenvalue p of particle x_1. The other term is an eigenfunction $\psi_p(x_2)$ corresponding to $-p$.

A measurement on x_1 of p determines the momentum state of x_2 $(-p)$ after they have separated. It is also possible to rewrite Eq. (12.5) as

$$\psi(x_1 x_2) = h \int \delta(x - x_2 + x_0)\, \delta(x_1 - x)\, dx, \tag{12.6}$$

the position eigenfunctions. We recognize $u_x(x_1) = \delta(x_1 - x)$ and $\psi_x(x_2) = \delta(x - x_2 + x_0)$ such that $x_2 = x + x_0$. This is consistent, since $x_1 - x_2$ and $p_1 + p_2$ commute. The choice of these commuting observables determines the proper $f_k(y)$ in the Schrödinger discussion of entanglements.

The answer to the dilemma may be that the two particles are one system having, even after interaction, the entangled $\psi(xy)$. They cannot be conceptually disentangled. To think of them apart is a fallacy. This point of view was emphasized by Bohr (Bohr, 1949; Einstein, 1949). Utilizing a micromaser cavity similar to that discussed in Chapter 11, Haroche (1998) has, by means of the Rabi oscillations, created entangled atom-field states:

$$|\psi\rangle = \frac{1}{\sqrt{2}} \left(|e, \alpha \exp i\phi\rangle + |g, \alpha \exp -i\phi\rangle \right), \tag{12.7}$$

e, g being the two-level atom state and α the coherent state of the cavity $n = |\alpha|^2$ photons (one to ten).

If we ignore the e, g for simplicity, these are "cat states" (Schrödinger, 1935a). The cat is built from entangling *macroscopic* nonorthogonal coherent states. Using the position representation of a coherent state $|\beta\rangle$, we have

$$\langle x \mid \beta\rangle = \pi^{-\frac{1}{4}} \exp \frac{1}{2}\left(\beta^2 - |\beta|^2\right) \exp\left[\frac{-1}{2}\left(x - \sqrt{2}\beta\right)^2\right].$$

The entangled "cat state" is then

$$\psi(x) = \pi^{-\frac{1}{4}} \frac{N}{\sqrt{2}} \exp\left[-\alpha^2 \sin^2\phi\right] \exp\left[\frac{1}{2}\alpha^2 \sin 2\phi\right]$$
$$\left\{ \begin{array}{c} \exp\left[\frac{-1}{2}\left(x - \sqrt{2}\alpha \exp(i\phi)\right)^2\right] \\ + \exp\left[-i\alpha^2 \sin(2\phi)\right] \exp\left[-\frac{1}{2}\left(x - \sqrt{2}\alpha \exp(-i\phi)\right)^2\right] \end{array} \right\},$$

where ϕ is the macroscopic angle in the $p - x$ plane between the symmetry center of $\langle x \mid \alpha \exp i\phi\rangle$ and the x-axis. Being coherent states, they are approximately macroscopic and distinguishable—in the cat paradox, the dead *and* alive cat. The entangled state is neither.

12.3 Entanglements: Q bits

The state representation of modern quantum computation is the Q bit (quantum bit), which is the two-level spin $\frac{1}{2}$ representation of the two-level atom model already discussed in detail in Chapter 2. We assume the reader is very familiar with this. Now we will consider the entanglement of two such states, of atom (spin) A and B. Consider first the direct product

$$|\psi\rangle_{AB} = |+1\rangle_A \otimes |+1\rangle_B \equiv |1, 1\rangle ,$$

whose density matrix operator

$$|+1\rangle_A \, |+1\rangle_B \, \langle +1|_A \, \langle +1|_B = \rho_{AB}$$

is pure $\rho_{AB}^2 = \rho_{AB}$.

We might be interested in a complete set of maximally entangled states where the Q bit states are $|1\rangle$ or $|0\rangle$ and the identity A, B is in the order

$$\begin{aligned} \left|\phi^{\pm}\right\rangle_{AB} &= \frac{1}{\sqrt{2}} \left(|00\rangle_{AB} \pm |11\rangle_{AB}\right) \\ \left|\psi^{\pm}\right\rangle_{AB} &= \frac{1}{\sqrt{2}} \left(|01\rangle_{AB} \pm |10\rangle_{AB}\right) . \end{aligned} \tag{12.8}$$

They are *not* factorable, but normalized.

The first pair determines the parity, and the second the phase of the entanglement. These are called Bell states (a compliment!). Yes–no information, $|+1\rangle$, $|0\rangle$, is carried in these states, but now it is hidden in the entanglement. We might operate on Q bit A with Pauli operator $\sigma_x \equiv \sigma_1$. This is a 90° y-axis rotation and causes the transformation of the Bell basis to

$$\begin{aligned} \left|\phi^{+}\right\rangle_{AB} &\to \left|\psi^{+}\right\rangle_{AB} \\ \left|\phi^{-}\right\rangle_{AB} &\to -\left|\psi^{-}\right\rangle_{AB} . \end{aligned} \tag{12.9}$$

A product of the unitary Hadamard transformation on a single Q bit is

$$H = \frac{1}{\sqrt{2}} (\sigma_1 + \sigma_3) \to \frac{1}{\sqrt{2}} \begin{bmatrix} 1 & 1 \\ 1 & -1 \end{bmatrix}, \tag{12.10}$$

and what is termed C_{not}, operating on two Q bits, is

$$C_{\text{not}} = \begin{pmatrix} 1 & 0 & 0 & 0 \\ 0 & 1 & 0 & 0 \\ 0 & 0 & 0 & 1 \\ 0 & 0 & 1 & 0 \end{pmatrix} \text{ where } \begin{matrix} |00\rangle \to & |00\rangle \\ |01\rangle \to & |01\rangle \\ |10\rangle \to & |11\rangle \\ |11\rangle \to & |12\rangle \\ |12\rangle \to & |10\rangle \end{matrix} . \tag{12.11}$$

These create Bell states from $|00\rangle_{AB}$, $|01\rangle_{AB}$, etc. We then have

$$|00\rangle_{AB} \stackrel{H}{\to} \frac{1}{\sqrt{2}}(|0\rangle_A + |1\rangle_A)\,|0\rangle_B \stackrel{\text{Cnot}}{\to} \left|\phi^+\right\rangle_{AB} \tag{12.12}$$
$$|01\rangle_{AB} \stackrel{H}{\to} \frac{1}{\sqrt{2}}(|0\rangle_A + |1\rangle_A)\,|1\rangle_B \stackrel{\text{Cnot}}{\to} \left|\psi^+\right\rangle_{AB}$$
$$|10\rangle_{AB} \stackrel{H}{\to} \frac{1}{\sqrt{2}}(|0\rangle_A - |1\rangle_A)\,|0\rangle_B \stackrel{\text{Cnot}}{\to} \left|\phi^-\right\rangle_{AB}$$
$$|11\rangle_{AB} \stackrel{H}{\to} \frac{1}{\sqrt{2}}(|0\rangle_A - |1\rangle_A)\,|1\rangle_B \stackrel{\text{Cnot}}{\to} \left|\psi^-\right\rangle_{AB}.$$

Being a unitary transformation, the inverse transformation reduces the Bell states to the factored ones. These unitary operations are commonly called gates and given diagrammatic circuit representations, which we shall not do. The product of these transformations is nonlocal. The H creates the two–Q bit entanglement.

As an illustration, consider the $\left|\phi^+\right\rangle_{AB}$ state. A measurement of $|0\rangle_A$ gives probability $\frac{1}{2}$. But now the B partner is in state $|0\rangle_B$. Similarly, a measurement of A in $|1\rangle_A$ implies B is in $|1\rangle_B$. The entanglement is apparent and destroyed by the measurement. The measurement of A and B separately exhibits 100% correlation between the results. Further, for this state, let us form

$$\rho_A = \text{Tr}_B \left|\phi^+\right\rangle_{AB}\ {}_{AB}\left\langle\phi^+\right| = \frac{1}{2} I_A$$

and also

$$\rho_B = \frac{1}{2} I_B.$$

In the measurement of A spin along *any* axis at all, we obtain probability $\frac{1}{2}$ for $|0\rangle_A$ state and $|1\rangle_A$ also. The spin is randomly oriented. To get more information we must use, not surprisingly, the other members of the Bell basis.

There are tests for the measured degree of entanglement (Kraus and Cirac, 2001). For the simple case of a pure state,

$$|\psi\rangle = \alpha\,|00\rangle + \beta\,|01\rangle + \gamma\,|10\rangle + \delta\,|11\rangle,$$

the quantity concurrence $c = 2\,|\alpha\delta - \beta\gamma|$ is a measure of entanglement. If and only if c is zero is the state separable. Maximal entanglement is $c = 1$. For the Bell states, $\alpha = \delta = \frac{1}{\sqrt{2}}$, and $\beta = \gamma$. Thus the Bell states are entangled maximally.

Another measure of entanglement is the Schmidt number E_s, which is the number of nonzero coefficients minus 1 in the bi-orthogonal expansion (Schmidt, 1907):

$$|\psi\rangle_{AB} = \sum_k c_k \left|\phi_k\right\rangle_A \left|\psi_k\right\rangle_B,$$

discussed by Schrödinger in his first paper. We shall discuss the properties of this in the appendix to this chapter.

Since the Bell states represent maximum entanglement, it is important theoretically (and possibly experimentally) to consider the observation after entanglement of these states, i.e. Bell state analysis. We will follow the discussion of Zeilinger (1998). The first thing to observe is that only $\left|\psi^{-}\right\rangle_{AB}$ is antisymmetric under interchange, whereas $\left|\psi^{+}\right\rangle_{AB}$, $\left|\phi^{+}\right\rangle_{AB}$ and $\left|\phi^{-}\right\rangle_{AB}$ are symmetric. We must also consider the spacial degrees of freedom $|x_A, x_B\rangle$, which can also be symmetric $|x_A, x_B\rangle_s$ or antisymmetric $|x_A x_B\rangle_a$. For a known two-boson case (two photons), the total wave function is then

$$\begin{aligned}
&\left|\psi^{-}\right\rangle_{AB} |x_A x_B\rangle_a \\
&\left|\psi^{+}\right\rangle_{AB} |x_A x_B\rangle_s \\
&\left|\phi^{+}\right\rangle_{AB} |x_A x_B\rangle_s \\
&\left|\phi^{-}\right\rangle_{AB} |x_A x_B\rangle_s \,.
\end{aligned}$$

Only in scattering do we observe an antisymmetric spacial state. We then identify the internal state as $\left|\psi^{-}\right\rangle_{AB}$. To distinguish $\left|\psi^{+}\right\rangle_{AB}$, $\left|\phi^{+}\right\rangle_{AB}$ and $\left|\phi^{-}\right\rangle_{AB}$, we must distinguish the internal states. In $\left|\psi^{+}\right\rangle_{AB}$, if the two Q bits have differing polarization, then $\left|\phi^{+}\right\rangle_{AB}$, $\left|\phi^{-}\right\rangle_{AB}$ have the same polarization. If we measure σ_3^A (or σ_3^B), does the other state then have the same spin direction? If it does not, we are finished, but if it does, we must distinguish $\left|\phi^{+}\right\rangle_{AB}$ from $\left|\phi^{-}\right\rangle_{AB}$. Now, as discussed, if we find on repeated measurement that $\rho_A = \frac{1}{2} I_A$ and $\rho_B = \frac{1}{2} I_B$, then we have $\left|\phi^{+}\right\rangle_{AB}$. The other possibility would be $\rho_A = \rho_B = 0$. (We will not do the Fermi case but leave it as a problem).

All this does not imply that such a scheme may be carried out experimentally. (However, see Boumeester and Zeilinger, 2000).

12.4 Entanglement consequences: quantum teleportation, the Bob and Alice story

The most remarkable effect of quantum entanglements is quantum teleportation, first suggested by Bennett and Wiesner (Bennett and Wiesner, 1992; Bennett, 1998). Quantum information may be sent with entangled states. As we have emphasized, the Bell entanglements hide the fundamental bits of which they are made. It is not possible to "eavesdrop" on messages in entangled pairs. Teleportation has recently been observed experimentally by Boumeester and Zeilinger.

There are three actors in an entanglement play: "Charlie," "Alice" and "Bob." We will speak in terms of entangled photons, since this is the first experimental system. "Alice" wants to transmit to "Bob" an arbitrary pure state, obtained from "Charlie," $|\psi\rangle_C = \alpha\,|1\rangle_C + \beta\,|0\rangle_C$. "Alice" has entangled states, as does Bob, having acquired them earlier. They have, for instance, together $\frac{1}{\sqrt{2}}\left(|{+1},\rangle_A\,|{+1},\rangle_B + |0\rangle_A\,|0\rangle_B\right) = \left|\phi^+\right\rangle_{AB}$. "Alice" performs a Bell analysis on the combined state $|\psi\rangle_C\left|\phi^+\right\rangle_{AB}$ and projects this onto the Bell basis $\left|\phi^\pm\right\rangle_{CA}, \left|\psi^\pm\right\rangle_{CA}$. *Two bits* of *classical information* result in the form of a *local* unitary transformation U_{ij} which is sent by "Alice" to "Bob." It is, for instance,

$$U_{00} = \begin{pmatrix} 1 & 0 \\ 0 & 1 \end{pmatrix} U_{01} = \begin{pmatrix} 1 & 0 \\ 0 & -1 \end{pmatrix}$$

$$U_{10} = \begin{pmatrix} 0 & -1 \\ -1 & 0 \end{pmatrix} U_{11} \begin{pmatrix} 0 & -1 \\ 1 & 0 \end{pmatrix}.$$

"Bob" then looks at his photon and finds by an inverse transformation on his bit α, β and $|\psi\rangle_C$. Thus a quantum state is teleported by two bits of classical information. To understand this, consider the Bell basis projection by "Alice" (see J. Preskill's clear discussion in *Lecture Notes for Physics 229*, California Institute of Technology, unpublished 1998; and also Jozsa [1998]):

$$\begin{aligned}
|\psi\rangle_C\left|\phi^+\right\rangle_{AB} &= \left(\alpha\,|0\rangle_C + \beta\,|1\rangle_C\right)\frac{1}{\sqrt{2}}\left(|00\rangle_{AB} + |11\rangle_{AB}\right) \\
&= \frac{1}{\sqrt{2}}\left(\alpha\,|000\rangle_{CAB} + \alpha\,|011\rangle_{CAB} + \beta\,|100\rangle_{CAB} + \beta\,|111\rangle_{CAB}\right),
\end{aligned}$$

which, upon using Eq. (12.8),

$$\begin{aligned}
&= \frac{1}{2}\alpha\left(|\phi+\rangle_{CA} + \left|\phi^-\right\rangle_{CA}\right)|0\rangle_B \\
&\quad + \frac{1}{2}\alpha\left(\left|\psi^+\right\rangle_{CA} + \left|\psi^-\right\rangle_{CA}\right)|1\rangle_B \\
&\quad + \frac{1}{2}\beta\left(\left|\psi^+\right\rangle_{CA} - \left|\psi^-\right\rangle_{CA}\right)|0\rangle_B \\
&\quad + \frac{1}{2}\beta\left(\left|\phi^+\right\rangle_{CA} - \left|\phi^-\right\rangle_{CA}\right)|1\rangle_B .
\end{aligned}$$

Collecting these, we have the Bell state representation:

$$\begin{aligned}
|\psi\rangle_C\left|\phi^+\right\rangle_{AB} &= \frac{1}{2}\left|\phi^+\right\rangle_{CA}\left(\alpha\,|0\rangle_B + \beta\,|1\rangle_B\right) \\
&\quad + \frac{1}{2}\left|\psi^+\right\rangle_{CA}\left(\alpha\,|1\rangle_B + \beta\,|0\rangle_B\right) \\
&\quad + \frac{1}{2}\left|\psi^-\right\rangle_{CA}\left(\alpha\,|1\rangle_B - \beta\,|0\rangle_B\right) \\
&\quad + \frac{1}{2}\left|\phi^-\right\rangle_{CA}\left(\alpha\,|0\rangle_B - \beta\,|1\rangle_B\right).
\end{aligned}$$

By the Bell state analysis on CA Q bits of "Alice," "Bob" may obtain one of these results with equal likelihood, and thus knowledge of α, β and thus $|\psi\rangle_C$. At this point no photon has been transmitted to "Bob." "Alice's" Bell state analysis of $|\psi\rangle_C \left|\phi^+\right\rangle_{AB}$ has caused "Bob" to become aware of $|\psi\rangle_C$ at the time of the wave function collapse in "Alice's" Bell state measurement. The classical information that is then sent by "Alice" to "Bob" is which of the Bell states, $\left|\phi^\pm\right\rangle_{CA}$ or $\left|\psi^\pm\right\rangle_{CA}$, was found expressed as a unitary U_{ij} transformation from $\left|\phi^+\right\rangle_{AB}$. We realize that in the measurement by "Alice," $|\psi\rangle_C \left|\phi^+\right\rangle_{AB}$ has been destroyed.

This is very remarkable. All that is needed is an arbitrary entangled state between "Alice" and "Bob" created at any time in the past. Then, by a Bell state analysis of an arbitrary state product with one of these, "Alice" (or "Bob") may, by means of a classical message, transmit this state precisely and instantly and over any distance to "Bob." Efforts at teleportation are reviewed by Boumeester and Zeilinger (2000). Why not simply transmit the original photon $|\psi\rangle_C$? Eavesdropping is more difficult, since knowledge of the classical message does not give another party the entanglement. In addition, the quality of the message is perfect, in principle, if the classical information is not garbled. Dense coding of the state $|\psi\rangle_C$ into classical information would, at best, give $\left|_B \langle\phi \mid \psi\rangle_A\right|^2 = \frac{2}{3}$.

12.5 Entanglement consequences: dense coding

Consider again the Bell states. As we have seen in Section 12.4, in order to switch from any one of the Bell states to any of the four, one must only manipulate one Q bit. For instance, if one begins with $\left|\psi^+\right\rangle$, then the operation of a phase shift of π *on* $\left|\psi^+\right\rangle_{AB}$, i.e. H, gives $\left|\psi^-\right\rangle$. We may also obtain $\left|\phi^+\right\rangle_{AB}$ and $\left|\phi^-\right\rangle_{AB}$ by unitary operations on $\left|\psi^+\right\rangle_{AB}$. Of course, the identity operator gives back $\left|\psi^+\right\rangle_{AB}$.

This classical coding of one bit gives any other desired Q bit of the four. This is more efficient than coding the two classical bits of quantum information $|0\rangle\,|1\rangle$ and so forth. Of course, B must have a Bell state analyzer to read this. It must be noticed that, in the past, "Alice" and "Bob" had built $\left|\psi^+\right\rangle_{AB}$. In an experimental realization of this with photons, it was possible to code $\log_2 3 = 58$ bits (Mattle *et al.*, 1996).

12.6 Entanglement consequences: quantum computation

Here we will discuss a simple algorithm showing that a quantum algorithm for computation is possible. This is due to Deutsch (1985). His was the first response to the call for such an algorithm by Feynman (1959). We will not outline the more difficult and useful factoring algorithm of Shor (1994), which is at the center of the focus to actually construct a quantum computer. This development is not the subject of our present discussion. An introduction to the Shor algorithm is given by Ekert (1998). An overview of the effort to produce the computer is in the Los Alamos scientific report already referred to and also in the article by Deutsch and Ekert (2000). The subject is proceeding so rapidly that any review is quickly out of date.

The key quantum elements which are potentially advantageous over a purely classical one are nicely outlined by Jozsa (2002). The classical computation is based on bits (yes, no) and the computation of functions. Quantum computation would transform vectors in a Hilbert space (Q bits in the present form) by means of unitary transformations. There are subtle advantages to the quantum calculations. The first advantage was termed quantum parallelism by Deutsch. We need not input only a single state $|a\rangle$ into the quantum computer U_f,

$$U_f |a\rangle \rightarrow |f(a)\rangle . \tag{12.13}$$

We might, by the linearity of quantum mechanics (superposition), input $\sum_{a\varepsilon A} |a\rangle$ so that

$$U_f \left| \sum_{a\varepsilon A} |a\rangle \right\rangle \rightarrow \sum_{a\varepsilon A} |a\rangle |f(a)\rangle . \tag{12.14}$$

In one operation a quantum unitary transformation has performed a parallel computation. Classical linearity is also possible, but quantum mechanics is more subtle. Eq. (12.14) may contain nonclassical entanglements. An example is the Hadamard gate mentioned in Eq. (12.10), operating on $|0\rangle$, $|1\rangle$ where

$$\begin{aligned} H|0\rangle &= \frac{1}{\sqrt{2}}(|0\rangle + |1\rangle) \\ H|1\rangle &= \frac{1}{\sqrt{2}}(|0\rangle - |1\rangle) . \end{aligned} \tag{12.15}$$

If we operate on a vector of n Q bits, $|0\rangle \ldots |0\rangle$, we obtain

$$\frac{1}{2^{\frac{n}{2}}}(|0\rangle + |1\rangle) \ldots (|0\rangle + |1\rangle) \tag{12.16}$$

and take a single output state $|0\rangle$, obtaining $|f\rangle$ with U_f operation:

$$|f\rangle = \frac{1}{2^{\frac{n}{2}}} \sum_{x\varepsilon A^n} |x\rangle |f(x)\rangle . \tag{12.17}$$

Our enthusiasm for the advantages of quantum calculation should be cautious, since the quantum theory of measurement (which we will discuss in the next chapter) does not allow us to know $|x\rangle$ and $f(x)$ in the entangled state. As emphasized by Jozsa, the quantum information is hidden and inaccessible. Certain global properties may be obtained, and this is the art of obtaining quantum algorithms, of which the Shor algorithm is a prime example.

Let us illustrate these comments by considering Deutsch's algorithm. Consider the space $|0\rangle, |1\rangle$ (B) and the map $f: B \to B$. The possible one-bit functions are

$$\begin{array}{cc} f(0)=0 & f(0)=1 \\ & \to \\ f(1)=0 & f(1)=1 \end{array} \tag{12.18}$$

and

$$\begin{array}{cc} f(0)=0 & f(0)=1 \\ & \to \\ f(1)=1 & f(1)=0. \end{array}$$

The second group has the "balanced" property, 0, 1, which appear both in input and output. The global object of the quantum calculation will be to determine, in one operation, whether the result is balanced or not.

We are given an "oracle," U_f, which is an unknown and inaccessible subroutine which computes one of Eq. (12.18), producing an output. It transforms

$$U_f |x\rangle |y\rangle \to |x\rangle |y \times f(x)\rangle \tag{12.19}$$

$$(\times \text{ means addition modulus 2}).$$

Now we start addition with input $|0\rangle$ and output $|0\rangle$. We apply the C_{not} operation to the output, and then H to both input and output. Recalling $C_{\text{not}} |0\rangle = |1\rangle$, we have the resulting input to U_f:

$$|0\rangle |1\rangle \to \left(\frac{|0\rangle + |1\rangle}{\sqrt{2}}\right)\left(\frac{|0\rangle - |1\rangle}{\sqrt{2}}\right) \equiv |\psi\rangle .$$

The result of H is to form *entangled* input. Now the oracle performs its function:

$$U_f |x\rangle \frac{(|0\rangle - |1\rangle)}{\sqrt{2}} \to (-1)^{f(x)} |x\rangle \frac{(|0\rangle - |1\rangle)}{\sqrt{2}}.$$

Thus we obtain

$$U_f |\psi\rangle = \pm \left[\frac{|0\rangle + |1\rangle}{\sqrt{2}}\right]\left[\frac{|0\rangle - |1\rangle}{\sqrt{2}}\right] \quad \text{for } f(0) = f(1)$$

$$U_f |\psi\rangle = \pm \left[\frac{|0\rangle - |1\rangle}{\sqrt{2}}\right]\left[\frac{|0\rangle - |1\rangle}{\sqrt{2}}\right] \quad \text{for } f(0) \neq f(1).$$

A subsequent Hadamard transformation to the input gives

$$HU_f\,|\psi\rangle = \pm\,|0\rangle\left[\frac{|0\rangle - |1\rangle}{\sqrt{2}}\right] \quad \text{for } f(0) = f(1) \tag{12.20}$$

$$HU_f\,|\psi\rangle = \pm\,|1\rangle\left[\frac{|0\rangle - |1\rangle}{\sqrt{2}}\right] \quad \text{for } f(0) \neq f(1) \quad \text{balanced.}$$

In this operation the entangled output is invariant. Now we ask one question: do we do *one* measurement to determine the two alternatives? The measurement of $(|0\rangle - |1\rangle)$ obtaining $\frac{|0\rangle - |1\rangle}{\sqrt{2}}$ would mean that it is balanced. The input is left in $\pm\,|1\rangle$ if f is balanced. This may be shown from $H \cdot H = I$. Thus we use the standard basis, not thc cntangled one, and look at the input to obtain the result. The analogous classical calculation would require two measurements. Thus there is a non-epsilon difference.

If we map instead $f\colon B^n \to B^1$, then the difference between a classical and quantum calculation becomes significant. Classically there are $0\,(2^n)$ questions to the oracle. In the quantum case, choosing $|0\rangle$ for n-dimensional input state superpositions, we choose the output state $\frac{1}{\sqrt{2}}\,(|0\rangle - |1\rangle)$, and we make one query to the oracle. The input state to U_f is the same as the previous example, except a product state is $\left[\frac{1}{\sqrt{2}}\,|0\rangle + |1\rangle\right]^n$. Transforming the output basis back to the standard basis, we have $\pm\,|0\rangle \ldots |0\rangle$ or $\pm\,|1\rangle \ldots |1\rangle$ for the constant or balanced result. The quantum algorithm requires $0\,(n)$ steps overall. This is the main result of the Deutsch algorithm. However, it has been shown that this difference disappears in the presence of noise in the quantum input (see Jozsa, 1998, 2002).

The quantum calculation in entangled Q bits has hidden information. It does not give us the elements of the oracle. For instance, in Eq. (12.20), it may tell us that we have the balanced case but not which two of the four. *Only proper global questions are possible in the quantum calculation.*

12.7 Decoherence: entanglement destruction

W. Zurek, in *Physics Today,* called attention to destruction of quantum correlations as the mechanism for the appearance of classical behavior (Zurek, 1991). See also Zurek (2003) for an extensive list of references. We have already discussed, in Chapters 2 and 5, the simple model introduced by Walls and Milburn (1985).

Recall, for an oscillator of the field, that

$$H = \hbar\omega a^+ a + a^+ a\Gamma,$$

the interaction with the environment being the second term representing phase damping. There is no energy damping, since $\left[a^+ a, H\right] = 0$. The number states

$|n\rangle$ are the so-called *pointer basis,* as termed by Zurek. In this eigenstate the environment interaction leaves this unchanged.

The reduced master equation for the system is exactly

$$\frac{\partial \rho}{\partial t} = \frac{\lambda}{2}\left(2a^{\dagger}a\rho a^{\dagger}a - \rho a^{\dagger}aa^{\dagger}a - a^{\dagger}aa^{\dagger}a\rho\right); \qquad t \geq 0. \tag{12.21}$$

It is completely positive, being of the Lindblad form as discussed in Chapter 5. The matrix elements, $\rho_{nn}(t)$, may be readily obtained. They are, again

$$\rho_{mn}(t) = \exp\left(-\lambda\left((n-m)^2\right)\frac{t}{2}\right)\rho_{mn}(0) \tag{12.22}$$

$$\dot{\rho}_{mm} = 0.$$

The correlations between the number basis decay as $\frac{t}{\tau_c}$ where $\tau_c = \frac{2}{\lambda}\frac{1}{(n-m)^2}$, which is more rapid than the decay of the diagonal elements which do not decay here at all. The quadratic dependence of the "distance" off the diagonal is rather characteristic.

Walls and Milburn also considered the damped harmonic oscillator after the results of Agarwal (1971). The Hamiltonian is

$$H = \hbar\omega a^{\dagger}a + \sum_j \hbar\omega_j a_j^{\dagger}aj \tag{12.23}$$

$$+ \hbar\sum_j\left[g_j a_j^{\dagger}\left(a + a^{\dagger}\right) + h.c.\right].$$

The Wigner function equation from the Born–Markov approximation master equation was discussed in previous chapters. For this oscillator, in interaction with a finite temperature environment, we have

$$\frac{\partial w(p,x)}{\partial t} = -\frac{\partial}{\partial x}\left[\frac{p}{m}w\right] + \frac{\partial}{\partial p}\left[\left(m\omega^2 x + 2kp\right)w\right] \tag{12.24}$$

$$+ 2m\hbar\omega k\left(\bar{n} + \frac{1}{2}\right)\frac{\partial^2 w(p,x)}{\partial p^2}.$$

Here $k = \pi f(\omega)\left|g(\omega)\right|^2$, $f(w)$ being the density of bath oscillators and $\bar{n}$ the non-interaction oscillator Planck distribution. This is the same result as that of Caldeira and Leggett at high temperature (Caldeira and Leggett, 1983) choosing the harmonic oscillator initially in a coherent state. Agarwal obtained the time-dependent solution to the Wigner function equation. The spacial entanglement is represented in the relation

$$w(xp,t) = \frac{1}{2\pi\hbar}\int \exp\left(\frac{ipy}{\hbar}\right)\left\langle x - \frac{1}{2}y\,|\rho|\,x + \frac{1}{2}y\right\rangle dy,$$

giving $\langle x - \frac{1}{2}y\,|\rho|\,x + \frac{1}{2}y\rangle$ by inverse transform. (See Chapter 4 for details.) It may be shown that the high-temperature bath destroys quantum correlations and at the final state $t \to \infty$ is

$$\left\langle x - \frac{1}{2}y\,|\rho|\,x + \frac{1}{2}y\right\rangle = N \exp\left[\frac{-x^2}{2\sigma_x^2}\right] \exp\left[\frac{-y^2}{2\sigma_y^2}\right], \tag{12.25}$$

where $\sigma_x^2 = \frac{kT}{m\omega^2}$ is a Gaussian mixture.

The time dependence of the entanglements (spacial correlations) is

$$\left\langle x - \frac{1}{2}y\,|\rho(t)|\,x + \frac{1}{2}y\right\rangle = N \exp\left[-\frac{[x - \langle x(t)\rangle]^2}{2\sigma_x^2(t)}\right] \exp\left[-\frac{[y - \langle y(t)\rangle]^2}{2\sigma_y^2(t)}\right], \tag{12.26}$$

where

$$\begin{aligned} \sigma_x^2 &= \frac{\hbar\bar{n}}{m\omega}(1 - \exp(-2kt)) + \frac{\hbar}{2m\omega} \\ \sigma_y^2 &= \frac{4\bar{n}m\omega}{\hbar}(1 - \exp(2kt)) + \frac{2m\omega}{\hbar}. \end{aligned} \tag{12.27}$$

Both the Gaussian spread of the coherent state (x dependence) and the spread of the coherence in y are seen here. For high temperature the off-diagonal correlations decay as $\frac{2kT}{\hbar\omega}(1 - \exp(-2kt))$. This is large for $\frac{kT}{\hbar\omega} >> 1$. The *width* of the diagonal spread in the coherent state also spreads by the same factor.

The difference between the off-diagonal time scale of change from that of the diagonal elements is the center of the decoherence time discussion. Zurek (2003) has argued, from examples and general considerations, that the master equation solution is of the form (at high temperature and $\hbar$ small)

$$\begin{aligned} \rho(xx', t) &= \rho(xx', 0) \exp\left(-\gamma t \frac{(x - x')^2}{\lambda_T}\right) \\ \rho(x, t) &= \rho(x, 0), \end{aligned} \tag{12.28}$$

where $\lambda_T = \frac{\hbar}{\sqrt{2mkT}}$ is the thermal de Broglie wavelength. The main point here is the loss of entanglement in a classical limit on a short time scale dependent on $(x - x')^2$, as we saw in the first model. Good estimates of these decay times are very model dependent. In the above formula, for $m = 1, T = 300K$ and $x - x' = 1cm$, the decoherence time is 10^{40} faster than γ^{-1}.

The fragile nature of entanglements, due to interaction with the environment, raises important questions concerning the use of these entanglements in quantum computation and other arenas. Let us take up this question for a bit (!) (Ekert *et al.*,

2002). These authors have modeled a Q bit interaction with the "environment" by means of a *reversible* unitary map $U(t)$ for the environment *plus* Q bit:

$$|0\rangle|E\rangle \overset{U(t)}{\rightarrow} |0\rangle|E_0(t)\rangle \qquad (12.29)$$
$$|1\rangle|E\rangle \overset{U(t)}{\rightarrow} |1\rangle|E_1(t)\rangle .$$

If the initial state is entangled, then

$$(a_0|0\rangle + a_1|1\rangle)\otimes|E\rangle \overset{U(t)}{\rightarrow} a_0|0\rangle|E_0(t)\rangle + a_1|1\rangle|E_1(t)\rangle . \qquad (12.30)$$

Decoherence is now viewed as a result of the environmental entanglement. The reduced Q bit density matrix is

$$\rho_Q(t) = \mathrm{Tr}_E \rho_{q+E} = \begin{vmatrix} |a_0|^2 & a_0 a_1^* \langle E_1 \mid E_0\rangle \\ a_1 a_0^* \langle E_0 \mid E_1\rangle & |a_1|^2 \end{vmatrix} . \qquad (12.31)$$

We really have no idea what the time dependence of $\langle E_1 \mid E_0\rangle$ is, but it is assumed to be exponential. Neither do we have a very good way of calculating this. Certain practical estimates have been made which are in the range of 10^4 to 10^{-12} s. However, the authors have raised a nice question concerning how this scales with the size of the computer. This we will now consider.

Now we model the bath as a system of harmonic oscillators in interaction with the two-level atom. The Hamiltonian is Eq. (12.23) with $\frac{1}{2\sigma_z\omega_0}$ replacing the first term and $a+a^\dagger$ replaced by σ_z. (See Chapter 2 in this connection.) Now $[\sigma_z, H] = 0$, and thus the two-level atom entangled state is a pointer state. Assuming the vacuum state of the bath to be coherent states $|\phi_k\rangle$, we may obtain, similar to the preceding discussion of the oscillator model,

$$|E_0\rangle = \Pi_k |-\phi_k\rangle \qquad (12.32)$$
$$|E_1\rangle = \Pi_k |\phi_k\rangle .$$

We assume short coherence length between the bath oscillators. Then for n Q bits,

$$|E_{k_1,k_2}\ldots\rangle = \Pi_{k_i=1}^n |E_{k_i}\rangle , \qquad (12.33)$$

and each Q bit decays independently exponentially, as in the one–Q bit model. We have

$$\rho_{111,\ldots,\,1;\,000,\ldots,\,0}(t) = \rho_{111,\ldots,\,1;\,000,\ldots,0}(0)\exp\frac{(-nt)}{\tau_c} . \qquad (12.34)$$

In this case the effective decoherence time scales are τ_c/n. This is not unexpected. In the opposite extreme of large oscillator coherence length, there is a collective decay of the Q bits, exactly as in super radiance (Nussenzweig, 1973). Then the decay is more rapid, depending on τ_c/n^2.

12.8 Decoherence correction (error correction)

How may we correct for the decoherence due to the environment? It may be viewed as a natural form of noise. From the classical point of view, we might create an ensemble of Q bits and make use of the expected $1/\sqrt{n}$ standard deviation law by using repetition. This is not a sophisticated view of error correction, but may it be done? Error correction is an enormous subject. See the introductions to quantum error correction of Macchiavello and Palma (2000) and of Knill *et al.* (2002.) The difficulty of creating an ensemble of *identical* entangled states is the no-cloning theorem (Wooters and Zurek, 1982). For the orthogonal quantum bits $|0\rangle_A\,|0\rangle_B$, $|1\rangle_A\,|1\rangle_B$, there is a unitary transformation, U:

$$|0\rangle_A\,|0\rangle_B \overset{U}{\rightarrow} |0\rangle_A\,|0\rangle_B$$

$$|1\rangle_A\,|0\rangle_B \overset{U}{\rightarrow} |1\rangle_A\,|1\rangle_B\,.$$

The associated entangled state $a\,|0\rangle_A + b\,|1\rangle_A$ becomes, under U,

$$(a\,|0\rangle_A + b\,|1\rangle_A)\,|0\rangle_B \overset{U}{\rightarrow} a\,|0\rangle_A\,|0\rangle_B + b\,|1\rangle_A\,|1\rangle_B\,.$$

The result is not a tensor product with the original. No unitary transformation can copy $|\psi\rangle$ and $|\phi\rangle$ if they are distinct and are not the same ray, and thus non-orthogonal.

The strategy to avoid this theorem was discovered by Shor (1995) and has been developed into error correction algorithms which appear practical and promising. Other possibilities are being explored, such as working in subspaces of the Hilbert space, for a given problem, which is identified as being nearly decoherence free. However, error correction is more developed and universal than classical experience with noise. Zurek briefly discusses these possibilities at the end of his article (Zurek, 2002).

To illustrate the quantum error correction routine, consider the following simple model (Macchiavello and Palma, 2000). This is a *three*-Q bit model which corrects errors on the system of interest, a *single* Q bit. This increase in the dimensionality is what obviates the cloning of the single Q bit. We adopt the model of Eq. (12.29). Due to the entanglement, there will be phase errors of the form

$$|0\rangle \rightarrow |0\rangle \tag{12.35}$$

$$|1\rangle \rightarrow -\,|1\rangle\,.$$

We choose the message *code words* of three entangled Q bits $|w_0\rangle$, $|w_1\rangle$:

$$|w_0\rangle = |000\rangle + |011\rangle + |101\rangle + |110\rangle \qquad (12.36)$$
$$|w_1\rangle = |111\rangle + |100\rangle + |010\rangle + |001\rangle .$$

Only one Q bit (any one) is taken as entangled with the environment by U. Now linear combinations

$$(a_0 |w_0\rangle + a_1 |w_1\rangle) |E\rangle \overset{U}{\to} \sum_{k=0}^{3} (a_0 |w_0\rangle_k + a_1 |w_1\rangle_k |E_k\rangle). \qquad (12.37)$$

The error states $|w_j\rangle_k$ being orthogonal,$_k \langle w_j \mid w_l\rangle_i = \delta_{kj}\delta_{jl}$, $j, l = 0, 1$ label the word, and $k, i = 0, 1, 2, 3$ label the Q bits. The Q bit 0 is no error, and 1, 2, 3 label the error (-1) on the relevant Q bit. The $|E_k\rangle$ are the environmental states with the associated Q bit error.

We may project the code space into the resulting error spaces identifying the errors. From this measurement on the error space, one now corrects the error by applying σ_z to the identified Q bit. We have measured in the error space, not the Q bit space. If $i = 0$, we do nothing. This iterated monitoring of $|w_0\rangle$ and $|w_1\rangle$ may continue without disturbing either, except to apply the appropriate σ_z. This illustrates the general case (which we will not go into, but leave it to the interested student). A nice, complete description of error-correcting methods in the quantum case, incorporating the classical methods, is given in the review by A. M. Steane (1998).

How effective is quantum error correction? Much work has been done recently with the perfection of threshold theorems (see Knill *et al.*, 2002). What are the possible tolerated error rates? This is of the order of 10^{-4} per computational step. For reviews of fault tolerant computation, see the book of Nielsen and Chuang (2000). It is a rather complete introduction to most of these topics. (See also Preskill's *Lecture Notes for Physics 229*, cited in Section 12.4 and available on the Internet: www.theory.caltech.edu/~preskill/ph219/topological.pdf).

Appendix 12A: entanglement and the Schmidt decomposition

In his fundamental introduction of entanglements, Schrödinger (1935a and 1935b) was apparently not aware of the potential mathematical basis due to Schmidt (1907). This is used in modern discussions of entanglements, and we will review it here.

Let $\mathfrak{H}_A$ and $\mathfrak{H}_B$ be Hilbert spaces with corresponding complete orthonormal basis, $|i\rangle_A$, $|j\rangle_B$. The joint Hilbert space $\mathfrak{H}_A \otimes \mathfrak{H}_B$ is $|i\rangle_A |j\rangle_B$. The state of the combined system is

$$\psi_{AB} = \sum_{ij} c_{ij} |i\rangle_A |j\rangle_B . \tag{12A.1}$$

We assume further

$$\sum_{ij} |c_{ij}|^2 = 1. \tag{12A.2}$$

If $|\psi\rangle_{AB}$ is *not* a direct product state $|\psi\rangle_A \otimes |\psi\rangle_B$, it is said to be *entangled*. The condition is obviously

$$c_{ij} = c_i^A c_j^B . \tag{12A.3}$$

Going further, we introduce the reduced density matrix ρ_A. As in Chapter 2,

$$\begin{aligned} \langle 0 \rangle_A &= \mathrm{Tr}_A(\mathrm{Tr}_B \, |\psi\rangle_{AB} \ {}_{AB}\langle \psi | \, 0_A) \\ &= \mathrm{Tr}_A \rho_A 0_A , \end{aligned}$$

where for pure states,

$$\rho_A = \mathrm{Tr}_B \, |\psi\rangle_{AB} \ {}_{AB}\langle \psi | . \tag{12A.4}$$

It is obvious that the product condition is true for pure states and

$$\rho_A^2 = \rho_A . \tag{12A.5}$$

As we know from Chapter 2, the inverse also is true.

Now let $|\phi_i\rangle_A$ be an eigenstate of ρ_A. We write

$$|\psi\rangle_{AB} = \sum_{ij} c_{ij} |\phi_i\rangle_A \langle j|_B , \tag{12A.6}$$

and introducing the state

$$\sum_j s_{ij} |j\rangle_B \equiv |\chi_i\rangle_B , \tag{12A.7}$$

$|\chi_i\rangle_B$ may be orthonormal. We have then

$$|\psi\rangle_{AB} = \sum_i d_i |\phi_i\rangle_A |\chi_i\rangle_B . \tag{12A.8}$$

d_i may be positive. This is a product representation of the entangled state $|\psi\rangle_{AB}$, Eq. (12A.1) of Schrödinger's discussion.

From Eq. (12A.8) entanglement is now apparent. Measurement of a single state $|\chi_i\rangle_B$ with certainty implies A is in state $|\phi_i\rangle_A$ with probability one, and the resulting $|\psi\rangle_{AB}$ is, on measurement, a product. Alternatively, if the measurement is not with certainty, then terms in Eq. (12A.8) are a succession of $|\phi_i\rangle_A$ and are obtained with probability $|d_i|^2$. $|\psi\rangle_{AB}$ is now a mixture, and the entanglement is again removed. It is classical. The measurement $|\chi_i\rangle_B$ may be done anytime, anywhere. The mixture is not unique, as discussed in Chapter 2.

The spacial Schrödinger dependence is quickly obtained in the $|x\rangle_A\,|y\rangle_B$ basis. Now the nonlocality of the results is seen, as has already been discussed in this chapter. This representation, Eq. (12A.8), and subsequent discussion are due, in physics, to Schrödinger.

References

Agarwal, G. S. (1971). *Phys. Rev. A* **4**, 739.

Bell, J. S. (1964). *Physics* **1**, 195.

Bennett, C. H. (1998). In *Proceedings of Nobel Symposium 104*, ed. E. B. Karlsson and E. Brändas, 210 (Stockholm, Physica Scripta).

Bennett, C. H. and Wiesner, S. J. (1992). *Phys. Rev. Lett.* **68**, 3121.

Bennett, C. H. and Wiesner, S. J. (1992). *Phys. Rev. Lett.* **69**, 2881.

Bohr, N. (1949). In *Albert Einstein: Philosopher and Scientist*, ed. P. A. Schlipp, 199 (Evanston, Library of Living Philosophers).

Boumeester, D. and Zeilinger, A. (2000). In *The Physics of Quantum Information,* ed. D. Boumeester, A. Ekert and A. Zeilinger, p. 1 (Berlin, Springer).

Boumeester, D., Pan, J. W., Weinfurter, H. and Zeilinger, A. (2000). *The Physics of Quantum Information*, ed. D. Boumeester and A. Zeilinger, p. 67 (Berlin, Springer).

Caldeira, A. O. and Leggett, A. J. (1983). *Physica A* **121**, 587.

Courant, R. and Hilbert, D. (1966). *Methods of Mathematical Physics* **1** (New York, Interscience).

Deutsch, D. (1985). *Proc. Roy. Soc. A* **400**, 97.

Deutsch, D. and Ekert, A. (2000). In *The Physics of Quantum Information,* ed. D. Boumeester, A. Ekert and A. Zeilinger (Berlin, Springer), p. 93.

Einstein, A. (1949). In *Albert Einstein: Philosopher and Scientist*, ed. P. A. Schlipp, 663 (Evanston, Library of Living Philosophers).

Einstein, A., Podolsky, B. and Rosen, N. (1935). *Phys. Rev.* **47,** 777.

Ekert, A. K. (1998). In *Proceedings of Nobel Symposium 104*, ed. E. B. Karlsson and E. Brändas, 218 (Stockholm, Physica Scripta).

Ekert, A. K., Palma, G. M. and Suomin, K. A. (2002). In *The Physics of Quantum Information,* ed. D. Boumeester, A. Ekert and A. Zeilinger, 222 (Berlin, Springer).

Feynman, R. (1959). Talk before the American Physical Society, California Institute of Technology.

Fry, E. S. (1998). In *Proceedings of Nobel Symposium 104*, ed. E. B. Karlsson and E. Brändas, 47 (Stockholm, Physica Scripta).

Haroche, S. (1998). In *Proceedings of Nobel Symposium 104*, ed. E. B. Karlsson and E. Brändas, 159 (Stockholm, Physica Scripta).

James, D. F. V. and Kwiat, P. C. (2002). Quantum state entanglements. *Los Alamos Science* **27**, 53.

Jammer, M. (1974). *The Philosophy of Quantum Mechanics* (New York, Wiley).

Jozsa, R. (1998). In *Introduction to Quantum Computation and Information,* ed. H. Lo, S. Popescu and T. Spiller (Singapore, World Scientific).

Jozsa, R. (2002). In *The Physics of Quantum Information,* ed. D. Boumeester, A. Ekert and A. Zeilinger, 104 (Berlin, Springer).

Knill, E., Laflamme, R., Barnum, H. H., Dalrit, D. A., Dziarmaga, J. J., Gubernatis, J. E., Gurvits, L., Ortiz, G., Viola, L. and Zurek, W. H. (2002). *Los Alamos Sci.* **27**, 2.
Kraus, B. and Cirac, J. I. (2001). *Phys. Rev. A* **63**, 1.
Macchiavello, C. and Palma, G. M. (2000). In *The Physics of Quantum Information,* ed. D. Boumeester, A. Ekert and A. Zeilinger, 232 (Berlin, Springer).
Mattle, K., Weinforter, H., Kwiat, P. C. and Zeilinger, A. (1996). *Phys. Rev. Lett.* **76**, 46, 56.
Nielsen, M. A. and Chuang, I. L. (2000). *Quantum Computation and Quantum Information* (London, Cambridge University Press).
Nussenzweig, H. M. (1973). *Introduction to Quantum Optics* (London, Gordon and Breach).
Schmidt, E. (1907). *Math. Ann.* **63**, 433.
Schrödinger, E. (1935a). *Naturwissenschaften* **23**, 807. Trans. J. D. Trimmer (1980), *Proc. Am. Phil. Soc.* **124**, 323.
Schrödinger, E. (1935b). *Proc. Camb. Phil. Soc.* **31,** 446.
Shor, P. W. (1994). In *Proceedings of the 36th Annual Symposium on Foundations of Computer Science*, ed. Caldwasser (Los Alamitos, I.E.E.E.) 124.
Shor, P. W. (1995). *Phys. Rev. A* **52**, 2493.
Steane, A. M. (1998). Quantum error correction. In *Introduction to Quantum Computation and Information*, ed. H. Huinkwang, S. Popesco and T. Spiller (Singapore, World Scientific).
Walls, D. F. and Milburn, G. J. (1985). *Phys. Rev. A* **31**, 2403.
Wooters, W. K. and Zurek, W. (1982). *Nature* **299**, 802.
Zeilinger, A. (1998). In *Proceedings of Nobel Symposium 104*, ed. E. B. Karlsson and E. Brändas, (Stockholm, Physica Scripta).
Zurek, W. (1991). *Phys. Today* **46**, 13.
Zurek, W. (2002). *Los Alamos Sci.* **27,** 1.
Zurek, W. (2003). *Rev. Mod. Phys.* **75**, 715.

13

Quantum measurement and irreversibility

13.1 Introduction

In the light of the preceding chapter, putting quantum measurement in some perspective is inescapable. This topic was begun by von Neumann (1955). We will see that the discussions in Chapter 12 of the consequences of entanglement take a very simplistic point of view, possibly leaving out important time scales.

To set the stage, let us first review what the postulates of quantum mechanics are, but not too mathematically rigorously:

1. The physical states $|\psi\rangle$ of a system are associated with a Hilbert space $\mathfrak{H}$ of normalized vectors. Physical observables, $\hat{O}$, are represented by these self-adjoint operators in $\mathfrak{H}$. The results of a measurement of $\hat{O}$ are the eigenvalues

$$\hat{O}\,|a_n\rangle = a_n\,|a_n\rangle\,, \tag{13.1}$$

assumed discrete and nondegenerate, for simplicity. a_n are real, and $|a_n\rangle$ normed and complete.

2. The time development of the state $|\psi\,(t)\rangle$ for the isolated system is given by the *linear* Schrödinger equation

$$i\hbar\frac{d}{dt}\,|\psi\,(t)\rangle = \hat{H}\,|\psi\,(t)\rangle\,. \tag{13.2}$$

$\hat{H}$ is the Hamiltonian operator in $\mathfrak{H}$. This is a reversible dynamic, as we emphasized in Chapter 5.

3. The probability of measuring a_n at time t is

$$P\,(a_n)\; = |\,\langle a_n|\,\psi\,|(t)\rangle\,|^2\,. \tag{13.3}$$

4.** The effect of measurement on the system is a *reduction of the state vector* from $|\psi\,(t)\rangle$ to $|a_n\rangle$:

$$\text{(before measurement)}\;\; |\psi\,(t)\rangle \rightarrow |a_n\rangle \;\;\text{(after measurement).} \tag{13.4}$$

This is a new dynamics. It is not often stated clearly nor agreed upon. Some do not even accept this as a postulate or problem.

The preceding outline may also be described in terms of the density operator, ρ :

1′. Assume the trace class, trace one, semipositive definite density operator

$$\rho = \sum_i P_i \left|\psi_i\right\rangle\left\langle\psi_i\right|. \tag{13.5}$$

$P_i = \frac{\mathfrak{N}_i}{\mathfrak{N}}$ is the weighting of state $\left|\psi_i\right\rangle$ in the ensemble $\mathfrak{N}$. As discussed in Chapter 2, we have either

$$\begin{aligned} \rho^2 &= \rho \quad \text{pure state} \\ \text{or } \rho^2 &\neq \rho \quad \text{mixture} \qquad \text{(entanglements!)}. \end{aligned} \tag{13.6}$$

2′. The time evolution of $\rho(t)$ is given by the linear reversible von Neumann equation

$$\frac{i\hbar d\rho(t)}{dt} = \left[\hat{H}, \rho(t)\right] \qquad -\infty \leq t \leq \infty, \tag{13.7}$$

$\hat{H}$ being the Hamiltonian operator in the commutator.

3′. The probability at time t of measuring $|a_n\rangle$ is

$$P[t, a_n] = \operatorname{Tr} P_n \rho(t) \tag{13.8}$$

where

$$P_n = |a_n\rangle\langle a_n|.$$

4′**. The measurement transforms ρ

$$\text{(before measurement) } \rho \to \frac{P_n \rho P_n}{\operatorname{Tr}[P_n \rho P_n]} \text{(after measurement)}. \tag{13.9}$$

This transformation leads to the wave packet reduction.

13.2 Ideal quantum measurement

Let us consider the ideal measurement of von Neumann, which leads to the so-called measurement problem. A deep and clear exposition is given by d'Espagnat (1971).

We must introduce the state of the operator that physically does the measurement. Call the state $|A\rangle$. It is macroscopic. There are possibly other degrees of freedom, called the environment or "rest of the universe." For the time being, we will ignore these degrees of freedom. The apparatus may be viewed as a "pointer" on the real line. Thus,

$$A|x\rangle = x|x\rangle; \qquad 0 < x \leq \infty.$$

This is, possibly, a position or a photographic plate. The state of the system plus the macro apparatus is then, initially, before measurement, $|A\rangle \otimes |a\rangle$. Now, to make the apparatus useful, we must correlate by means of repetitive separate measurement the apparatus value with a_n. The process is ideal, and we ignore errors due to noise (classical!) and back reaction. These may be taken into account (see Bassi and Ghirardi, 2003; also Wigner, 1963; Margenau and Park, 1967).

Many examples of quantum measurement are treated by D. Bohm in his prophetic book (Bohm, 1951). We urge the student *not* to leave this unread. For instance, he emphasizes that classical measurements may be made arbitrarily weak, whose errors may be corrected for by classical dynamics. However, quantum errors cannot be so simply discussed. See the error correction discussion of the previous chapter.

Now, once the perfect correlations between the apparatus and the system have been made, we may dispense with the system coordinates. Measurement is a recording of the *apparatus* coordinates. The total Hamiltonian of system, plus apparatus with interaction, governs the measurement with the associated Schrödinger equation. We find that

$$|a_n\rangle \otimes |A_0\rangle \overset{U(t)}{\rightarrow} |a_n\rangle \otimes |A_n\rangle . \tag{13.10}$$

The macro $|A_n\rangle$ implies the system state $|a_n\rangle$, which has not changed in the reversible measurement. There is no trouble with a complete set of commuting observables, which may be similarly treated. Now, what if the initial system state is entangled?

$$|a\rangle = \frac{1}{\sqrt{2}} (|a_n\rangle + |a_l\rangle) .$$

The analysis now becomes

$$|a\rangle \otimes |A_0\rangle \rightarrow \frac{1}{\sqrt{2}} [|a_n\rangle \otimes |A_n\rangle + |a_l\rangle \otimes |A_l\rangle] . \tag{13.11}$$

The *macroscopic* pointer must read two separate distinct values. That is an absurdity. This necessitates the idea of a "collapse" where the reading is $|A_n\rangle$ or $|A_l\rangle$, not each with probability of 1/2. But what is the mechanism or dynamics of such a "collapse"? This is the problem.

As pointed out by von Neumann, the argument may be carried further in a hierarchical fashion. At a first stage we measure

$$|a_n\rangle \otimes |A_0\rangle \rightarrow |a_n\rangle \otimes |A_n\rangle .$$

Next, instrument A interacts with macroscopic instrument B,

$$|A_n\rangle\,|B_0\rangle \rightarrow |A_n\rangle\,|B_n\rangle\,,$$

and also B with C,

$$|B_n\rangle\,|C_0\rangle \rightarrow |B_n\rangle\,|C_n\rangle\,, \qquad \text{etc.}$$

The correlation is transferred through a succession of macroscopic measuring instruments. The argument with entangled states may also be made successively. We may imagine A, B, C, D etc. to be a succession of larger (more complex) macro systems in use by the experimenter, such as a developed image, photo plate, scanner, computer. All these then contain the information $|a_n\rangle$. When the hierarchy is terminated, by definition the measurement has occurred. Is this satisfactory?

These considerations have led to an enormous body of debate ranging from "there is no problem!" to "hidden variables," the "many universe" interpretation, etc. We will not review these. Bassi and Ghirardi have given a compact recent review, with many references, as well as a helpful road map through the interesting jungle as an introduction to their personal contributions. To a large extent the contributions to the measurement problem are an effort to modify and enlarge on the above rules of quantum mechanics and in a sense to create a "new quantum mechanics."

13.3 Irreversibility: measurement master equations

The suggestion that irreversibility plays a key role goes back in time to Szilard (1929) and von Neumann, whose work caused von Neumann to contend that it was impossible to formulate a consistent theory of measurement without reference to human consciousness. Thus the above hierarchy is broken (see Jammer, 1974). The collapse of the wave function appears analogous to the *Stosszahlansatz* of Boltzmann (see Chapter 6). Jordan (1949) asserted that an element of the wave function collapse was irreversible, as in "thermodynamical" statistics. Misra, Prigogine and Courbage have pointed out that the general entropy principle would lead conceptually to a solution of the measurement issue, although it was not carried out in detail (Misra *et al.*, 1979).

The system, plus macro measuring devices, is inescapably in interaction with the environment and thus represents an open system, the subject of our book. Open system dynamics is irreversible, at least in reasonable approximation, governed by master equations of the type already discussed in many early chapters. An alternative generalization has been made by Ludwig (1953) in his attempt to create a new Hilbert space formalism to properly define macro observables consistent with quantum mechanics and the classical world. The idea was to consider the apparatus variables as in a metastable state evolving under a perturbation by the small system to a stable state. A cloud chamber is a system illustrating this. N. G.Van Kampen (1962) coarse-grained the Hilbert space, introducing coarse-grained macro quantum observables and derived by qualitative argument the Pauli equation governing the coarse-grained irreversible dynamics.

The subsequent work of the Trieste school is summarized in the long recent review of Bassi and Ghirardi (2003). They term this approach dynamic reduction. It introduces, in place of the Schrödinger equation, a nonlinear stochastic modification. A spontaneous localization is achieved continuously on the particle coordinates. The resulting master equation is a semi-group equation of the Lindblad form. For a single particle it is

$$\frac{d\rho}{dt} = \frac{-i}{\hbar}\left[H_1 \rho(t)\right] - \lambda\left(\rho(t) - T\left[\rho(t)\right]\right),$$

where

$$T\left[\rho(t)\right] = \sqrt{\frac{\alpha}{\pi}} \int_{-\infty}^{+\infty} dx \exp\left(\frac{-\alpha}{2}(q-x)^2\right) \rho \exp\frac{\alpha}{2}(q-x)^2 .$$

In between the localization disturbance, the system evolution has the Schrödinger form. The spontaneous hitting of the particle is Poissonian, having a probability λdt of occurrence in dt. This spontaneous localization is in a sense ad hoc. To maintain quantum mechanics on a micro scale, they choose $\lambda_{\text{micro}} = 10^{-16}\ \text{sec}^{-1}$. The localization distance is $1/\sqrt{\alpha}$ taken as 10^{-5} cm. Consider a macroscopic entangled state $\psi = \psi_1 + \psi_2$ at position "1" and "2," a distance larger than $1/\sqrt{\alpha}$. The spontaneous localization transforms ψ into a statistical mixture of ψ_1 and ψ_2.

We will *not* adopt this approach now in this chapter, but rather, first, take an alternative viewpoint called environment-induced superselection, which restricts the class of observables by means of the interaction of the system plus pointer with the environment. This is the open system master equation approach to measurement strongly argued by Zurek (Dineri *et al.,* 1962; Jauch, 1964; Zurek, 1991). Emphasis on the open system master equation approach will allow us to treat a simple model in detail, illustrating the point of view due to Walls (Walls *et al.,* 1985; see also Walls and Milburn, 1994). Many of the things now discussed were also covered in the section on decoherence in Chapter 12.

The ambiguity as to which pointer state the macro measuring device is in may be noted in a different fashion from Eq. (13.11). If the system is initially in the state $|\psi\rangle = \sum_i c_i |a_i\rangle$, then on measurement,

$$|\psi\rangle \otimes |A_0\rangle \rightarrow \sum_i c_i |A_i\rangle \otimes |a_i\rangle . \tag{13.12}$$

The reduced density matrix of the system is the mixture

$$\rho_s = \sum_i \mid c_i^2 \mid |a_i\rangle \langle a_i| . \tag{13.13}$$

However, this is not unique. The complete meter basis may be transformed, assuming the meter states are complete, as

$$|A_i\rangle = \sum_j \langle B_j \mid A_i\rangle \, |B_j\rangle . \tag{13.14}$$

Then Eq. (13.12) becomes

$$\sum_i c_i \, |A_i\rangle \otimes |a_i\rangle = \sum_j d_j \, |B_j\rangle \otimes |b_j\rangle , \tag{13.15}$$

where

$$d_j \, |b_j\rangle = \sum_k c_k \langle B_j \mid A_j\rangle \, |a_j\rangle . \tag{13.16}$$

Are we measuring $\sum_j d_j \, |b_j\rangle$ or $\sum_j c_j \, |a_j\rangle$? We have

$$\rho_s = \sum_j |d_j|^2 \, |d_j\rangle\langle d_j| . \tag{13.17}$$

The mixture may be made unique if there is selection by the environment of a *preferred basis*. Call it the pointer basis. We choose that basis for which

$$\left[\hat{O}, H_A + H_{AE}\right] = 0. \tag{13.18}$$

$\hat{O}$ is that special class of observables for which Eq. (13.18) holds. H_{AE} is the apparatus–environmental interaction, H_E the environment Hamiltonian, and E_E its energy. For a large environment (a thermal bath, for instance) for small systems, we have approximately $\left[\hat{O}(t), H_A + H_E\right] = 0$. Eq. (13.18) ensures no "back reaction" between the macro apparatus and the environment. The state $|A, E\rangle$ is macro in nature, and

$$(H_A + H_E)\,|A, E\rangle = (E_A + E_E)\,|A, E\rangle \tag{13.19}$$

are then the diagonal representation of $\hat{O}$. Approximately, these $\hat{O}(t)$ are constant and unchanging, even with the apparatus–environment interaction. The $|AE\rangle$ are the pointer basis. Zurek (1982) has given a long discussion of the pointer basis. By introducing the environment, we are no longer dealing with the reversible Schrödinger equation but rather with irreversible master equations for open systems. The time scales have already been discussed in Chapters 3, 4, 5, 6 and 11. By a selection rule, the so-called quantum measurement problem is answered. However, it is not clear how the macro nature of $|A\rangle$ appears in this approach, nor are the time scales of the measurement very explicit. To see this, we must turn to another model.

13.4 An open system master equation model for measurement

We have already discussed the first model in Chapter 2. We take the apparatus to be a harmonic oscillator and the interaction with the environment to be of the phase-damping form

$$H_{AE} = b^{\dagger} b \Gamma. \tag{13.20}$$

The oscillator energy is conserved, but the phase is changed by E. The density operator for the apparatus obeys the irreversible master equation

$$\frac{d\rho}{dt} = \frac{\gamma}{2}\left[2b^{\dagger}b\rho b^{\dagger}b - \left(b^{\dagger}b\right)^2\rho - \rho\left(b^{\dagger}b\right)^2\right]. \tag{13.21}$$

It is of the Lindblad form, as already noted in Chapter 6. In the energy eigenstate $|n\rangle$, we had

$$\begin{aligned} \rho_{mn}(t) &= \exp\left(-\gamma\,(m-n)^2\,t\right)\rho_{mn}(0) \\ \rho_{mn}(t) &= \rho_{mn}(0)\,. \end{aligned} \tag{13.22}$$

The off-diagonal apparatus correlations rapidly decay. The macro observable $b^{\dagger}b$ obeys

$$\left[b^{\dagger}b, H_{AE}\right] = 0 \tag{13.23}$$

and is the pointer operator O. $|n\rangle$ are the pointer states, now macroscopic. The interaction of the apparatus with the general system may be used to correlate $|n\rangle$ with the system states $|a\rangle$, thus performing the measurement.

This is a general model being restricted by the form of Eq. (13.21). The states $|n\rangle$ may be taken to be coherent states $|\alpha\rangle$. Thus, in an appropriate limit, the apparatus becomes apparently classical. The apparatus correlations have decayed rapidly on a time scale $(\gamma)^{-1}$. This is the collapse time scale and the apparatus decoherence time scale. Thus we see, implicit in the measurements discussed in the previous chapter, that there are apparatus–environmental time scales. The effect of this is not clearly seen in such discussions.

Another model which illustrates this in more detail is the following: assume the apparatus is a harmonic oscillator in interaction with a system harmonic oscillator with the Hamiltonian

$$H_{SA} = \frac{\hbar}{2} a^{\dagger} a \left(bE^* + b^{\dagger}E\right). \tag{13.24}$$

$b^\dagger, b$ are, as before, the apparatus operators, and E a classical driving field. The apparatus is coupled to the environment by a more realistic interaction:

$$H_{AE} = b\Gamma^\dagger + b^\dagger\Gamma. \tag{13.25}$$

Now we will find that $b^\dagger b$ is an approximate pointer operator.

The system plus apparatus master equation is

$$\begin{aligned}\frac{d\rho}{dt} &= \frac{1}{2}\left[\left(Eb^\dagger - E^* b\right)a^\dagger a, \rho\right] \\ &\quad + \frac{\gamma}{2}\left(2b\rho b^\dagger - b^\dagger b\rho - \rho b^\dagger b\right).\end{aligned} \tag{13.26}$$

We assume the environment is at zero temperature and take

$$\rho(0) = \sum_{nm} \rho_{nm} |n\rangle\langle m| \otimes |0\rangle\langle 0|, \tag{13.27}$$

the apparatus being in the ground state $|0\rangle\langle 0|$, and $\rho_{nm} \equiv \langle n|\rho_S(0)|m\rangle$.

To solve Eq. (13.26), Walls (Walls *et al.*, 1985) utilized the characteristic function transformation

$$\chi_{nm}(t) = \mathrm{Tr}\left[\exp\left(\lambda b^\dagger - \lambda^* b\right)\rho_{nm}(t)\right] \tag{13.28}$$

with

$$\chi_{nm}(0) = \sum_{\alpha\beta} N_{nm}(\alpha\beta)\exp\left(-\frac{1}{2}|\lambda|^2 + \lambda\beta^* - \lambda^*\alpha\right).$$

Obtaining the partial differential equation for $\chi_{mn}(\lambda, t)$,

$$\frac{\partial\chi_{mn}}{\partial t}(t) = \frac{\gamma}{2}\left\{\begin{array}{c} -|\lambda|^2 - \frac{1}{2}(n+m)\left[\frac{E}{\gamma}\lambda^* - \frac{E^*}{\gamma}\lambda\right] \\ -\left[\lambda - (n-m)\frac{E}{\gamma}\right]\frac{\partial}{\partial\lambda} - \left[\lambda^* - (n-m)\frac{E^*}{\lambda^*}\right]\frac{\partial}{\partial\lambda^*}\end{array}\right\}\chi_{mn}(t). \tag{13.29}$$

We leave it as a problem for the student to solve this and show that

$$\begin{aligned}\rho(t) = \sum_{nm} \rho_{nm} \exp\left\{\frac{|E|^2}{\gamma^2}(n-m)^2\left[1 - \frac{\gamma t}{2} - \exp\left(-\frac{\gamma t}{2}\right)\right]\right\} \\ \times |n\rangle\langle m| \underset{S}{|} \otimes \frac{|\alpha_n(t)\rangle\langle\alpha_m(t)|}{\langle\alpha_m(t)\,|\,\alpha_n(t)\rangle}\underset{A}{|},\end{aligned} \tag{13.30}$$

where the time-dependent apparatus coherent states are

$$|\alpha_n(t)\rangle = \frac{En}{\gamma}\left(1 - \exp\left(\frac{-\gamma t}{2}\right)\right). \tag{13.31}$$

From Eq. (13.21) we see that the apparatus irreversibly decays rapidly on a time scale $\left(\frac{\gamma}{2}\right)^{-1}$ and goes to the coherent state, the pointer basis. The amplification in the "meter" reading is seen in the classical magnitude of the field E. Being a coherent state, this is approximately classical macroscopic for large amplitudes and is orthogonal.

Consider now the solution, Eq. (13.30). The master equation is not valid at short time, as we have discussed in earlier chapters. The t dependence is not physical. For sufficiently long time then,

$$\begin{aligned} \rho &\to \sum_{nm} \exp\left[\left(\frac{E^2}{\gamma^2}(n-m)^2\right)\left(1-\frac{\gamma t}{2}\right)\right] |n\rangle\langle m| \underset{S}{|} \otimes |\alpha_n\rangle\langle\gamma_m| \underset{A}{|} \\ &\to 0 \qquad\qquad n \neq m \\ &\to |n\rangle\langle n|_s \otimes |\alpha_n\rangle\langle\alpha_n|_A \qquad\qquad n = m \end{aligned} \tag{13.32}$$

if $\frac{\gamma t}{2} \gg 1$ and the decoherence of the off diagonal elements is of the form $\exp\frac{-E^2}{2\gamma}(n-m)^2 t$. Thus the intensity of the classical field amplifies the decoherence rate. Eq. (13.32) is a statement of the wave function collapse naturally appearing in the solution. It is not an ad hoc postulate here but appears as a result of the irreversible open system dynamics governed by a master equation. Of course, the solution depends on a particular model, but the qualitative suggestion is general.

13.5 Stochastic energy based collapse

In the previous section we have discussed some of the historical development of ideas related to the process by which a linear superposition of wave functions makes a transition to a mixture of pure states, for which each is defined by an eigenvalue of some self-adjoint operator characterizing the outcome of a measurement process. As we have explained, one might think that there is no problem, since the experimental consequences of the quantum theory, well verified, are consistent with the computation of the probability of some outcome according to the absolute square scalar product of the initial wave function with the wave function of the final state sought by the apparatus. This leaves open, however, the question of how this transition takes place, and that is the subject of many discussions that have appeared in the literature. It is clear that the mechanism for this reduction—or collapse, as it is sometimes called—of the wave function cannot be generated by the linear action of a one-parameter unitary group such as the action of the normal evolution through the ordinary Schrödinger equation. Jauch (1968) has discussed carefully three apparent paradoxes that illustrate the philosophical difficulties involved in the reduction process (Schrödinger, 1935; Einstein *et al.*, 1935; Wigner, 1962), all of which involve the destruction of coherence in the construction of the linear superposition of wave functions. Thus, the application of possible mechanisms of disturbance, such as interaction with a random environment, provides an effective way of looking at these difficulties, as we have discussed above. These mechanisms, however, are generally called upon to accomplish this task without a complete specification of their nature and without accounting for their apparent universality. In recent years, a mechanism has been introduced which is both universal and mathematically clear and rigorous, and which can therefore bear careful investigation to the extent of model building within the framework of known physical theory. We shall discuss here some further details of this mechanism, which we shall call stochastic reduction, referred to in Section 13.3.

The basic structure of this mechanism seems to have been first introduced by Gisin (1984) and Diósi (1988) and was brought to a level that has been useful for detailed calculations by Ghirardi, Pearle and Rimini (Ghirardi *et al.*, 1990) and Hughston (1996). Much of the large literature that has developed is recorded and referred to in the work of Bassi and Ghirardi (2003) and in the book of Adler (2004), which embeds the idea into a framework provided by a new form of the dynamics of quantum field theory. This last is an interesting example of the deeper investigations of the underlying physical processes that can now be carried out given this relatively recently developed, well-defined, structural model, and illustrates its connection with statistical mechanics in a fundamental way.

To describe this model, we write an extended Schrödinger equation in the form

$$\begin{aligned} d\,|\psi(t)\rangle = -iH\,|\psi(t)\rangle\,dt - \frac{\sigma^2}{8}(H - H_t)^2\,|\psi(t)\rangle\,dt \\ + \frac{\sigma}{2}(H - H_t)\,|\psi(t)\rangle\,dW(t)\,, \end{aligned} \tag{13.33}$$

where $H_t = < \psi(t)\,|H|\,\psi(t) >$, σ is a parameter characterizing the reduction time scale, and $W(t)$ is a standard Wiener process describing Brownian fluctuations, satisfying the relation

$$dW(t)^2 = dt. \tag{13.34}$$

The first term on the right side of Eq. (13.33) corresponds to the usual Schrödinger evolution, and the last is a stochastic contribution to the evolution law; both the second and last terms are nonlinear, since they depend, through the expectation value, on the state $|\psi(t)\rangle$ itself.

Using the rules of the Itô calculus, based on Eq. (13.34), it is straightforward to prove that the evolution law Eq. (13.33) preserves the norm of the wave function, which we take to be unity (Itô, 1950; see Malliavin, 1997, for a discussion of properties and applications of these techniques). At this point the student might jump ahead and consider Section 14.2 of the following chapter for the derivation of the quantum Langevin equation.

In Eq. (13.33) one applies the idea to the wave function of a quantum state. It is this Brownian motion (Einstein, 1926; van Kampen, 1983) which represents the fluctuations that may be induced by quantum fields or an "environment" and supplies a mathematically rigorous basis for calculations as well as posing, in a well-defined framework, deep physical questions for further investigation.

The stochastic variable dW has the property that its expected value $E\,[dW\,(t)]$ under the Brownian distribution is zero. Making use of the Itô calculus, one sees that the expectation value of H is given by

$$H_t = H_{t=0} + \sigma \int_0^t dW\,(s)\,V_s, \tag{13.35}$$

where

$$V_t = \langle\psi\,(t)|\,(H - H_t)^2\,|\psi\,(t)\rangle \tag{13.36}$$

is the variance of the energy in the state $|\psi\,(t)\rangle$. The expected value of H_t, under the stochastic distribution $E\,[H_t]$, is therefore conserved (Hughston, 1996).

Furthermore, using the Itô calculas again, one easily finds that

$$dV_t = -\sigma^2 V_t^2 dt_\sigma \beta_t dW\,(t)\,, \tag{13.37}$$

where $\beta = \langle\psi\,(t)|\,(H - H_t)^3\,|\psi\,(t)\rangle$ is the third moment of the deviation of H, and that therefore (Hughston, 1996; Ghirardi *et al.,* 1990; Adler and Horwitz, 2000, 2003)

$$E\,[V_t] = E\,[V_{t=0}] - \sigma^2 \int_0^t ds E\,[V_s]^2\,. \tag{13.38}$$

This is the essential result of the stochastic reduction theory. Since $E\,[V_t]$ must be positive, the integral must converge as $t \to \infty$, and therefore $E\,[V_t] \to 0$. The physical state, therefore, approaches a state in which the dispersion of the Hamiltonian operator goes to zero, and hence it must be an eigenstate (Hughston, 1996) (for the nondegenerate case). We have therefore described a process in which the system starts in some arbitrary state of the system, for example, a linear superposition of energy eigenstates, and under the evolution Eq. (13.33) it necessarily goes over to an eigenstate. It was shown in Ghirardi *et al.* (1990; Hughston (1996); and Adler and Horwitz, 2003) that the probabilities for convergence to each of the eigenstates obeys the Born rule, i.e. they are equal to the squared modulus of the scalar product of the initial state with the corresponding eigenstate. The collapse mechanism described by Eq. (13.33) is therefore consistent with the required results of the quantum theory.

It is clear that during this process of collapse, the initial pure state goes over to a density matrix, since the outcome is a mixture of pure states with *a priori* probabilities, given by the Born rule, i.e. one finds one or the other of the final states, not a linear superposition. In fact, under the stochastic expectation, the pure density matrix obtained by the direct product $|\psi(t)\rangle\langle\psi(t)|$ becomes a state which, under stochastic expectation (all terms linear in $dW(t)$ vanish), evolves under the Lindblad equation, of the type we have discussed above, with well-defined coefficients (Ghirardi *et al.,* 1990; Adler and Horwitz, 2003; Adler, 2004).

For the collapse of a system of two spins in a spin zero state, involving a linear superposition of up–down and down–up states to the experimentally detected up–down state, as occurs in the E.P.R. experiment (Silman *et al.,* 2008), it was found that the nonlinear structure of the evolution law accounts for correlation between measurements, even though the model for the Hamiltonian is a simple sum.

It appears that there will be interesting physics in the further exploration of the methods of stochastic reduction of the type described here.

References

Adler, S. L. (2004). *Quantum Theory as an Emergent Phenomenon* (Cambridge, Cambridge University Press).
Adler, S. L. and Horwitz, L. P. (2003). *J. Math. Phys.* **41**, 2485.
Bassi, A. and Ghirardi, G. (2003). Dynamic reduction models, in *Quantum Physics* **2,** 1.
Bohm, D. (1951). *Quantum Theory* (New York, Prentice Hall).
d'Espagnat, B. (1971). *Conceptual Foundations of Quantum Mechanics* (Menlo Park, W. Benjamin).
Dineri, A., Loinger, A. and Prosperi, G. M. (1962). *Nucl. Phys.* **33**, 297.
Diósi, L. (1988). *J. Phys. A: Math. Gen.* **21,** 2885; *Phys. Lett. A* **129** 419; see also (1989). *Phys. Rev. A* **40**, 1165.
Einstein, A. (1926). *Investigations on the Theory of Brownian Movement,* ed. R. Furth (London, Methuen).
Einstein, A., Podolsky B. and Rosen N. (1935). *Phys. Rev.* **47,** 777.
Ghirardi, G. C., Pearle, P. and Rimini, A. (1990). *Phys. Rev. A* **42**, 78.
Gisin, N. (1984). *Phys. Rev. Lett.* **52**, 1657.
Hughston, L. P. (1996). *Proc. Roy. Soc. Lond. A* **452,** 953.
Itô, K. (1950). *Nagoya Math. J.* **3**, 55.
Jammer, M. (1974). *The Philosophy of Quantum Mechanics* (New York, Wiley).
Jauch, J. M. (1964). *Helv. Phys. Acta* **37**, 293.
Jauch, J. M. (1968). *Foundations of Quantum Mechanics* (Reading, Addison-Wesley).
Jordan, P. (1949). *Phil. Sci.* **16**, 269.
Ludwig, G. (1953). *Z. Phys.* **152**, 98.
Malliavin, P. O. (1997). *Stochastic Analysis* (Berlin, Springer).
Margenau, H. and Park, J. L. (1967). *Delaware Seminar in the Foundations of Physics,* ed. M. Bunge (New York, Springer).
Misra, B., Prigogine, I. and Courbage, M. (1979). *Proc. Natl. Acad. Sci. USA* **76**, 4768.
Schrödinger, E. (1935). *Naturwiss* **48**, 52.

Silman, J., Machnes, S., Shnider, S., Horwitz, L. P. and Belenkiy, A. (2008). *J. Phys. A: Math Gen.,* in print.
Szilard, L. (1929). *Z. Phys.* **53**, 840.
Van Kampen, N. G. (1962). In *Fundamental Problems in Statistical Mechanics*, ed. E. G. D. Cohen (Amsterdam, North Holland).
Van Kampen, N. G. (1983). *Stochastic Processes in Physics and Chemistry* (Amsterdam, North Holland).
von Neumann, J. (1955). *Mathematical Foundations of Quantum Mechanics,* trans. R. B. Beyer (Princeton, Princeton University Press).
Walls, D. F. and Milburn, G. J. (1994). *Quantum Optics* (New York, Springer).
Walls, D. F., Collet, M. J. and Milburn, G. J. (1985). *Phys. Rev. D* **32**, 3208.
Wigner, E. (1962). *The Scientist Speculates*, ed. I. J. Good (London, Heinemann).
Wigner, E. P. (1963). *Am. J. Phys.* **31**, 6.
Zurek, W. (1982). *Phys. Rev. D* **26**, 1822.
Zurek, W. (1991). *Phys. Today,* Oct., 36.

14

Quantum Langevin equation and quantum Brownian motion

14.1 Introduction

We will now consider a continuation of the topic of quantum reservoir damping begun with the master equation description of Chapter 3 and continued in the chapter on quantum optics, Chapter 10. The Heisenberg equation approach, utilizing the Langevin equation type description, will contain elements of approximations already made in those chapters.

The operator Langevin equation description is interesting in that it sheds new light on the physical elements of the discussion, if not new results. We could have derived the Langevin equation from the previous results, but it is profitable to start from the beginning in the Heisenberg quantum description. Haken, in his detailed theory of the laser (Haken, 1984), adopted this point of view. Senitzky (1960) early discussed the quantum damped harmonic oscillator. Many of the elements of this are quite general. It is an interesting paper to be read profitably by the student.

The classical Brownian motion equation,

$$\begin{aligned}\frac{d\mathbf{v}}{dt} &= -\gamma\mathbf{v} + \mathbf{F}(t) + \Gamma(t) \\ &\equiv a(x,t) + b(x,t)\,\zeta(t),\end{aligned} \tag{14.1}$$

is Newton's second law with damping, $-\gamma\mathbf{v}$. $\mathbf{F}(t)$ is an external driving force, and $\Gamma(t)$ a classical random stochastic force. See Gardiner (1983) and Wax (1954) for the original Ornstein–Uhlenbeck theory. $\mathbf{v}(t)$ is a random variable also assuming the Markov property for continuous in time random processes. Examining

$$v(t) = \int_0^t dt'\zeta(t), \tag{14.2}$$

it may be shown that the average, over the ensemble of random processes, is

$$\langle v(t+\Delta t) - v_0 \rangle = 0$$
$$\left\langle [v(t+\Delta t) - v_0]^2 \right\rangle = \Delta t.$$

We assume that

$$\int_0^t \zeta(t')\, dt' = W(t) \qquad (14.3)$$
$$\text{or} \quad dW(t) = \zeta(t)\, dt$$

is a Wiener random process called Brownian motion in one dimension with

$$\langle \mathbf{W}(t) \rangle = \mathbf{W}_0 \qquad (14.4)$$
$$\left\langle [W_i(t) - W_{0i}]\left[W_j(t) - W_{0j}\right] \right\rangle = (t - t_0)\,\delta_{ij}.$$

Eq. (14.4) indicates that the sample paths are highly irregular. They are, in addition, nondifferentiable, although $W(t)$ is continuous. Examining the solution for $X(t)$, we have

$$x(t) - x(0) = \int_0^t a(x, s)\, ds + \int_0^t b(x(s), s)\, dW(s). \qquad (14.5)$$

This is a stochastic Stieltjes integral over the sample path $W(t)$. It is the source of much discussion and the origin of the Itô stochastic integral, and also that of Stratonovich (see Gardiner, 1983). These interpretations also appear in the quantum case, to be discussed here.

Markov assumptions lead to the property for the classical stochastic forces,

$$\left\langle \Gamma_i(t), \Gamma_j(t_0) \right\rangle = G_{ij}\delta(t - t_0). \qquad (14.6)$$

The important point, in the quantum case, is that there are additional conditions to be applied to the random operator "forces."

14.2 Quantum Langevin equation

We will, largely, follow the paper of Gardiner and Collett (1985). Let us consider the idealized Hamiltonian

$$H = H_S + H_{SB} + H_B, \qquad (14.7)$$

where the reservoir $H_B = \hbar \int d\omega \omega b^\dagger(\omega)\, b(\omega)$ is a system of bosons $\left[b(\omega), b^\dagger(\omega')\right] = \delta(\omega - \omega')$ and the interaction with the system H_S is taken as

$$H_{SB} = i\hbar \int_{-\infty}^{+\infty} d\omega K(\omega)\left[b^\dagger(\omega)\, c - c^\dagger b(\omega)\right]. \qquad (14.8)$$

c is a system interaction operator left rather general. H_S need not be specified any further now. Two things should be said: (a) the rotating wave approximation is implicit in the simple choice of H_{SB} (note Chapter 10); and (b) the integral on ω is taken to $-\infty$ and will lead to $\delta\left(t-t'\right)$. These assumptions facilitate the treatment.

As in Chapter 2, we may obtain the Heisenberg equations for an operator of the system and also of b. They are immediately

$$\dot{b}\,(\omega, t) = -ib\,(\omega, t) + K\,(\omega)\,c\,(\omega, t) \tag{14.9}$$

and

$$\dot{a}\,(t) = \frac{-i}{\hbar}\,[a, H_S] + \int_{-\infty}^{+\infty} d\omega K\,(\omega)\left\{b^{\dagger}\,(\omega, t)\,[a, c] - \left[a, c^{\dagger}\right] b\,(\omega, t)\right\}. \tag{14.10}$$

We may formally solve Eq. (14.9) for $b\,(\omega, t)$:

$$b\,(\omega, t) = \exp\left(-i\omega\,(t-t_0)\right) b_0\,(\omega) + K\,(\omega)\int_{t_0}^{t} (-i\omega\,(t-\tau))c\,(\tau)\,d\tau) \qquad t \geq t_0, \tag{14.11}$$

where $b_0\,(0) = b\left(t-t'=0\right)$. We use it to eliminate $b\,(\omega)$ on the right side of Eq. (14.10), obtaining closed non-Markovian equations for operator $a\,(t)$ of the system:

$$\dot{a} = \frac{-i}{\hbar}\,[a, H_S] + \int_{-\infty}^{+\infty} d\omega K\,(\omega)\begin{Bmatrix} \exp\left(i\omega\,(t-t_0)\right) b_0^{\dagger}\,(\omega)\,[a, c] \\ -\left[a, c^{\dagger}\right]\exp\left(-i\omega\,(t-t_0)\right) b_0\,(\omega) \end{Bmatrix} \tag{14.12}$$
$$+ \int_{-\infty}^{+\infty} d\omega K^2\,(\omega)\int_{t_0}^{t} d\tau \begin{Bmatrix} \exp\left(i\omega\,(t-\tau)\right) c^{\dagger}\,(\tau)\,[a, c] \\ -\left[a, c^{\dagger}\right]\exp\left(-i\omega\,(t-\tau)\right) c\,(\tau) \end{Bmatrix}.$$

In our discussion of spontaneous emission in Chapter 10, the second term gave the fluctuations, and the third the radiation reaction. There, an important point was the necessity of adopting an ordering in system and reservoir operators and maintaining it.

Recall here the derivation of the generalized master equation in Chapter 3. Note, in principle, that the commutation laws of a and c are known. Now we make the equivalent assumption to the Born–Markov approximation of Chapter 3. (The derivation of the Pauli equation was discussed there.) We have $K^2\,(\omega) = \gamma/2\pi$, where the memory function in the resulting equation is $2\gamma\delta\,(t)$. Eq. (14.12) becomes

$$\dot{a} = \frac{-1}{\hbar}\,[a, H_S] - \begin{Bmatrix} \left[a, c^{\dagger}\right]\left[\frac{\gamma}{2}c - \sqrt{\gamma}b_{\text{in}}\,(t)\right] \\ -\left[\frac{\gamma}{2}c^{\dagger} + \sqrt{\gamma}b_{\text{in}}^{\dagger}\,(t)\right][a, c] \end{Bmatrix} \qquad t > t_0, \tag{14.13}$$

where

$$b_{\text{in}}\,(t) \equiv \frac{1}{\sqrt{2\pi}}\int_{-\infty}^{+\infty} d\omega \exp\left(-i\omega\,(t-t_0)\right) b_0\,(\omega)\,.$$

We have not included any explicit external time dependence. For the harmonic oscillator system $c \to a$,

$$\dot{a} = -i\omega_0 a - \frac{\gamma}{2}a - \sqrt{\gamma} b_{\text{in}}(t); \quad t > t_0. \tag{14.14}$$

We may call the last term the fluctuating operator force $F(t) = -\sqrt{\gamma} b_{\text{in}}(t)$. We note that the condition $t \geq t_0$ appears on the formal integration of Eq. (14.11), just as in the generalized master equation of Chapter 3, also discussed in Chapter 5. The Langevin equation is irreversible.

There is a time reversed Langevin equation which is obtained by the replacement

$$\begin{aligned} \sqrt{\gamma} &\to \sqrt{\gamma} \\ \frac{\gamma c}{2} &\to -\frac{\gamma c}{2} \\ b_{\text{in}} &\to b_{\text{out}}, \end{aligned} \tag{14.15}$$

where

$$b_{\text{out}} = \frac{1}{\sqrt{2\pi}} \int_{-\infty}^{+\infty} d\omega \exp\left(-i\omega\left(t - t'\right)\right) b_1(\omega).$$

We have

$$b_{\text{out}}(t) - b_{\text{in}}(t) = \sqrt{\gamma} c(t). \tag{14.16}$$

The time dependent commutation laws are

$$\begin{aligned} \left[a(t), b_{\text{in}}\left(t'\right)\right] &= -\theta\left(t - t'\right)\sqrt{\gamma}\left[a(t), c\left(t'\right)\right] \\ \left[a(t), b_{\text{out}}\left(t'\right)\right] &= \theta\left(t - t'\right)\left[a(t), c\left(t'\right)\right], \end{aligned} \tag{14.17}$$

θ being the Heaviside function reflecting the semi-group behavior of the forward and backward equations.

We have not yet characterized the noise structure of the bath dynamics. There is already a noise in $b_{\text{in}}(t)$, since there are vacuum fluctuations having effects in spontaneous emission and in the Lamb shift. These are not yet stochastic equations in the classical sense. Let us now define a *quantum* Wiener process. Let, for the operators,

$$B(t, t_0) = \int_{t_0}^{t} b_{\text{in}}\left(t'\right) dt' \quad \text{(a Heisenberg operator)} \tag{14.18}$$

for an ensemble of *operator* inputs. This is a natural generalization of the c-number $W(t)$. Two ensemble averages are

$$\begin{aligned} \left\langle B^{\dagger}(t, t_0) B(t, t_0)\right\rangle &= \bar{N}(t - t_0) \\ \left\langle B(t, t_0) B^{\dagger}(t, t_0)\right\rangle &= \left(\bar{N} + 1\right)(t - t_0), \end{aligned} \tag{14.19}$$

and the commutator is

$$\left[B(t,t_0), B^\dagger(t,t_0)\right] = t - t_0,$$

where $\bar{N} = \frac{1}{(\exp K - 1)}$ for $B(t,t_0)$ quantum Gaussian where

$$\rho(t,t_0) = (1 - \exp(-K)) \exp\left\{\frac{\bar{K} B^\dagger(tt_0) B(tt_0)}{t - t_0}\right\}.$$

We may make a further idealization to quantum white noise. The input assumption is then

$$\left\langle b_0^\dagger(\omega) b_0(\omega')\right\rangle = \bar{N}\delta(\omega - \omega'),$$

and thus

$$\left\langle b_{\mathrm{in}}^\dagger(t') b_{\mathrm{in}}(t)\right\rangle = \bar{N}\delta(t - t'), \tag{14.20}$$

in which $\bar{N}$ is constant.

For a two-level atom ω_0 in interaction with thermal radiation in weak coupling, one obtains in the continuum approximation

$$\left\langle E(t) E(t')\right\rangle = \frac{\hbar}{\pi}\int_{-\infty}^{+\infty} d\omega \left(-\omega + \omega \coth\left(\frac{\hbar\omega}{kT}\right)\right) \exp i\omega (t - t').$$

Assuming a resonance interaction at $\pm\omega_0$, the integrand is removed and evaluated at these points. We have $\delta(t - t')$. This gives the white noise result, approximately, in a narrow range of ω_0.

If we integrate the white noise force correlation function, in the case of $a(t)$ being that of the harmonic oscillator in Eq. (14.14), $F = -\gamma B_{\mathrm{in}}(t)$, we obtain

$$\gamma = \bar{N}^{-1} \int_{-\infty}^{+\infty} dt \left\langle F^\dagger(t) F(0)\right\rangle. \tag{14.21}$$

This is the quantum fluctuation dissipation theorem relating γ to the dissipative force fluctuation. We may carry this further. We write, for a general a_i,

$$\dot{a}_i = D_i(t) + F_i(t), \tag{14.22}$$

and

$$\left\langle F_i(t') F_j(t)\right\rangle = 2\left\langle D_{ij}\right\rangle \delta(t' - t).$$

One may show, by expansion around $t = 0$,

$$\left\langle a_i(t) F_j(t)\right\rangle = \frac{1}{2}\int_{-\infty}^{+\infty} dt' \left\langle F_i(t') F_j(t)\right\rangle = \left\langle D_{ij}\right\rangle \tag{14.23}$$

because of causality and assuming the noise to be stationary in time. Utilizing $\langle F_i(t)\, a_j(t)\rangle = \langle D_{ij}\rangle$, we obtain from Eq. (14.22)

$$2\langle D_{ij}\rangle = -\langle a_i D_j\rangle - \langle D_j a_i\rangle + \frac{d}{dt}\langle a_i a_j\rangle . \tag{14.24}$$

This is the Einstein formula relating a diffusion coefficient, D_{ij}, to the drift coefficient and is a manifestation of a quantum fluctuation-dissipation (Gardiner, 1991).

Further, we may easily obtain the quantum regression theorem of Lax (1967). We consider, from Eq. (14.22),

$$\frac{d}{dt}\langle a_i(t)\, a_j(t')\rangle = \langle D_i a_j(t')\rangle + \langle F_i(t)\, a_j(t')\rangle; \qquad t' < t.$$

The process is Markovian and causal. $a_j(t)$ cannot be affected by the future noise, so

$$\frac{d}{dt}\langle a_i(t)\, a_j(t')\rangle = \langle D_i(t)\, a_j(t')\rangle . \tag{14.25}$$

The two-time system correlation function obeys the same equation of motion as the single-time $a_j(t)$ Heisenberg equation.

Now, what is the meaning of such operator stochastic integrals? We define the Itô stochastic integral as

$$I\int_{t_0}^{t} g(t')\, dB(t') = \lim_{i\to\infty}\sum_i g(t_i)\left[B(t_{i+1}, t_0) - B(t_i, t_0)\right] \tag{14.26}$$

(see Gardiner, 1983). $g(t)$ is any Heisenberg system operator. The Itô increments may be shown to commute with $g(t')$. $I(da)$ may also be shown to be equivalent to the quantum Langevin equation, because $Id(ab) = adb + bda + dadb$.

The Stratonovich operator integral is defined as

$$S\int_{t_0}^{t} g(t')\, dB(t') = \lim_{i\to\infty}\sum_i \frac{g(t_i + t_{i+1})}{2}\left[B(t_{i+1}, t_0) - B(t_i, t_0)\right]. \tag{14.27}$$

$dB(t')$ does not commute with $g(t')$. In fact, we have the general result

$$S\int_{t_i}^{t} g(t')\, dB(t') - S\int_{t_i}^{t} dB(t')\, g(t') = \frac{\sqrt{\gamma}}{2}\int_{t_0}^{t} dt'\left[g(t'), c(t')\right]. \tag{14.28}$$

From this we may show

$$S\int_{t_0}^{t} g(t')\, dB(t') = I\int_{t_0}^{t} g(t')\, dB(t') + \frac{1}{2}\sqrt{\gamma}\bar{N}\int_{t_0}^{t}\left[g(t'), c(t')\right] dt', \tag{14.29}$$

and similarly for $dB(t')\, g(t')$ and $g(t')\, dB^{\dagger}(t')$ and also $dB^{\dagger}(t')\, g(t')$.

We may show that the quantum Stratonovich stochastic equation is equivalent to the Itô quantum stochastic equation and they are both of the quantum Langevin form. For instance,

$$S(da) = \frac{-i}{\hbar}[a, H_S]\,dt - \frac{\gamma}{2}\left\{\left[a, c^{\dagger}\right]c - c^{\dagger}[a, c]\right\}dt \\ - \sqrt{\gamma}\left[a^{\dagger}c^{\dagger}\right]dB(t) + \sqrt{\gamma}\,dB^{\dagger}(t)[a, c]\,dt. \tag{14.30}$$

In addition, for the Stratonovich case, ordinary *noncommuting* calculus is true for two arbitrary Heisenberg operators,

$$S(d(ab)) = adb + dab. \tag{14.31}$$

Gardiner and Collett have made a succinct comparison of these two definitions. The consequence is that the Stratonovich view is useful for formulating physical problems, since it maintains the ordering rule of ordinary calculus. However, the definition Eq. (14.27) is difficult to utilize theoretically. Theoretically, the Itô view is more useful. We may use either, depending on the problem at hand.

From the Langevin equation, we may obtain the master equation of Chapters 3 and 10. Assume initially, at $t_0 = t$,

$$\rho = \rho_S(0) \otimes \rho_B(0).$$

Then, for a given operator,

$$\langle a(t)\rangle = \operatorname*{Tr}_S(a(0)\,\rho(t))$$

with exactly

$$\rho(t) = \operatorname*{Tr}_B\left\{U^{\dagger}(t, 0)\,\rho_S(0) \otimes \rho_B(0)\,U(t, 0)\right\}.$$

The Itô stochastic differential equation is

$$I(da) = \frac{i}{\hbar}[a, H_S]\,dt + \frac{\gamma}{s}\left(\bar{N} + 1\right)\left[2c^{\dagger}ac - ac^{\dagger}c - c^{\dagger}ca\right]dt \\ + \frac{\gamma}{2}\bar{N}\left(2cac^{\dagger} - acc^{\dagger} - cc^{\dagger}a\right)dt \\ - \sqrt{\gamma}\left[a, c^{\dagger}\right]dB(t) + \sqrt{\gamma}\,dB^{\dagger}[a, c]\,dt. \tag{14.32}$$

From this we have the average equation

$$\frac{d\langle a\rangle}{dt} = \operatorname*{Tr}_S\left\{a\frac{d\rho}{dt}\right\},$$

where we identify the master equation for the density operator ρ,

$$\frac{d\rho}{dt} = \frac{i}{\hbar}[\rho, H_S] + \frac{\gamma}{2}\left(\bar{N}+1\right)(2c\rho c^\dagger - c^\dagger c\rho - \rho c^\dagger c) + \frac{\gamma}{2}\bar{N}\left(2c^\dagger\rho c - cc^\dagger\rho - \rho cc^\dagger\right). \tag{14.33}$$

This is in the Lindblad form already discussed in Chapters 3, 6 and 10. Hence Eq. (14.33), for the density operator, is physically equivalent to the quantum Langevin equation, Eq. (14.13). Since the Langevin equation is impossible to solve, it is better to use the master equation approach.

In Eq. (14.33), for the two-level atom, let $H_S = \frac{1}{2}\hbar\omega_0\sigma_z$ and $c \to \sigma^-$. It is left as an exercise for the student to write down the appropriate quantum Langevin equations.

14.3 Quantum Langevin equation with measurement

Let us return in this section to measurement. This work has already been mentioned in Chapter 13, particularly Section 13.5. Because of the relation to the Langevin approach, we reconsider it here.

The aim of the work of Ghirardi is to replace the isolated system Schrödinger equation with a stochastic Langevin-type equation in Hilbert space which incorporates measurement and thus the wave function collapse (Bassi and Ghirardi, 2003). Consider the assumed linear Itô equation for the ensemble of wave function $|\psi\rangle$ which obeys

$$I\left(d\,|\psi\rangle\right) = Cdt + \mathbf{A}\cdot\mathbf{dB}\,|\psi\rangle\,. \tag{14.34}$$

$\mathbf{A}$ are a set of operators, C an operator, $C - C^\dagger = -\frac{i}{\hbar}H$, and $\mathbf{dB}$ is a set of real Wiener processes such that

$$\langle dB_i\rangle = 0 \tag{14.35}$$

and

$$\left\langle dB_i dB_j\right\rangle = \gamma\delta_{ij}dt.$$

This does not preserve the norm $\|\psi(t)\|^2$. The $\langle\rangle$ indicate ensemble averages of random processes generated by $\mathbf{dB}$. Define $P_\psi = |\psi|^2$ (not normalized!), $|\psi\rangle$ being a solution to Eq. (14.34) and

$$d\,(\text{Phy}) = P_\psi\,\|\psi\|^2 = \text{“physical” probability} = \|\phi\|^2 \tag{14.36}$$

$$\int d\,(\text{Phy}) = 1.$$

We assume a new ensemble such that $d\left\langle \|\psi\|^2 \right\rangle = 0$, giving the conditions

$$C + C^{\dagger} = -\gamma \mathbf{A}^{\dagger} \cdot \mathbf{A} \tag{14.37}$$

and thus

$$d\,\|\psi\|^2 = \left\langle \psi \left| \left(\mathbf{A} + \mathbf{A}^{\dagger}\right) \right| \psi \right\rangle \cdot d\mathbf{B}. \tag{14.38}$$

We may obtain an Itô equation for a state $|\phi\rangle$, giving the probability, Eq. (14.36), from the new ensemble. It obeys

$$\begin{aligned} I\,(d\,|\phi\,(t)\rangle) &= \left[C - C^{\dagger} - \frac{1}{2}\gamma\,(\mathbf{A} - \mathbf{R})^2\,dt + (\mathbf{A} - \mathbf{R}) \cdot d\mathbf{B}\right] |\phi\,(t)\rangle \\ R &= \langle \phi\,|A|\,\phi\rangle \end{aligned} \tag{14.39}$$

for $A = A^{\dagger}$. Implicit in this now nonlinear stochastic operator equation is the calculation of the physical average, Eq. (14.36). But now the probability is obtained with $|\phi\rangle$ by the usual rule. A similar equation has been proposed by Gisin, Pearle and Diosi (Gisin, 1984a and b; Pearle, 1984; Diosi, 1988, 1989).

We may also carry this out in a Stratonovich way. Assuming **A** are self-adjoint and that the ensemble is Gaussian white noise, the linear Stratonovich equation corresponding to Eq. (14.34) is

$$S\frac{d}{dt}\,|\psi\,(t)\rangle = \left[C - C^{\dagger} + \mathbf{A} \cdot \mathbf{V}\,(t) - \gamma \mathbf{A}^2\right] |\psi\,(t)\rangle\,, \tag{14.40}$$

where $\langle V\,(t)\rangle = 0$, and

$$\left\langle V_i(t_1) V_j\,(t_2)\right\rangle = \gamma \delta_{ij} \delta\,(t_1 - t_2)\,. \tag{14.41}$$

The physical probability is

$$\mathrm{Phy}\,[\psi\,(t)] = P_{\psi}\,\|\psi\,(t)\|^2 \equiv \||\phi\,(t)\rangle|^2\,. \tag{14.42}$$

From this the nonlinear Stratonovich equation for $|\phi\,(t)\rangle$ is

$$S\frac{d}{dt}\,|\phi\,(t)\rangle = \left[\begin{array}{c} C - C^{\dagger} + (\mathbf{A} - \mathbf{R}) \cdot \mathbf{V}\,(t) \\ -\gamma\,(\mathbf{A} - \mathbf{R})^2 + \gamma\left(\mathbf{Q}^2 - \mathbf{R}^2\right) \end{array}\right] |\phi\,(t)\rangle\,, \tag{14.43}$$

where $\mathbf{R} = \langle \phi\,|\mathbf{A}|\,\phi\rangle$ and $Q^2 = \left\langle \phi\left|\mathbf{A}^2\right|\phi\right\rangle$. We choose a single A and assume a Wiener process with no sample path memory,

$$B\,(t) = \int_0^t d\tau V\,(\tau)\,. \tag{14.44}$$

We have a nonlinear Brownian process for the state vector $|\phi\,(t)\rangle$. Instead of solving this nonlinear equation, we solve the linear equation, Eq. (14.40). We assume a two-level state given by the α, β eigenvalues of A. Taking initially

$\psi(0) = P_\alpha |\psi(0)\rangle + P_\beta |\psi(0)\rangle$ and neglecting $C - C^\dagger$, the Hamiltonian, the solution to the linear Stratonovich equation is

$$|\psi(t)\rangle = \exp\left(\alpha B(t) - \alpha^2 \gamma t\right) P_\alpha |\psi(0)\rangle + \exp\left(\beta B(t) - \beta^2 \gamma t\right) P_\beta |\psi(0)\rangle, \tag{14.45}$$

where $P_\alpha = |\alpha\rangle\langle\alpha|$ and $P_\beta = |\beta\rangle\langle\beta|$. Since $V(t)$ is a Gaussian ensemble, we then have, from a Fokker–Planck equation solution,

$$\mathrm{Phy}[\omega(t)] = \|P_\alpha |\psi(0)\rangle\|^2 \frac{1}{\sqrt{2\pi\gamma t}} \exp\left(\frac{-1}{2\gamma t}\left[B(t) - 2\gamma\alpha t\right]^2\right) + \left\|P_\beta |\psi(0)\rangle\right\|^2 \frac{1}{\sqrt{2\pi\gamma t}} \exp\frac{-1}{2\gamma t}\left[B(t) - 2\gamma\beta t\right]^2. \tag{14.46}$$

This is classical Brownian motion in a state space $|a\rangle$, which are eigenfunctions of A. The ensemble is sampled by $B(t)$. There are no interference terms from $|\psi(t)\rangle$, since $\mathrm{Phy}[\psi(t)] = |\phi|^2$ represents the collapse of the wave function to $|\alpha\rangle$ or $|\beta\rangle$. The effective state space diffusion coefficient is $\gamma/2$.

$\mathrm{Phy}[\psi(t)]$ must be normalizable so that as $t \to \infty$ the ensemble $B(t \to \infty)$ must be contained in a width $\sqrt{\gamma t}$ near either $2\gamma\alpha t$ or $2\gamma\beta t$. Note that as $t \to \infty$, the value of $\sqrt{\gamma t}$ to $2(\alpha - \beta)\gamma t$ tends to zero. The rate of collapse, $(2\gamma)^{-1}$, is determined by the white noise constant γ. All this seems interesting, but what is the source of this continuous white noise (in this case) which is appended to the dynamics of the Schrödinger equation, thus leading to a Langevin quantum dynamics via the nonlinear Itô equation, Eq. (14.39)? This is the point of much discussion (Bassi and Ghirardi, 2003). We shall not take it up here, as our purpose is to *introduce* the reader to this interesting equation in Hilbert space.

Finally, we remark that, by means of the Itô equation and the definition of $\mathrm{Phy}[\psi(t)]$, we may obtain an equivalent density operator $\rho(t)$ in the same fashion as in the previous section. It is

$$\frac{d\rho}{dt} = C - C^\dagger + \gamma \mathbf{A}\rho(t)\mathbf{A}^\dagger - \frac{\gamma}{2}\left[\mathbf{A}^\dagger \cdot \mathbf{A}, \rho(t)\right]_+; \qquad t > 0. \tag{14.47}$$

Interestingly, it is of the Lindblad form also, so ρ obeys a completely positive semi-group equation. A good question is whether or not this is the most general quantum Brownian motion equation.

References

Bassi, A. and Ghirardi, G. C. (2003). *Arcive Quantum Physics* (Trieste, A. Salam Institute).

Collett, M. J. and Gardiner, C. W. (1984). *Phys. Rev. A*, 1386.

Diosi, L. (1988). *Phys. Lett A* **129**, 419.

Diosi, L. (1989). *Phys. Rev. A* **40**, 1165.
Gardiner, C. W. (1983). *Handbook of Stochastic Methods* (Berlin, Springer).
Gardiner, C. W. (1991). *Quantum Noise* (Berlin, Springer).
Gardiner, C. W. and Collett, M. J. (1985). *Phys. Rev.* **31**, 3761.
Gisin, N. (1984a). *Phys. Rev. Lett.* **52,** 1657.
Gisin, N. (1984b). *Phys. Rev. Lett*. **53**, 1776.
Haken, H. (1984). *Laser Theory* (Berlin, Springer).
Lax, M. (1967). *Phys. Rev.* **157**, 213.
Pearle, P. (1984). *Phy. Rev. Lett.* **53**, 1775.
Senitzky, I. R. (1960). *Phys. Rev.* **119**, 670.
Senitzky, I. R. (1961). *Phys. Rev.* **124**, 642.
Wax, N. (1954). *Selected Papers on Noise and Stochastic Processes* (New York, Dover).

15

Linear response: fluctuation and dissipation theorems

15.1 Introduction

Linear response is the perturbative steady state and temporal description of a system in interaction with a reservoir, thermal and/or mechanical. We have already discussed this, implicitly, in Chapter 6, on dissipation. There the first topic was the thermodynamic description of linear response and the introduction of transport coefficients as well as the Onsager symmetries (Onsager, 1931). Chapter 6 also dealt with the results of the quantum Boltzmann kinetic equation approach, particularly in terms of the Chapman–Enskogg solution leading to the steady transport laws in gases (Chapman and Cowling, 1939).

Linear response theory, a parallel approximate description of system–reservoir interactions leading to "exact" closed equations for the transport coefficients, will be discussed in this chapter in detail. The related topic is steady fluctuations and their connection to the "dissipative" behavior due to the system–reservoir coupling. This leads to a general form of the fluctuation-dissipation theorems, which we will obtain. Let us now consider the simple classical origins of this.

Einstein, early in his treatment of Brownian motion, obtained the diffusion constant of the form (Einstein, 1905, 1910)

$$D = \frac{kT}{m\gamma}, \tag{15.1}$$

considering the diffusion current with the linear law

$$j(x) = -D\frac{\partial n(x)}{\partial x} + u_d n(x), \tag{15.2}$$

$n(x)$ being the concentration and u_d the drift velocity where

$$u_d = \frac{-1}{m\gamma} \times \frac{dV}{dx} \tag{15.3}$$

is the potential $V(x)$, and $m\gamma$ the dissipative friction constant. In equilibrium the two terms compensate each other. This leads to the Einstein relation Eq. (15.1); thus fluctuation and dissipation are apparently related. It can be seen more clearly by examining the classical Langevin equation already met in Chapter 14. Assume the stochastic Brownian motion equation for the particle velocity $u(t)$,

$$m\dot{u}(t) = -m\gamma u + F(t), \tag{15.4}$$

$F(t)$ being the stochastic random force. Assuming short correlation for the elements of the ensemble,

$$\langle F(t_1) F(t_2)\rangle = 2\pi G\, \delta(t_1 - t_2). \tag{15.5}$$

We may write a Fokker–Plank equation for the stochastic classical and random ensemble probability, $W(u_0, t_0; u, t)$,

$$\frac{\partial W}{\partial t} = \frac{\partial}{\partial u}\left(D\frac{\partial}{\partial u} + \gamma u\right) W, \tag{15.6}$$

where initially

$$W(u, t_0; u_0, t_0) = \delta(u - u_0). \tag{15.7}$$

The bath is at thermal equilibrium. Thus,

$$W(u_0 t_0; u\infty) = \text{const}\ \exp\left(-\frac{1}{2}\frac{mu^2}{2}\right) \tag{15.8}$$

and Eq. (15.1) follows. The process has been assumed to be Gaussian. Further, it may be shown heuristically that

$$D = \frac{1}{m^2}\int_0^\infty \langle F(t_0) F(t_0 + t)\rangle\, dt \tag{15.9}$$

and is an integral of the force time correlation function depending on the force *fluctuation* at equilibrium. There is a nice review by Callen (1985) of the calculation of the *equilibrium* fluctuations begun by Einstein (1910), utilizing the Boltzmann formula

$$d\Omega(u) = \exp\left[\frac{S(u)}{k}\right] du. \tag{15.10}$$

We will not discuss this here but focus on the non-equilibrium aspects (see, however, Chapter 7). First we will consider steady state linear response and then temporal. All of this is closely connected to dissipation, as discussed in Chapter 6. The reader should remind him- or herself of the results, particularly the entropy production theorem.

15.2 Quantum linear response in the steady state

Let us consider the statistical mechanics of an open system in the steady state, due to the work by McLennan (1959; see also Zubarev, 1975). We take the total Hamiltonian to be

$$H_T = H + H_R + V, \tag{15.11}$$

H being the system Hamiltonian with a time-independent interaction V with the surroundings, which we call the reservoir, H_R. The von Neumann equation for the universe, H_T, is again

$$\frac{\partial \rho}{\partial t} + \frac{1}{i}[\rho, H_T] = 0; \qquad \hbar = 1. \tag{15.12}$$

Let

$$f = \mathrm{Tr}_R \rho, \tag{15.13}$$

then

$$\frac{\partial f}{\partial t} + \frac{1}{i}\left[f, H\right] + \frac{1}{i}\mathrm{Tr}_R\left[\rho, V\right] = 0. \tag{15.14}$$

Now define X by

$$\rho = fX. \tag{15.15}$$

The X operator will be identified later, thermodynamically. We assume f near equilibrium initially at $t = -\infty$,

$$f = f_0\left(1 + \eta\right), \tag{15.16}$$

where

$$f_0 = z^{-1}\exp\left(-\beta H + \beta\mu N\right),$$

μ being the chemical potential and N the number of particles. The system is close to grand canonical equilibrium initially and will be changed slowly in time from that state by V. We may, to the lowest perturbation order, obtain an equation for $\eta\left(t\right)$:

$$\frac{\partial \eta}{\partial t}\left(t\right) + \frac{1}{\hbar}\left[\eta, H\right] = h \tag{15.17}$$

$$h = \frac{-1}{i f_0}\mathrm{Tr}_R\left[f_0 X, V\right].$$

This may be formally integrated to long time at $t = 0$:

$$\eta = \int_{-\infty}^{0} \exp\left(\varepsilon t'\right) h\left(t'\right) dt' \tag{15.18}$$

$$h\left(t\right) = \exp\left(iHt\right) h_0 \exp\left(-iHt\right).$$

The parameter $\exp(\varepsilon t')$ achieves a slow adiabatic turning on of the interaction to achieve a steady state. $1/\varepsilon$ is large compared with the time necessary to approach this steady state. This is admittedly not a rigorous discussion in the spirit of Spohn and Lebowitz (1978) used in Chapter 6. The critical reader should return to that discussion and allow us to proceed more physically here. $h\,(t)$ is a system Heisenberg operator,

$$h\,(t) = \exp\,(iHt)\,h_0 \exp\,(-iHt)\,.$$

The steady state system ensemble is

$$f = f_0 \left[1 + \int_{\infty}^{0} dt' \, \exp \varepsilon t' h_0\left(t'\right)\right]. \tag{15.19}$$

Now $h\,(t)$ must be related to thermodynamic forces discussed in Chapter 6. Let there be r reservoirs, $\left(R = \sum r\right)$, for which the system may be in grand canonical equilibrium;

$$f \to f_r = Z^{-1} \exp\left(-\beta_r H + \beta_r \mu_r N\right).$$

To obtain a linear thermodynamic description, we then take *initially*

$$f_0 = \left[1 - \left(\beta_r - \beta\right) H - \left(\beta_r \mu_r - \beta\mu\right)\right] f_r. \tag{15.20}$$

This imposes a condition on X_r:

$$\mathrm{Tr}_r \left[f_0 X_r, V_r\right] = 0. \tag{15.21}$$

Hence we write

$$h = \sum_r \left[\left(\beta - \beta_r\right) q_r - \left(\mu - \beta_r \mu_r\right) j_r\right], \tag{15.22}$$

where

$$\begin{aligned} q_r &= \frac{1}{i} \mathrm{Tr}_r \left[H, V_r\right] X_r \\ j_r &= \frac{1}{i} \mathrm{Tr}_r \left[N, V_r\right] X_r. \end{aligned} \tag{15.23}$$

Here we have identified the energy flux q_r from the reservoir r to the system and also the particle flux j_r. This is a general microscopic expression for the assumptions of linear irreversible thermodynamics (see Chapter 6).

The external force also may be added. In this case we take $\mathrm{Tr}_R\,[X, V] = 0$. The contribution of mechanical forces to h is

$$h = \frac{-1}{i f_0} \mathrm{Tr}_R \left[f_0 V\right] X. \tag{15.24}$$

Assuming $[H, V]$ commutes with H, we obtain the additional term for the power input from the external device:

$$h = -\beta W$$
$$W = -\frac{1}{i}\mathrm{Tr}\,[H, V]\,X. \tag{15.25}$$

We further observe, in Eq. (15.22), that X_r must be interpreted as a general macroscopic thermodynamic force. We write macroscopically

$$h = \frac{1}{k}\sum_r X_r \langle J_r \rangle \tag{15.26}$$

and identify the ensemble average

$$\langle J_r \rangle = \left\langle J_r \left[1 + \frac{1}{k}\sum_\alpha X_\alpha \int_{-\infty}^{0} \exp \varepsilon t\, J_\alpha dt \right] \right\rangle_0 . \tag{15.27}$$

The fluxes vanish at equilibrium, so

$$\langle J_\beta \rangle_0 = \sum_\alpha L_{\beta\alpha} X_\alpha, \tag{15.28}$$

and

$$L_{\beta\alpha} = \frac{1}{k}\int_{-\infty}^{0} \exp \varepsilon t \left\langle J_\beta(-\infty)\, J_\alpha(t) \right\rangle_0 . \tag{15.29}$$

$\langle\rangle_0$ indicates the ensemble average with respect to f_0. Eq. (15.28) is the general linear response steady state statement. The generalized flux is the result of the thermodynamic linear forces, X_α. We now have a further result from the microscopic theory, a general formula for the transport coefficients $L_{\beta\alpha}$ which is a microscopic flux time correlation function. This is the Green–Kubo formula (Green, 1954; Kubo *et al.*, 1957). We leave it to the student to prove, from Eq. (15.29) and the mechanical equations of motion, that the Onsager symmetry follows (see Chapter 6):

$$L_{\beta\alpha} = L_{\alpha\beta}. \tag{15.30}$$

We may also prove the entropy production theorem of Chapter 6. The positive integral

$$I = \left\langle \left[\int_{-\infty}^{0} \exp \varepsilon t\, J_\alpha(t) \right]^2 \right\rangle_0 \geq 0 \tag{15.31}$$

may be rewritten

$$I = \int_{-\infty}^{0} \exp(\varepsilon t)\, dt \int_{-\infty}^{0} dt' \left\langle J_\alpha(-\infty)\, J_\alpha\left(t' - t\right)\right\rangle_0 \tag{15.32}$$

$$= \int_{-\infty}^{0} \exp(\varepsilon t)\, dt \int_{-\infty}^{-t} ds \exp \varepsilon (s + t) \left\langle J_\alpha(-\infty)\, J_\alpha(s)\right\rangle_0 . \tag{15.33}$$

Doing a partial integration, this gives

$$I = \frac{1}{\varepsilon} \int_{-\infty}^{0} \exp \varepsilon t \left\langle J_\alpha(-\infty)\, J_\alpha(t)\right\rangle dt. \tag{15.34}$$

Thus,

$$L_{\alpha a} = \frac{\varepsilon}{k} I \geq 0. \tag{15.35}$$

The diagonal elements are positive in any representation. Hence the thermodynamic entropy production is

$$\sigma = \sum_{\alpha,\beta} X_\alpha L_{\alpha\beta} X_\beta \geq 0.$$

This result of the Green–Kubo form has already been discussed in Chapter 6.

15.3 Linear response, time dependent

We will now *not* turn on the linear response adiabatically but be interested, in particular, in the time-frequency dependence of the weak response near initial equilibrium (see Kubo, 1969; Chester, 1969). Again we assume the system is in equilibrium initially, now $t = 0$, and we have

$$\rho_0(0) = Z^{-1} \exp(-\beta H + \beta\mu N).$$

Take the potential V to be a perturbation turned on, not necessarily slowly, at $t = 0$. Expanding near $t = 0$, we have

$$\rho(t) = \rho_0 + \rho',$$

where

$$\rho'(t) = -i \int_0^t dt' \exp -iH\left(t - t'\right)\left[V, \rho_0\right] \exp +iH\left(t - t'\right). \tag{15.36}$$

The ρ_0 appears in the right side by iteration around $t = 0$, assuming ρ' to be small due to V. Thus we have linear response. V may be explicitly time dependent. Assume the form $V = BF(t)$. B is a Hermitian operator. $F(t)$, a c-number, contains the form of the time dependence.

Let us now calculate the response of $\langle A(t)\rangle$ to this. A is another hermitian operator. From Eq. (15.36), we have

$$\langle A(t)\rangle = \mathrm{Tr}\left(\rho', A\right) = -i\int_0^t \mathrm{Tr}\left\{A\left[B\left(t'-t\right), \rho_0\right]F\left(t'\right)\right\},$$

where

$$B(t) = \exp(+iHt)\,B\exp(-iHt)$$

is the unperturbed Heisenberg representation and $\langle C\rangle_0 = \mathrm{Tr}(\rho_0 C)$, since the average is on ρ_0. Changing $t - t' \rightarrow \tau$ and utilizing the cyclic trace property, we have

$$\langle A(t)\rangle = -i\int_0^t d\tau\,\langle[A(\tau), B]\rangle_0\,F(t-\tau). \tag{15.37}$$

We define

$$\phi_{AB} = -i\,\langle[A(\tau), B]\rangle_0 \tag{15.38}$$

to be the response function of A to B. The arguments made in Chapter 6 are necessary to remove the non-Markovian memory and extend the limit to $t = \infty$. We assume this to be so. Then

$$\langle A(t)\rangle = -i\int_0^\infty d\tau\,\langle[A(\tau), B]\rangle_0\,F(t). \tag{15.39}$$

We need only mention that Eq. (15.37) is an almost periodic function of time, and it would seem necessary to use the thermodynamic limit here also. Eq. (15.37) is certainly irreversible (see Chapter 5). The simplification is that $\langle\rangle_0$ is an equilibrium ensemble average. Kubo was the first to derive such an equation (Kubo, 1957). We have already met these results in the previous section. A simple result may be obtained with external forces (-βW earlier). We choose $V = -\mathbf{P}\cdot\mathbf{E}$, $\mathbf{P}$ being the polarization, $\mathbf{E}$ the external electric field. We choose to consider the response of the electric current J_α. Assuming $F(t)$ is constant,

$$\langle J_\alpha\rangle = i\sum_\beta\int_0^\infty d\tau\left\langle\left[J_\alpha, P_\beta\right]\right\rangle_0 E_\beta.$$

We identify the conductivity tensor as

$$\sigma_{\alpha\beta} = i\int_0^\infty d\tau\left\langle\left[J_\alpha, P_\beta\right]\right\rangle_0,$$

a Green–Kubo formula for the coefficient $\sigma_{\alpha\beta}$.

Such an equation as Eq. (15.37) can be written in another form. We use the identity

$$\left[B, \rho_0\right] = \rho_0 \int_0^\beta d\lambda \left[B\left(i\lambda\right), H\right] \tag{15.40}$$

(the student should prove this) to obtain

$$\langle A \rangle = -i \int_0^t d\tau \int_0^\beta d\lambda \left\langle A \dot{B}\left(\tau - t + i\lambda\right)\right\rangle_0, \tag{15.41}$$

having a complex time evolution. The response function is then

$$\phi_{AB} = \int_0^\beta d\lambda \left\langle \dot{B}\left(t - i\lambda\right) A\right\rangle_0 \tag{15.42}$$

$$= -\int_0^\beta d\lambda \left\langle B \dot{A}\left(-t + i\lambda\right)\right\rangle_0. \tag{15.43}$$

The latter useful form utilizes the translational invariance.

For frequency-dependent electrical conductivity, we take $E = E_0 \exp\left(i\omega t\right)$. Then $\dot{A} = -J$, and $B = J/V$. We have

$$\left\langle \frac{J}{V} \right\rangle = \sigma\left(\omega\right) E\left(t\right) \tag{15.44}$$

and

$$\sigma\left(\omega\right) = \int_0^\infty dt \exp\left(i\omega t\right) \int_0^\beta d\lambda \frac{1}{V} \left\langle J\left(-t + i\lambda\right) J\right\rangle.$$

This is Kubo's formula for frequency-dependent conductivity (Kubo *et al.*, 1957; see also Kubo *et al.*, 1995).

As we have implied in Section 15.2, intrinsic transport is more difficult to deal with. What are the stimulus and response? There are two interesting tricks. Montroll (1959) pointed out that diffusion can take place either by an internal gradient or by an external gravitational interaction. The result is easy to see for an external gravitational force. Then $F = -mgz$. Thus, $A = z$ and $B = v_z$. $\langle v_z \rangle = \mu F$, and μ is the mobility. Now, from the response function,

$$\mu = \beta \int_0^\infty dt \left\langle v_z\left(t\right), v_z\right\rangle_0. \tag{15.45}$$

The diffusion coefficient may be defined as the autocorrelation function:

$$D = \int_0^\infty d\tau \left\langle v_z\left(\tau\right) v_z\left(0\right)\right\rangle. \tag{15.46}$$

Thus

$$D = kT\mu, \tag{15.47}$$

which is the Einstein relation mentioned above.

A similar remark may be made concerning the shear viscosity. Montroll also suggested that Feynman realized that an incompressible flow pattern may be generated by a suitable boundary perturbation. By canonical transformation this may be recast in a Hamiltonian time-dependent perturbation form with fixed boundaries, and treated by the methods mentioned here. The result is η, the shear viscosity, which may be written as

$$\eta = \beta \int_0^{\infty} dt \left\langle \frac{1}{2} \left[F_{xy}(0), F_{xy}(t) \right]_+ \right\rangle_0 .$$

F_{xy} is a volume integrated momentum flux. The thermal conductivity is another matter but was treated by methods resulting from the discussion in Section 15.2 (McLennan, 1960, 1989). McLennan showed on a relevant time scale, classically, that the frequency-dependent thermal conductivity may be written exactly:

$$\lambda(\omega) = \frac{1}{VkT^2} \int_{-\infty}^{0} \exp(i\omega t)\, dt\, \langle S(0)\, S(t) \rangle_0 . \tag{15.48}$$

S is the total energy flux. By means of the Chapman–Enskogg methods, this may be shown to give the Boltzmann answer discussed in Chapters 4 and 6. Such formulas, as Eq. (15.45) and Eq. (15.46), have been obtained by McLennan much more systematically, but that is too lengthy to discuss here (McLennan, 1989).

From this discussion it is clear that transport coefficients may be written exactly and in a somewhat independent way by means of the linear response approach. However, their evaluation requires solutions to kinetic equations or the knowledge of Green's function solutions, which will be discussed in the next chapter.

15.4 Fluctuation and dissipative theorems

The term "dissipative" might seem to indicate that the present section is closely related to the discussion of Chapter 6. There it was stated that the transport laws are fundamentally dissipative, as emphasized by the separation in the Chapman–Enskogg procedure of the local equilibrium hydrodynamic quantities from the dissipative part D_{ij}^* S_i^* in Eq. (6.12). This is related to the entropy production $\sigma \geq 0$, since from Eq. (6.20),

$$\sigma = \lambda T^{-2} (\Delta T)^2 + 2T^{-1} \eta \left(D_{ij} \right)^2$$

is positive because λ and η are positive. This is the local entropy production and may be time dependent through local equilibrium variables such as $T(x, t)$.

Now we will turn to a somewhat related topic of fluctuation dissipation theorems. This is a misnomer, in a sense, since such theorems may also be true for nondissipative systems. There is a variety of these theorems, possibly the first being due to Nyquist (1928), which had to do with electrical circuits, as we shall see. The first generalization is a general susceptibility fluctuation theorem due to Callen and Welton (1951). They showed that the mean square of the fluctuating force $\langle V^2 \rangle$ may be related to $R(\omega)$ such that

$$\langle V^2 \rangle = \left(\frac{2}{\pi}\right) \int_0^\infty R(\omega)\, E(\omega, T)\, d\omega,$$

where $R(\omega)$ is the resistance and

$$E(\omega, T) = \left\{ \hbar\omega + \hbar\omega \left[\exp\left(\frac{\hbar\omega}{kT}\right) - 1 \right]^{-1} \right\}.$$

See also the book by Landau and Lifshitz (1980).

To consider such relations in general, we introduce the Fourier transform of the response function (Kubo, 1969), Eq. (15.38):

$$\chi_{BA}(\omega) \equiv \int_0^\infty dt \exp(-i\omega t)\, \phi_{BA}(t), \tag{15.49}$$

where

$$\phi_{BA}(t) = i^{-1} \langle [A(0), B(t)]_- \rangle_0$$
$$\text{or } = -\int_0^\beta i\lambda \langle B\dot{A}(-t + i\lambda) \rangle_0 .$$

We call χ_{BA} a generalized susceptibility. Kubo (1969) considered such correlations and their symmetries, defining

$$\langle X; Y \rangle_0 \equiv \frac{1}{\beta} \int_0^\beta d\lambda \mathrm{Tr}\rho_0 \exp(\lambda H)\, X \exp(-\lambda H)\, Y, \tag{15.50}$$

and wrote Eq. (15.49) in the diagonal representation of H and also considered the symmetrized equilibrium correlation

$$\langle [AB(t) + B(t)A] \rangle_0 \equiv \langle [A, B(t)]_+ \rangle_0$$

in the diagonal representation. A term-by-term comparison shows that

$$\chi_{BA} = \chi'_{BA} + i\chi''_{BA} \tag{15.51}$$

(see Chapter 16). We have

$$\chi''_{BA}(\omega) = \hbar^{-1} \left(\tanh \frac{1}{2}\beta\hbar\omega\right) \langle [A, B(t)]_+ \rangle_0 . \tag{15.52}$$

These relations between the imaginary parts of the susceptibilities to the *equilibrium* correlation function are called the general fluctuation dissipation theorem.

Case (1971) has given a critical review of such relations. He pointed out, because of $\tanh 0 = 0$, the inversion of Eq. (15.52) is

$$\left\langle [B(t), A]_+ \right\rangle_0 = \hbar \coth\left(\frac{1}{2}\beta\hbar\omega\right) \chi''_{BA}(\omega) + C\delta(\omega) . \tag{15.53}$$

The δ function arises because of such a term in the expansion of χ'' A principal part is also present in χ' in this series. C is arbitrary. Thus the inversion is *not* unique. However, physical results may be obtained. Landau and Lifshitz (1980) discuss in detail $\chi'_{BA}(\omega)$ and $\chi''_{BA}(\omega)$ and their symmetry.

Let us briefly write the symmetry properties of both $\left\langle [X(0), Y(t)]_+ \right\rangle$ and $\langle X; Y \rangle$. The second case is the same as the first:

1. Stationarity-equilibrium:

$$\left\langle [X(0), Y(t)]_+ \right\rangle = \left\langle [X(t_0), Y(t_0 + t)]_+ \right\rangle .$$

2. If X, Y are hermitian, $\left\langle X^2 \right\rangle \geq 0$.
3. Time inversion:

$$\left\langle [X(0), Y(t)]_+ \right\rangle = \left\langle [Y(0), X(-t)]_+ \right\rangle .$$

With time reversal, let H be a classical external magnetic field and $\varepsilon_X = \pm 1$, for even $(+1)$ and for odd momentum (-1) dependence, we have

$$\left\langle [X(0), Y(t)]_+ \right\rangle_H = \varepsilon_X \varepsilon_Y \left\langle [X(0), Y(-t)]_+ \right\rangle_{-H} = \varepsilon_X \varepsilon_Y \left\langle [Y(0), X(t)]_+ \right\rangle_{-H} .$$

Now, $\phi_{BA}(t)$ has been said to be dissipative in the sense of Landau and Lifshitz. The expression for energy dissipation is proportional to $\chi''_{BA'}$, the imaginary part of χ. It may, however, be complex. Let us consider this. We take the rate of work on the system by an external "force" to be

$$\frac{dW}{dt} = X \frac{df}{dt} . \tag{15.54}$$

For a harmonic driving, which is real,

$$f(t) = \frac{1}{2}\left(f_0 \exp(-i\omega t) + f_0^* \exp(i\omega t)\right) . \tag{15.55}$$

We have

$$\bar{X} = \frac{1}{2}\left[\chi(\omega) f_0 \exp(-i\omega t) + \chi(-\omega) f_0^* \exp(i\omega t)\right] , \tag{15.56}$$

$\chi(\omega)$ being the susceptibility. The average time rate of work is

$$\frac{d\bar{W}}{dt} = \frac{1}{4} i\omega \left(\chi^* - \chi\right) |f_0|^2 = \frac{1}{2} \omega \chi'' |f_0|^2 . \tag{15.57}$$

If this work is dissipative, with it being turned only into entropy change, then the condition χ'' is positive. This association of χ'' to *system* dissipation is not compelling, and the word "dissipation" should not be used in this context. An additional fact should be added. There are dispersion relations relating $\chi'(\omega)$ and $\chi''(\omega)$. They are general. These Kramers–Kronig relations are derived in the text of Kubo (Kubo *et al.*, 1995). We will also prove them in the next chapter. They are a result of the Plemelj formulas of complex integration (see Balescu, 1963). They are

$$\chi'(\omega) = \chi^{\infty} + \frac{1}{\pi}\int_{-\infty}^{+\infty} d\omega' \frac{P}{\omega' - \omega}\chi''(\omega') \tag{15.58}$$

$$\chi''(\omega) - \frac{-1}{\pi}\int_{-\infty}^{+\infty} d\omega' \frac{P}{\omega' - \omega}[\chi'(\omega') - \chi^{\infty}].$$

Let us illustrate this further with the simple example of electrical transport where the susceptibility χ_{BA} is the frequency-dependent electrical tensor $\sigma_{\mu\nu}(\omega)$ (Kubo, 1969). Take this as

$$\sigma_{\mu\nu}(\omega) = \int_{0}^{\infty} \phi_{\mu\nu}(t) \exp(-i\omega t). \tag{15.59}$$

We write a Fourier transform, $f_{\mu\nu}(\omega)$:

$$\phi_{\mu\nu}(t) = \int_{-\infty}^{+\infty} d\omega' f_{\mu\nu}(\omega') \exp(-i\omega' t). \tag{15.60}$$

Now, by time symmetry, $f^{*}_{\mu\nu}(\omega) = f_{\nu\mu}(\omega)$. The tensor $\sigma_{\mu\nu}$ is divided into symmetric and anti-symmetric pieces. We further utilize

$$\int_{0}^{\infty} dt \exp(i\omega t) = \pi\delta(\omega) + i\frac{P}{\omega} \tag{15.61}$$

and find

$$\begin{aligned} f^{*s}_{\mu\nu}(\omega) &= f^{s}_{\mu\nu}(-\omega) && \text{real} \\ f^{a}_{\mu\nu}(\omega) &= -f^{a}_{\mu\nu}(-\omega) && \text{purely imaginary.} \end{aligned}$$

Now Eq. (15.59) is

$$\sigma_{\mu\nu}(\omega) = \frac{1}{2}f_{\mu\nu}(\omega) + \frac{-i}{2\pi}\int_{-\infty}^{+\infty} d\omega' \frac{P}{\omega' - \omega} f_{\mu\nu}(\omega'). \tag{15.62}$$

As before, let $\sigma'_{\mu\nu}$ be the real and $\sigma''_{\mu\nu}$ the imaginary part of the conductivity susceptibility. The results are then

$$\sigma'^{s}_{\mu\nu}(\omega) = \frac{1}{2} f^{s}_{\mu\nu}(\omega) \tag{15.63a}$$

$$\sigma''^{s}_{\mu\nu}(\omega) = -\frac{1}{2} \int_{-\infty}^{+\infty} d\omega' \frac{P}{\omega' - \omega} f^{s}_{\mu\nu}(\omega') \tag{15.63b}$$

$$\sigma'^{a}_{\mu\nu}(\omega) = -\frac{1}{2} \int_{-\infty}^{+\infty} d\omega' \frac{P}{\omega' - \omega} f^{a}_{\mu\nu}(\omega') \tag{15.63c}$$

$$\sigma''^{a}_{\mu\nu}(\omega) = -\frac{1}{2} f^{a}_{\mu\nu}(\omega). \tag{15.63d}$$

Taking the inverse transform, we may then write Eqs. (15.63a) and (15.63d) in terms of the response function as

$$\sigma'^{s}_{\mu\nu}(\omega) = \frac{1}{2} \int_{-\infty}^{+\infty} dt \phi^{s}_{\mu\nu}(t) \cos\omega t \tag{15.64}$$

$$\sigma'^{a}_{\mu\nu}(\omega) = \frac{1}{2} \int_{-\infty}^{+\infty} dt \phi^{a}_{\mu\nu}(t) \sin\omega t.$$

In this pair of equations, we have extended $\phi_{\mu\nu}(t)$ to negative time using $\phi_{\mu\nu}(t) = \phi_{\nu\mu}(-t)$, and thus $\phi^{s}_{\mu\nu}(-t) = \phi^{s}_{\mu\nu}(t)$, $\phi^{a}_{\mu\nu}(-t) = -\phi^{a}_{\mu\nu}(t)$. For the electrical conductivity we have

$$\phi^{s}_{\mu n}(t) = \frac{1}{2} \left\langle \left[J_{\mu}(0), J_{\nu}(t) \right]_{+} \right\rangle, \tag{15.65}$$

and thus Eqs. (15.64) are time-dependent Green–Kubo formulas. Eq. (15.63a) and Eq. (15.63d) are a form of the Nyquist–Callen–Welton theorem. It must be remembered that there are yet relationships of the form of Eq. (15.63b) and Eq. (15.63c) which may be utilized. To obtain the Nyquist theorem, we consider the symmetrized time correlation function $\phi^{s}_{\mu\nu}(t)$ and show that the Fourier transform, now called $\phi^{s}_{\mu\nu}(\omega)$, is again

$$\phi^{s}_{\mu\nu}(\omega) = \frac{\hbar\omega}{2} + \frac{\hbar\omega}{\exp(\beta\hbar\omega) - \frac{1}{2}} \tag{15.66}$$

$$\phi^{s}_{\mu\nu}(\omega) = \frac{\hbar\omega}{2} \coth\left(\frac{\beta\hbar\omega}{2}\right) f_{\mu\nu}(\omega).$$

Thus,

$$\phi^{s}_{\mu\nu}(t) = \frac{\hbar}{\pi} \int_{0}^{\infty} d\omega\omega \coth\left(\frac{\beta\hbar\omega}{2}\right) f^{s}_{\mu\nu}(\omega) \cos\omega t. \tag{15.67}$$

The factor $\frac{\hbar\omega}{2}\coth\left(\frac{\beta\hbar\omega}{2}\right)$ was obtained by Nyquist. Since this relates $\phi^{s}_{\mu\nu}(t)$ to the symmetric part of $f^{s}_{\mu\nu}(\omega)$, it is called a fluctuation (symmetric) dissipation theorem.

15.5 Comments and comparisons

Comparison of Chapters 3, 4 and 6 with this chapter indicates a considerable difference in the derivation of transport coefficients, or what we may now term susceptibilities. In earlier chapters, on the kinetic description, the transport coefficients appear as a result of the solution to the *irreversible* transport equations by methods such as that of Chapman and Enskogg. These dissipative kinetic equations are obtained from either the B.B.G.Y.K. hierarchy or the generalized master equation by reduction procedures. It was emphasized that the method of Bogoliubov is an "exact" reduction. No ad hoc coarse-graining or stochastic *Stosszahlansatz* is employed. The procedure provides the form of the transport coefficient as well as the necessary solution.

In the linear response theory, an apparently exact formula for the susceptibility is immediately obtained. After the initial system equilibrium ensemble assumption, and as with linear thermodynamics, a truncation linear in the external field is obtained. It is, surprisingly, a reversible result depending on initial *equilibrium* correlations $\langle[A(t), B]\rangle_0$. The derivation is irreversible. This is, in fact, the same symmetry that exists in the Onsager derivation. (Some comments were made in Chapter 6.) No method of solution of this correlation function is given, and then, at the next stage, one must use the equivalent of the kinetic method or Green's function to obtain results. In some sense the two methods overlap. However, the conditions for the strong initial equilibrium assumption are not clear, nor is there a method for examining this basic assumption within the theory itself. Van Kampen (1971) has questioned the linear response approach. Kubo (Kubo *et al.*, 1995) has offered a rejoinder. We invite the student to look into this matter. The Green's function approach will be considered in the next chapter.

Balescu (1961) has bridged the gap between the two views to some extent. He introduced, classically (the quantum version has not been carried through), an external field in the exact Liouville equation by H^e:

$$H = H^i + H^e,$$

and assuming, just as in the linear response theory,

$$f^s_N(x, p, 0) = \alpha \exp\left(-\beta H^i\right).$$

Then, using the causal Liouville Green's function,

$$L\Gamma\left(xpt \mid \alpha' p' t'\right) = \theta\left(t - t'\right)\delta\left(x - x'\right)\delta\left(p - p'\right).$$

$L = \{H, \ \}$, and θ is the Heaviside function. The response of the electric current to this is exactly

$$\mathbf{J}(t) = e\sum_m \int dxdp \int dx'dp'\mathbf{v}_m\Gamma^i\left(xpt \mid x'p't'\right) f_N^0(x, p.0),$$

which we may show to have a Kubo equation form *linear in the field*:

$$J(\mathbf{t}) = -e^2\beta\sum_m\sum_n\int_0^t dt'\int dxdp\int dx'dp'$$
$$\mathbf{v}_m\Gamma^i\left(xpt \mid x'p't'\right)\mathbf{E}\left(t'\right)\cdot\mathbf{v}_n' f_0^N\left(x'p'\right).$$

The temporal response Γ of $\mathbf{J}$ to $\mathbf{E}$ is averaged over the equilibrium ensemble f_0^N. The Liouville equation approach has given the linear response. A quantum version of this is expected to be similar; it is just more complicated because of the necessity of utilizing the Wigner function Liouville equation.

There is an additional important point. Utilizing the Fourier representation of Prigogine and his colleagues (Prigogine, 1967), we may write the Laplace transform of $\mathbf{J}(t)$ as

$$\mathbf{J}(z) = -e^2\beta\sum_{m,n}\sum_k\int dpv_m\mathbf{E}(z)\left\langle 0\left|R^i(z)\right|k\right\rangle v_n f_k^0(p).$$

For a time-independent field, $E(z) = E\left(\frac{1}{-iz}\right)$, and the $\left\langle k\left|R(z)\right|k'\right\rangle$ is the Laplace and Fourier transform of the "resolvent" of the system Liouville Green's function:

$$\left\langle xp\left|R^i(z)\right|x'p'\right\rangle = \int_0^\infty dz\exp(-iz\tau)\Gamma^i\left(xp \mid x'p', \tau\right).$$

In the steady state, only $\left\langle 0\left|R^i(z)\right|0\right\rangle$ appears in $\mathbf{J}(z)$. By the analytic properties of $\langle 0\,|R(z)|\,0\rangle$, we may show

$$\left\langle 0\left|R^i(z)\right|0\right\rangle = \left(\frac{1}{2}\right)\frac{1}{z - i\psi(z)},$$

where $\psi(z)$ is a holomorphic operator. It is a p space differential operator in the upper half z-plane. The details may be expressed by perturbation theory. The steady transport properties depend on $\psi(z)$ and not on the full irreversible operator as expressed by the Γ^i Green's function. Thus, a tool for the calculation of the $J(z)$

or $J(t)$ is in our hands, as well as a formula for the transport. The final formula expressing this is

$$\mathbf{J} = \mathbf{E}e^2\beta \sum_m \sum_n \int dp\mathbf{v}_m \frac{1}{-\psi(0)} \mathbf{v}_n f_N^0(p),$$

where

$$\langle 0 | R(z) | 0 \rangle = \frac{1}{\psi(0)}.$$

These generalities do not answer the question of whether or not the Green–Kubo type linear response formula gives the same answer as the kinetic equation approach. Mori (Mori *et al.*, 1961) was the first to show that they were the same for the dilute gas in the Chapman–Enskogg approach. Résebois has extended the previous results to inhomogeneous systems utilizing the classical diagrammatic methods of Severne (Résebois, 1964; Severne, 1965). In a very elaborate calculation, Résebois showed that in a dense gas the kinetic approach and that of linear response gave the same answer.

In his derivation of the kinetic transport coefficients to higher order in the density, McLennan (1989), has shown that the Green–Kubo formulas hold. However, there are limitations to this in the failure of the formulation due to long time effects (the "tails"). McLennan discusses this in some detail.

The reader, reconsidering Chapter 6, will rightly accuse us of "glossing over" the question of time scales. Balescu (1961) has discussed this to some extent. The alert reader will rightly suggest that H^i contains the interaction with a reservoir leading to dissipation as well as H^e. This has not been clearly discussed, but the reader should return to the comments of Spohn and Lebowitz (1978) for an overview of the more rigorous considerations of the approach to a transport steady state.

We conclude by reminding the reader that the response theory is general, being valid for *small* reversible quantum systems. It is useful in discussing the time-dependent susceptibility phenomena. We will comment extensively on small systems (particularly resistance) in Chapter 19.

References

Balescu, R. (1961). *Physica* **27**, 693.
Balescu, R. (1963). *Statistical Mechanics of Charged Particles* (New York, Interscience).
Callen, H. (1985). *Thermodynamics* (New York, Wiley).
Callen, H. and Welton, T. A. (1951). *Phys. Rev.* **83**, 34.
Case, K. M. (1971). *Trans. Th. Stat. Phys.* **2**, 129.
Chapman, S. and Cowling, T. G. (1939). *The Mathematical Theory of Non-uniform Gases* (Cambridge, Cambridge University Press).
Chester, G. B. (1969). In *Many Body Problems* (New York, W. A. Benjamin).

Einstein, A. (1905). *Ann. Phys. Lpz.* **17**, 549.
Einstein, A. (1910). *Investigations on the Theory of Brownian Movement* (New York, Dover).
Green, M. S. (1954). *J. Chem. Phys.* **20**, 1281 and **22**, 398.
Kubo, R. (1957). *J. Phys. Soc. Jpn.* **12**, 570.
Kubo, R. (1969). In *Many Body Problems* (New York, W. A. Benjamin).
Kubo, R., Yokota, M. and Nakajima, S. (1957). *J. Phys. Soc. Jpn.* **12**, 1203.
Kubo, R., Toda, M. and Hashitsume, N. (1995). *Statistical Physics II* (New York, Springer).
Landau, L. and Lifshitz, E. (1980). *Statistical Physics,* 3rd edn. trans. J. B. Sykes and M. J. Kearsley (Oxford, Butterworth-Hermann).
McLennan, J. A. (1959). *Phys. Rev.* **115**, 1405.
McLennan, J. A. (1960). *Phys. Fluids* **3**, 493.
McLennan, J. A. (1989). *Introduction to Non-equilibrium Statistical Mechanics* (Englewood Cliffs, N.J., Prentice Hall).
Montroll, E. (1959). In *Rend. Cont. della Scuela Institute di Fisica Enrica Fermi, Varenna* and *Lectures in Theoretical Physics,* Boulder **3**, 221.
Mori, H., Opopenheim I. and Ross, J. (1961). *Studies in Statistical Mechanics* (New York, Interscience).
Nyquist, H. (1928). *Phys. Rev.* **22**, 110.
Onsager, L. (1931). *Phys. Rev.* **37**, 405 and **38**, 2265.
Prigogine, I. (1967). *Non-equilibrium Statistical Mechanics* (New York, Interscience).
Résebois, P. (1964). *J. Chem. Phys.* **41**, 2979.
Severne, G. (1965). *Physica* **31**, 877.
Spohn, H. and Lebowitz, J. L. (1978). *Adv. Chem. Phys.* **38**, 109, ed. S. A. Rice and I. Prigogine.
Van Kampen, N. G. (1971). *Phys. Nouv.* **5**, 279.
Zubarev, D. N. (1974). *Non-equilibrium Statistical Thermodynamics,* trans. P. J. Shepherd (New York, Consultants Bureau).

16

Time-dependent quantum Green's functions

16.1 Introduction

Mathematically, given a linear differential operator $L_{\mathbf{x}}$,

$$L_{\mathbf{x}} = a_0(\mathbf{x}) + a_1(\mathbf{x})\frac{\partial}{\partial x_1} + a_{11}(\mathbf{x})\frac{\partial^2}{\partial x_1^2} + a_{12}(\mathbf{x})\frac{\partial^2}{\partial x_1 \partial x_2} \ldots a_{nn}\frac{\partial^2}{\partial x_n^2}$$
$$+ \cdots + a_{n\ldots n}(\mathbf{x})\frac{\partial^p}{\partial x_n^p},$$

one encounters the solution to the inhomogeneous differential equation

$$L_{\mathbf{x}}\phi(\mathbf{x}) = -\rho(\mathbf{x}). \tag{16.1}$$

Here $\rho(\mathbf{x})$ is a given source function. For a given boundary condition, we assume a solution to exist. The solution can be reduced to a simpler problem. Let

$$L_x G(\mathbf{x}, \mathbf{y}) = -\delta(\mathbf{x} - \mathbf{y}). \tag{16.2}$$

$G(\mathbf{x}, \mathbf{y})$ is the Green's function. This is a function of $\mathbf{x}$ with $\mathbf{y}$ a parameter. Take $G(\mathbf{x}, \mathbf{y})$ to satisfy the same boundary conditions as $\phi(\mathbf{x})$. Then

$$\phi(\mathbf{x}) = \int d\mathbf{y} G(\mathbf{x}, \mathbf{y}) \rho(\mathbf{y}), \tag{16.3}$$

since

$$\begin{aligned} L_{\mathbf{x}}\,\phi(\mathbf{x}) &= \int L_{\mathbf{x}} G(\mathbf{x}, \mathbf{y}) \rho(\mathbf{y})\, d\mathbf{y} \\ &= -\int \delta(\mathbf{x} - \mathbf{y}) \rho(\mathbf{y})\, d\mathbf{y} = -\rho(\mathbf{x}). \end{aligned}$$

An example of $L_{\mathbf{x}}$ is, of course, the Schrödinger operator

$$L_{\mathbf{x},t} = \left(-i\hbar\frac{\partial}{\partial t} + \frac{\hbar^2}{2m}V\right).$$

We take $\rho(\mathbf{x}) = V(\mathbf{x})\,\psi(\mathbf{x},t)$, $\psi(\mathbf{x},t)$ being the wave function and $V(\mathbf{x})$ the potential operator.

We are interested in Green's functions taken over from the techniques of quantum field theory (Schweber, 1961; Lifshitz and Petaevskii, 1981). We will concern ourselves particularly with one- and two-time Green's functions, since our principal interest is to show a connection to the calculations of linear response theory (Chapter 15) as well as to quantum kinetic equations. Then we wish to compare the methods with those described in Chapter 4. In this we will follow the work of L. P. Kadanoff and G. Baym (1962) and of L. V. Keldysh (1965) and also Zubarev (1974). We will not discuss equilibrium statistical mechanics utilizing Green's function techniques for many-body problems. The literature is exhaustive (see Abrikosov *et al.,* 1963; Fetter and Walecka, 1971). A good general introduction is the book by G. D. Mahan (2000).

16.2 One- and two-time quantum Green's functions and their properties

Let us introduce the creation operator $\psi^\dagger(r,t)$ and annihilation operator $\psi(r,t)$ of the second quantization formalism (see Schweber, 1961). They have the equal time commutation rules for Bose and Fermi particles:

$$\begin{aligned}
\left[\psi(r,t),\psi\left(r',t\right)\right]_\pm &= 0 \qquad + \ F.D. \\
\left[\psi^\dagger(r,t),\psi^\dagger\left(r',t\right)\right]_\pm &= 0 \qquad - \ B.E. \\
\left[\psi(r,t),\psi^\dagger\left(r',t\right)\right]_\pm &= \delta\left(r-r'\right).
\end{aligned} \tag{16.4}$$

The Hamiltonian operator for the particles is

$$\begin{aligned}
H = &\int dr \frac{\nabla\psi^\dagger(r,t)\,\nabla\psi(r,t)}{2m} \\
&+ \frac{1}{2}\int dr dr' \psi^\dagger(r,t)\,\psi^\dagger\left(r',t\right) V\left(\left|r-r'\right|\right)\psi\left(r',t\right)\psi(r,t),
\end{aligned} \tag{16.5}$$

and the number density of particles at $\mathbf{r}t$ is the operator

$$n(r,t) = \psi^\dagger(r,t)\,\psi(r,t). \tag{16.6}$$

Note that $r = (r_1 \ldots r_N)$ for $(1,2,3,\ldots,N)$ and $V\left(\left|r-r'\right|\right)$ is the pair potential depending, as in Chapter 4, on the scalar distance between the particles.

Now we define the one-particle time-dependent Green's function as

$$G\left(1,1'\right) = -i\left\langle T\psi(r_1t_1)\,\psi^\dagger\left(r_1,t_1'\right)\right\rangle. \tag{16.7}$$

Here the Wick chronological operator, for two operators A and B, is

$$TA(t)\,B\left(t'\right) = \theta\left(t-t'\right)A(t)\,B\left(t'\right) + \eta\theta\left(t'-t\right)B\left(t'\right)A(t)$$

where $\eta = \pm 1$, and the Heaviside function is

$$\theta(t) = 1 \quad t > 0$$
$$= 0 \quad t < 0.$$

The two-particle time-dependent Green's function is

$$G_2\left(12, 1'2'\right) = i^2 \left\langle \begin{array}{c} T\psi(r_1, t_1)\,\psi(r_2, t_2) \\ \times\,\psi^\dagger\left(r_2', t_2'\right)\psi^\dagger\left(r_1', t_1'\right) \end{array} \right\rangle. \qquad (16.8)$$

Of course, there is a hierarchy of these. Here,

$$\langle A \rangle = \mathrm{Tr}\left[\exp\left(-\beta\left(H - \mu N\right)\right) A\right]$$

means a grand canonical ensemble with μ the chemical potential. In addition, T is again the time-ordering operator of Wick (or chronological operator), and

$$T\psi(1)\,\psi^\dagger\left(1'\right) = \psi(1)\,\psi^\dagger\left(1'\right) \qquad \text{for } t_1 > t_1' \qquad (16.9)$$
$$= \pm\psi^\dagger\left(1'\right)\psi(1) \quad \text{for } \mathrm{t}_1 < t_1'.$$

The earliest time appears on the right, and the later time on the left with the introduction of ± 1 for Fermi particles. Here the $\pm$ depends on the evenness or oddness of the permutation of the original order.

These Green's functions may be further generalized to the retarded Green's function:

$$G_r^\pm\left(t, t'\right) = \theta\left(t - t'\right)(-i)\left\langle\left[\hat{A}(t), \hat{B}\left(t'\right)\right]_\pm\right\rangle \qquad (16.10)$$
$$(\theta(t) = 1 \text{ for } t > 0 \text{ and } 0 \text{ for } t < 0),$$

where the Heaviside function introduces a causality. $[\ ]_\pm$ are the anticommutator, commutator brackets and $\hat{A}$, $\hat{B}$ are arbitrary operator functions of ψ and $\psi^\dagger$. We also define an advanced Green's function,

$$G_a^\pm\left(t, t'\right) = i\theta\left(t' - t\right)\left\langle\left[\hat{A}(t), \hat{B}\left(t'\right)\right]_\pm\right\rangle, \qquad (16.11)$$

and we see the Green's functions in a special form of Eq. (16.10) and Eq. (16.11), which may be written as the causal Green's function (the correlation function),

$$G_c\left(t, t'\right) = -i\left\langle\psi(t), \psi^\dagger\left(t'\right)\right\rangle \equiv G^>\left(t, t'\right). \qquad (16.12)$$

(no commutator brackets!)

$$G\left(t, t'\right) = G^>\left(t, t'\right) \quad t > t'$$
$$G\left(t, t'\right) = G^<\left(t, t'\right) \quad t < t'.$$

It should be noted that as $\beta \to \infty$, these particle Green's functions go over to the field theoretic ones averaged over the vacuum. The double time-temperature

Green's functions were first introduced by Bonch-Bruevich (1956, 1957) and Bogoliubov and Tyablikov (1959) and reviewed in detail by Zubarev (1960, 1974).

The equations of motion for all these Greens's functions are easily obtained from the Heisenberg equations of motion for $\hat{A}(t)$, $\hat{B}(t)$ and the fact that

$$\frac{d}{dt}\left[\pm\theta\left(\pm\left(t-t'\right)\right)\right]=\delta\left(t-t'\right).$$

The equation for $G_r^{\pm}\left(t,t'\right)$ is

$$i\frac{dG_r^{\pm}}{dt}\left(t,t'\right)=\delta\left(t-t'\right)\left\langle\left[\hat{A},\hat{B}\right]_{\pm}\right\rangle+\theta\left(t-t'\right)\left\langle\frac{1}{i}\left[\hat{A}\left(t\right),H\right],\hat{B}\left(t'\right)]_{\pm}\right\rangle. \tag{16.13}$$

The Hamiltonian operator is assumed to be time independent. The right side contains new double time Green's functions for which equations of motion may be formulated, then the whole process repeated, forming a hierarchy of the appropriate Green's functions. This is not unexpected in the light of the quantum B.B.G.Y.K. hierarchy. Here the set of equations is supplemented by boundary conditions. This hierarchy must be uncoupled by supplemental assumptions. More will be said about this in Section 16.5.

16.3 Analytic properties of Green's functions

As emphasized by Landau (1958), it is the analytic properties of the Green's function approach which are important. Let us turn to this function approach now. We term this the spectral representation. Let $\mathfrak{H}\phi_{\nu}=E_{\nu}\phi_{\nu}$. The Fourier transform of the retarded Green's function is (dropping $\pm$ now),

$$G_r\left(t-t'\right)=\int_{-\infty}^{+\infty}G_r\left(E\right)\exp\left(-iE\left(t-t'\right)\right)dE \tag{16.14}$$

and the inverse

$$G_r\left(E\right)=\frac{1}{2\pi}\int_{-\infty}^{+\infty}G_r\left(t\right)\exp\left(iEt\right)dt.$$

Here, because of the equilibrium average, $G_r\left(t-t'\right)$ has the time dependence of the familiar time equilibrium correlation function already met in Chapter 14:

$$C_{BA}\left(t-t'\right)=\left\langle B\left(t'\right)A\left(t\right)\right\rangle. \tag{16.15}$$

We use Wick's time ordering to write for the $\pm$ commutator

$$G_r\left(t,t'\right)=-i\theta\left(t-t'\right)\left\{\left\langle A\left(t\right)B\left(t'\right)\right\rangle-\eta\left\langle B\left(t'\right)A\left(t\right)\right\rangle\right\};\qquad\eta=\pm1. \tag{16.16}$$

Thus,

$$G_r(E) = \frac{1}{2\pi i}\int_{-\infty}^{+\infty} dt \exp\left(iE\left(t-t'\right)\right)\theta\left(t-t'\right) \\ \times \left\{\left\langle A(t)\,B\left(t'\right)\right\rangle - \eta\left\langle B\left(t'\right)A(t)\right\rangle\right\}. \tag{16.17}$$

Following Zubarev (1974), we assume that ϕ_v are complete and ν discrete. We then may write the correlation functions in Eq. (16.17) as

$$\left\langle B\left(t'\right)A(t)\right\rangle = Z^{-1}\sum_{\nu,\mu}\left(\phi_\nu^* B(0)\,\phi_\mu\right)\left(\phi_\mu^* A(0)\,\phi_\nu\right) \\ \times \exp\left(-\frac{E_\nu}{\theta}\right)\exp\left\{i\left(E_\mu - E_\nu\right)\left(t-t'\right)\right\},$$

and similarly for $\left\langle A(t)\,B\left(t'\right)\right\rangle$. By interchanging the indices μ, ν and comparing these two expressions, we find

$$\left\langle B\left(t'\right)A(t)\right\rangle = \frac{1}{2\pi}\int_{-\infty}^{+\infty} J_{BA}(\omega)\exp i\omega\left(t'-t\right)d\omega \tag{16.18a}$$

and

$$\left\langle A(t)\,B\left(t'\right)\right\rangle = \frac{1}{2\pi}\int_{-\infty}^{+\infty} J_{BA}(\omega)\exp(\beta\omega)\exp\left(i\omega\left(t'-t\right)\right)d\omega, \tag{16.18b}$$

where

$$J_{BA}(\omega) = 2\pi Z^{-1}\sum_{\mu\nu}\left(\phi_\mu^* B(0)\,\phi_\nu\right)\left(\phi_\nu^* A(0)\,\phi_\mu\right) \\ \times \exp\left(-\beta E_\mu\right)\delta\left(E_\mu - E_\nu - \omega\right). \tag{16.19}$$

Eq. (16.18a) and (16.18b) are the spectral representations of time correlation functions introduced by Callen and Welton (1951), as previously mentioned. Now

$$J_{AB}(-\omega) = J_{BA}(\omega)\exp\beta\omega. \tag{16.20}$$

We note that because $\left\langle B\left(t'\right)A(t)\right\rangle$ depends on the time difference, the first equation is a statement of a Fourier transform. We leave it as an exercise for the student to prove that the second follows immediately. If now the limit exists, $\lim\left|t-t'\right| \to \infty$, then

$$\left\langle A(t)\,B\left(t'\right)\right\rangle = \langle A(t)\rangle\left\langle B\left(t'\right)\right\rangle.$$

If $\langle A\rangle = 0$, the right side is zero. We may then write

$$\left\langle B\left(t'\right)A(t)\right\rangle - \langle B\rangle\langle A\rangle = \frac{1}{2\pi}\int_{-\infty}^{+\infty} J_{BA}(\omega)\exp\left(i\omega\left(t'-t\right)\right)d\omega. \tag{16.21}$$

Since for finite systems the states are almost periodic functions, the above equation is true only in the thermodynamic limit ($N \to \infty$, N/V = constant). Now, by the Riemann–Lebesgue lemma, the right side of Eq. (16.21) is zero. These matters have already been met in Chapters 5 and 6.

Let us return to the Green's function. Using Eq. (16.17) we have the Fourier transform, $G_r(\omega)$:

$$G_r(\omega) = \frac{1}{2\pi}\int_{-\infty}^{+\infty} d\omega' J_{BA}(\omega')\left(\exp(\beta\omega') - \eta\right) \times (-i)\int_{-\infty}^{+\infty} dt \exp\left(i(\omega - \omega')t\right)\theta(t).$$

Using the representation

$$\delta(t) = \frac{1}{2\pi}\int_{-\infty}^{+\infty} \exp(-ixt)\,dx \tag{16.22}$$

$$\theta(t) = \frac{i}{2\pi}\int_{-\infty}^{+\infty} \frac{\exp(-ixt)}{x + i\varepsilon}dx, \tag{16.23}$$

we obtain, for $\varepsilon \to +0$, the retarded Green's function:

$$G_r(\omega) = \frac{1}{2\pi}\int_{-\infty}^{+\infty} \left(\exp\beta\omega' - \eta\right) J_{BA}(\omega')\frac{d\omega'}{\omega - \omega' + i\varepsilon}. \tag{16.24}$$

The advanced one is with $i\varepsilon \to -i\varepsilon$ in Eq. (16.24). $G(\omega)$ may be viewed as a function of the complex variable ω. Consider

$$G_r - G_a = G(\omega + i\varepsilon) - G(\omega - i\varepsilon) = \frac{1}{2\pi}\int_{-\infty}^{+\infty} d\omega' \left(\exp\left(\beta\omega' - \eta\right) J_{BA}(\omega')\right) \times \left\{\frac{1}{\omega - \omega' + i\varepsilon} - \frac{1}{\omega - \omega' - i\varepsilon}\right\}. \tag{16.25}$$

Now we use

$$\delta\left(\omega - \omega'\right) = \lim_{\varepsilon\to 0}\frac{1}{2\pi i}\left[\frac{1}{\omega - \omega' - i\varepsilon} - \frac{1}{\omega - \omega' + i\varepsilon}\right].$$

We have

$$G(\omega + i\varepsilon) - G(\omega - i\varepsilon) = \frac{1}{i}\left(\exp\ (\beta\omega - \eta)\right) J_{BA}(\omega) \equiv \frac{1}{i}f(\omega); \quad \omega \text{ real}. \tag{16.26}$$

The $G(\omega)$ has a discontinuity on the real axis. We shall call this the first Plemelj formula. $G(\omega)$ is a sectionally regular function if we assume the Hölder condition:

$$\|f(\omega_2) - f(\omega_1)\| \le A(\omega_2 - \omega_1)^{\mu} \text{ for } A > 0, \quad 0 \le \mu \le 1$$

(Muskelishvilli, 1953). We may also add, using

$$\frac{1}{\omega - \omega' \pm i\varepsilon} = \mathfrak{P}\frac{1}{\omega - \omega'} \mp i\pi\delta\left(\omega - \omega'\right), \tag{16.27}$$

$$G\left(\omega + i\varepsilon\right) + G\left(\omega - i\varepsilon\right) = \frac{1}{\pi i}\int_{-\infty}^{+\infty} d\omega' \frac{f\left(\omega'\right)}{\omega - \omega'}.$$

This is the second Plemelj formula. We obtain

$$G_r\left(\omega\right) = \frac{1}{2\pi i}\mathfrak{P}\int_{-\infty}^{+\infty} f\left(\omega'\right)\frac{d\omega'}{\omega - \omega'} + \frac{1}{2}f\left(\omega\right) \tag{16.28}$$

$$G_a\left(\omega\right) = \frac{1}{2\pi i}\mathfrak{P}\int_{-\infty}^{+\infty} f\left(\omega'\right)\frac{d\omega'}{\omega - \omega'} - \frac{1}{2}f\left(\omega\right). \tag{16.29}$$

These are the fundamental formulas. From them we obtain the dispersion relations already mentioned in Chapter 15. The above discussion is the proof. Also,

$$\text{Re } G_r\left(\omega\right) = \frac{\mathfrak{P}}{\pi}\int_{-\infty}^{+\infty} \frac{\text{Im } G_r\left(\omega'\right)}{\omega' - \omega} d\omega' \tag{16.30}$$

$$\text{Re } G_a\left(\omega\right) = \frac{-\mathfrak{P}}{\pi}\int_{-\infty}^{+\infty} \frac{\text{Im } G_a\left(\omega'\right)}{\omega' - \omega} d\omega'.$$

From these considerations we see that we may analytically continue (for instance, $G_r(\omega)$) into the upper half plane, providing $f(\omega)$ may be continued. The continuation of $G_r(\omega)$ to the upper half plane and $G_a(\omega)$ to the lower half plane creates two Riemann surfaces which intersect on the real ω axis.

Finally, in this section, the causal Green's function $G_c(\omega)$ may also be Fourier analyzed. It has the form

$$G_c\left(\omega\right) = \frac{1}{2\pi}\int_{-\infty}^{+\infty} J\left(\omega'\right) d\omega' \left\{\frac{\exp\left(\beta\omega\right)}{\omega - \omega' + i\varepsilon} - \frac{\eta}{\omega - \omega' - i\varepsilon}\right\}.$$

Now one may prove

$$\text{Re } G_c\left(\omega\right) = \frac{\mathfrak{P}}{2\pi}\int_{-\infty}^{+\infty} d\omega' \left(\exp\left(\beta\omega'\right) - \eta\right) J\left(\omega'\right)\frac{d\omega'}{\omega - \omega'} \tag{16.31a}$$

and

$$\text{Im } G_c\left(\omega\right) = \frac{1}{2}\left(\exp\left(\beta\omega\right) + \eta\right) J\left(\omega\right). \tag{16.31b}$$

Landau (1958) first obtained such relations.

16.4 Connection to linear response theory

According to Eq. (15.38) and Eq. (15.39), we may generalize the response function ϕ_{ik} to a Green's function,

$$\phi_{ik}\left(t-t'\right) = -i\theta\left(t-t'\right)\mathrm{Tr}\rho_0\left[a_i, a_k\right]. \tag{16.32}$$

Here the operators $\hat{A}$, $\hat{B}$ are replaced by $\hat{a}_i$, $\hat{a}_k$. Then the susceptibility is again

$$\chi_{ik}(\omega) = \int_{-\infty}^{+\infty} \phi_{ik}(t)\exp(i\omega t).$$

From the spectral representation for the Green's function, we may immediately have

$$\chi_{ik}(\omega) = -\frac{1}{2\pi}\int_{-\infty}^{+\infty}\left(\exp\left(\beta\omega'\right)-1\right)J_{a_k a_i}\left(\omega'\right)\frac{d\omega'}{\omega-\omega'+i\varepsilon} \tag{16.33}$$

$$= \frac{1}{2\hbar}\left(\exp(\beta\omega)-1\right)J_{a_k a_i}(\omega) - \frac{1}{2\pi}\mathfrak{P}\int_{-\infty}^{+\infty}\left(\exp\left(\beta\omega'-1\right)\right)J_{a_k a_i}\left(\omega'\right)\frac{d\omega'}{\omega-\omega'}.$$

In the case of a symmetrized time correlation function,

$$\left[a_k, a_i(t)\right] \rightarrow \left[a_k, a_i(t)\right]_+ \tag{16.34}$$

and

$$J_{a_k a_i}(\omega) = \frac{1}{2}\left[J_{a_k a_i}(\omega) + J_{a_i a_k}(-\omega)\right] \tag{16.35}$$

but also

$$J_{a_k a_i}(\omega) = J_{a_i a_k}(-\omega)\exp(-\beta\omega) \tag{16.36}$$

$$= \frac{1}{2}J_{a_k a_i}(\omega)\left(1+\exp\beta\omega\right). \tag{16.37}$$

Thus we have

$$\chi_{ik}(\omega) = -\frac{1}{\pi}\int_{-\infty}^{+\infty}\tanh\frac{\beta\omega'}{2}J_{a_k a_i}\left(\omega'\right)\frac{d\omega'}{\omega-\omega'+i\varepsilon}. \tag{16.38}$$

This is the Callen–Welton result already obtained in Chapter 15. From this it follows, using

$$J^*_{a_k a_i}\left(\omega'\right) = J_{a_i a_k}\left(\omega'\right),$$

that for the symmetric case,

$$\mathrm{Im}\,\chi^s_{a_k}(\omega) = \tanh\frac{\beta\omega}{2}\,\mathrm{Re}\,J_{a_i a_k}\left(\omega'\right), \tag{16.39}$$

and for the antisymmetric case,

$$\mathrm{Re}\,\chi^a_{ik} = \tanh\frac{\beta\omega}{2}\,\mathrm{Im}\,J_{a_i a_k}(\omega). \tag{16.40}$$

These useful results relate χ_{ik} to the spectral density $J_{a_i a_k}(\omega)$. The point here is to emphasize the connection of the spectral properties of the generalized susceptibility χ_{ik} to the spectral properties of the retarded Green's function of this chapter, as expected from Eq. (16.31).

16.5 Green's function hierarchy truncation

Let us return to the few-body Green's function of Eq. (16.7) and Eq. (16.8) with the purpose of deriving a kinetic equation from the hierarchy outlined earlier. First we will follow the earliest development in the book of Kadanoff and Baym (1962). We should say that the Green's function hierarchy was studied extensively by means of the diagrammatic techniques originated by Feynman (Feynman, 1949). This is principally focused on the equilibrium time-independent many-body phenomena. For more on this topic, see Abrikosov *et al.*, 1963. We will consider the Keldysh time-dependent theory in Section 16.6.

To analyze the causal one-particle Green's function (correlation function),

$$\begin{aligned} G\left(1,1'\right) &= -i\left\langle T\psi\left(1\right),\psi^{\dagger}\left(1'\right)\right\rangle \\ &= -i\left\langle\left\{\theta\left(t_1-t_1'\right)\psi\left(1\right)\psi^{\dagger}\left(1'\right)+\eta\theta\left(t_1'-t_1\right)\psi^{\dagger}\left(1'\right)\psi\left(1\right)\right\}\right\rangle, \end{aligned} \tag{16.41}$$

we observe that

$$\left[-i\frac{d}{dt_1}+\left[H\left(1\right)\right]_-\right]G\left(1,1'\right)=\delta\left(t_1-t_1'\right)\delta\left(r_1-r_1'\right), \tag{16.42}$$

which is similar to Eq. (16.13), $H(1)$ being the one-particle Hamiltonian. A similar equation may be written for $G\left(1,2,1',2'\right)$. It is

$$G\left(1,2,1',2'\right)=i^2\left\langle T\psi\left(1\right)\psi(2)\psi^{\dagger}\left(2'\right)\psi^{\dagger}\left(1'\right)\right\rangle.$$

From Eq. (16.41) we may define the correlation functions which play a central role here and in the subsequent section:

$$\begin{aligned} G^{>}\left(1,1'\right) &\equiv -i\left\langle\psi\left(1\right)\psi^{\dagger}\left(1'\right)\right\rangle \qquad t_1>t_1' \\ G^{<}\left(1,1'\right) &\equiv -i\eta\left\langle\psi^{\dagger}\left(1'\right),\psi\left(1\right)\right\rangle \quad t_1<t_1' \\ \eta &= \pm 1. \end{aligned} \tag{16.43}$$

Note that we have not included the Heaviside function of the previous analysis in Eq. (16.43).

Now consider boundary conditions. We note that for the spectral function of the correlation function,

$$J_{AB}\left(\omega\right)=\exp\left(\beta\omega\right)J_{BA}\left(\omega\right). \tag{16.44}$$

A and B are Hermitian (Kubo *et al.*, 1992). Then from

$$\left\langle A\left(t'\right) B\left(t\right)\right\rangle = \int_{-\infty}^{+\infty} \exp\left[i\omega\left(t-t'\right)\right]\exp\left(\beta\omega\right) J_{BA}\left(-\omega\right),$$

we find

$$\left\langle A\left(t\right) B\left(t'\right)\right\rangle = \left\langle A\left(t-i\hbar\beta\right) B\left(t'\right)\right\rangle. \tag{16.45}$$

This suggests introducing the temperature Green's function because of the analog of $-i\hbar\beta$ with a time. The analog of the unitary time operator $\exp\left(-iH\frac{t}{\hbar}\right)$ is $\exp\left(+\beta H\right) = \exp i\left(-i\beta H\right)$ and was first developed by Matsubara (1955). We shall not dwell on this formalism but refer the reader to Kubo *et al.* (1992). The causal Green's function (the correlation) does not have the simple analytic properties of the retarded Green's function. From Eq. (16.45) we may show, for the single-particle causal Green's function,

$$G^{<}\left(t_1, t_1'\right) = \pm\exp\beta\mu G^{>}\left(1, 1'\right)|_{t_1=-i\beta}. \tag{16.46}$$

Here we are extending the Green's function and T to a complex time domain, a time contour. Other paths than this may be simpler for the purpose of diagrammatic analysis. We shall, in a subsequent section, consider the choice of the Keldysh contour (Keldysh, 1965). We restrict imaginary t_1 to the range $0 < it_1 \leq \beta$. The farther down the imaginary time axis, the "later" it is. The $\pm$ in Eq. (16.46) has come from the Wick theorem in the imaginary time domain. Now the boundary conditions are obtained. We have

$$G\left(1, 1'\right)|_{t_1=0} = G^{<}\left(1, 1'\right)|_{t_1=0}, \tag{16.47}$$

since $it_1 = 0 < it_1'$ for all t_1', and

$$G\left(1, 1'\right)|_{t_1=-i\beta} = G^{>}\left(1, 1'\right)|_{t_1=-i\beta},$$

since $\beta = it_1 > it_1'$ for all t_1'. By direct computation (a homework problem for the student), we may also show for imaginary time from Eq. (16.46),

$$G\left(1, 1'\right)|_{t_1=0} = \pm\exp\left(\beta\mu\right) G\left(1, 1'\right)|_{t_1=-i\beta}. \tag{16.48}$$

Also, by Eq. (16.47), for the imaginary time causal two-particle Green's function $G\left(12, 1'2'\right)$, we have

$$G\left(12, 1'2'\right)|_{t_1=0} = \pm\exp(\beta\mu) G\left(12, 1'2'\right)|_{t_1=-i\beta}. \tag{16.49}$$

This is the boundary condition to be imposed on the causal Green's function in the imaginary time domain. Here analyticity is maintained in the "time" range $\mathrm{Re}\left(i\left(t_1-t_1'\right)\right) > 0, -\beta + \mathrm{Im}\left(t_1-t_1'\right) > 0$. To do this we will utilize a rather special Fourier series. Assuming time translational invariance, which is possible since this is an equilibrium average, we write in momentum space

$$G\left(p, t-t'\right) = (-i\beta)^{-1} \sum_{\nu} \exp\left(-i z_{\nu}\left(t-t'\right)\right) G\left(p, z_{\nu}\right) \tag{16.50}$$

and the inverse

$$G\left(p, z_{\nu}\right) = \int_{0}^{-i\beta} dt \exp i\left\{\left[\frac{\pi\nu}{-i\beta} + \mu\right]\left(t-t'\right)\right\} G\left(p, t-t'\right), \tag{16.51}$$

where

$$0 \le it \le \beta, \quad 0 \le it' \le \beta.$$

The boundary condition of Eq. (16.49) requires

$$1 = \pm \exp \beta\left(\mu - z_{\nu}\right) \tag{16.52}$$

$$\text{or } z_{\nu} = \mu \mp \frac{\pi\nu i}{\beta},$$

where

$$\begin{aligned} \nu &= \text{even} \quad &+ \text{ Bose–Einstein} \\ &= \text{odd} \quad &- \text{ Fermi–Dirac.} \end{aligned}$$

Utilizing Eq. (16.51) we may write the Hilbert transform:

$$G\left(p, z_{\nu}\right) = \frac{1}{2\pi} \int_{-\infty}^{+\infty} \frac{A(p,\mu)}{z_{\nu} - \mu} \text{ at } z_{\nu} = \frac{\pi\nu}{-i\beta} + \mu \tag{16.53}$$

$$\text{and } A(p\omega) = \lim_{\varepsilon\to 0} i\left[G(p, \omega + i\varepsilon) - G(p, \omega - i\varepsilon)\right]. \tag{16.54}$$

For free particles,

$$\left(z_{\nu} - \frac{p^2}{2m}\right) G\left(p, z_{\nu}\right) = 1, \tag{16.55}$$

and then we have

$$A(p\omega) = 2\pi\delta\left(\omega - \frac{p^2}{2m}\right).$$

The equations of motion will now be considered. In the Heisenberg picture the operator equation of motion is

$$i\frac{\partial\psi}{\partial t}(r, t) = \left[\psi(\mathbf{r}, t), H(t)\right].$$

As in Eq. (16.5), we take

$$H(t) = \int dr \frac{-\nabla\psi^{\dagger}(r,t)\cdot\nabla\psi(r,t)}{2m} + \frac{1}{2}\int dr_1 dr_2 V(r_1 - r_2)\,\psi^{\dagger}(r_1,t)\,\psi^{\dagger}(r_2,t)\,\psi(r_2,t)\,\psi(r_1,t), \tag{16.56}$$

and thus, by commutation laws,

$$[\psi(rt), H(t)] = \frac{-\nabla^2}{2m}\psi(r,t) + \int dr'\psi^{\dagger}(r't)\,V(r'-r)\,\psi(r',t)\,\psi(r,t), \tag{16.57}$$

and similarly for $\left[\psi^{\dagger}(r,t), H(t)\right]$.

Now consider $G\left(r_1t_1, r_1't_1'\right)$. We form, using Wick's theorem,

$$i\frac{\partial}{\partial t_1}\left[T\psi(1)\,\psi^{\dagger}(1')\right] = i\delta\left(t_1 - t_1'\right)\delta\left(r_1 - r_1'\right) \tag{16.58}$$

$$\pm\,\theta\left(t_1' - t_1\right)\psi^{\dagger}(1')\,i\frac{\partial\psi(1)}{\partial t} + \theta\left(t_1 - t_1'\right)i\frac{\partial\psi(1)}{\partial t_1}\psi^{\dagger}(1')$$

$$= i\delta\left(1 - 1'\right) \pm \int dr_2 V(r_2 - r_1)\left[T\psi(1)\,\psi(2)\,\psi^{\dagger}(2^{+})\,\psi^{\dagger}(1')\right]_{t_2=t_1} - \frac{\nabla_1^2}{2m}\left[T\psi(1)\,\psi(1')\right]. \tag{16.59}$$

The notation $\left(2^{+}\right)$ requires $t_2^{+} > t_2$ infinitesimally, and the $t_2 = t_1$ reminds us that there is a one-time variable t_1. Carrying T through the time derivative introduces a $\delta\left(t_1 - t_1'\right)$. Time ordering does not commute with T.

The result is

$$\left(i\frac{\partial}{\partial t_1} + \frac{\nabla_1^2}{2m}\right)G\left(1,1'\right) = \delta\left(1-1'\right) \pm i\int dr_2 V(r_1 - r_2)\,G_2\left(12, 1'2^{+}\right)|_{t_2=t_1} \tag{16.60}$$

(Kadanoff and Baym, 1962). We rewrite Eq. (16.60) as

$$i\left(\frac{\partial}{\partial t_1} + \frac{\nabla_1^2}{2m}\right)G\left(1,1'\right) = \delta\left(1-1'\right) + \int dr_2\Sigma(1,2)\,G\left(21'\right), \tag{16.61}$$

introducing the self energy $\Sigma(1,2)$. Diagrammatic perturbation theory defines this. A re-summation of diagrams gives it a formal solution, a Dyson equation:

$$G\left(1,1'\right) = G^0\left(1,1'\right) + \int dr_2 \int dr_3 G^0(1,2)\,\Sigma(2,3)\,G\left(3,1'\right). \tag{16.62}$$

See the discussion and proof in Abrikosov *et al.* (1963). The $\Sigma(12)$ are introduced variationally by Kadanoff and Baym (1962).

Eq. (16.60) is the beginning of the hierarchy for G implied earlier. Again,

$$G\left(12, 1'2'\right) = -i^2 \left\langle T\psi\left(1\right)\psi\left(2\right)\psi^{\dagger}\left(2'\right)\psi^{\dagger}\left(1'\right)\right\rangle . \qquad (16.63)$$

The truncation of the hierarchy is the difficult point, as it is with the B.B.G.Y.K. hierarchy discussed in Chapter 4 on the derivation of kinetic equations. To illustrate this in the simplest way, we adopt the Hartree approximation, which is to factor the $G_2\left(12, 1'2^{+}\right)$:

$$G_2(12, 1'_1 2^{+}) \to G\left(1, 1'\right) G\left(2, 2^{+}\right). \qquad (16.64)$$

Now introduce the one-particle time/position-dependent density

$$n\left(\mathbf{r}, t\right) = \psi^{\dagger}\left(\mathbf{r}, t\right)\psi\left(\mathbf{r}, t\right) \qquad (16.65)$$

and the average

$$\begin{aligned} G\left(2, 2^{+}\right) &= \pm\frac{1}{i}\left\langle\psi^{\dagger}\left(2^{+}\right)\psi\left(2\right)\right\rangle \\ &= \pm\frac{1}{i}\left\langle n\left(\mathbf{r}\right)\right\rangle . \end{aligned} \qquad (16.66)$$

Thus,

$$\left[i\frac{\partial}{\partial t_1} + \frac{\nabla_1^2}{2m}\right] G\left(1, 1'\right) = \delta\left(1 - 1'\right) + \int dr_2 V\left(\left|r_2 - r_1\right|\right)\left\langle n\left(r_2\right)\right\rangle G\left(1, 1'\right). \qquad (16.67)$$

The self-consistency of this equation for the "reduced" one-body $G\left(1, 1'\right)$ is apparent as in the Vlasov equation in Chapter 3. This is a Hartree self-consistent Green's function equation. What is the justification? That is not clear, just as in the case of the B.B.G.Y.K. hierarchy. The identity of the particles is not maintained. To do this, one must introduce the Hartree–Fock approximation and add an additional term to the factorization:

$$G\left(12, 1'2'\right) \to G\left(1, 1'\right) G\left(2, 2'\right) \pm G\left(1, 2'\right) G\left(2, 1'\right). \qquad (16.68)$$

Such equations are not truly quantum kinetic equations but proto-quantum operator equations. To proceed further, as with the B.B.G.Y.K. hierarchy, one must introduce phase space distributions. The Wigner function of Chapter 4 has been used by Kubo (Kubo *et al.*, 1992) to obtain, in an elegant way, the quantum Vlasov equation. We refer the reader to this development. Eq. (16.67) does not lead to the Boltzmann equation, as we would expect. Kadanoff and Baym (1962) have derived the Born approximation quantum Boltzmann equation in their book. It is not simple, but we refer the reader to it for the discussion.

Now, proceeding further and in a simpler way, let us consider the work of G. D. Mahan (2000) in his detailed book introducing Green's functions as applied to condensed matter. He has carried the analysis of the hierarchy further. See also his earlier article (Mahan, 1987). He obtained coupled operator Green's function equations for $G^{<}(k, \omega, RT)$, k, ω being the time-space Fourier transform of the two-body relative coordinates, $\mathbf{r}, t$ and $\mathbf{R}, T$ being the center of mass position and time. Let us outline this, following his work. We go to center of mass space-time coordinates

$$(\mathbf{r}, t) = x_1 - x_2, \quad (R, T) = \frac{1}{2}(x_1 + x_2)$$

and have

$$G^{<}(x_1, x_2) = i\left\langle \psi^{\dagger}\left(R - \frac{1}{2}r, T - \frac{t}{2}\right)\psi\left(R + \frac{1}{2}r, T + \frac{1}{2}t\right)\right\rangle.$$

We introduce $G^{<}(k, \omega, RT)$, the space-time correlation function, and recognize that the Fourier transformed Wigner function is

$$w(k, \omega, RT) = -iG^{<}(k, w, RT). \tag{16.69}$$

Now we utilize Eq. (16.60) for $G^{<}, G^{a}, G^{>}, G^{r}$ and complex conjugates. These four equations form a matrix equation for $\mathbf{G} = \begin{vmatrix} G^{>} & G^{a} \\ G^{r} & G^{<} \end{vmatrix}$, a form similar to that which we shall meet in the following section. We add and subtract complex conjugate pairs to obtain equations for the relative and center of mass motion.

Terms are dropped which represent system spacial temporal inhomogeneity. Further, to obtain gauge invariance in the electromagnetic interaction case, a transformation is made. Note that Ω is the center of mass energy, and $\mathbf{E} = -\nabla\phi + \nabla \cdot \mathbf{A} \equiv E_s + E_v$. The transformation is

$$\begin{aligned} \Omega + e\mathbf{E}_s \cdot \mathbf{R} &\rightarrow \omega \\ \mathbf{q} + eE_vT &\rightarrow \mathbf{k} \\ \nabla_R &\rightarrow \nabla_R + e\mathbf{E}_s\frac{\partial}{\partial\omega} \\ \frac{\partial}{\partial T} &\rightarrow \frac{\partial}{\partial T} + e\mathbf{E_v}\cdot\nabla_k. \end{aligned} \tag{16.70}$$

Then the "hydrodynamic" part of the added and subtracted equations becomes

$$\left[\omega - \omega_k + \frac{1}{8m}\left(\nabla_R + e\mathbf{E}\frac{\partial}{\partial\omega}\right)^2\right]G^{>}(k, \omega, RT)$$

and

$$i\left[\frac{\partial}{\partial T} + \mathbf{v}_k \cdot \nabla_R + e\mathbf{E}\left(\nabla_k + \mathbf{r}_k\frac{\partial}{\partial\omega}\right)\right]G^{>}(k, \omega, RT).$$

The second expression is similar to the classical Boltzmann equation.

The right side of Eq. (16.61) must now be transformed also. (This is similar to what has been done in Chapter 4.) To obtain the Wigner function form, we go to the center of mass variables, and, for instance, the arguments $\left(y, X + \frac{z}{2}\right)$ and $\left(z, X - \frac{y}{2}\right)$ appear. Write the preceding expressions in Fourier transforms, obtaining

$$\left[\omega - \varepsilon_k + \frac{1}{8m}\left(\nabla_R + eE\frac{\partial}{\partial\omega}\right)^2\right] G^< (k, \omega, RT) \tag{16.71}$$
$$= \frac{1}{2}\int dz \exp(-iqz) \int dy \exp(-iqy)$$
$$\left[\Sigma_t G^< - \Sigma^< G_{\bar{t}} + G_t \Sigma^< - G^< \Sigma_{\bar{t}}\right].$$

$\bar{t}$ here means anti-time ordering, which we will discuss later. Thus, also,

$$i\left[\frac{\partial}{\partial T} + v_k \cdot \nabla_R + eE\left(\nabla_k + v_k\frac{\partial}{\partial\omega}\right)\right] G^< (k, \omega, RT) \tag{16.72}$$
$$= \int dz \exp(-izq) \int dy \exp(-iqy) \left[\Sigma_t G^< - \Sigma^< G_{\bar{t}} - G_t \Sigma^< + G^< \Sigma_{\bar{t}}\right],$$

and there is a similar equation pair for G_r in terms of Σ_r. These are the first equations of a hierarchy introduced by $\Sigma^<$ etc.; they are a form of Wigner function hierarchy. In the preceding equations,

$$G_t + G_{\bar{t}} = G^> + G^<, \tag{16.73}$$
$$G_t - G_{\bar{t}} = 2\,\mathrm{Re}\,[G_r].$$

Also,

$$G^> = G^< - iA \tag{16.74}$$
$$\Sigma^> = \Sigma^< - 2i\Gamma,$$

A being the non-equilibrium spectral function.

We now do a gradient in space-time expansion in the center of mass variables about (R, T), as in Chapter 4. Assuming spacial-temporal homogeneity, the equations for $G_r(k, \omega, RT)$ become, in the static E approximation,

$$[\omega - \varepsilon k - \Sigma_r]\, G_r = 1 \tag{16.75}$$
$$ie\mathbf{E} \cdot \left[\left(1 - \frac{\partial \Sigma_r}{\partial\omega}\right)\nabla_k + (v_k + \nabla_k \Sigma_r)\frac{\partial}{\partial\omega}\right] G_r = 0.$$

The solution is to $0\left(E^2\right)$,

$$G_r(k\omega) = (\omega - \varepsilon_k - \Sigma_r(k\omega))^{-1}. \tag{16.76}$$

Finally, the equation for $G^<$ becomes, for the spacial-temporal homogeneous case,

$$\begin{aligned} & eE\left[\left(1-\frac{\partial\Sigma_r}{\partial\omega}\right)\nabla_k+(v_k+\nabla_k\Sigma_r)\frac{\partial}{\partial\omega}\right]G^< \\ & -ieE\cdot\left[\frac{\partial\Sigma^<}{\partial\omega}\nabla_k\operatorname{Re}[G_r]-\frac{\partial G_r}{\partial\omega}\nabla_k\Sigma^<\right] \\ & =\Sigma^<A-2\Gamma G^<. \end{aligned} \tag{16.77}$$

Further, as in the Chapman–Enskogg procedure, the quantities on the "left" side (hydrodynamic) may be taken to be the equilibrium values. Now assume

$$A=\frac{2\Gamma}{\sigma^2+\Gamma^2}, \tag{16.78}$$

where

$$\begin{aligned} \sigma &= \omega-\varepsilon_k-\operatorname{Re}[\Sigma_r] \\ \Gamma &= -\operatorname{Im}\Sigma_r. \end{aligned}$$

We obtain a proto-Boltzmann equation for the electrons:

$$A^2(k\omega)\frac{\partial n_F}{\partial\omega}e\mathbf{E}\cdot\left[\begin{array}{c}(\mathbf{v}_k+\nabla_k\operatorname{Re}\Sigma_r)\Gamma \\ +\sigma\nabla_k\Gamma\end{array}\right]=\Sigma^>G^<-\Sigma^<G^>. \tag{16.79}$$

This is still not uncoupled from a hierarchy on the right. This will now be discussed for the special case of dilute impurity scattering of the electrons. $\Sigma^>$ and $\Sigma^<$ contain the effects of scattering, creating the correlation. A general thing to do would be to form equations for the two-particle Wigner functions, as in the hierarchy discussion earlier in this chapter and in Chapter 4. This, however, can be "finessed" by the diagrammatic analysis. Eq. (16.79) was applied to electron scattering by dilute impurities. In this case,

$$\Sigma_r(k,\omega)=n_iT_{kk}(\omega), \tag{16.80}$$

$T_{kk}(\omega)$ being the off shell scattering matrix. Now

$$\Sigma^{<,>}=n_i\int\frac{d_3p}{(2\pi)^3}\left|T_{pk}(\omega)\right|^2G^{<,>}. \tag{16.81}$$

The nonlinear structure and scattering form of the right side of Eq. (16.81) is now apparent. For further discussion we refer the reader to Mahan's book (Mahan, 2000).

Let's compare this derivation with a similar derivation from the hierarchy in Chapter 4. There are equivalent assumptions. Eq. (16.79) is a Wigner function hierarchy. The density expansion is not explicitly done until a final step, whereas in Chapter 4 this is done initially. This complicates the Green's function approach. In addition, here the Markovianization is done within the perturbation analysis. The dependence upon the perturbation diagrammatic analysis to do this, as well as to obtain irreversible equations, which Eq. (16.77) is, does not add to the clarity of the logic. The derivation of the kinetic equation in Chapter 4 does not depend on the diagrammatic analysis.

For analytically extending the quantum operator Boltzmann equations, the diagrammatic methods are advantageous. Danielewicz (1984) has done this, to an extent, in his derivation. However, Hawker and others have extended the Boltzmann equation to include the further gradient terms in the expansion around the local values. They are termed collisional transfer corrections and have physical importance. The Waldman–Snider equation is an example (Waldman, 1957; Snider and Sanctuary, 1971). See also the book by McLennan (1989) for further references.

16.6 Keldysh time-loop path perturbation theory

Let us turn to the Keldysh (1965) analysis, suggested by Schwinger (1961), of a Green's function in time-contour perturbation theory. The main point is to do a diagrammatic resummation of the one-body Green's function terms in the perturbation, obtaining by standard equilibrium techniques a closed Dyson equation. The Dyson equation has the form

$$G(q) = G_0(q) + G_0(q)\,\Sigma(q)\,G(q),$$

where $G_0(q)$ is the unperturbed Green's function, $G(q)$ the *exact* Green's function, and $\Sigma(q)$ the self energy function. This has the same structure as the resolvent equation met earlier in our discussions. The reader should consult the book of Abrikosov (Abrikosov *et al.,* 1963), as well as the citations for the equilibrium discussion. This equation is the starting point of the analysis of a quantum kinetic theory in condensed matter applications. It is not exact.

We have already met the chronological ordering operator T, which arises in the field theory interaction representation for the expression

$$\left\langle S(-\infty,+\infty)\,T\left\{\hat{A}(t)\,\hat{B}(t')\ldots S(+\infty,-\infty)\right\}\right\rangle.$$

Here,

$$S(t,-\infty) = T\exp\left(-i\int_{-\infty}^{t} H_i(\tau)\,d\tau\right) \qquad t > -\infty \tag{16.82}$$

is a generalized many-body scattering matrix. $H_i(t)$ is turned on adiabatically from $t=-\infty$ and off at $t=+\infty$, much as in the discussion of the derivation of linear response in Chapter 15. Keldysh apparently generalized this for the case when reservoirs or irreversible absorption and emission are present, as in the Gamov vector discussion of Chapter 17. Then,

$$S(+\infty, -\infty)\,\phi_0 \neq \exp(i\alpha)\,\phi_0.$$

He takes

$$\left\langle S(-\infty, +\infty)\, T\left\{\hat{A}(t)\,\hat{B}(t')\,S(+\infty, -\infty)\right\}\right\rangle = \left\langle T_c\left\{\hat{A}(t)\,\hat{B}(t')\ldots S_c\right\}\right\rangle. \tag{16.83}$$

Here $\langle\rangle = \mathrm{Tr}\rho_0$.

T_c is an ordering operator on a new multi-time contour c running from $-\infty$ to $+\infty$ along a positive increasing time branch and then returning along a negative time oriented branch to $-\infty$. In Eq. (16.83) $t,\ t'$ are on the positive branch. The complete S matrix is $S_c = S(-\infty, +\infty)\,S(+\infty, -\infty)$. $S(-\infty, +\infty)$ is the positive branch, and $S(+\infty, -\infty)$ is the negative branch. Ordering on c means return branch times are *later* than the positive branch times. The path ordering with time loop was introduced by Schwinger (1961). See the history and many references in Rammer and Smith's review (Rammer and Smith, 1986). The Keldysh paths are not "exact," having omitted the initial correlation decay as well as non-Markovian contributions.

There are four one-particle Green's functions between the t_+ (plus branch) and t_- (minus branch):

$$G^<\left(t_+, t'_-\right) = i\left\langle\psi^\dagger\left(t'_+\right)\psi\left(t_-\right)\right\rangle \tag{16.84}$$

$$G^>\left(t_-, t'_+\right) = -i\left\langle\psi\left(t_-\right)\psi^\dagger\left(t'_+\right)\right\rangle \tag{16.85}$$

$$G^c\left(t_+, t'_+\right) = -i\left\langle T\psi\left(t_+\right)\psi^\dagger\left(t'_+\right)\right\rangle \tag{16.86}$$

$$\tilde{G}^c\left(t_-, t'_-\right) = -i\left\langle\tilde{T}\psi\left(t_-\right)\psi^\dagger\left(t'_-\right)\right\rangle. \tag{16.87}$$

G^+ and G^- in Keldysh's original paper are frequently called $G^<$ and $G^>$ respectively in the literature, which we shall follow from now on. The $\tilde{T}$ ordering on the minus branch is

$$\begin{aligned} \tilde{T}\psi(t)\,\psi^+\left(t'\right) &= \psi(t)\,\psi^+\left(t'\right) \qquad & t < t' \\ &= -\psi^+\left(t'\right)\psi(t) \qquad & t > t'. \end{aligned}$$

This use of Wick's ordering theorem is only true for fermions and bosons. Here T products do decompose into sums of T products taken pairwise. The four cases for G_0 in Eq. (16.84) through Eq. (16.87) may represent lines in a graphical Feynman pictorial decomposition. The line in Eq. (16.84) goes from the minus to the plus branch. The line in Eq. (16.85) from the plus to the minus branch, etc.

The diagram summation is equivalent to a time integration along c and thus is an integration from $-\infty$ to $+\infty$ *and* a summation over subscripts $+$, $-$. To do the latter, we introduce a 2×2 matrix. We define a Green's function matrix $\mathbf{G}$ of the four possibilities:

$$\mathbf{G} = \begin{pmatrix} G^c & G^< \\ G^> & \tilde{G}^c \end{pmatrix} \tag{16.88}$$

We must note that $G^c, G^>, G^<$ are related. $G^c + \tilde{G}^c = G^> + G^< \equiv G_K$, the Keldysh Green's function.

The standard procedures of equilibrium for re-summation are then made by forming the Dyson equation (Abrikosov *et al.*, 1963: Mahan, 1987). This will be illustrated in some detail in the next chapter.

$$\mathbf{G}\left(rt; r't'\right) = G^0\left(rt; r't'\right) + \int G^0\left(rt; r't'\right) \Sigma\left(rt, r't'\right) \times \mathbf{G}\left(rt, r't'\right) dt dt', \tag{16.89}$$

where the 2×2 self energy matrix is

$$\Sigma = \begin{pmatrix} \Sigma^c & \Sigma^- \\ \Sigma^+ & \tilde{\Sigma}^c \end{pmatrix}. \tag{16.90}$$

Note also that $\Sigma^c + \tilde{\Sigma}^c = -\left(\Sigma^> + \Sigma^<\right)$. This is a one-body Green's function equation with the electron interactions incorporated in Σ. We will examine it shortly. It is again a proto-kinetic equation and a hierarchy. Alternative transformations of $\mathbf{G}$ have been employed, and we have also dropped the explicit t_+, t_- notation. See the review by Rammer and Smith (1986). If we first transform $\mathbf{G} \Rightarrow \sigma_3 \mathbf{G} = \mathbf{G}'$ and then perform the rotation, we obtain

$$\mathbf{G} = LG'L^\dagger = \begin{pmatrix} G_r & G_k \\ 0 & G_a \end{pmatrix}.$$

This has advantages. Rammer and Smith discuss the Feynman graph rules, and we refer the reader to this required review at this point.

In the diagrammatic analysis, $x = \mathbf{r}t$, we have the irreducible summation

$$\Sigma_{ij}\left(x, x'\right) = \int \gamma^k_{ii'} G_{i',j'}\left(x, x_1\right) \Gamma^{h'}_{j'j}\left(x_1 x'; y\right) \times D_{k'k}\left(y, x\right) d_4 x_1 d_4 y. \tag{16.91}$$

x is the incoming electron line, and x' the outgoing; y is an external phonon line. Because of the matrix form, the matrix σ_z enters;

$$\gamma^k_{ij} = \delta_{ij}\left(\sigma_z\right)_{jk}.$$

The plus part of c corresponds to $+1$, and the minus part to -1. Subscripts are electron lines and superscripts the phonon line. D is the matrix for the Bose particles

Green's function. We are quoting Keldysh, who presented these aspects, but not in much detail (see Mahan, 2000).

The vacuum of field theory (ϕ_0) has been replaced by a trace over ρ_0, an initial distribution. This causes the Dyson equation to depend on initial ρ_0, which is somewhat inconsistent with the Green's function approach. However, this may be handled with a transformation by a differential operator, which transforms the equation and does not contain ρ_0. The solution is unique up to the solution of a homogeneous equation. The uniqueness of the solution is, however, proved by Keldysh.

A different canonical rotation used by Keldysh of the **G** matrix and also Σ and **D** is employed:

$$\begin{aligned} \mathbf{G} &= \frac{1}{2}\left(I - i\sigma_y\right)\mathbf{G}\frac{1}{2}\left(I + i\sigma_y\right), \\ &= \begin{pmatrix} 0 & G_a \\ G_r & G_K \end{pmatrix}, \end{aligned} \tag{16.92}$$

where

$$\begin{aligned} G_a &= G^c - G^> \\ G_r &= G^c - G^< \\ G_K &= G^c + \tilde{G}^c = G^> + G^<, \end{aligned} \tag{16.93}$$

and

$$\Sigma \to \begin{pmatrix} \Sigma_K & \Sigma_r \\ 0 & \Sigma_a \end{pmatrix} \tag{16.94}$$

$$\begin{aligned} \Sigma_a &= \Sigma^c + \Sigma^< \\ \Sigma_r &= \Sigma^c + \Sigma^> \\ \Sigma_K &= \Sigma^c + \tilde{\Sigma}^c. \end{aligned} \tag{16.95}$$

G_a and G_r are advanced and retarded Heisenberg Green's functions, previously met in Eq. (16.10) and Eq. (16.11):

$$\begin{aligned} G_r\left(x, x'\right) &= i\Theta\left(t - t'\right)\left\langle\left[\psi\left(x\right), \psi^{\dagger}\left(x'\right)\right]_+\right\rangle \\ G_a\left(x, x'\right) &= -i\Theta\left(t' - t\right)\left\langle\left[\psi\left(x\right), \psi^{\dagger}\left(x'\right)\right]_+\right\rangle, \end{aligned} \tag{16.96}$$

and the new

$$G_K\left(x, x'\right) = -i\left\langle\left[\psi\left(x\right), \psi^{\dagger}\left(x'\right)\right]_-\right\rangle. \tag{16.97}$$

G_K is called a causal Green's function. In this representation of **G**, it represents one of its components. The perturbation summed equation is a matrix equation for all three one-body Green's functions together. Remember G_a and G_r are not independent. The presence of these elements together in **G** is an interesting feature of the Keldysh theory.

In weak coupling (to g^2 order in Σ), the hierarchy uncouples, and an independent equation for G_K or $G^<$ may be obtained. This localizes the equation in $\mathbf{x} - \mathbf{x}''/2$. It is for $G^>$,

$$\left\{\frac{\partial}{\partial t} + \mathbf{v}\cdot\nabla + e\mathbf{E}\cdot\nabla_p\right\} G^>(xp) = g^2\left[\Sigma^>(xp)\,G^<(xp) - \Sigma^<(xp)\,G^>(xp)\right]. \tag{16.98}$$

This is a one-body birth–death weak coupling gain–loss equation of a familiar form. x is (x_s, t_x), the center of mass position and time, and can be transformed to an equation for a single-particle Wigner function.

S. Datta (1989) has further examined the equation for $G^<(x, p)$ similar to Eq. (16.97), for the special case of a steady state, where $\tau = t_2 - t_1$ is constant. In addition, he assumes the oscillator reservoir to be in equilibrium, interacting with the electron with a $\delta(r_1 - r_2)$ potential. He considers, then, one-phonon weak coupling process and obtains the approximate self energies $\Sigma^>$, $\Sigma^<$. The result is

$$\Sigma^>(r_1 r_2, E) = \frac{i\hbar}{\tau^>(r_1, E)}\delta(r_1 - r_2) \tag{16.99}$$

$$\Sigma^<(r_1 r_2, E) = \frac{i\hbar}{\tau^<(r_1, E)}\delta(r_1 - r),$$

where

$$\frac{1}{\tau^>} = \frac{2\pi}{\hbar}\int dE' F\left(r, E' - E\right) p\left(r E'\right) \tag{16.100}$$

$$\frac{1}{\tau^<} = \frac{2\pi}{\hbar}\int dE' F\left(r, E - E'\right) n\left(r_1 E\right),$$

when $n(r, E)$ and $p(r, E)$ are equilibrium electron and hole densities. $n(r, t)$ and $p(r, t)$ are diagonal and thus have one-particle Green's functions. The weak coupling birth–death structure of Eq. (16.98) is now apparent. $F(r, E)$ is effectively the phonon equilibrium distribution function at temperature T, and $1/\tau^{\gtrless}$ are the rates of scattering of the electrons by phonons. Thus we have a relaxation time model of the Boltzmann-type picture. Here $1/\tau \approx \lambda^2$ is the electron–phonon interaction constant.

Datta (1989) obtains from Eq. (16.99) a coupled equation set for the nondiagonal $G^<(r_1 r_2; E)$ and $G^>(r_1 r_2; E)$ Green's functions. They may be formally solved beginning with

$$G^<(r_1 r_2, E) = i\hbar \int d_3 r \frac{G_r(r_1 r_3, E)\, G_a(r_3 r_2, E)}{\tau^<(r_3, E)}.$$

These are nondiagonal operator equations, not Boltzmann equations. More will be said concerning the use of the Keldysh theory in the discussion of tunneling in Chapter 19.

References

Abrikosov, A. A., Gorkov, L. P. and Dejaloshinskii, I. E. (1963). *Methods of Quantum Field Theory in Statistical Physics* (New York, Prentice Hall).
Bogoliubov, N. N. and Tyablikov, S. V. (1959). *Sov. Phys. Dokl.* **4**, 589.
Bonch-Bruevich, V. L. (1956). *J. Exp. Theor. Phys.* **31**, 522.
Bonch-Bruevich, V. L. (1957). *Sov. Phys. J. Exp. Theor. Phys.* **4**, 457.
Callen, H. and Welton, T. A. l (1951). *Phys. Rev.* **34**, 83.
Danielewicz, P. (1984). *Ann. Phys.* (N.Y.) **152**, 239.
Datta, S. (1989). *Phys. Rev. B* **40**, 583.
Fetter, A. L. and Walecka, J. D. (1971). *Quantum Theory of Many Particle Systems* (New York, McGraw-Hill).
Feynman, R. P. (1949). *Phys. Rev.* **76,** 749 and 769.
Kadanoff, L. P. and Baym, G. (1962). *Quantum Statistical Mechanics* (New York, W. A. Benjamin).
Keldysh, L. V. (1965). *Sov. Phys. J. Exp. Theor. Phys.* ***20***, 1018.
Kubo, R., Todus, M. and Hashitsume, H. (1992). *Statistical Physics* (New York, Springer).
Landau, L. D. (1958). *Sov. Phys. J. Exp. Theor. Phys.* **7**, 182.
Lifshitz, E. and Petaevskii, L. P. (1981). *Physical Kinetics* (New York, Pergamon).
Mahan, G. D. (1987). *Phys. Rep.* **5**, 251.
Mahan, G. D. (2000). *Many Particle Physics*, 3rd edn. (New York, Kluwer).
Matsubara, T. (1955). Prog. Theor. Phys. 14, 351.
McLennan, J. A. (1989). *Introduction to Non-equilibrium Statistical Mechanics* (Englewood Cliffs, N.J., Prentice Hall).
Muskelishvilli, N. I. (1953). *Singular Integral Equations* (Groning, Noordhoff).
Rammer, J. and Smith, H. (1986). *Rev. Mod. Phys.* **58**, 323.
Schweber, S. S. (1961). *An Introduction to Quantum Field Theory* (New York, Harper & Row).
Schwinger, J. (1961). *J. Math. Phys.* (N.Y.) **2**, 407.
Snider, R. F. and Sanctuary, B. C. (1971). *J. Chem. Phys.* **55**, 1555.
Waldman, L. (1957). *Z. Naturforsh* **12A**, 660.
Zubarev, D. N. (1960). *Sov. Phys. Uspekhi* **3**, 320.
Zubarev, D. N. (1974). *Non-equilibrium Statistical Thermodynamics*, trans. D. J. Shepherd (New York, Consultants Bureau).

17

Decay scattering

17.1 Basic notions and the Wigner–Weisskopf theory

Although the notions of bound states, scattering and quantum transitions are well defined in the quantum theory, the description of an unstable system, involved in the process of decay, has remained an outstanding issue for many years. The problem is fundamental, since it concerns the nature of irreversible processes, one of the most important issues in statistical mechanics and the theme that is central to this book.

The theory of decay is intimately connected with scattering theory and necessarily contains mathematical ideas and methods. We shall try to explain these points carefully as we get to them.

We treat elsewhere in the book the ideas of Boltzmann, Van Hove and Prigogine on irreversible phenomena. The tools that are developed there are basically approximate, although very useful. One can argue that the basic rigorous characteristic of an irreversible process is that, as represented in terms of the evolution of a state in the Hilbert space of the quantum theory, it must be a semigroup. This type of evolution, resulting in an operation $Z(t)$ on a state ψ, should satisfy the property

$$Z(t_2)\, Z(t_1) = Z(t_1 + t_2)\,. \tag{17.1}$$

The argument is as follows. If the system evolves in time t_1 and is stopped, then evolves further at time t_2, since the process has no memory, the total evolution should be as if the system evolved from the initial state to a state at $t_1 + t_2$ independently of the fact that it was done in two stages (Piron, 1976). Since the process is irreversible, the operator $Z(t)$ may have no inverse. Such an evolution is called a *semigroup*. As we shall see, it is not possible to obtain such an evolution law in the framework of the standard quantum theory (Horwitz *et al.*, 1971), but recently much work has been done, and methods have been developed, based on ideas of

Sz.-Nagy and Foias (1976), such as the theory of Lax and Phillips (1967) and its extension to the quantum theory (Strauss *et al.*, 2000) in which semigroup evolution can be achieved. In Chapter 18, we discuss in detail the structure of the Liouville space (a linear space of operators in the Hilbert space containing the density matrices, and isomorphic to a larger Hilbert space defined through the trace norm), which also provides an important framework for the realization of these recently developed methods for the description of unstable systems and resonances.

We start by describing some of the history of the subject in the framework of the standard quantum theory. In 1928, Gamow made the first striking application of quantum theory to the α-decay of nuclei (Gamow, 1928). From a simple classical point of view, one thinks of a collection of N unstable nuclei with a probability Γ (per unit time, per particle) to decay by the emission of an α-particle. The rate of change of the number of nuclei in the original state is described by

$$\frac{dN}{dt} = -\Gamma N \tag{17.2}$$

with solution

$$N = e^{-\Gamma t} N_0, \tag{17.3}$$

where N_0 is the original number of nuclei. To achieve such a result in the framework of the quantum theory, Gamow assumed the form

$$i\frac{\partial \psi}{\partial t} = \left(E - i\frac{\Gamma}{2}\right)\psi \tag{17.4}$$

for the Schrödinger equation, i.e. that the state ψ is an eigenfunction of the Hamiltonian operator (usually taken to be self-adjoint) with complex eigenvalue. The solution of this equation,

$$\psi_t = e^{-i(E-i\frac{\Gamma}{2})}\psi_0, \tag{17.5}$$

has the property that

$$N_t = e^{-\Gamma t} N_0, \tag{17.6}$$

where we have taken $|\psi_0|^2$ as the probability to find N_0 particles undecayed initially (obtained by multiplying the usual normalized probability to find a particle by N_0) and $N_0|\psi_t|^2$ as the probability to find N_t undecayed particles remaining at time t. The formula of Gamow satisfies the semigroup property and has been very useful in describing experimental results. We shall return to this important point later.

We remark that the Laplace transform, well defined for $\Gamma > 0$,

$$\int_0^\infty e^{izt}\psi_t dt = \frac{-i}{z - (E - i\frac{\Gamma}{2})}\psi_0, \tag{17.7}$$

has a simple pole in the lower half plane. We shall see some of these characteristics emerge from much more sophisticated theories of unstable systems, and in fact, the exponential law Eq. (17.5) has been shown to give a very precise representation of the data (Winstein *et al.,* 1997) in its two-channel generalization, a parametrization more recently proposed by Lee, Oehme, and Yang (1957) and Wu and Yang (1964) for the description of neutral K meson decay.

An obvious objection to the form Eq. (17.5) given by Gamow, however, is that the momentum of a free particle is proportional to the square root of the Hamiltonian. Such a momentum would be, in this case, complex and gives rise to an exponential divergence of the wave function.

Weisskopf and Wigner, in a fundamental work (Weisskopf and Wigner, 1930), provided a possible theory for the description of unstable systems on a more fundamental level, using a proper self-adjoint Hamiltonian in a form consistent with the standard structure of the quantum theory, and obtained, nevertheless, an exponential decay law in good approximation. We shall describe their method in the following section. (This has also been discussed in previous chapters.)

Their method, which we shall refer to as the Wigner–Weisskopf method (following the nomenclature used in much of the literature on this subject), starts with the general Schrödinger equation for the evolution of a quantum system

$$i\frac{\partial\psi}{\partial t} = H\psi, \tag{17.8}$$

with H a self-adjoint Hamiltonian with (exact) solution

$$\psi_t = e^{-iHt}\psi_0. \tag{17.9}$$

Weisskopf and Wigner then proceed to assume that the initial state ψ_0 represents an unstable system, and that its evolution Eq. (17.8) induces a *decay* of that system. Note that Eq. (17.8), in the framework of the quantum theory, describes the evolution of the system represented by ψ; the assumption that this evolution corresponds to a decay of the system from some initial type of system to another, as a strong physical assumption, is the basis for the Wigner–Weisskopf model. Examples are the decay of a discrete state of some characterizing (unperturbed) Hamiltonian, such as the state of a neutron, to the set of states with continuous spectrum, such as the proton, electron, antineutrino final state. Other examples are the excited atom decaying into a ground state with the emission of a photon, or the excited nucleus decaying to a nucleus in a lower level with the emission of electromagnetic radiation or an α-particle, as in Gamow's application. We emphasize that this idea is not a natural consequence of the general structure of quantum theory, for which the evolution generated by Eq. (17.8) constitutes a continuous, probability-preserving change in the state of a given system, but involves an additional explicit assumption that the nature of the system itself is undergoing a change in structure. In its corresponding formulation in quantum field theory – where, for example, one can assume an interaction consisting of the annihilation operator for a neutron and the product of creation operators for the proton, electron and antineutrino – the evolution still, through the action of unitary evolution, follows a continuous transition subject to the criticisms which we shall describe in Sections 17.4 and 17.5. As we shall see, if this change is of an irreversible nature, the applicability of the Wigner–Weisskopf formulation, in terms of evolution in the usual Hilbert space of states, can only be approximate, and in some cases is not adequate to serve even approximately as a basic theory. In succeeding sections, we shall discuss formulations capable of describing irreversible processes more accurately.

It is remarkable, however, that the analytic structure of the resolvent (or Green's function) for the standard quantum evolution associated with the Wigner–Weisskopf formulation, which we shall describe below, is a very robust feature of the analysis. The primary difficulties arise in the representations of the evolution in terms of quantum states, and it will be our purpose in this chapter to describe some of the techniques that have been developed to deal with this problem. In Chapter 18 we will discuss the extension of these ideas to statistical mechanics.

17.2 Wigner–Weisskopf method: pole approximation

We shall start with a rather general analysis of this underlying analytic structure, in the standard Wigner–Weisskopf framework. Consider the amplitude, according to the Wigner–Weisskopf model, for which the state of the system remains in its initial (undecayed) state,

$$A(t) = \langle \psi_0 | e^{-iHt} | \psi_0 \rangle, \tag{17.10}$$

often called the survival amplitude (Misra and Sudarshan, 1977). Although the original calculation of Weisskopf and Wigner (1930) was done in first-order perturbation theory, we shall follow a somewhat different method here. Consider the Laplace transform, for $\operatorname{Im} z > 0$:

$$iR(z) \equiv \int_{-\infty}^{\infty} e^{izt} \langle \psi_0 | e^{-iHt} | \psi_0 \rangle = i \langle \psi_0 | \frac{1}{z - H} | \psi_0 \rangle. \tag{17.11}$$

Since the Hamiltonian is a self-adjoint operator, it has a spectral resolution of the form

$$H = \int \lambda dE(\lambda) \tag{17.12}$$

(von Neumann, 1955; Riesz and Sz.-Nagy, 1955; Reed and Simon, 1979), where $E(\lambda)$ is a spectral family of projections satisfying

$$E(\lambda)\,E(\mu) = E(\min(\lambda,\mu)) \tag{17.13}$$
$$dE(\lambda)\,dE(\lambda') = \begin{Bmatrix} 0 & \text{if } \lambda \neq \lambda' \\ dE(\lambda) & \text{if } \lambda = \lambda' \end{Bmatrix},$$

and λ, μ correspond to the spectrum of H.

If we assume that the operator H is absolutely continuous, so that $E(\lambda)$ is differentiable, we may write the spectral representation as in Dirac's book (Dirac, 1947), in terms of bras and kets:

$$dE(\lambda) = |\lambda\rangle\langle\lambda|d\lambda \tag{17.14}$$

The bra-ket combination corresponds to the derivative of $E(\lambda)$.

If there is a discrete spectrum, for example a point eigenvalue at λ_0, then $dE(\lambda_0)$ is infinite (there is a jump in the spectral function), but the integral in the neighborhood of λ_0 is finite and projection-valued:

$$\int_{\lambda_0-\varepsilon}^{\lambda_0+\varepsilon} \lambda dE(\lambda) = \lambda_0 P_0, \tag{17.15}$$

where $P_0 = \lim_{\varepsilon\to 0} E(\lambda_0+\varepsilon) - E(\lambda_0-\varepsilon)$ is a simple projection operator, i.e. $P_0^2 = P_0$, and it is self-adjoint. If H had a totally discrete spectrum, it could be expressed in the familiar form

$$H = \sum_i \lambda_i P_i.$$

We shall not discuss here the third case, of singular continuous spectrum, which does not have the property Eq. (17.14). It is defined by the fact that $E(\lambda)$ is not the integral (with endpoint λ) of some operator valued function. As an example, to see how such a construction could come about, one may think of a point spectrum which is imbedded in a continuum, i.e. there is an absolutely continuous spectrum between the points; then consider taking a limit in which the density of points becomes so high that the derivative is no longer defined.

We shall assume for our present purposes that H has an absolutely continuous spectrum. The discrete eigenstates of an "unperturbed" operator H_0, where $H = H_0 + V$, may be used to characterize the initial states of the system. The operator V here induces the decay, corresponding to a transition to the continuous spectrum of H_0.

Now, due to Eq. (17.13),

$$H^2 = \int \lambda^2 dE(\lambda),$$

and, generally,

$$H^n = \int \lambda^n dE(\lambda).$$

Therefore, for any function that can be formed as a sequence of polynomials (finite or infinite),

$$f(H) = \int f(\lambda)\, dE(\lambda). \tag{17.16}$$

It then follows that Eq. (17.11) can be written as

$$R(z) = \langle\psi_0| \int \frac{dE(\lambda)}{z-\lambda} |\psi_0\rangle, \tag{17.17}$$

from which it is clear that, if the Hamiltonian has spectrum $\lambda \geq 0$, the function $R(z)$ is analytic in the cut plane excluding the positive real line. The inverse transform is given by

$$A(t) = \langle\psi_0|e^{-iHt}|\psi_0\rangle = \frac{1}{2\pi i}\int_C R(z)\, e^{-izt} dz, \tag{17.18}$$

where C is a contour running slightly above the real line on the z plane from $+\infty$ to zero and then, going around the branch point, from zero back to $+\infty$ slightly below the real line. The proof of this statement can be achieved by reversing the order of integration in Eq. (17.18):

$$\frac{1}{2\pi i}\int_C \langle\psi_0| \int \frac{dE(\lambda)}{z-\lambda}|\psi_0\rangle e^{-izt} dz = \frac{1}{2\pi i}\langle\psi_0| \int dE(\lambda) \int_C \frac{e^{-izt}}{z-\lambda}|\psi_0\rangle. \tag{17.19}$$

For each fixed λ, the integral on the contour C can be pinched down to a small circle around λ, which just gives a residue $2\pi i e^{-i\lambda t}$. The completion of the integral, after cancellation of the factor $2\pi i$, is then, according to Eq. (17.16),

$$\langle\psi_0| \int dE(\lambda)\, e^{-i\lambda t}|\psi_0\rangle = \langle\psi_0 \mid e^{-iHt} \mid \psi_0\rangle. \tag{17.20}$$

We are, however, interested in utilizing Eq. (17.18) to obtain an approximate result, since the exact explicit calculation of this expression is, in general, difficult. To do this, we first note that one may deform the part of the contour C from the branch point to $+\infty$ below the real line to an integral along the imaginary axis from the branch point to $-i\infty$. This can be done, since the line integral along the quarter circle arc in the lower half plane vanishes in the limit that the radius goes to ∞ (the exponent e^{-izt} decreases exponentially with the radius). The part of the contour above the real line must then be deformed through the cut to the second Riemann sheet of $R(z)$, to bring it to the negative imaginary half line as well. This can be done as follows.

We wish to construct a complex analytic function which is defined in the lower half plane and is continuously and differentially connected to $R(z)$ in the upper half plane. Such a function is identified as the extension of $R(z)$ to the second Riemann sheet. Consider the difference of $R(z)$ immediately below the real line (the analytic continuation of the function $R(z)$ defined in the upper half plane around the branch point to the lower half plane, all on the first Riemann sheet) and the function $R(z)$ evaluated immediately above the real line. Using the spectral form Eq. (17.17), we see that

$$R(\mu + i\varepsilon) - R(\mu - i\varepsilon) = \int |\langle\lambda \mid \psi_0\rangle|^2 \left(\frac{1}{\mu + i\varepsilon - \lambda} - \frac{1}{\mu - i\varepsilon - \lambda}\right) d\lambda, \tag{17.21}$$

where we have used the form Eq. (17.14) applicable to a Hamiltonian with absolutely continuous spectrum, and the fact that $< \psi_0|\lambda\rangle\langle\lambda|\psi_0 >= |\langle\lambda \mid \psi_0\rangle|^2$. With the well-known result of the theory of distributions,

$$\lim_{\varepsilon\to 0} \frac{1}{x + i\varepsilon} = P\left(\frac{1}{x}\right) - i\pi\delta(x), \tag{17.22}$$

we obtain

$$\lim_{\varepsilon\to 0} R(\mu + i\varepsilon) - R(\mu - i\varepsilon) = -2\pi i|\langle\mu \mid \psi_0\rangle|^2. \tag{17.23}$$

If we assume that $|\langle\mu \mid \psi_0\rangle|^2$ is the boundary value on the real axis of a function $W(z)$ analytic in some region of the lower half plane, we see that the continuous differentiable extension we were looking for is given by

$$R^{II}(z) = R(z) - 2\pi i W(z). \tag{17.24}$$

It is clear that in the limit as z goes to the real line from below, by Eq. (17.23), $R^{II}(z)$ approaches the limit of $R(z)$ onto the real line from above, smoothly. We shall show in Section 17.3 that there are models, such as the Lee–Friedrichs model (Lee, 1954, Friedrichs, 1950, to be discussed later in this chapter), for which the assumptions we made are justified. Furthermore, it can occur that the function $W(z)$ has a pole in the lower half plane in the extension of its domain of analyticity, a situation which we shall argue for in the framework of these models. Let us assume for now that such a simple pole exists in $W(z)$ and return to our construction of the approximate form for the reduced evolution, Eq. (17.10).

The part of the contour which remained above the real line can now be distorted by rotation downward, where the integration is now on the second sheet function $R^{II}(z)$. This line can, by the same argument given above, be rotated down to the negative imaginary axis, curving above the branch point into the line integral obtained earlier on the first Riemann sheet. In moving this line downward, we encounter the pole that we have assumed, say, at $z_0 = E_0 - i\frac{\Gamma}{2}$, resulting in the following exact form:

$$A = \frac{1}{2\pi i}\int_C e^{-izt} R(z)\, dz = \frac{1}{2\pi i}\int_{C_1} e^{-izt} R(z)\, dz - 2\pi i e^{-iz_0 t}\, \mathrm{Res}\, W(z_0), \tag{17.25}$$

where C_1 corresponds to the contour around the negative imaginary axis (the left part in the first sheet and the right part in the second sheet), and Res $W(z_0)$ is the residue of the function $W(z)$ at the pole position z_0. These integrals carry the factor e^{-izt} for z in the lower half plane, and for $t > 0$ and not too small, one can consider neglecting these contributions. These terms are called "background" contributions. The remaining part, proportional to $e^{-iz_0 t}$, is the principal contribution for this time range (t not too small and not too large) and is called the "pole approximation." Actually, part of the integration along C_1 has a weaker time decrease than this pole contribution, but it is generally of higher order in some small coupling constant (Bleistein *et al.*, 1977).

Thus,

$$A(t) \cong -2\pi i e^{-iz_0 t}\, \mathrm{Res}\, W(z_0). \tag{17.26}$$

For t very large, the pole term decreases, of course, exponentially, and the integral on C_1 in the neighborhood of the branch cut, where $|\,\mathrm{Im}\, z|$ is small, will dominate the integral. This usually gives rise to an inverse polynomial dependence on t, that is, t^{-n}, where n is the space dimension of the problem (Bleistein *et al.*, 1977; Höhler, 1958).

For t very small, the integral on C_1 cannot be neglected, and the best path of integration (minimum descent path) (Bleistein *et al.*, 1977) is along the real axis, where the expression for $A(t)$ can be expanded in a power series. This results in a very simple form for the survival amplitude:

$$A(t) \cong \left\langle \psi_0 \left| 1 - iHt - \frac{1}{2}H^2t^2 + \cdots \right| \psi_0 \right\rangle = 1 - i\langle\psi_0|H|\psi_0\rangle - \frac{1}{2}\langle\psi_0|H^2|\psi_0\rangle t^2 + \cdots \tag{17.27}$$

The absolute square is (to order t^2) the survival probability

$$p(t) = |A(t)|^2 \cong \left(1 - \frac{1}{2}\langle\psi_0|H^2|\psi_0\rangle t^2\right)^2 + \langle\psi_0|H|\psi_0\rangle^2 t^2 \cong 1 - t^2 \Delta H^2, \tag{17.28}$$

where

$$\Delta H^2 = \langle\psi_0|H^2|\psi_0\rangle - \left(\langle\psi_0|H|\psi_0\rangle\right)^2, \tag{17.29}$$

the dispersion of the Hamiltonian operator in the state $|\psi_0\rangle$.

A very important consequence of this calculation is that for small t, $p(t)$ does not go linearly in t, as a pure exponential dependence (semigroup evolution) would, but only quadratically.

In addition to destroying the possibility that the Wigner–Weisskopf method could give rise to a semi-group, this so-called Zeno effect (Misra and Sudarshan, 1977) results in an apparent paradox, called the Zeno paradox. These effects are related in the sense that if the decay were of semigroup form, there would be no Zeno effect. The observation of a Zeno effect is a consequence of *reversible* evolution. It has been observed (Itano *et al.*, 1990; Wilkinson *et al.*, 1997; Fischer *et al.*, 2001) under conditions that minimize radiation and inelastic collisions.

Let us first describe this latter phenomenon before going on to the consequences of the failure of the theory to provide a semigroup law of decay. If one thinks of a series of measurements to extract, by a filter, the initial state from the beam, after each filtering, the evolution process we have described here must be started again. Done at fairly short time, one would find a quadratic decay law for this short time, followed by another quadratic decay, followed by another, and so on. The envelope of this curve would look approximately exponential (see articles of E. Joos and H. D. Zeh in Giuliani *et al.* (1996)), accounting for exponential decay in the Wigner–Weisskopf model as a result of successive interference by an "environment" (selective scattering, with the effect of a filtering measurement). Such efforts have been largely replaced by the use of stochastic terms in the Schrödinger evolution, a fundamental idea previously discussed in the chapter on measurement.

If the frequency of selective filtering measurements becomes very high, it is clear that the sequence of quadratic decays converges to a constant occupancy for the initial state (Misra and Sudarshan, 1977), i.e. in spite of a perturbation inducing decay, the state is completely stabilized. This is the so-called Zeno paradox (associated with this Zeno effect) and has been used by Aharonov (Aharonov and Vardi, 1980) to theoretically stabilize an unstable state, and even to guide its evolution macroscopically.

The Zeno effect, as seen from the expansion in Eq. (17.28), is an inevitable consequence of the application of Hilbert space techniques to calculate transition amplitudes under Hamilton (or any one-parameter group) evolution.

A serious consequence of the $O\left(t^2\right)$ decay law, as we have pointed out above, is the obstruction it forms to the property of semigroup evolution, which is a fundamental property of irreversible processes.

The exponential, or pole approximation, of the Wigner–Weisskopf method would have this property for the single channel, or one decay mode case at intermediate times, but as we have seen, at long or short times, this approximation is not valid.

For the two-channel case, studied experimentally very carefully for the neutral K meson decay, it has been shown (Winstein *et al.*, 1997) that the two-dimensional generalization (Lee *et al.*, 1957; Wu and Yang, 1964) of Gamow's formula provides an extremely accurate description, while even in the pole approximation, the Wigner–Weisskopf method predicts results that disagree with the experiments. Therefore, for the two- (or more) channel case, if there is no decoupling due to symmetry, the Wigner–Weisskopf method is not suitable. We demonstrate this result in a soluble model in the next section.

A fundamental theory, based on the scattering approach of Lax and Phillips (1967), has recently been developed which provides an exact semigroup evolution, and therefore a theoretical basis for the Gamow construction (Flesia and Piron, 1984; Horwitz and Piron, 1993; Eisenberg and Horwitz, 1997; Strauss, 2003, 2005a, 2005b, 2005c). We shall discuss this theory in Section 17.7.

17.3 Wigner–Weisskopf method and Lee–Friedrichs model with a single channel

Before describing these developments, let us return to a quantitative discussion of the Wigner–Weisskopf method in the framework of the soluble Lee–Friedrichs model (Friedrichs, 1950; Lee, 1954), where we shall be able to make precise statements as well as to introduce in a simple way the notion of the rigged Hilbert space, or Gel'fand triple (Bailey and Schieve, 1978; Baumgartel, 1978: Bohm, 1978, 1980; Horwitz and Sigal, 1980; Parravicini *et al.*, 1980; Bohm and Gadella, 1989; Bohm and Kaldass, 2000), which has been widely used to obtain an exact semigroup behavior. (We shall discuss the Gel'fand triple approach in Section 17.6.)

Although the Gel'fand triple states provide exponential evolution, there is, in general, no scalar product defined in such spaces (they are Banach spaces, not Hilbert spaces), and therefore properties such as expectation values of observables, for example, for the spatial dispersion of a resonant state are not available. There are, however, many robust properties of these theories which have their counterparts in the more complete physics contained in the Lax–Phillips type of approach, and therefore these theories are important and worth studying.

The Lee–Friedrichs model for the decay of an unstable system is defined by a Hamiltonian of the form

$$H = H_0 + V, \tag{17.30}$$

where H_0 has absolutely continuous spectrum $\{\lambda \geq 0\}$, with spectral function $dE(\lambda) = |\lambda><\lambda|d\lambda$, and a discrete eigenvalue λ_0 embedded in this continuum with eigenstate $|\psi_0\rangle$, which we shall identify with the initial (unstable) state.

The perturbation V has the property, essential for the model, that for all λ, λ',

$$\langle\lambda|V|\lambda'\rangle = 0. \tag{17.31}$$

The nonvanishing matrix elements are $\langle\lambda|V|\psi_0\rangle$ and its conjugate $\langle\psi_0|V|\lambda\rangle$. A nonvanishing expectation value $\langle\psi_0|V|\psi_0\rangle$ would contribute a shift to λ_0 in all resulting expressions and may be taken as zero as well.

Historically, Lee (1954) formulated this model in the framework of nonrelativistic quantum field theory. The special structure of the interaction terms permits the problem to be decomposed to sectors involving one unstable particle which decays into two final particles, or two unstable particles which decay into two pairs of final particles, and so on. The problem in each sector is identical to that of the first sector. It is therefore equivalent to the original quantum mechanical form given by Friedrichs (1950).

In this construction, the eigenfunction for the discrete state corresponds to, as noted above, the unstable system, and the continuum corresponds to the final states of the decayed system. (In the quantum mechanical form, there is no reference to the number of particles in the final state, so long as it is in a single degenerate continuum.)

There are examples of decaying systems for which a multiplicity of continua occur with a sequence of distinct thresholds (lower bounds on each continuum), as in molecular physics. The analytic continuation that we shall carry out in our discussion is complicated by the occurrence of these nondegenerate continua. If the potential is an analytic function of coordinates, it is possible to carry out what is known as rotation of spectra, which effectively separates the many Riemann sheets occurring in the lower half plane. This is done by carrying out a unitarily induced dilation, and then using the fact that a one-parameter unitary transformation is an analytic function of the parameter. All discrete parts of the spectrum (including resonance poles) are left invariant (independent of the value of the real parameter), but the continuum rotates. The method was originally developed by Aguilar, Balslev, Combes and Simon (Aguilar and Combes, 1971; Balslev and Combes, 1971; Simon, 1972). We shall not discuss this method further here, but refer the reader to the excellent discussions in the literature.

With this model, let us again consider the general identity (often called the second resolvent equation or just the resolvent equation):

$$G(z) = G_0(z) + G_0(z)\,VG(z)\,, \tag{17.32}$$

where

$$G(z) = \frac{1}{z - H} \qquad G_0(z) = \frac{1}{z - H_0}.$$

The identity is easily proven by factoring out $G_0(z)$ to the left and $G(z)$ to the right:

$$\begin{aligned} G(z) &= G_0(z)[z - H + V]G(z) \\ &= G_0(z)[z - H_0]G(z) \equiv G(z). \end{aligned}$$

Now we consider the expectation value

$$R(z) = \langle\psi_0|G(z)|\psi_0\rangle, \tag{17.33}$$

as in Eq. (17.11). With the resolvent equation, we see that

$$\begin{aligned} R(z) &= \langle\psi_0|G_0(z)|\psi_0\rangle + \langle\psi_0|G_0(z)VG(z)|\psi_0\rangle \\ &= \frac{1}{z - \lambda_0} + \frac{1}{z - \lambda_0}\langle\psi_0|VG(z)|\psi_0\rangle, \end{aligned} \tag{17.34}$$

where ψ_0 is a discrete eigenstate of H_0. Furthermore, since the operator V connects ψ_0 only to the continuum $|\lambda\rangle$ (we have assumed $\langle\psi_0|V|\psi_0\rangle = 0$), Eq. (17.34) becomes

$$(z - \lambda_0) R(z) = 1 + \int_0^\infty \langle\psi_0|V|\lambda\rangle\langle\lambda|G(z)|\psi_0\rangle d\lambda. \tag{17.35}$$

It is then necessary for us to consider $\langle\lambda|G(z)|\psi_0\rangle$. Using the resolvent Eq. (17.32) again, we obtain

$$\langle\lambda|G(z)|\psi_0\rangle = \frac{1}{z - \lambda}\langle\lambda|V|\psi_0\rangle\langle\psi_0|G(z)|\psi_0\rangle, \tag{17.36}$$

since, again, the operator V connects $\langle\lambda|$ only to $|\psi_0\rangle$. This is the essential point of the Lee–Friedrichs model. Substituting Eq. (17.36) into Eq. (17.35), we obtain

$$(z - \lambda) R(z) = 1 + \int \frac{\langle\lambda|V|\psi_0\rangle^2}{z - \lambda} d\lambda R(z),$$

or

$$\left[(z - \lambda_0) - \int_0^\infty \frac{\omega(\lambda)}{z - \lambda} d\lambda\right] R(z) = 1, \tag{17.37}$$

where the spectral weight function $\omega(\lambda)$ for the Lee–Friedrichs model is given by

$$\omega(\lambda) = |\langle\lambda|V|\psi_0\rangle|^2. \tag{17.38}$$

We write,

$$h(z) = z - \lambda_0 - \int_0^\infty \frac{\omega(\lambda)}{z - \lambda} d\lambda, \tag{17.39}$$

and the condition, Eq. (17.37),

$$h(z)\,R(z) = 1, \tag{17.40}$$

implies that if $h(z)$ goes to zero at some value $z \to z_0$, then $R(z)$ will have a pole at z_0. It is easy to see, with some simple assumptions, that there is no zero of $h(z)$ in the cut plane.

For z on the negative real axis, say, $z = -E$, $E > 0$, we would have to satisfy

$$-E - \lambda_0 + \int \frac{\omega(\lambda)}{E+\lambda} d\lambda = 0. \tag{17.41}$$

Since

$$\int_0^\infty d\lambda \frac{\omega(\lambda)}{E+\lambda} \leq \int_0^\infty \frac{\omega(\lambda)}{\lambda} d\lambda,$$

if $\omega(\lambda)$ vanishes as $\lambda \to 0$, so that the integral on the right side is defined (vanishing of the spectral weight at the threshold for decay), then for sufficiently small coupling, measured by the norm

$$\int |\langle\lambda|V|\psi_0\rangle|^2 d\lambda = \|V\psi_0\|^2,$$

the zero in Eq. (17.41) cannot be achieved for some finite λ_0 ($\lambda_0 + E \geq \lambda_0$).

We now consider complex z. Taking the imaginary part of Eq. (17.39), the vanishing of $h(z)$ at some point $\mathrm{Im}\, z \neq 0$ would imply

$$\begin{aligned} 0 &= \mathrm{Im}\, z + \int \frac{\omega(\lambda)}{|z-\lambda|^2} \mathrm{Im}\, z d\lambda \\ &= \mathrm{Im}\, z \left(1 + \int \frac{\omega(\lambda)\, d\lambda}{|z-\lambda|^2}\right). \end{aligned}$$

Since the second factor on the right is positive, this zero cannot be achieved for any z in the cut plane.

As we have described in our discussion of the general case, in Eq. (17.25), we must now consider the analytic continuation of $R(z)$ to the second Riemann sheet. From Eq. (17.40) we see that the second sheet function $R(z)^{II}$ is defined by the analytic continuation of $h(z)$ through the cut, evident in Eq. (17.39), on the real positive axis. The technique described in Eq. (17.21) through Eq. (17.24) can be applied directly to $h(z)$. Let us compare $h(\mu + i\varepsilon)$ and $h(\mu - i\varepsilon)$ in the first sheet, for μ real and positive and ε small, to obtain a function in the second Riemann sheet which is the analytic continuation of $h(z)$ above the cut into the lower half plane. Consider

$$h(\mu + i\varepsilon) - h(\mu - i\varepsilon) = -\int_0^\infty \omega(\lambda) \left\{\frac{1}{\mu + i\varepsilon - \lambda} - \frac{1}{\mu - i\varepsilon - \lambda}\right\} d\lambda$$

in the limit $\varepsilon \to 0$. Then

$$h(\mu + i\varepsilon) - h(\mu - i\varepsilon) = 2\pi i \int_0^\infty \omega(\lambda)\,\delta(\mu - \lambda)\,d\lambda = 2\pi i\omega(\mu). \tag{17.42}$$

We thus have the relation

$$h(\mu + i\varepsilon) = h(\mu - i\varepsilon) + 2\pi i\omega(\mu), \tag{17.43}$$

the second term corresponding to the "jump" across the cut. We now wish to make a further assumption, namely, that $\omega(\mu)$ is the boundary value, on the real line, of a function analytic in some sufficient domain in the lower half plane. Calling this function $\omega(z)$, it follows from Eq. (17.43) that

$$h^{II}(z) = h(z) + 2\pi i\omega(z) \tag{17.44}$$

satisfies the conditions for the second sheet continuation of $h(z)$ across the cut. As $z \to \mu - i\varepsilon$, this function smoothly approaches the value of $h(z)$ just above the cut.

Now let us examine again the imaginary part of $h^{II}(z)$ for z in the lower half plane:

$$\operatorname{Im} h^{II}(z) = \operatorname{Im} z \left(1 + \int_0^\infty \frac{\omega(\lambda)\,d\lambda}{|z - \lambda|^2}\right) + 2\pi\omega(z). \tag{17.45}$$

In a region for which $\operatorname{Im} z$ is small, $\omega(z)$ must be predominantly real and positive; it goes smoothly to $\omega(\mu)$ on the real line. Since $\operatorname{Im} z < 0$, it is quite reasonable to assume that $\operatorname{Im} h^{II}(z)$ defined in Eq. (17.45) vanishes at some value of z in the lower half plane (close to the real axis). If the real part vanishes as well, then $h^{II}(z)$ becomes zero at this point, implying that $R^{II}(z)$ has a singularity. There are simple examples for which these assumptions are valid.

Assuming, then, that $R^{II}(z)$ has a pole at some point z_0 for $\operatorname{Im} z_0 < 0$ (and small), the contour integral Eq. (17.18) takes on the form

$$A(t) = e^{-iz_0 t} \operatorname{Res} R^{II}(z)\,|_{z_0} + \text{background contribution}, \tag{17.46}$$

where the first term dominates for t not too large and not too small. Since

$$R^{II}(z) = \frac{1}{h^{II}(z)} \cong \frac{1}{z - z_0} \frac{1}{h^{II}(z_0)'}$$

in the neighborhood of the pole, the residue is the inverse of

$$h^{II}(z_0)' = 1 + \int_0^\infty d\lambda \frac{\omega(\lambda)}{(z_0 - \lambda)^2} + 2\pi i\omega'(z_0). \tag{17.47}$$

Since we have assumed that $\omega(\lambda)$ is the boundary value of a function analytic in the lower half plane down to the neighborhood of Im z_0, at least, we may make an estimate of the integral by distorting the contour below the real axis in some neighborhood of Re z_0. Calling the complex value of the variable on the contour ζ, we have that

$$\int_0^\infty d\lambda \frac{\omega(\lambda)}{(z_0-\lambda)^2} = \int_C \frac{\omega(\zeta)\,d\zeta}{(z_0-\zeta)^2},$$

where C is a small deviation of the real line (holding the origin $\lambda = 0$ fixed) below the real axis. Continuing below to cross the pole position (the sense of encirclement is negative), we obtain

$$-2\pi i \omega'(z_0)$$

as the contribution of the pole. This term is canceled by the third term on the right of Eq. (17.47), and what remains of the integration is expected to be a well-bounded contribution of second order in the coupling

$$\sim |\langle \text{Re } z_0|V|\psi_0\rangle|^2.$$

The residue is, therefore, very close to unity (for weak coupling).

We therefore conclude that to a very good approximation, for t not too small or too large,

$$p(t) = |A(t)|^2 \cong e^{-\Gamma t} \tag{17.48}$$

for $\Gamma = |\text{Im } z_0|$.

In a similar way, an estimate can be made for the decay width (Im z_0) if it is small (a small width is characteristic of a resonance, which is almost a bound state). Returning to Eq. (17.45), we see that the vanishing of Im $h^{II}(z)$ at $z = z_0$ implies that

$$\text{Im } z_0 \left(1 + \int_0^\infty \frac{\omega(\lambda)\,d\lambda}{|z_0-\lambda|^2}\right) + 2\pi\omega(z_0) = 0. \tag{17.49}$$

For Im z_0 small,

$$\frac{1}{(\text{Im } z_0)^2 + (\text{Re } z_0 - \lambda)^2} \cong \frac{\pi}{|\text{Im } z_0|}\delta(\text{Re } z_0 - \lambda), \tag{17.50}$$

where we have used the relation

$$\lim_{\varepsilon\to 0} \frac{\varepsilon}{\varepsilon^2 + x^2} = \pi\delta(x), \tag{17.51}$$

approximately true, without taking the limit, for ε small. Thus Eq. (17.49) becomes

$$-|\text{Im } z_0|\left(1 + \frac{\pi}{|\text{Im } z_0|}\omega(\text{Re } z_0)\right) + 2\pi\omega(z_0) \cong 0,$$

or

$$|\text{Im } z_0| \cong \pi\omega(\text{Re } z_0), \tag{17.52}$$

where we have approximated $\omega(z_0) \cong \omega(\text{Re } z_0)$. The result of Eq. (17.52) coincides with the first Born approximation (the Golden Rule) for the transition rate $|\langle\psi_0|V|\psi_0\rangle|^2$, the result of the original perturbation calculation of Weisskopf and Wigner (1930).

This very useful result of the paper of Weisskopf and Wigner, in the so-called pole approximation, a form first postulated by Gamow, appeared to provide a fundamental theory describing the decay law for an unstable system.

17.4 Wigner–Weisskopf and multichannel decay

There remain two fundamental difficulties, related to the fact that the amplitude $A(t)$ does not satisfy a semigroup law. The first is the vanishing of the decay at very short times, and the second is that *even in pole approximation,* the N-channel $(N \geq 2)$ decay law that follows from the Wigner–Weisskopf method does not obey the semigroup law, although the pole approximation in the one-channel case does, to a good approximation.

For the N-channel case, we consider a Hamiltonian H_0 with a continuous spectrum of multiplicity N. We assume for our present discussion that the lower bounds on all of these spectra are at zero; the case of differing thresholds (onset values of the final decay channels) slightly complicates the discussion of analyticity (Aguilar and Combes, 1971; Balslev and Combes, 1971; Simon, 1972), as mentioned above. We furthermore assume that there are N discrete states embedded in these continua, but we admit coupling between the different channels, since this possibility gives rise to the well-known CP violation effects and other similar physical phenomena involving symmetry breakdown in decay processes. The physical idea is that we have several types of initial resonant states which decay into a set of continuum final states.

To formulate this problem, we consider an initial state in the finite dimensional subspace spanned by the N discrete eigenstates of H_0:

$$|\psi_0\rangle = \sum_{\alpha=1}^{N} \alpha_\alpha \varphi_a. \tag{17.53}$$

According to the Wigner–Weisskopf method, the probability of decay can be described as follows.

One can argue that into any channel α, the probability of decay is given by

$$p_\alpha^D(t) = \int_0^\infty d\lambda |\langle\lambda, \alpha|e^{-iHt}|\psi_0\rangle|^2 \tag{17.54}$$

(a similar formulation is discussed in Antoniou *et al.*, 1993), and the total decay into all channels is

$$p^D(t) = \sum_\alpha \int_0^\infty d\lambda |\langle\lambda, \alpha| e^{-iHt} |\psi_0\rangle|^2 \tag{17.55}$$
$$= 1 - \sum_\alpha |\langle\varphi_a| e^{-iHt} |\psi_0\rangle|^2,$$

since the set $\{| \varphi_\alpha >, |\lambda_\alpha\rangle\}$ is complete. Therefore,

$$\sum_\alpha |\langle\varphi_\alpha| e^{-iHt}|\psi_0\rangle|^2 + \sum_\alpha \int_0^\infty d\lambda |\langle\lambda, \alpha| e^{-iHt}|\psi_0\rangle|^2 = \|e^{-iHt} \mid \psi_0\rangle\|^2 = 1.$$

Since ψ_0 is given by Eq. (17.53), what we must study in order to evaluate Eq. (17.54) and Eq. (17.55) are the matrix elements $\langle\varphi_\alpha|e^{-iHt}|\varphi_\beta\rangle$. This finite matrix can be thought of as the evolution e^{-iHt} of the system restricted to the subspace spanned by $\{|\varphi_\alpha\rangle\}$, sometimes called the reduced evolution. It is this evolution law which is expected to satisfy the semigroup law for an irreversible process. We shall show in the following that this cannot be true in the Wigner–Weisskopf method, and moreover, even in the pole approximation (which does satisfy this requirement to a very good approximation in the single-channel case), deviations from the semigroup law can be very large.

Let us consider again the reduced resolvent matrix obtained by the Laplace transform of

$$\langle\varphi_\alpha|e^{-iHt}|\varphi_\beta\rangle,$$

obtained as in Eq. (17.11):

$$R_{\alpha\beta}(z) = \left\langle \varphi_\alpha \left| \frac{1}{z - H} \right| \varphi_\beta \right\rangle. \tag{17.56}$$

This matrix is a set of functions of the complex variable z analytic (as seen from the representation Eq. (17.17)) in the cut plane. Using the same methods employed to obtain Eq. (17.24), we may define the second sheet continuation of $R_{\alpha\beta}(z)$ to construct $R^{II}_{\alpha\beta}$. It is then convenient to define a matrix $W^{II}_{\alpha\beta}(z)$ in terms of $R^{II}_{\alpha\beta}$, so that as a matrix equation,

$$R^{II}(z) = \frac{1}{z - W^{II}(z)}. \tag{17.57}$$

The poles of this matrix-valued function occur at values of z for which it is equal to an eigenvalue of the matrix $W^{II}(z)$. We shall argue that the $N \times N$ matrix residues at different pole values are, in general, not orthogonal, and therefore that the semigroup property is not obeyed even in pole approximation.

Let us use the fact that (almost) every finite matrix, Hermitian or not, has a set of right and left eigenvectors with eigenvalues ω_α, with α running from one to N. Denoting the left and right eigenvectors of $W^{II}(z)$, respectively, by ${}_L\langle\alpha, z|$, $|\alpha, z\rangle_R$, where we take into account the explicit z dependence of the matrix $W^{II}(z)$, we have

$$W^{II}(z)\,|\alpha, z\rangle_R = \omega_\alpha(z)\,|\alpha, z\rangle_R \tag{17.58}$$

and

$${}_L\langle\alpha, z|W^{II}(z) = \omega_\alpha(z)\,{}_L\langle\alpha, z|. \tag{17.59}$$

Clearly,

$$\begin{aligned} {}_L\langle\beta, z|W^{II}(z)\,|\alpha, z\rangle_R &= \omega_\alpha(z)\,{}_L\langle\beta, z \mid \alpha, z\rangle_R \\ &= \omega_\beta(z)\,{}_L\langle\beta, z \mid \alpha, z\rangle_R, \end{aligned} \tag{17.60}$$

which can be valid only if

$${}_L\langle\beta, z \mid \alpha, z\rangle_R = 0 \tag{17.61}$$

for $\omega_\alpha(z) \neq \omega_\beta(z)$. Taking into account this orthogonality, we see that we can construct a finite dimensional spectral representation, at any point z, for the non-Hermitian matrix

$$W^{II}(z) = \sum_\alpha \omega_\alpha(z)\,Q_\alpha(z),$$

where the

$$Q_\alpha(z) = \frac{|\alpha, z\rangle_R\;{}_L\langle\alpha, z|}{{}_L\langle\alpha, z \mid \alpha, z\rangle_R}$$

satisfy

$$Q_\alpha(z)\,Q_\beta(z) = Q_\alpha(z)\,\delta_{\alpha\beta}.$$

The reduced resolvent may be represented in the form

$$R^{II}(z) = \sum_\alpha \frac{Q_\alpha(z)}{z - \omega_\alpha(z)}, \tag{17.62}$$

for which poles occur at the points satisfying $z = \omega_\alpha(z)$. If this condition is true at some point z for some α, it is generally true that $z \neq \omega_\beta(z)$ for $\beta \neq \alpha$. If we search for a second pole, we may find one at some other point in the complex plane, e.g. z', for which we may suppose, for example, that $z' = \omega_\beta(z')$. This process may be continued until we have located all of the poles of the reduced resolvent. The residue of the first pole at $z = \omega_\alpha(z)$ has a residue proportional to $Q_\alpha(z)$, and the residue of the second pole has residue proportional to $Q_\beta(z')$. While $Q_\alpha(z)\,Q_\beta(z)$ is zero for $\omega_\alpha(z) \neq \omega_\beta(z)$ (two poles at the same point z), there is no reason why $Q_\alpha(z)\, Q_\beta(z')$ must be zero, and therefore the pole residues from different channels in the pole approximation will not, in general, be mutually orthogonal. Extracting the pole contributions from the inverse Laplace transform, as we did for the single-channel case,

$$\frac{1}{2\pi i}\int_C e^{-izt} R(z)\, dz \cong \sum_\alpha \frac{e^{-iz_\alpha t}}{1-\omega'_\alpha(z_a)} Q_\alpha(z_\alpha), \tag{17.63}$$

where $\{z_\alpha\}$ are the pole positions, we see that (assuming, for weak interactions, $\omega'_\alpha(z_\alpha) << 1$, for example) the reduced evolution has the matrix valued form

$$U_{red}(t) \simeq \sum_\alpha e^{-iz_\alpha t} Q_\alpha(z_\alpha). \tag{17.64}$$

Repeated application of this reduced evolution is then

$$U_{red}(t_2)\, U_{red}(t_1) \cong \sum_{a\beta} e^{-iz_\alpha t_2} e^{-iz_\beta t_1} Q_\alpha(z_\alpha)\, Q_\beta(z_\beta). \tag{17.65}$$

Although $Q_\alpha(z_\alpha)^2 = Q_\alpha(z_\alpha)$, as we have pointed out, $Q_\alpha(z_\alpha)\, Q_\beta(z_\beta) \neq 0$, so that even in the pole approximation, for more than one channel, the semi-group property is not valid. In the following section, we estimate this effect in a multichannel Lee–Friedrichs model.

17.5 Wigner–Weisskopf method with many-channel decay: the Lee–Friedrichs model

Let us use a Lee–Friedrichs model for an N-channel system to illustrate this point. In this model, we may estimate the departure from the semigroup property and show that in at least one careful set of experiments—in particular, in the two-channel case—the deviation predicted is larger than the experimental error.

The Lee–Friedrichs model, as in Eq. (17.30) and Eq. (17.31), is defined for the N-channel case by excluding all final state interactions. For

$$H = H_0 + V, \tag{17.66}$$

the matrix elements

$$\langle\lambda\alpha|V|\lambda'\beta\rangle = 0 \tag{17.67}$$

for all α, β, and we shall take, as before (in the subspace),

$$\langle\varphi_\alpha|V|\varphi_\beta\rangle = 0 \tag{17.68}$$

for all α. β, i.e. V does not connect the bound states in different channels as well. (If we did not assume Eq. (17.68), we would have a mass matrix instead of a shift;

the channels could then be chosen to be their unperturbed mass eigenstates; we would then have only shifts in each channel.)

Again, using the resolvent equation identity, the reduced resolvent in this model takes on the form

$$
\begin{aligned}
R_{\alpha\beta}(z) &= \left\langle \varphi_\alpha \left| \frac{1}{z-H} \right| \varphi_\beta \right\rangle \\
&= \left\langle \varphi_\alpha \left| \frac{1}{z-H_0} \right| \varphi\beta \right\rangle + \left\langle \varphi_\alpha \left| \frac{1}{z-H_0} V \frac{1}{z-H} \right| \varphi\beta \right\rangle \qquad (17.69) \\
&= \frac{1}{z-\lambda_\alpha}\delta_{\alpha\beta} + \frac{1}{z-\lambda_\alpha}\sum_\gamma \int d\lambda \langle \varphi_\alpha | V | \lambda\gamma \rangle \left\langle \lambda\gamma \left| \frac{1}{z-H} \right| \varphi_\beta \right\rangle .
\end{aligned}
$$

We therefore need

$$
\begin{aligned}
\left\langle \lambda\gamma \left| \frac{1}{z-H} \right| \varphi_\beta \right\rangle &= \left\langle \lambda\gamma \left| \frac{1}{z-H_0} V \frac{1}{z-H} \right| \varphi_\beta \right\rangle \qquad (17.70) \\
&= \frac{1}{z-\lambda}\sum_{\gamma'} \langle \lambda\gamma | V | \varphi_{\gamma'} \rangle R_{\gamma'\beta}(z) .
\end{aligned}
$$

Note that in Eq. (17.70) we have used the fact that, by our assumption, the continuous spectrum of H_0 is N-fold degenerate.

Substituting Eq. (17.70) into Eq. (17.69), we find

$$
R_{\alpha\beta} = \frac{1}{z-\lambda_\alpha}\delta_{\alpha\beta} + \frac{1}{z-\lambda_\alpha}\sum_\gamma \int d\lambda \frac{\langle \varphi_\alpha | V | \lambda\gamma \rangle \langle \lambda\gamma | V | \varphi_{\gamma'} \rangle}{z-\lambda} R_{\gamma'\beta}(z) . \qquad (17.71)
$$

Defining the matrix

$$
\omega_{\alpha\beta}(\lambda) = \sum_\gamma \langle \varphi_\alpha | V | \lambda\gamma \rangle \langle \lambda\gamma | V | \varphi_\beta \rangle, \qquad (17.72)
$$

it follows from Eq. (17.71) that

$$
\sum_\gamma h_{\alpha\gamma}(z) R_{\gamma\beta}(z) = \delta_{\alpha\beta}, \qquad (17.73)
$$

where

$$
h_{\alpha\gamma}(z) = (z-\lambda_\alpha)\,\delta_{\alpha\gamma} - \int d\lambda \frac{\omega_{\alpha\gamma}(\lambda)}{z-\lambda}. \qquad (17.74)
$$

The proofs that there are no zeros of the determinant of $h_{\alpha\beta}(z)$ for z in the first sheet (cut plane) are similar to those of the single-channel case. The analytic continuation of $h_{a\beta}(z)$ to the second sheet follows the same method as in Eq. (17.42), i.e.

$$h_{\alpha\beta}(\mu + i\varepsilon) - h_{\alpha\beta}(\mu - i\varepsilon) = -\int \omega_{\alpha\beta}(\lambda) \left\{ \frac{1}{\mu + i\varepsilon - \lambda} - \frac{1}{\mu - i\varepsilon - \lambda} \right\} d\lambda$$
$$= 2\pi i \omega_{\alpha\beta}(\mu).$$

Again, assuming the elements $\omega_{\alpha\beta}(\mu)$ are boundary values on the real line of a set of functions analytic in the lower half plane (in some region containing the singularities), i.e. a matrix-valued analytic function $\omega(z)$, we can define the second sheet function in the same way as in Eq. (17.44):

$$h^{II}_{\alpha\beta}(z) = h_{\alpha\beta}(z) + 2\pi i \omega_{\alpha\beta}(z). \tag{17.75}$$

The second sheet reduced resolvent is then defined in terms of the second sheet function Eq. (17.75) as

$$\sum_{\gamma} h^{II}_{\alpha\gamma}(z) R^{II}_{\gamma\beta} = \delta_{\alpha\beta}, \tag{17.76}$$

and $R^{II}_{\alpha\beta}$ will have a singular determinant where the determinant of $h_{\alpha\beta}(z)$ vanishes. This vanishing can be expressed in terms of a matrix $W_{\alpha\beta}(z)$ defined by

$$h^{II}_{\alpha\beta}(z) = z\delta_{\alpha\beta} - W^{II}_{\alpha\beta}(z); \tag{17.77}$$

the singularities in $R^{II}(z)$ then occur on condition that the eigenvalue $\omega_\alpha(z)$ of $W^{II}_{\alpha\beta}(z)$ has the property

$$z_\alpha = \omega_\alpha(z)\,|_{z=z_\alpha}, \tag{17.78}$$

that is, at the point z where the matrix has an eigenvalue equal to that value of z. As we have pointed out, this condition may be satisfied at several different values of z (or none, in which case there will be no resonance poles), and although the spectral factors associated with each eigenvalue at a point are orthogonal, the spectral factors associated with the eigenvalues at *different* points z will not, in general, be orthogonal.

Let us consider in detail the case of $N = 2$ and use the formulas of Eq. (17.72), Eq. (17.74) and Eq. (17.75) with $\lambda_1 = \lambda_2 \equiv \lambda_0$ to estimate the lack of orthogonality of $Q_1(z_1)$ and $Q_2(z_2)$, and thus the violation of the semigroup property in pole approximation. This case is of particular interest in the investigation of the decay of the neutral K meson system. The neutral K meson system consists of two particle (resonance) types, the K^0 and $\bar{K}^0$, a meson and its antiparticle which differ by the sign of a quantum number called hypercharge in the $SU(3)$ classification of the scalar meson octet (eight-dimensional representation of $SU(3)$). (See, for example Ne'eman, 1967; Gell-Mann and Ne'eman, 1964.) This system decays into two π mesons (uncharged or one positive, one negative), or into three π mesons (of zero total charge). The π mesons have zero hypercharge, a quantity not conserved in weak interactions, but both the K mesons and the π mesons have negative parity in the sense that the state vectors in the quantum mechanical Hilbert space representing these particles change sign under space reflection. Thus, the decay into two π mesons appears to violate parity conservation in the interaction that induces the decay. It was hoped before 1964 (see Cahn and Goldhaber, 1989; Martin and Shaw, 1997, for further discussion) that the combination of charge conjugation, by which the signs of all charges are inverted, including the sign of the hypercharge, and parity reflection, called CP, would be a symmetry obeyed by this interaction. Although the deviations are small, of order one in 10^3, the CP symmetry is not obeyed (Christenson *et al.,* 1964). This admits a non-orthogonality of the matrices corresponding to the pole residues, and therefore a failure of the semi-group property in K meson decay (Horwitz and Marchand, 1969; Horwitz and Mizrachi, 1974; Winstein *et al.,* 1997; Cohen and Horwitz, 2001; estimates based on Hagiwara *et al.*, 2002).

The experiments carried out at Fermilab (Winstein *et al.,* 1997) are accurate enough to rule out the applicability of the Wigner–Weisskopf model as a description for the decay process. This experiment represents a very fundamental difficulty in the application of the theory. It provides a simple and direct demonstration that irreversible processes actually exist on a fundamental level in particle decay to a very high accuracy and do not constitute just an effective macroscopic idealization. Furthermore, it demonstrates that although Wigner–Weisskopf theory may well apply on a rigorous quantum mechanical level to reversible processes governed by Hamiltonian evolution in the standard Schrödinger theory, for which one sees the short-time Zeno effect (Wilkinson *et al.,* 1997; Fischer *et al.,* 2001), it does not apply to the irreversible process involved in weak nonleptonic particle decay for which the time dependence can be experimentally studied. It will be therefore necessary to develop some new theoretical techniques, some of which we shall describe in the next sections.

To estimate the non-orthogonality predicted by the Wigner–Weisskopf model, we use the Lee–Friedrichs model and specialize our formulas to two channels. We then proceed to estimate the pole residues and the magnitude of their non-orthogonality.

According to Eq. (17.76), the poles of $R^{II}_{\alpha\beta}$ are determined by the zeros of the determinant of $h^{II}_{\alpha\beta}$, now a 2×2 matrix. In the neighborhood of one of the poles (say, z_p for $p = 1, 2$), let us write

$$\det h^{II}(z) = \left(z - z_p\right) \Delta\left(z_p\right), \tag{17.79}$$

where $\Delta\left(z_p\right)$ is the derivative of the determinant, and using the easily derived formula

$$\begin{pmatrix} \alpha & b \\ c & d \end{pmatrix}^{-1} = \frac{1}{(ad - bc)} \begin{pmatrix} d & -b \\ -c & a \end{pmatrix},$$

we see that the residue of the pole in R^{II} is

$$g_p(z_p) = \frac{1}{\Delta(z_p)} \begin{pmatrix} h_{22}^{II} & -h_{12}^{II} \\ -h_{21}^{II} & h_{11}^{II} \end{pmatrix}. \tag{17.80}$$

The quantity Δ can be estimated as follows. One can apply the well-known formula

$$\delta \ln \det X = \mathrm{Tr}\,(X^{-1}\delta X),$$

easily seen as follows. By definition,

$$\det (X - \delta X) - \det X = \delta \det X,$$

so that

$$\delta \det X = \det\left(1 + X^{-1}\delta X\right) - \det X,$$

and the result then follows from

$$\det\left(1 + X^{-1}\delta X\right) = 1 + \mathrm{Tr}\left(X^{-1}\delta X\right) + O\left(\delta X^2\right).$$

The derivative of the determinant is then

$$\begin{aligned} \Delta(z) &= \frac{\partial}{\partial z} \det h^{II} \mathrm{Tr}\left(h^{II^{-1}} \frac{\partial h^{II}}{\partial z}\right) \\ &= \mathrm{Tr}\left[\begin{pmatrix} h_{22}^{II} & -h_{12}^{II} \\ h_{21}^{II} & h_{11}^{II} \end{pmatrix}\left(\frac{\partial h^{II}}{\partial z}\right)\right], \end{aligned} \tag{17.81}$$

where $\det h^{II}$ has canceled. Since, by Eq. (17.75),

$$h_{\alpha\beta}^{II}(z) = (z - \lambda_0)\,\delta_{\alpha\beta} + \int d\lambda \frac{\omega_{\alpha\beta}}{z - \lambda} + 2\pi i \omega_{\alpha\beta}(z), \tag{17.82}$$

we have

$$\frac{\partial h_{\alpha\beta}^{II}}{\partial z} = \delta_{\alpha\beta} + \int d\lambda \frac{\omega_{\alpha\beta}}{(z - \lambda)^2} + 2\pi i \frac{\partial \omega_{\alpha\beta}(z)}{\partial z}. \tag{17.83}$$

By distorting the contour $(0, \infty)$ of the λ integration to the lower half plane, the residue remaining in passing the double pole is $-2\pi i \frac{\partial \omega_{\alpha\beta}(z)}{\partial z}$, canceling the third term (Horwitz and Mizrachi, 1974; Cohen and Horwitz, 2001).

The smooth contribution of the background integral that remains can be estimated by taking into account orders of magnitude of the matrix elements appearing in

$$\omega_{\alpha\beta}(\lambda) = \sum_{\gamma} \langle \varphi_\alpha | V | \lambda\gamma \rangle \langle \lambda\gamma | V | \varphi_\beta \rangle,$$

i.e.

$$\begin{aligned}
\langle \lambda 1|V|\varphi_1\rangle &= O\,(g) \\
\langle \lambda 2|V|\varphi_1\rangle &= O\,(g\alpha) \\
\langle \lambda 1|V|\varphi_2\rangle &= O\,(g\alpha) \\
\langle \lambda 2|V|\varphi_2\rangle &= O\,(g)\,,
\end{aligned} \tag{17.84}$$

where g is the weak decay coupling constant (or order 10^{-6} in units of eV) and α measures the relative CP violation $\left(\alpha^2 \sim 10^{-3}\right)$ (Charpak and Gourdin, 1967).

Thus,

$$\frac{\partial h^{II}_{\alpha\beta}}{\partial z} = 1 + O\left(g^2\right), \tag{17.85}$$

so, from Eq. (17.81), at each pole position,

$$\Delta\left(z_p\right) = \mathrm{Tr}h^{II}\left(z_p\right) = O\left(g^2\right), \tag{17.86}$$

and the pole residues are

$$g_p\left(z_p\right) = \frac{1}{\mathrm{Tr}h^{II}} \begin{pmatrix} h^{II}_{22} & -h^{II}_{12} \\ h^{II}_{21} & h^{II}_{11} \end{pmatrix}. \tag{17.87}$$

In contrast, as we have pointed out, by Eq. (17.76) and Eq. (17.77), these poles occur at the eigenvalues of $W^{II}\,(z)$ when z satisfies the condition Eq. (17.78), here, for $\alpha = 1$ or 2. We define the right and left eigenfunctions $|K_{L\alpha}\,(z)\rangle$ and $|K_{R\alpha}\,(z)\rangle$ for $\alpha = 1, 2$ at any z, and the (non-Hermitian) projections

$$Q_\alpha\,(z) = |K_{R\alpha}\,(z)\rangle\langle K_{R\alpha}\,(z)\,|. \tag{17.88}$$

The reduced resolvent can then be written in the spectral form

$$\frac{1}{z - W^{II}\,(z)} = \sum_{\alpha=1,2} \frac{1}{z - \omega_a\,(z)} Q_\alpha\,(z) \tag{17.89}$$

at any z in the lower half plane. At the poles, $z = w_\alpha\,(z)$, in particular, at z_1 and z_2. Also note that, in general (e.g. for the CP-violating case), at z_1, $z_1 \neq w_2\,(z_1)$, and at z_2, $z_2 \neq z_1\,(z_2)$. The pole residues are then at z_1

$$\frac{1}{1 - w'\,(z_1)} Q_1\,(z_1) \tag{17.90}$$

and at z_2

$$\frac{1}{1 - w_2'\,(z_2)} Q_2\,(z_2)\,. \tag{17.91}$$

We now proceed to estimate the non-orthogonality of the pole residues. The matrix W^{II} is given by

$$W^{II}_{\alpha\beta} = \delta_{\alpha\beta} z - h^{II}_{\alpha\beta} = \lambda_0 \delta_{\alpha\beta} + \int d\lambda \frac{\omega_{\alpha\beta}(\lambda)}{1-\lambda} + 2\pi i w_{\alpha\beta}(z),$$

and therefore

$$W'^{II}_{\alpha\beta}(z) = -\int d\lambda \frac{\omega_{\alpha\beta}(\lambda)}{(z-\lambda)^2} - 2\pi i \omega'_{\alpha\beta}(z).$$

The matrix W'^{II} is therefore second order in the weak coupling constant, and its eigenvalues are very small compared with unity. The residues Eq. (17.88) and Eq. (17.89) are therefore well approximated by the non-Hermitian projections $Q_1(z_1)$ and $Q_2(z_2)$ alone. The eigenstates of $W^{II} = z \cdot 1 - h^{II}$ are the eigenstates of h^{II}.

We shall relate these eigenstates, which we shall call $|K_S\rangle$ and $|K_L\rangle$, corresponding to the short- and long-lived states, respectively (the $CP+$ state, admitting a two-pion decay is short lived relative to the $CP-$ state, which corresponds to a three-pion decay), to the unperturbed eigenstates of $CP\pm$, which we shall call, respectively, $|K_1\rangle$ and $|K_2\rangle$. The latter states have the property that for these, $\omega^0_{12} = \omega^0_{21} = 0$ (for a V^0 which contains no CP violation); the corresponding matrix $h^{0\ II}$ is then diagonal:

$$\begin{aligned} h^{0\ II}_{11} &= z - \lambda_0 - \int d\lambda \frac{\omega_{11}{}^0}{z-\lambda} - w\pi i \omega_{11}{}^0 \\ h^{0\ \ II}_{12} &= h^{0\ \ II}_{21} = 0 \\ h^{0\ \ II}_{22} &= z - \lambda_0 - \int d\lambda \frac{\omega_{22}{}^0}{z-\lambda} - 2\pi i \omega_{22}{}^0. \end{aligned} \tag{17.92}$$

The states $|K_1\rangle$ and $|K_2\rangle$ corresponding to the eigenvalues $h^{0\ II}_{11}$ and $h^{0\ II}_{22}$ are, in our two-state representation,

$$\begin{pmatrix}1\\0\end{pmatrix} \text{ and } \begin{pmatrix}0\\1\end{pmatrix}. \tag{17.93}$$

The eigenstates of the interacting form of the matrix

$$h^{II} = \begin{pmatrix} h_{11}{}^{II} & h_{12}{}^{II} \\ h_{21}{}^{II} & h_{22}{}^{II} \end{pmatrix}$$

for null eigenvalue are given by

$$\begin{pmatrix}u\\v\end{pmatrix} = \begin{pmatrix}1\\ -\frac{h_{11}{}^{II}}{h_{12}{}^{II}}\end{pmatrix} = \begin{pmatrix}1\\ -\frac{h_{21}{}^{II}}{h_{22}{}^{II}}\end{pmatrix} \tag{17.94}$$

since the determinant is zero at the pole positions. These eigenvectors obviously are linear combinations of the two states given in Eq. (17.93). Since, as we have seen, $h_{11}{}^{II}$ and $h_{22}{}^{II}$ determine the linear combination

$$|K_S\rangle = \frac{|K_1\rangle + \varepsilon_S|K_2\rangle}{\sqrt{1+|\varepsilon_S|^2}} \tag{17.95}$$

and

$$|K_L\rangle = \frac{|K_2\rangle + \varepsilon_L|K_1\rangle}{\sqrt{1+|\varepsilon_L|^2}}, \tag{17.96}$$

where we have written the result in accordance with the standard notation of Wu and Yang (1964). In our case, at the corresponding pole positions,

$$\varepsilon_S = -\frac{h_{21}{}^{II}(S)}{h_{22}{}^{II}(S)} \tag{17.97}$$

and

$$\varepsilon_L = -\frac{h_{12}{}^{II}(L)}{h_{11}{}^{II}(L)}, \tag{17.98}$$

where we have again used the relation $h_{11}^{II}h_{22}^{II} = h_{12}^{II}h_{21}^{II}$ at the pole positions, and written the eigenfunctions in a way making explicit the small quantities. The magnitudes of the ε's are $\frac{O(g^2\alpha)}{O(g^2)} = O(\alpha)$, where α is the order of the CP-violating decay amplitude (it appears only linearly in the off-diagonal elements).

In the same way, we can compute the left eigenvectors of h^{II} to obtain

$$\begin{aligned}\langle K_S| &= \frac{\langle K_1| + \tilde{\varepsilon}_S\langle K_2|}{\sqrt{1+|\tilde{\varepsilon}_S|^2}}\\ \langle K_L| &= \frac{\langle K_2| + \tilde{\varepsilon}_L\langle K_1|}{\sqrt{1+|\tilde{\varepsilon}_L|^2}},\end{aligned} \tag{17.99}$$

where

$$\tilde{\varepsilon}_S = -\frac{h_{12}^{II}(S)}{h_{22}^{II}(S)} \tag{17.100}$$

and

$$\tilde{\varepsilon}_L = -\frac{h_{21}{}^{II}(L)}{h_{11}{}^{II}(L)}. \tag{17.101}$$

We are now in a position to evaluate the non-orthogonality of the residues. As we have seen in our discussion of the residues, they correspond to the matrix coefficients of $z - z_P$ at the pole positions $P = S, L$ and have the form (from Eq.(17.87))

$$g_P = \frac{1}{\Delta}\begin{pmatrix} h_{22}{}^{II}(P) & -h_{12}{}^{II}(P) \\ -h_{21}{}^{II}(P) & h_{11}{}^{II}(P) \end{pmatrix} \tag{17.102}$$

at the pole positions $P = S, L$. Factoring out $h_{22}{}^{II}(S)$ from the matrix for g_S and $h_{11}{}^{II}(L)$ from the matrix for g_L, these residue functions can be written as

$$g_S = \frac{h_{22}{}^{II}(S)}{\Delta(S)}\begin{pmatrix} 1 & \tilde{\varepsilon}_S \\ \varepsilon_S & \varepsilon_S\tilde{\varepsilon}_S \end{pmatrix} \tag{17.103}$$

and

$$g_L = \frac{h_{11}{}^{II}(L)}{\Delta(L)}\begin{pmatrix} \varepsilon_L\tilde{\varepsilon}_L & \varepsilon_L \\ \tilde{\varepsilon}_L & 1 \end{pmatrix}. \tag{17.104}$$

Now, multiplying the residues, we obtain

$$g_S g_L = \frac{h_{11}{}^{II}(L)\,h_{22}{}^{II}(S)}{\Delta(S)\,\Delta(L)}\left(\varepsilon_L + \tilde{\varepsilon}_S\right)\begin{pmatrix} \tilde{\varepsilon}_L & 1 \\ \varepsilon_S\tilde{\varepsilon}_L & \varepsilon_S \end{pmatrix} \tag{17.105}$$

and

$$g_L g_S = \frac{h_{11}{}^{II}(L)\,h_{22}{}^{II}(S)}{\Delta(S)\,\Delta(L)}\left(\tilde{\varepsilon}_L + \varepsilon_S\right)\begin{pmatrix} \varepsilon_L & \tilde{\varepsilon}_L\varepsilon_S \\ 1 & \tilde{\varepsilon}_S \end{pmatrix}. \tag{17.106}$$

In both of these expressions, the matrices contain a term of order one, and the coefficients

$$\frac{h_{11}{}^{II}(L)\,h_{22}{}^{II}(S)}{\Delta(S)\,\Delta(L)}$$

are of order unity. It therefore remains to estimate $(\varepsilon_L + \tilde{\varepsilon}_S)$ and $(\tilde{\varepsilon}_L + \varepsilon_S)$. Let us consider the first of these:

$$\left(\varepsilon_L + \tilde{\varepsilon}_S\right) = \varepsilon_S\left(1 + \frac{\tilde{\varepsilon}_S}{\varepsilon_L}\right) = \varepsilon_L\left[\frac{h_{12}{}^{II}(z_S)\,h_{11}{}^{II}(z_L)}{h_{12}{}^{II}(z_L)\,h_{22}{}^{II}(z_S)} + 1\right]. \tag{17.107}$$

Since $|z_S - z_L| = O\left(g^2\right)$,

$$h_{12}{}^{II}(z_S) - h_{12}{}^{II}(z_L) = O\left(g^4\alpha\right),$$

and therefore,

$$\frac{h_{12}{}^{II}(z_S)}{h_{12}{}^{II}(z_L)} = \frac{h_{12}{}^{II}(z_L) + O\left(g^4\alpha\right)}{h_{12}{}^{II}(z_L)} = 1 + O\left(g^2\right). \tag{17.108}$$

Finally, we estimate the second factor as follows. For the imaginary part of $z_{L,S}$ small compared with the real part, we can approximate the formulas for $h_{11}{}^{II}(z_L)$ and $h_{22}{}^{II}(z_S)$ by

$$h_{11}{}^{II}(z_L) \cong (z_L - \lambda_0) - P\int \frac{\omega_{11}(\lambda')}{\lambda_L - \lambda'} d\lambda' + \pi i \omega_{11}(\lambda_L)$$

and

$$h_{11}{}^{II}(z_S) \cong (z_S - \lambda_0) - P\int \frac{\omega_{22}(\lambda')}{\lambda_S - \lambda'} d\lambda' + \pi i \omega_{22}(\lambda_S),$$

where $\lambda_{S,L}$ are the real parts of the pole values. Furthermore, from the eigenvalue conditions and the formulas for the matrix elements of the h^{II}'s, we see that

$$\begin{aligned} z_L &= \lambda_0 + P\int \frac{\omega_{22}(\lambda')}{\lambda_L - \lambda'} d\lambda' - \pi i \omega_{22}(\lambda_L) + O\left(g^2\alpha^2\right) \\ z_S &= \lambda_0 + P\int \frac{\omega_{11}(\lambda')}{\lambda_S - \lambda'} d\lambda' - \pi i \omega_{11}(\lambda_S) + O\left(g^2\alpha^2\right), \end{aligned} \tag{17.109}$$

where we have used $\varepsilon_{S,L} = O(\alpha)$. Hence

$$\begin{aligned} h_{11}{}^{II}(z_L) &\cong P\int \frac{\omega_{22}(\lambda') - w_{11}(\lambda')}{\lambda_L - \lambda'} d\lambda' + \pi i\left(\omega_{11}(\lambda_L) - \omega_{22}(\lambda_L)\right) + O\left(g^2\alpha^2\right) \\ h_{22}{}^{II}(z_S) &\cong P\int \frac{\omega_{22}(\lambda') - \omega_{11}(\lambda')}{\lambda_S - \lambda'} d\lambda' + \pi i\left(\omega_{11}(\lambda_S) - \omega_{22}(\lambda_S)\right) + O\left(g^2\alpha^2\right). \end{aligned} \tag{17.110}$$

Since $|\lambda_S - \lambda_L| = O\left(g^2\right)$, we have

$$h_{11}{}^{II}(z_L) + h_{22}{}^{II}(z_S) = O\left(g^2\alpha^2\right) + O\left(g^4\right),$$

or

$$\frac{h_{11}{}^{II}(z_L)}{h_{22}{}^{II}(z_S)} = -1 + O\left(\alpha^2\right). \tag{17.111}$$

Putting these results together, we see that

$$\varepsilon_L + \tilde{\varepsilon}_S = \varepsilon_L\left(1 + \frac{h_{11}{}^{II}(z_L)\, h_{12}{}^{II}(z_S)}{h_{22}{}^{II}(z_S)\, h_{12}{}^{II}(z_L)}\right).$$

The first factor in the second term is $\left(-1 + O\left(\alpha^2\right)\right)$, and the second is $\left(1 + O\left(\alpha^2\right)\right)$, so that (since ε_L is $O(\alpha)$)

$$\varepsilon_L + \tilde{\varepsilon}_S = O\left(a^3\right). \tag{17.112}$$

A similar argument can be made for $\tilde{\varepsilon}_L + \varepsilon_S$. We see that the non-orthogonality of g_S and g_L is far from negligible; the products depend only on the CP violation amplitude, and not on the weak coupling constant.

The idempotence properties for the residues can be similarly checked and are valid up to $O\left(g^2\right)$.

Some of the crucial experiments with K meson decay proceed as follows. The beam of K^0 mesons produced by collisions in an accelerator decay rapidly to two pions, corresponding to the K_S component of the linear superposition making up the beam. The remaining K_L component decays primarily into three pions, but due to CP violation, there is a small component of two-pion decays (the volume of phase space for three-pion decays is smaller than that for two-pion decays). The beam is passed through a block of material containing heavy nuclei, such as copper, and the special linear combination of K^0 and $\bar{K}^0$ making up the K_L component is disturbed by the scattering of the K^0 and $\bar{K}^0$ components on the nuclei, which involves phase shifts corresponding to the strong interactions sensitive to the opposite sign of hypercharge of these states. Thus, an admixture of K_S is regenerated (and the apparatus is called a regenerator). The short-lived decay of this K_S beam then interferes coherently with the CP-violating two-pion decay of the older K_L beam, which was unaffected by the regeneration.

This process of regeneration takes place at a sequence of the order of 10^8 atomic scattering sites, and between each scattering the beam is presumed to develop according to the reduced evolution law. The deviations from semigroup behavior are brought into evidence at each scattering center when the reduced evolution starts to flow from the new initial conditions. The sum of these deviations add up over all these scatterings and give rise to a disagreement with experiment of the order of a few percent. In contrast, the model of Wu and Yang (1964), which is an exact semigroup, since it is constructed of the exponential of a non-self-adjoint effective 2×2 matrix Hamiltonian, leads to a very accurate fit to the data (Winstein *et al.,* 1997). Details of these calculations are in papers by Horwitz, Mizrachi and Cohen (Horwitz and Mizrachi, 1974; Cohen and Horwitz, 2001; numerical estimates based on Hagiwara, 2002), where further estimates are made for the deviations from experiment in the exit channel of the regenerator due to the non-orthogonality of the residues.

In view of the inadequacy of the Wigner–Weisskopf model, even in pole approximation, for the more than one channel decay system, the development of a theory for unstable systems which contains the property of exact semigroup evolution becomes very important. In the following sections, we describe two relatively recent developments for achieving the semigroup property for the evolution of an unstable system. These developments have fundamental importance, since, as we have pointed out, the semigroup property corresponds to irreversible evolution. It was for this reason that we gave considerable detail to the computation of the estimates of the deviation from the semigroup law in the framework of the approach commonly used in the application of quantum theory to the description of unstable systems.

17.6 Gel'fand triple

We have shown in the previous sections that the Wigner–Weisskopf theory does not lead to a semigroup property for the decay law of an unstable system. To achieve a semigroup law, one may seek a function in some space which corresponds to the complex pole, and for which the evolution has exact exponential decay.

An important development in this direction is the introduction of the Gel'fand triple construction, for which an element of a generalized linear space larger than the original Hilbert space represents the resonance, and the extension of the unitary evolution to this element results in an exact exponential decay. As we shall see, this functional is, in general, not useful for the description of a quantum mechanical problem, since scalar products and expectation values of physical observables are not available in this generalized space. It is, however, an important concept, and in some mathematical formulations, the space of generalized states has turned out to be useful. To illustrate the construction, we shall again utilize the Lee–Friedrichs model discussed in the previous section.

We have seen that the reduced resolvent $R(z)$, defined in Eq. (17.33) may acquire a pole in its analytic continuation to the second Riemann sheet. Since this function is the expectation value of the resolvent operator

$$G(z) = \frac{1}{z - H}, \tag{17.113}$$

the process of analytic continuation appears to provide a complex eigenvalue for the Hermitian operator H; this is, of course, not possible in the Hilbert space but indicates that some type of extension could be consistent with a complex eigenvalue. We show in the following section how this can happen, most simply in the rank one Lee–Friedrichs model (Bailey and Schieve, 1978; Baumgartel, 1978; Horwitz and Sigal, 1980) discussed in the previous section.

As we have remarked, the occurrence of a pole in $G(z)$ suggests the existence, in some sense, of a complex eigenvalue for H. A self-adjoint operator in a Hilbert space cannot, of course, have a complex eigenvalue; we shall see in the following discussion precisely in what sense this statement can be true.

Let us search for an eigenfunction $f(z)$, a vector parametrized by the complex number z, for which

$$H|f(z)\rangle = z|f(z)\rangle. \tag{17.114}$$

Since $H = H_0 + V$, where in this model V does not connect the discrete eigenstate $|\psi_0\rangle$ of H_0 with itself and has no continuum–continuum matrix elements, the continuum part of Eq. (17.114) is:

$$\begin{aligned}\langle\lambda|H_0 + V|f(z)\rangle &= \lambda\langle\lambda \mid f(z)\rangle + \langle\lambda|V|\psi_0\rangle\langle\psi_0 \mid f(z)\rangle \\ &= z\langle\lambda \mid f(z)\rangle,\end{aligned} \tag{17.115}$$

from which it follows that

$$\langle \lambda \mid f(z) \rangle = \frac{1}{z-\lambda} \langle \lambda | V | \psi_0 \rangle \langle \psi_0 \mid f(z) \rangle. \tag{17.116}$$

We now consider the discrete component of Eq. (17.114). Taking the scalar product with $\langle \psi_0 |$, and remembering that $\langle \psi_0 \mid \lambda \rangle = 0$, we obtain

$$\int d\lambda \langle \psi_0 | V | \lambda \rangle \langle \lambda \mid f(z) \rangle + \lambda_0 \langle \psi_0 \mid f(z) \rangle = z \langle \psi_0 \mid f(z) \rangle,$$

and substituting the result Eq. (17.116), we find the condition

$$\left\{ (z - \lambda_0) - \int d\lambda \frac{|\langle \lambda | V | \psi_0 \rangle|^2}{z - \lambda} \right\} \langle \psi_0 \mid f(z) \rangle = 0. \tag{17.117}$$

The coefficient of $\langle \psi_0 \mid f(z) \rangle$ is exactly what we called $h(z)$ in Eq. (17.39); the condition that $h(z)$ vanish corresponds to the resonant pole condition and can be achieved only by analytic continuation to the second Riemann sheet. In Eq. (17.114) the factor $\langle \psi_0 \mid f(z) \rangle$ cannot vanish; according to Eq. (17.116) this would imply vanishing of the continuum part of $f(z)$ as well, so that the vector f itself would be zero. Hence, the condition for the complex eigenvalue is the same as that for the complex pole of the reduced resolvent. We must continue the eigenvalue equation to the second Riemann sheet in order to achieve this. The analytic continuation of Eq. (17.114) is not a trivial matter; a vector in the Hilbert space defined by a relation such as Eq. (17.116) can be analytically continued only in the sense of considering its scalar product with all other vectors in the Hilbert space, and the integration over λ required for carrying out the scalar product then admits the analytic continuation to the second Riemann sheet. Let us examine this expression, taking the scalar product with some vector g:

$$\langle g | H | f(z) \rangle = z \langle g \mid f(z) \rangle. \tag{17.118}$$

We have studied two parts of this scalar product, consisting of the continuum and the discrete parts. For z in the upper half plane, it is easy to see that $f(z)$ is a very well-defined vector with finite norm in the Hilbert space. However, to satisfy the "eigenvalue" condition, we had to carry out an analytic continuation. The analytic continuation of $\langle g \mid f(z) \rangle$ is carried out as follows:

$$\begin{aligned} \langle g \mid f(z) \rangle &= \langle g \mid \psi_0 \rangle \langle \psi_0 \mid f(z) \rangle + \int d\lambda \langle g \mid \lambda \rangle \langle \lambda \mid f(z) \rangle \qquad (17.119) \\ &= \langle g \mid \psi_0 \rangle \langle \psi_0 \mid f(z) \rangle + \int d\lambda \langle g \mid \lambda \rangle \frac{\langle \lambda | V | \psi_0 \rangle}{z - \lambda} \langle \psi_0 \mid f(z) \rangle. \end{aligned}$$

Following the procedure of Eq. (17.42), it is clear that analytic continuation through the cut on the positive half line will pick up a factor of $\langle g \mid \lambda \rangle$ upon crossing the cut, and carry the function defined as its analytic continuation (assuming it is the boundary value on the real line of a function analytic in at least part of the lower half plane) into the lower half plane.

The Riesz theorem (Riesz and Sz.-Nagy, 1955) states that if there is a bounded linear functional $L(g)$ mapping vectors of the Hilbert space into the complex numbers, for which

$$L(ag) = aL(g), \qquad |L(g)| \leq K\|g\|, \tag{17.120}$$

for *all* g in the Hilbert space $\mathfrak{H}$, then there is a unique f in the Hilbert space for which

$$L(g) = \langle f \mid g \rangle. \tag{17.121}$$

The functionals L are called *dual,* since they constitute also a linear space; it is a special property of Hilbert spaces that they are *self-dual;* the f's of Eq. (17.121) are elements of the Hilbert space themselves. However, in the case we have studied, although the analytic continuation of $\langle g \mid f(z) \rangle$ is linear in g, this procedure cannot be carried out for all g, since the function $\langle g \mid \lambda \rangle$ is the boundary value of an analytic function only for a subset of g's in $\mathfrak{H}$. Hence, the analytic continuation of $\langle g \mid f(z) \rangle$ does not define a vector in the original Hilbert space. If the value of this function is known for all $g \varepsilon \mathfrak{D}$, where $\mathfrak{D} \subset \mathfrak{H}$, a proper subspace, then Eq. (17.121) defines an element of a larger space (Gel'fand and Shilov, 1967) which we may call $\bar{\mathfrak{H}}$, so that

$$\mathfrak{D} \subset \mathfrak{H} \subset \bar{\mathfrak{H}}.$$

This construction is called a Gel'fand triple, and we see that the "state" corresponding to the resonance, the continuation of $f(z)$, is an element of a Gel'fand triple. According to Eq. (17.118), continued to the location of the pole in the lower half plane (where it can be satisfied), the Hamiltonian operator (now extended to operate on such elements) has a complex eigenvalue, and the extension of the evolution operator $\exp -iHt$ would induce an exact exponential decay (semigroup) on this "state" (Bohm, 1978; Parravicini *et al.,* 1980; Bohm, 1980, 1981; Bohm and Gadella, 1989; Bohm *et al.,* 1989; Bohm and Kaldass, 2000).

We have therefore succeeded in finding a representation of the unstable state which evolves according to an exact semigroup property, but the "state" defined in this way is not an element of a Hilbert space. It is an element of a Banach space, in which a norm can be defined (in this case related to a maximum modulus norm on a complex variable) (Baumgartel, 1978; Horwitz and Sigal, 1980), but no scalar products or expectation values can be defined in general; only bilinear forms such as the continuation of $\langle g \mid f(z) \rangle$ can be defined, where g belongs to a subset of elements of $\mathfrak{H}$ for which $\langle g \mid \lambda \rangle$ are the boundary values of analytic functions in λ at least in a region of the lower half plane including the resonance pole. There are, however, cases in which the Gel'fand triple states lie in a Hardy class of states (Bohm and Gadella, 1989), which are Hilbert spaces (of a special type which we shall describe later, after Eq. (17.169)), and in such cases, more information may become available. Hence we cannot deduce properties of the resonant state, such as its locality in coordinate or momentum space.

There are important physical problems for which this information is essential. For example, an electron on a conducting sheet of material with embedded electrodes that create a potential well (a so-called quantum dot) may pass the region of the well as a resonance. It is possible to detect the resulting current and find signatures of smooth or chaotic behavior. For example, Altshuler (Agam *et al.*, 1995, 1996) has conjectured that if the resonance is not localized in momentum space (so that it may be localized in coordinate space), one may find chaotic behavior, based on the analog of a classical billiard, a localized object in a closed boundary (of irregular shape). It therefore is important to find a description of a resonance which can be represented in a Hilbert space, for which one can define observables in terms of self-adjoint operators, and scalar products and expectation values are well defined.

In the next section we develop the framework for such a theory, directly applicable to some important problems, and for which current research indicates extension to a wider class of applications (Strauss, 2003, 2005a, 2005b, 2005c; Baumgartel, 2006).

17.7 Lax–Phillips theory

In the previous sections we have pointed out the importance of describing the state of a resonance as a vector in a Hilbert space, for which the evolution is of the form of a semigroup. Such a theory was developed by Lax and Phillips in 1967 for application to classical wave equations, such as electromagnetic or acoustic waves (Lax and Phillips, 1967). We shall describe this theory in this section and show that the structure can be extended to the quantum theory. The development of this theory in the quantum framework is under rapid development at present and shows indications of reaching wide classes of non-equilibrium physical phenomena.

In their study of wave equations, Lax and Phillips (1967) showed how to write second-order wave equations in a way that defines an effective Hamiltonian evolution. Given such an evolution, it was possible to characterize the system in terms of invariant subspaces of a Hilbert space. The notion of invariant subspaces is fundamental to the development of theories for unstable systems which admit a semigroup evolution law.

There has been considerable effort in recent years in the development of the theoretical framework of Lax–Phillips scattering theory for the description of quantum mechanical systems (Horwitz and Piron, 1993; Eisenberg and Horwitz, 1997; Strauss, 2003, 2005a, 2005b, 2005c; Baumgartel, 2006).

As an example of how an effective Hamiltonian evolution can be introduced into the study of wave equations, consider the wave equation in three dimensions, valid in some exterior domain G (Lax and Phillips, 1967):

$$u_{tt}(x,t) - \Delta u(x,t) = 0,$$

where Δ corresponds to the Laplacian, for all x on some bounded region $G(R)$, for which $|x| < R$. Subscripted letters indicate partial derivatives (e.g. $u_x(x,t) \equiv \partial_x u(x,t)$), and we suppose that $u(x,t)$ satisfies the boundary condition $u(x,t) = 0$ on the boundary ∂G of this region. The boundaries of this region define an "exterior" problem, where the wave motion is outside of regions of varying density or speed of propagation. The "free" problem corresponds to a similar structure, but with such unperturbed waves everywhere. We shall return to this point in our discussion of the translation representations below, and its analog in the quantum mechanical problem in Eq. (17.138) and what follows. Lax and Phillips (1967; see also Reed and Simon, 1979) show that there is a conserved "energy norm" which can be used to define a scalar product on a Hilbert space of solutions, and that the evolution can be described in terms of the action of a one-parameter unitary group acting on this space. This example forms a prototype of the structure of the general Lax–Phillips theory, and we describe it briefly here.

The energy norm is defined as

$$E(u(t), R) = \frac{1}{2}\int_{G(R)} \{|u_x(x,t)|^2 + |u_t(x,t)|^2\} dx.$$

By multiplying the wave equation by u_t and integrating over the space-time region $|x| < R + T - t, 0 < t < T$, we can show that

$$E(u(T), R) \le E(u(0), R+T),$$

and by reversing the direction of time, that

$$E(u(T), R) \ge E(u(0), R-T).$$

Thus, if the initial values for u vanish in the ball $|x| < R$, then $u(x,t)$ vanishes in the region $|x| < R - t$, and if the total energy is finite, it has the same energy for all time.

Let us now express this problem in terms of a pair of complex functions $f \equiv \{f_1, f_2\}$, defined on the same space domain G, and write the energy norm as

$$\|f\|_E^2 = \frac{1}{2}\int_G \left\{|\partial_x f_1|^2 + |f_2|^2\right\} dx.$$

We then define a Hilbert space H as the (closed) set of functions in this norm with compact support in G. Note that then both the derivative (gradient) of f_1 squared and f_2 must be finitely integrable.

We now define the operator

$$A = \begin{pmatrix} 0 & I \\ \Delta & 0 \end{pmatrix},$$

with domain of definition $D(A)$ the set of all f treated as a two-dimensional vector under the action of this matrix such that Af lies in H, i.e. has finite energy norm. We now show that this matrix generates a unitary group $U(t)$ on H, and that if f is in $D(A)$, when $f_1 = u(x,t)$ and $f_2 = u_t(x,t)$, then $u(x,t)$ satisfies the second-order wave equation written above.

From the energy norm formula, we can write a scalar product

$$\langle f \mid g \rangle = \frac{1}{2} \int_G \left\{ \partial_x f_1^* \partial_x g_1 + f_2^* g_2 \right\} dx,$$

and therefore, by multiplying out A on the (two-dimensional) function f and integrating by parts in G, we find

$$\langle Af \mid g \rangle = \frac{1}{2} \int_G \left[\partial_x f_2^* \partial g_1 - \partial_x f_1^* \partial_x g_2 \right] dx,$$

so that

$$\langle Af \mid g \rangle = -\langle f \mid Ag \rangle,$$

and A is therefore a skew-Hermitian operator. Thus, A generates a unitary group of operators $\{U(t)\}$ for which

$$\frac{d}{dt} U(t) f = AU(t) f,$$

and it is easy to see that the second component of this equation, for $f_1 = u(x,t)$ and $f_2 = u_t(x,t)$, assures that the wave equation is satisfied under this evolution.

In addition to the formulation of wave equations in the form of unitary evolution, another essential feature of Lax–Phillips theory is that one can construct *translation representations* for the unitary evolution, for which there are *invariant subspaces* under its action, called *ingoing* and *outgoing* subspaces. This result is not surprising in view of the Huygens principle governing wave equations in odd-dimensional spaces. To construct these subspaces, we choose $\rho > 0$ so that $|x| < \rho$ contains ∂G in its interior, and set $D_+^\rho = U_0(\rho) D_+$ and $D_-^\rho = U_0(-\rho) D_-$, where $D_\pm$ are the incoming and outgoing subspaces of the solutions in free space (the unperturbed problem referred to above), and $U_0(t)$ is the unitary propagation on the full unperturbed space. By construction, $\left[U(t) f \right](x)$, for f in $D_\pm^\rho$, vanishes in the forward (backward) cones

$$|x| < t + \rho \, (t > 0)$$
$$|x| < -t + \rho \, (t < 0),$$

respectively. Then

$$U(t) D_+^\rho \subset D_+^\rho \text{ for } t > 0$$
$$\cap U(t) D_+^\rho = \{0\}$$
$$\overline{\cup U(t) D_+^\rho} = H;$$

that is, D_+^ρ is an outgoing subspace, defined by successive inclusion. Similarly, D_-^ρ is an incoming subspace; these two subspaces are orthogonal.

As we shall show below, these properties provide the Lax–Phillips theory with the power to describe resonances as irreversible phenomena, and the resonant state as a state in a Hilbert space. Clearly, these properties would be desirable for the description of quantum mechanical systems. The Huygens principle to assure the outgoing, incoming and translation properties is not available for quantum mechanical systems, but as we shall show below, similar arguments comparing free and perturbed motion, with the additional assumption of the existence of wave operators, provide sufficient structure to develop the corresponding quantum theory.

The quantum Lax–Phillips theory (Strauss, 2003, 2005a, 2005b, 2005c; Baumgartel, 2006), constructed by embedding the quantum theory into the original Lax–Phillips scattering theory (Lax and Phillips, 1967), describes a resonance as a state in a Hilbert space, and therefore it is possible, in principle, to calculate all measurable properties of the system in this state. Moreover, the quantum Lax–Phillips theory provides a framework for understanding the decay of an unstable system as an irreversible process. It appears, in fact, that this framework may be categorical for the description of irreversible processes on a fundamental level.

In the following discussion, we distinguish the abstract Hilbert space and *representations* in terms of L^2 (square integrable) functions, since we shall be using explicitly different types of representations. In preparation for this idea, we remind the reader that, for a given physical state of the system, there corresponds a vector—say, ψ in a Hilbert space—and this vector can be represented as a wave function $\langle x \mid \psi \rangle \equiv \psi(x)$ or equally well as a function in momentum space, $\langle p \mid \psi \rangle \equiv \psi(p)$. Other complete sets may be used as well to represent the state vector, such as energy and angular momentum eigenstates. These representations have special properties. For example, in the x-representation, the position operator X is diagonal, in the sense that $\langle x|X|\psi\rangle = x\langle x \mid \psi\rangle$, with similar properties for the operator P in the p-representation, or energy and angular momentum operators $\left(\text{e.g. } H, J^2 \text{ and} J_z\right)$ in a representation provided by energy and angular momentum eigenstates. In conventional quantum mechanical scattering theory, one thinks in terms of the evolution of a set of incoming *states* to a set of outgoing *states,* and in terms of these (asymptotic) states, to describe the scattering with the so-called scattering S-matrix. The Lax–Phillips theory defines incoming and outgoing *representations* and defines an analogous S-matrix which relates these representations, defined in terms of the property of translations along a line. With these ideas in mind, we may proceed to define the basic ideas of the Lax–Phillips scattering theory.

The scattering theory of Lax and Phillips (1967) assumes the existence of a Hilbert space $\overline{\mathfrak{H}}$ of physical states in which there are two distinguished orthogonal subspaces, $\mathfrak{D}_+$ and $\mathfrak{D}_-$, and a unitary evolution $U(\tau)$, a function of the physical laboratory time, with the properties

$$\begin{aligned} U(\tau)\,\mathfrak{D}_+ \subset \mathfrak{D}_+ \quad & \tau > 0 \\ U(\tau)\,\mathfrak{D}_- \subset \mathfrak{D}_- \quad & \tau < 0 \\ \bigcap_\tau U(\tau)\,\mathfrak{D}_\pm = \{0\} & \\ \overline{\bigcup_\tau U(\tau)\,\mathfrak{D}_\pm} = \overline{\mathfrak{H}}\,, & \end{aligned} \tag{17.122}$$

where $\cap$ and $\cup$ correspond to the intersection and union of the subspaces (here evolved by $U(\tau)$). Thus, the subspaces $\mathfrak{D}_\pm$ are assumed to be stable under the action of the full unitary dynamical evolution $U(\tau)$, for positive and negative times τ respectively; over all τ, the evolution operator generates a dense set in $\overline{\mathfrak{H}}$ from either $\mathfrak{D}_+$ or $\mathfrak{D}_-$. Note that we are here discussing *abstract* vectors, (pure) *states* in the sense of Dirac (1947), in $\overline{\mathfrak{H}}$ without reference to a particular representation.

We shall call $\mathfrak{D}_+$ the *outgoing subspace* and $\mathfrak{D}_-$ the *incoming subspace* with respect to the group $\{U(\tau)\}$.

A theorem of Sinai (Cornfeld *et al.,* 1982) then assures that $\overline{\mathfrak{H}}$ can be represented as a family of Hilbert spaces obtained by foliating $\overline{\mathfrak{H}}$ along a real line, which we shall call $\{s\}$, in the form of a direct integral

$$\overline{\mathfrak{H}} = \int_\oplus \mathfrak{H}_s, \tag{17.123}$$

where the set of auxiliary Hilbert spaces $\mathfrak{H}_s$ are all isomorphic. This foliated structure does not rule out the possibility that there are operators on $\overline{\mathfrak{H}}$ which have matrix elements between different values of s.

Representing these spaces in terms of square-integrable functions, we define the norm in the direct integral space (we use Lesbesgue measure ds on the line s) as

$$\langle f \mid f\rangle \equiv \|f\|^2 = \int_{-\infty}^{\infty} ds \|f_s\|_H^2, \tag{17.124}$$

where $f \in \overline{H}$ represents a vector in $\overline{\mathfrak{H}}$, and $f_s \in H$, the L^2 function space representing $\mathfrak{H}_s$ for any s. (It then represents the vector f in $\overline{\mathfrak{H}}$ in the L^2 function space $L^2(-\infty, \infty,) H$.) Thus, f_s is an L^2 valued function in some representation (e.g. $f_s(x)$, square integrable in x; we do not specify the variables of the measure space of the auxiliary spaces H at this stage). The Sinai theorem furthermore asserts that there are representations for which the action of the full evolution group $U(\tau)$ on $L^2(-\infty, \infty, H)$ is translation by τ units. Given $D_{\pm}$ (the L^2 spaces representing $\mathfrak{D}_{\pm}$), there is such a representation, called the *incoming translation representation* (Lax and Phillips, 1967), for which functions in D_- have support in $L^2(-\infty, 0, H)$, and another called the *outgoing translation representation,* for which functions in D_+ have support in $L^2(0, \infty, H)$.

We remark that the foliation variable s, whose existence is asserted by Sinai, might be thought of as the laboratory time *at which the state of the unstable system could potentially be observed* as an unstable system, i.e. with some probability of existing at that value of s. The notion that occurs here is a difficult concept to grasp but also occurs in standard Floquet theory (Zhang and Feng, 1995), a method developed to manage the Schrödinger evolution with a time-dependent Hamiltonian in a systematic way. The idea is that there is a physical *state* at laboratory time τ which predicts not only the space distribution, say, of anticipated events, but also the times s at which such events are to be observed. The difficulty in grasping this notion is that the time s corresponds to the time of the measurement, a value of τ itself. Thus, the theory, as in the Floquet case, predicts what will (or would be) found if one did the measurement at a time different, in general, from the time at which this state is extant. As in any quantum mechanical statement, the state of the system provides an *a priori* prediction, but here we have the prediction at time τ for what would be seen at a different time $\tau' = s$ if the measurement were done then. That prediction, of course, varies with the time τ at which the state is determined, since the distribution and therefore the prediction changes at each τ. Such concepts also occur in relativistic quantum theory, where the state of a system evolving in τ (in this case a universal world time that is reflected on the laboratory clock) is a distribution on space-time (Stueckelberg, 1941; Schwinger, 1951; Feynman, 1948, 1950; Horwitz and Piron, 1973; Fanchi, 1979; Kyprionides, 1986 and references therein), a structure consistent with the covariance of special relativity (see Chapter 10). The events occurring in that space-time occur, when measured, as times on the laboratory clock. It is this expanded space (the space of variables on which the wave functions are represented) for the representation of the state that enables one to define projection operators into subspaces that are partly determined by time boundaries (Horwitz and Piron, 1993). It is a fundamental result, as we shall see, that the evolution operator of the system, projected into such subspaces, may become non-self-adjoint, with complex spectrum, and thus capable of describing exact semigroup evolution.

Lax and Phillips (1967) show that there are unitary operators $W_\pm$, called wave operators, which map elements in $\overline{\mathfrak{H}}$, respectively, to the incoming and outgoing representations. For example, suppose f is an element of $\overline{\mathfrak{H}}$; then $|W_\pm f\rangle_s$, an element of $L^2(-\infty, \infty, H)$, has the property that

$$U(\tau)\,|W_\pm f\rangle_s = |W_\pm f\rangle_{s-\tau}, \tag{17.125}$$

where we have used the notation that $U(\tau)$ is considered as an operator on the L^2 space as well. That is, we write, as is usual in quantum theory literature,

$$|Ag\rangle \equiv A|g\rangle$$

for A, some operator defined on g. Eq. (17.125) corresponds to translation of $|f\rangle_s$ forward by τ units. These representations furthermore have the property that, for $f_+ \in \mathfrak{D}_+$, $|f_+\rangle_s$ vanishes for s ≤ 0; it has support only on the positive half line and therefore is an element of $L^2(0, \infty, H)$. For $f_- \in \mathfrak{D}_-$, $|f_-\rangle_s$ similarly vanishes for $s \geq 0$ and is therefore an element of $L^2(-\infty, 0, H)$.

Lax and Phillips then define an S-matrix,

$$S = W_+ W_-^{-1}, \tag{17.126}$$

which connects these representations; it is unitary, commutes with translations, and maps $L^2(-\infty, 0)$ into itself. The singularities of the S-matrix, in what we shall define as the *spectral representation,* obtained by means of Fourier transform, correspond to the spectrum of the generator of the exact semigroup characterizing the evolution of the unstable system. This very striking result is analogous to the approximate treatment of resonances in standard scattering theory (Taylor, 1972; Newton, 1976).

With the assumptions stated above on the properties of the subspaces $\mathfrak{D}_+$ and $\mathfrak{D}_-$, Lax and Phillips (1967) prove that the family of operators

$$Z(\tau) \equiv P_+ U(\tau) P_- \qquad (\tau \geq 0), \tag{17.127}$$

where $P_\pm$ are projections into the orthogonal complements of $\mathfrak{D}_\pm$, respectively, is a contractive, continuous semigroup. This operator annihilates vectors in $\mathfrak{D}_\pm$ and carries the space

$$\mathfrak{K} = \overline{\mathfrak{H}} \ominus \mathfrak{D}_+ \ominus \mathfrak{D}_- \tag{17.128}$$

into itself, with norm tending to zero for every element in $\mathfrak{K}$. It is clear that therefore

$$Z(\tau) = P_{\mathfrak{K}} U(\tau) P_{\mathfrak{K}} \tag{17.129}$$

as well.

In the following, we use the L^2 function space (square integrable functions) representing $\overline{\mathfrak{H}}$ and denote the corresponding subspaces of functions as $D_\pm$ and K. The semigroup property of the operator $Z(\tau)$ of Eq. (17.127) may then be proven as follows (Lax and Phillips, 1967).

$Z(\tau)$ clearly vanishes on the subspace D_-, and by the stability of D_+ under $U(\tau)$ for $\tau \geq 0$, it vanishes on D_+ as well. It is therefore nonzero only on the subspace K, and on such vectors, the operator P_- can be omitted. We must then consider, on a vector $|f_K\rangle$ in K,

$$Z(\tau_1)\, Z(\tau_2)\, |f_K\rangle = P_+ U(\tau_1)\, P_- P_+ U(\tau_2)\, |f_K\rangle.$$

But $U(\tau_2)\;|f_K\rangle$ lies in $K+D_+$, and P_+ annihilates the part in D_+. The projection operator P_- can then be omitted, since K is orthogonal to D_-. We then have

$$\begin{aligned} Z(\tau_1)\, Z(\tau_2)\, |f_K\rangle &= P_+ U(\tau_1)\, P_+ U(\tau_2)\, |f_K\rangle \\ &= P_+ U(\tau_1)\left[1 - \left[1 - P_+\right]\right] U(\tau_2)\, |f_K\rangle; \end{aligned}$$

the second term in brackets makes no contribution, since it is a projection into D_+, and $U(\tau_1)$ leaves D_+ invariant (it is then annihilated by the first factor P_+). We are then left with

$$\begin{aligned} Z(\tau_1)\, Z(\tau_2)\, |f_K\rangle &= P_+ U(\tau_1)\, U(\tau_2)\, |f_K\rangle \\ &= P_+ U(\tau_1 + \tau_2)\, |f_K\rangle \\ &= Z(\tau_1 + \tau_2)\, |f_K\rangle, \end{aligned} \tag{17.130}$$

completing the proof of the semigroup property of $Z(\tau)$.

The outgoing subspace D_+ is defined, in the outgoing representation, in terms of support properties (this is also true for the incoming subspace in the incoming representation). One can then easily understand that the fundamental difference between Lax–Phillips theory and the standard quantum theory, as pointed out, lies in this property. The subspace defining the unstable system in the standard theory is usually defined as the eigenstate of an unperturbed Hamiltonian and does not correspond to an interval on a line associated with evolution. The subspaces of the Lax–Phillips theory correspond, generally, to semibounded intervals (e.g. the positive and negative half lines in the outgoing and incoming representations). The generator of the semigroup (the generator of the full group restricted to the subspace K) is then, in general, not self-adjoint. To see this, consider the action of the semigroup on a vector $|f_K\rangle$ in K. In the outgoing translation representation (Horwitz and Piron, 1993),

$$P_+ U(\tau)\, P_- |f_K\rangle_s = P_+ U(\tau)\, |f_K\rangle_s = \theta(-s)\, |f_K\rangle_{s-\tau}, \tag{17.131}$$

since K is in the complement of D_-, and therefore

$$P_+ K | f_K\rangle_s = i\theta\,(-s)\,\frac{\partial\ |f_K\rangle_{s-\tau}}{\partial s}\ |_{\tau\to 0_+}\,, \tag{17.132}$$

where $|f_K\rangle_s$ is a vector-valued function, as described above, and K is the self-adjoint generator associated with $U\,(\tau)$. (We use the symbol K for the generator as well as for the subspace orthogonal to $D_\pm$; there should be no confusion in context.) If we then compute the scalar product of the vector given in Eq. (17.131) with a vector $|g\rangle$, we find that

$$\int_{-\infty}^{\infty} ds\ {}_s\langle g|P_+K|f_K\rangle_s = i\delta\,(s)\ {}_0\langle g\mid f\rangle_0 + \int_{-\infty}^{\infty} ds\ {}_s\langle P_+Kg)\mid f_K\rangle_s. \tag{17.133}$$

The generator is therefore symmetric but not self-adjoint. It is through this mechanism that the Lax–Phillips theory, as remarked above, provides a description that has the semigroup property for the evolution of an unstable system (Horwitz and Piron, 1993; see also Eisenberg and Horwitz, 1997).

Comparing with the formula of Wigner and Weisskopf discussed in Section 17.1, let us consider a vector ψ_0 in $\mathfrak{K}$, corresponding to an "unstable state," and evolve it under the action of $U\,(\tau)$; the projection back into the original state is, using the form Eq. (17.129),

$$\begin{aligned} A\,(\tau) &= \langle\psi_0|U\,(\tau)\,|\psi_0\rangle \\ &= \langle\psi_0|P_K U\,(\tau)\,P_K|\psi_0\rangle \\ &= \langle\psi_0|Z\,(\tau)\,|\psi_0\rangle, \end{aligned} \tag{17.134}$$

so that the survival amplitude of the Lax–Phillips theory, analogous to that of the Wigner–Weisskopf formula, Eq. (17.10), has the exact exponential behavior. The difference between this result and the corresponding expression given in Section 17.1 for the Wigner–Weisskopf theory can be accounted for by the fact that there are translation representations for $U\,(\tau)$ and that the definition of the subspace K is related to the support properties along the foliation axis on which these translations are induced. As a consequence of this structure, as made explicit in Eq. (17.133), the generator of $Z(\tau)$ in Eq. (17.134) is not self-adjoint. The presence of complex eigenvalues then implies the possibility of the (often experimentally observed) exponential decay law for $A\,(\tau)$.

Functions in the space $\overline{H}$, representing the elements of $\mathfrak{H}$, depend on the variable s as well as the variables of the auxiliary space H. The measure space of this Hilbert space of states is one dimension larger than that of a quantum theory represented in the auxiliary space alone. Identifying this additional variable with an *observable time* (in the sense of a quantum mechanical observable), we may understand this representation of a state as a *virtual history*. The collection of such histories forms a quantum ensemble; the absolute square of the wave function corresponds to the probability that the system would be found, as a result of measurement, at time s in a particular configuration in the auxiliary space (in the state described by this wave function), i.e. an element of one of the virtual histories. For example, the expectation value of the position variable x at a given s is, in the standard interpretation of the auxiliary space as a space of quantum states,

$$\langle x \rangle_s = \frac{\langle \psi_s | x | \psi_s \rangle}{\|\psi_s\|^2}. \tag{17.135}$$

The full expectation value in the physical Lax–Phillips state, according to Eq. (17.124), is then

$$\int ds |\langle \psi_s | x | \psi_s \rangle|^2 = \int ds \|\psi_s\|^2 \langle x \rangle_s, \tag{17.136}$$

(Strauss *et al.*, 2000), so we see that $\|\psi_s\|^2$ corresponds to the probability to find a signal which indicates the presence of the system at the time s.

The generator of the full evolution restricted to the subspace $\mathfrak{K}$ has a family of complex eigenvalues, if there are resonances, in the lower half plane. We can easily see this by writing the action of the semigroup (in the subspace $\mathfrak{K}$) as $Z(\tau) = e^{-iB\tau}$ for B non-self-adjoint. Then the Laplace transform is given by

$$\int_0^\infty d\tau e^{-i(\mu - B)\tau} = -\frac{i}{\mu - B}$$

for μ in the lower half plane, where the integral is well defined. There may be poles, however, if B has a discrete spectrum, for which

$$B|f\rangle = \mu|f\rangle.$$

(Lax and Phillips, 1967).

Then, in the outgoing representation,

$$\langle s | e^{-iB\tau} | f \rangle_{\text{out}} = \langle s + \tau \mid f \rangle_{\text{out}} = e^{-i\mu\tau} \langle s \mid f \rangle_{\text{out}}$$

for $s + \tau > 0$. For, in particular, $s = 0$, $\langle \tau \mid f \rangle_{\text{out}} = e^{-i\mu\tau} \langle 0 \mid f \rangle_{\text{out}}$ and zero for τ negative; i.e. the eigenfunctions are

$$\langle \tau \mid f \rangle_{\text{out}} = \begin{cases} e^{-i\mu\tau} n & \tau \geq 0 \\ 0 & \tau < 0, \end{cases} \tag{17.137}$$

where $n \equiv \langle 0 \mid f \rangle_{\text{out}}$ is some vector in the auxiliary space. This is a very fundamental result, displaying the *eigenvector of a resonance* in a Hilbert space.

The preceding ideas were worked out by Lax and Phillips (1967) for classical wave equations, where the Huygens principle could be used to provide illustrations of the existence of the invariant subspaces $\mathfrak{D}_\pm$. In the following discussion we show

how the theory can be extended to the quantum case, for which the construction of these subspaces is the central problem.

The fundamental step in bringing the Lax–Phillips framework to the quantum theory is to construct the incoming and outgoing subspaces. The definition of these subspaces plays a crucial role, as we have explained, in achieving the required semigroup law for the description of irreversible processes. To achieve this (Strauss *et al.*, 2000), we consider the evolution operator to be composed of an unperturbed part and a small perturbation in the form

$$K = K_0 + V, \tag{17.138}$$

and assume that there are *wave operators* that intertwine K and K_0. We may then construct $D^0_\pm$ in terms of support properties on the spectrum of K_0 and lift the result to the subspaces $D_\pm$ for the full structure of the generator K. The notion of the wave operator is very important in conventional scattering theory as well, and we therefore review the idea briefly in the following.

The foundation of standard scattering theory (Jauch, 1958; Taylor, 1972; Newton, 1976) lies in the following statement. Under the usual Schrödinger evolution

$$\psi_t = e^{-iHt}\psi_0 \tag{17.139}$$

for large positive or negative t, ψ_t approaches asymptotically a state ϕ_t, called an asymptotic state, which evolves according to

$$\phi_t = e^{-iH_0t}\phi_0, \tag{17.140}$$

where $H = H_0 + V$ and V is a "small" operator, relatively bounded with respect to H_0 (Jauch, 1958; Taylor, 1972; Newton, 1976). We may write this condition as

$$\lim_{t\to\pm\infty} \|\psi_t - \phi_t^\pm\| = 0, \tag{17.141}$$

where we have distinguished the incoming and outgoing asymptotic states with the sign $\pm$.

Since the norm taken in Eq. (17.141) is invariant under multiplication by a unitary operator, it can be rewritten as

$$\lim_{t\to\pm\infty} \|\psi_0 - e^{iHt}e^{-iH_0t}\phi_0^\pm\| = 0. \tag{17.142}$$

In this relation, the norm of the difference between ψ_0 and a sequence of vectors in t converges to zero, so that we may write

$$\psi_0 = \lim_{t\to\pm\infty} e^{iHt}e^{-iH_0t}\phi_0^\pm. \tag{17.143}$$

Since the operators multiplying $\phi_0^\pm$ are bounded for every t, if these asymptotic states form a dense set in the Hilbert space, the operator

$$\lim_{t\to\pm\infty} e^{iHt} e^{-iH_0 t} \tag{17.144}$$

can be defined everywhere. These limits are defined as the *wave operators*

$$\Omega_\pm = \lim_{t\to\pm\infty} e^{iHt} e^{-iH_0 t} \tag{17.145}$$

and have some remarkable properties. Although they are constructed of a sequence of products of unitary operators, the limit is not necessarily unitary, although the wave operators are isometric. The exceptional situation occurs when either H or H_0 has bound states; we refer the reader to discussions in Taylor or Newton (Taylor, 1972; Newton, 1976; Jauch, 1958) for the general case. (In the specific example of the Stark effect, which we shall treat below, the wave operators are unitary.) The convergence of Eq. (17.145) will certainly not occur on a dense set if one of the elements in the Hilbert space is an eigenfunction of H_0, and of the inverse, if one of the elements is an eigenfunction of H, so these subspaces must be excluded in the definition. One may, in fact, think of the convergence as an interference in phases between the two factors.

A property that we will use for the wave operator is that of *intertwining,* obtained by differentiating the product before taking the limit. If the limit indeed converges, then the limit of the derivative must be zero. One finds in this way that

$$H\Omega_\pm = \Omega_\pm H_0. \tag{17.146}$$

(In this computation one must take care that H does not commute, in general, with H_0; the result is obtained by bringing H to the left and H_0 to the right when differentiating.) Then, when $\Omega_\pm^{-1}$ exists,

$$H = \Omega_\pm H_0 \Omega_\pm^{-1}, \tag{17.147}$$

a simple relation between the unperturbed and perturbed Hamiltonians. This property is precisely what we need for the construction of the quantum Lax–Phillips theory.

We remark that, looking at the derivative before the limit in another way (bringing H to the right and H_0 to the left of the exponents), it is easy to see that the convergence implies that

$$\| V e^{-iH_0 t} \phi_0^\pm \| \to 0 \tag{17.148}$$

for $t \to \pm\infty$; that is, for example, for a local potential of the form $\langle x|V|x'\rangle = V(x)\,\delta\left(x - x'\right)$, the free evolution must bring the function $\phi_0^\pm(x)$ out of the range of $V(x)$ faster than the tail of the wave packet spreads. If this condition is met on a dense set, the wave operator exists. There are, of course, many other beautiful properties of scattering theory (Kato, 1966; Reed and Simon, 1979), but this brief discussion will suffice for our purposes.

We now return to the Lax–Phillips theory and the definition of the invariant subspaces $\mathfrak{D}_\pm$ in the quantum case. To do this, we assume that there is an evolution operator K_0 for an "unperturbed" system in the Lax–Phillips Hilbert space $\overline{\mathfrak{H}}$, and a full perturbed operator K of the form $K = K_0 + V$, and seek a translation representation for the unperturbed evolution, i.e., a solution of the differential equation

$$K_0 \chi_s = -i \frac{\partial}{\partial s} \chi_s, \tag{17.149}$$

obtained from the requirement that $e^{-iK_0 s}$ acts as translation on χ_s (extracted for infinitesimal s). Note that this is *not* a Schrödinger equation for the evolution of a state χ_s in "time" s, but an actual translation along the s-axis of this function which is square integrable (on the absolute square) over all s (clearly, wave packets must be constructed). The Schrödinger evolution of a state in quantum theory has a finite norm but is not square integrable over the time, since the norm is invariant in time in the standard nonrelativistic theory, and t commutes with all observables. Eq. (17.149) is completely analogous to the translation in space of a wave function $\psi(x)$ in the sense that $e^{ipa}\psi(x) = \psi(x + a)$, where p is the canonical momentum operator, and our problem is to find the analog of the momentum representation for K_0 for translations in s. Eq. (17.149) defines the operator K_0 as $-i\frac{\partial}{\partial s}$ on this representation. Let us now choose a set of functions χ_s^+ with support in $(0, +\infty)$, and another set of functions χ_s^- with support in $(-\infty, 0)$. These form the two invariant subspaces for the unperturbed problem, $\mathfrak{D}_\pm^0$. Now suppose that there are wave operators defined in an analogous way to the wave operators of standard scattering theory,

$$\Omega_\pm = \lim_{\tau \to \pm\infty} e^{iK\tau} e^{-iK_0\tau}. \tag{17.150}$$

Then by the intertwining relation discussed above, we see that

$$f_s^\pm = \Omega_\pm \chi_s^\pm \tag{17.151}$$

form the functions that lie in $\mathfrak{D}_\pm$ for the full interacting evolution $e^{-iK\tau}$. In general, this lifting does not complete the space $\overline{H}$ with $D_\pm$; the deficit is just the subspace K, spanned by the resonance states. We have thus arrived at a procedure for the construction of the incoming and outgoing representations for the full quantum Lax–Phillips theory (Strauss *et al.*, 2000). The possibility of carrying through this construction depends, of course, on the existence of the wave operators defined in Eq. (17.150).

A model that has been widely used, and discussed in this context by Flesia and Piron (1984), is to take

$$K = E + H, \tag{17.152}$$

where E is represented by $i\frac{\partial}{\partial t}$ in the "physical" coordinate-time representation $\langle x, t \mid f\rangle$ of elements f in $\overline{H}$, as will be discussed below, as the form of the Lax–Phillips evolution operator. This structure was also used by Howland (1974, 1979) in his discussion of how to deal with time-dependent Hamiltonians in the framework of the standard theory, both classically and quantum mechanically, and is the basis for the mathematical formulation of Floquet theory, as discussed above (Zhang and Feng 1995). Taking for K_0 the form

$$K_0 = E + H_0, \tag{17.153}$$

it is clear that, since $K - K_0 = H - H_0$, there is a good likelihood that the wave operator defined in Eq. (17.150) can be found in particular applications. Furthermore, the time derivative term assures that the spectra of K_0 and of K are unbounded from below, even though H_0 and H may be semibounded. Thus, a translation representation may be accessible. We shall discuss in detail toward the end of this section an important example in which these operators exist and can be computed explicitly. Note that the variable t does not, in general, coincide with the foliation variable, whose existence is asserted by Sinai (Cornfeld *et al.*, 1982). In any model for K, K_0, formulated in terms of functions on x, t (for a one-particle system, x may correspond to the position of the particle at a given t), one has to find representations in which these operators act as translations, involving the construction of unitary transformations.

The translation representation for a state f may be written as $\langle s, \xi \mid f\rangle$, where the variables of the auxiliary space ξ associated with the foliated description along s do not, in general, coincide with the "physical" variables x (corresponding to the coordinate description of the configuration space). The transformations to the translation representations generally require that the new variables s, ξ are functions of t, x. For a one-particle system, for example, x may correspond just to the position of the particle. For X, the position operator, we have not only

$$\langle t, x|X|f\rangle = x\langle t, x \mid f\rangle \tag{17.154}$$

and, as usual, the canonically conjugate relation (related by Fourier transform)

$$\langle t, x|P|f\rangle = -i\frac{\partial}{\partial x}\langle t, x \mid f\rangle, \tag{17.155}$$

where P is the canonical momentum operator, but also the relation

$$\langle t, x|E|f\rangle = i\frac{\partial}{\partial t}\langle x, t \mid f\rangle. \tag{17.155'}$$

In the *translation* representation associated with K_0, what we might call the *free* translation representation, we have

$$\langle s, \xi | K_0 | f \rangle = -i \frac{\partial}{\partial s} \langle s, \xi \mid f \rangle. \tag{17.156}$$

We can now see that the model posed in Eq. (17.152) and Eq. (17.153) does not admit a unique definition of the translation representation. We shall show, however, that unitarity requirements are sufficient to resolve the ambiguity up to a phase. Nevertheless, important new developments in the subject have arisen, motivated by the problem of dealing directly with a Hamiltonian with support on the half line (Strauss, 2003, 2005a, b; Strauss *et al.*, 2006; see also Strauss, 2005c).

Let us write out the condition Eq. (17.156) in terms of the transformation functions $\langle s, \xi \mid t, x \rangle$. To do this, we write

$$\int dtdx \langle s, \xi \mid x, t \rangle \langle x, t \mid (E + H_0) \mid f \rangle = i \frac{\partial}{\partial s} \int dtdx \langle s, \xi \mid x, t \rangle \langle x, t \mid f \rangle \tag{17.157}$$

and use the properties of the operator E of Eq. (17.155') and that of H_0:

$$\langle x, t | H_0 | f \rangle = \hat{H}_0 \langle x, t \mid f \rangle, \tag{17.158}$$

where $\hat{H}_0$ is now an operator on x alone—for example, $-\frac{\partial^2}{\partial x^2}$. The equation (17.157) then becomes

$$\int dtdx \langle s, \xi \mid x, t \rangle \left(i \frac{\partial}{\partial t} + \hat{H}_0 \right) \langle x, t \mid f \rangle = i \frac{\partial}{\partial s} \int dtdx \langle s, \xi \mid x, t \rangle \langle x, t \mid f \rangle. \tag{17.159}$$

Assuming that the differential operator $\hat{H}_0$ is even in derivatives with respect to x, we may integrate both terms by parts so that the operator on the left side acts only on the transformation function $\langle s, \xi \mid x, t \rangle$. We may then extract from this equation the (arbitrary) function $\langle x, t \mid f \rangle$ to obtain the condition

$$\left\{ -i \frac{\partial}{\partial t} + \hat{H}_0 \right\} \langle s, \xi \mid x, t \rangle = i \frac{\partial}{\partial s} \langle s, \xi \mid x, t \rangle. \tag{17.160}$$

Bringing the term $-i \frac{\partial}{\partial t} \langle s, \xi \mid x, t \rangle$ to the right, we obtain

$$i \left(\frac{\partial}{\partial t} + \frac{\partial}{\partial s} \right) \langle s, \xi \mid x, t \rangle = \hat{H}_0 \langle s, \xi \mid x, t \rangle, \tag{17.161}$$

which provides a condition which involves the transformation function $\langle s, \xi \mid x, t \rangle$ only as a function of $s + t$ and does not determine the dependence on $s - t$. It must satisfy, however, the normalization and orthogonality conditions

$$\int dxdt \langle s, \xi \mid x, t \rangle \langle s' \xi' \mid x, t \rangle^* = \delta \left(s - s' \right) \delta \left(\xi - \xi' \right) \tag{17.162}$$

and

$$\int dsd\xi \langle s, \xi \mid x, t \rangle \langle s\xi \mid x', t' \rangle^* = \delta\left(x - x'\right)\delta\left(t - t'\right). \tag{17.163}$$

Let us define

$$u = \frac{s+t}{2} \qquad v = \frac{s-t}{2} \tag{17.164}$$

and

$$\langle s, \xi \mid x, t \rangle = f\left(u, v; \xi, x\right). \tag{17.165}$$

It then follows from Eq. (17.162) and Eq. (17.163) that the unitarity and completeness relations can be satisfied for

$$f\left(u, v; \xi, x\right) = e^{iv\alpha} g\left(u; \xi, x\right), \tag{17.166}$$

for any constant α. The main results of the theory, such as the S-matrix poles and eigenstates for the resonances, to be discussed below, do not depend on the choice of α, and in spite of this ambiguity, the theory is therefore predictive, even in the case of Hamiltonians H_0 which are semibounded and treated in the context of the model proposed in Eq. (17.152). Although this form for the embedding of standard Hamiltonian theory into the Lax–Phillips structure is not necessary, it is a very natural generalization from the point of view of Zhang and Feng (1995) and Howland (1974, 1979). It also occurs in the nonrelativistic limit of a covariant relativistic theory (Horwitz *et al.,* 1981).

As we shall see, this ambiguity does not occur in the Stark model that we shall treat below, since the H_0 that we choose in that case naturally has spectrum $(-\infty, \infty)$, without the addition of a time derivative. In the relativistic quantum theory (Stueckelberg, 1941; Schwinger, 1951; Feynman, 1948, 1950; Horwitz and Piron, 1973; Fanchi, 1979; Kyprionides, 1986 and references therein), the generator of motion has spectrum $(-\infty, \infty)$ quite generally; in place of the Hamiltonian H_0, usually of the form of Laplacian, one has a d'Alembert differential operator, containing only second derivatives in both space and time, and therefore the transformation function of the Lax–Phillips theory is well defined. An analysis of the Lax–Phillips theory in the relativistic treatment of the neutral K meson decay was carried out by Strauss and Horwitz (2000), and formulas given for the semigroup evolution in agreement with the Yang–Wu phenomenological formulation (Lee *et al.,* 1957; Wu and Yang, 1964).

We shall assume in the following discussion that the translation representations are well defined; each particular example must be examined to determine how this can be done.

As in the transition to momentum space from coordinate space, let us define the *free spectral* representation in terms of the Fourier transform

$$\langle \sigma\xi \mid g\rangle = \int e^{-i\sigma s} \langle s\xi \mid g\rangle ds, \tag{17.167}$$

where $\langle \sigma\xi \mid g\rangle$ satisfies

$$\langle \sigma\xi | K_0 | g\rangle = \sigma \langle \sigma\xi \mid g\rangle. \tag{17.168}$$

Here $|g\rangle$ (the abstract vector, or Dirac ket) is an element of $\overline{\mathfrak{H}}$ and ξ remains as the set of variables (measure space) of the auxiliary space associated to each value of σ, which with σ comprise a complete spectral set. The functions may be thought of as a set of functions of the variables ξ indexed on the variable σ in a continuous sequence of auxiliary Hilbert spaces, as described in the book of Lax and Phillips (1967), isomorphic to H. Clearly, K_0 acts as the generator of translations in the representation $\langle s\xi \mid g\rangle$.

We now proceed to define the incoming and outgoing subspaces $\mathfrak{D}_\pm$. Let us consider, as remarked above, the sets of functions with support in $L^2\,(0, \infty)$ and in $L^2\,(-\infty, 0)$, and call these subspaces $D_0^\pm$. The Fourier transform back to the free spectral representation provides the two sets of functions,

$$_f\langle \sigma\xi \mid g_0^\pm\rangle = \int e^{-i\sigma s} \langle s\xi \mid g_0^\pm\rangle ds \in H_\pm, \tag{17.169}$$

for $g_0^\pm \in D_0^\pm$. Since these functions are defined as Fourier transforms on half lines, they are *Hardy class* functions, i.e. functions satisfying square integrable conditions as functions of a complex variable.

We remind the reader that an analytic function $f\,(x + iy)$ is said to belong to a Hardy class in the upper half plane if

$$\int_{-\infty}^{+\infty} |f\,(x + iy)\,|^2 dx < \infty$$

for all $y > 0$. A Fourier transform with weight factor $e^{i\sigma s}$ of a square integrable function $g\,(s)$ with support on the positive real line has the property that, as a function of complex σ, it is of Hardy class in the upper half plane.

We may now define the subspaces $\mathfrak{D}_\pm$ in the Hilbert space of states $\overline{\mathfrak{H}}$ in the energy representation. To do this, we first map these Hardy class functions in $\overline{H}$ to $\overline{\mathfrak{H}}$, i.e. we define the subspaces $\mathfrak{D}_0^\pm$ by

$$\int d\xi d\sigma |\sigma\xi\rangle_f \; _f\langle \sigma\xi \mid g_0^\pm\rangle \in \mathfrak{D}_0^\pm. \tag{17.170}$$

We shall now make use of the wave operators, discussed above, which intertwine K_0 with the full evolution K. (We shall explicitly construct these wave operators in the case of the Stark effect that we shall study as an example here.) The construction of $\mathfrak{D}_\pm$ is then completed with the help of the wave operators. We define these subspaces by

$$\begin{aligned}\mathfrak{D}_+ &= \Omega_+ \mathfrak{D}_0^+ \\ \mathfrak{D}_- &= \Omega_- \mathfrak{D}_0^- .\end{aligned} \tag{17.171}$$

We remark that these subspaces are not produced by the same unitary map. This procedure is necessary to realize the Lax–Phillips structure nontrivially. If a single unitary map were used, then there would exist a transformation into the space of functions on $L^2(-\infty, \infty, H)$ which has the property that all functions with support on the positive half line represent elements of $\mathfrak{D}_+$, and all functions with support on the negative half line represent elements of $\mathfrak{D}_-$ in the same representation. The resulting Lax–Phillips S-matrix would then be trivial.

The requirement that $\mathfrak{D}_+$ and $\mathfrak{D}_-$ be orthogonal is not an immediate consequence of our construction. Since the functions ${}_f\langle sx \mid g_0^\pm\rangle$ have support on, respectively, the positive and negative half lines, and the orthogonality of $\mathfrak{D}_\pm$ is determined by the integral of the product of these functions with an operator valued kernel $\mathbf{S}\left(s - s'\right)$ (to be defined below), one sees that suitable analyticity properties of the transformed kernel $S(\sigma)$ assure that these subspaces will be orthogonal. This analyticity property (upper half plane analyticity) is true in the Stark model that we shall treat below.

The wave operators defined by Eq. (17.150) intertwine K and K_0:

$$K\Omega_\pm = \Omega_\pm K_0. \tag{17.172}$$

We may therefore construct the outgoing (incoming) spectral representations from the free spectral representation. Since

$$\begin{aligned}K\Omega_\pm|\sigma\xi\rangle &= \Omega_\pm K_0|\sigma\xi\rangle \\ &= \sigma\Omega_\pm|\sigma\xi\rangle,\end{aligned} \tag{17.173}$$

we may identify

$$|\sigma\xi\rangle_{\substack{\text{out}\\ \text{in}}} = \Omega_\pm|\sigma\xi\rangle. \tag{17.174}$$

The Lax–Phillips S-matrix is defined as the operator on H which carries the incoming to outgoing translation representations of the evolution operator K. Supposing g is an element of $\overline{\mathfrak{H}}$, its incoming spectral representation, according to Eq. (17.174), is

$${}_{\text{in}}\langle\sigma\xi|g\rangle = \langle\sigma\xi|\Omega_-^{-1}g\rangle. \tag{17.175}$$

Let us now act on this function with the Lax–Phillips S-matrix in the free spectral representation, and require the result to be the outgoing representer of g:

$$
\begin{aligned}
{}_{\text{out}}\langle\sigma\xi|g\rangle &= \langle\sigma\xi|\Omega_+^{-1}g\rangle \\
&= \int d\sigma' d\xi' \langle\sigma\xi|\mathbf{S}|\sigma'\xi'\rangle\langle\sigma'\xi'|\Omega_-^{-1}g\rangle, \qquad (17.176)
\end{aligned}
$$

where $\mathbf{S}$ is the Lax–Phillips S-operator (defined on $\overline{\mathfrak{H}}$). Transforming the kernel to the free translation representation with the help of Eq. (17.167), i.e.

$$
\langle s\xi|\mathbf{S}|s'\xi'\rangle = \frac{1}{(2\pi)^2}\int d\sigma d\sigma' e^{i\sigma s'} e^{-i\sigma' s'} \langle\sigma\xi|\mathbf{S}|\sigma'\xi'\rangle, \qquad (17.177)
$$

we see that the relation Eq. (17.176) becomes, after using Fourier transform in a similar way to transform the *in* and *out* spectral representations to the corresponding *in* and *out* translation representations,

$$
\begin{aligned}
{}_{\text{out}}\langle s\beta|g\rangle = \langle s\beta|\Omega_+^{-1}g\rangle &= \int ds' d\xi' \langle s\xi|\mathbf{S}|s'\xi'\rangle\langle s'\xi'|\Omega_-^{-1}g\rangle \\
&= \int ds' d\xi' \langle s\xi|\mathbf{S}|s'\xi'\rangle_{\text{in}}\,\langle s'\xi'|g\rangle. \qquad (17.178)
\end{aligned}
$$

Hence the Lax–Phillips S-matrix is given by

$$
S = \left\{\langle s\xi|\mathbf{S}|s'\xi'\rangle\right\} \qquad (17.179)
$$

in free translation representation. It follows from the intertwining property Eq. (17.172) that

$$
\langle\sigma\xi|\mathbf{S}|\sigma'\xi'\rangle = \delta\left(\sigma-\sigma'\right) S^{\xi\xi'}\left(\sigma\right). \qquad (17.180)
$$

This result can be expressed in terms of operators on $\overline{\mathfrak{H}}$. Let

$$
w_-^{-1} = \left\{\langle s\xi|\Omega_-^{-1}\right\} \qquad (17.181)
$$

be a map from $\overline{\mathfrak{H}}$ to $\overline{H}$ in the incoming translation representation—the same sense that $\{\langle x|\}$ provides a map from the vectors $\{\Psi\}$ in an abstract Hilbert space to the functions $\langle x \mid \Psi\rangle$ in a space of square integrable functions in the x representation. Similarly, let

$$
w_+^{-1} - \left\{\langle s\xi|\Omega_+^{-1}\right\} \qquad (17.182)
$$

be a map from $\overline{\mathfrak{H}}$ to $\overline{H}$ in the outgoing translation representation. It then follows from Eq. (17.178) that

$$
S = w_+^{-1} w_- \qquad (17.183)
$$

is a kernel (integral operator) on the free translation representation. This kernel is understood to operate on the representer of a vector g in the incoming representation and map it to the representer in the outgoing representation.

17.8 Application to the Stark model

In preparation for the application of the Lax–Phillips theory to the Stark model (Ben-Ari and Horwitz, 2004), let us first study the Wigner–Weisskopf approach to this problem. The results will be useful in writing out the solutions to the Lax–Phillips analysis as well. The potential for the Stark effect problem, of the form $-Ex$, is unbounded (on the full space; we shall work in one space dimension). For a model of the form (Friedrichs and Rejto, 1962)

$$H = -Ex + \lambda P_0, \tag{17.184}$$

where λ is real and P_0 is a rank one projection operator, it will be convenient to consider $-Ex \equiv H_0$ as the unperturbed Hamiltonian and the second term, $\lambda P_0 \equiv V$ as the perturbation. (The resolvent is, of course, unaffected by this choice, but the form of the perturbation theory is very different.) In this case, therefore, neither H nor H_0 has bound states, and the resulting wave operators are unitary. We study in this section the Wigner–Weisskopf description of the resonance, and in the next section embed this analysis in the Lax–Phillips Hilbert space.

Let us choose for P_0 the form

$$\langle x|P_0|x'\rangle = \left(\frac{2}{\pi}\right)^{\frac{1}{2}} e^{-(x^2+x'^2)}. \tag{17.185}$$

The resolvent satisfies the identity (second resolvent equation)

$$G = G_0 + G_0 V G, \tag{17.186}$$

where, as before, $G = (z - H)^{-1}$, and $G_0 = (z - H_0)^{-1}$, defined for z in the upper half plane, where $H_0 = -Ex$. The x, x' matrix element of G is therefore

$$\langle x|G|x'\rangle = \frac{1}{z+Ex}\delta\left(x-x'\right) + \frac{1}{z+Ex}\lambda \int_{-\infty}^{\infty} \langle x|P_0|x''\rangle\langle x''|G|x'\rangle dx''. \tag{17.187}$$

Let us define

$$f\left(z,x'\right) = \int dx'' e^{-x''^2} \langle x''|G|x'\rangle dx''. \tag{17.188}$$

It then follows from Eq. (17.187) that

$$f\left(z,x'\right) = \int dx \frac{e^{-x^2}}{z+Ex} + \lambda \int dx \frac{e^{-x^2}}{z+Ex}\sqrt{\frac{2}{\pi}} e^{-(x^2+x''^2)} \langle x''|G|x'\rangle dx''.$$

We can write this as

$$f\left(z,x'\right) = \frac{e^{-x'^2}}{z+Ex'} + \lambda\sqrt{\frac{2}{\pi}} F(z)\, f\left(z,x'\right)$$

or

$$f\left(z, x'\right) = \frac{1}{a - \lambda\sqrt{\frac{2}{\pi}}F\left(z\right)}\frac{e^{-x'^2}}{z + Ex'}, \tag{17.189}$$

where

$$F\left(z\right) = \int dx \frac{e^{-2x^2}}{z + Ex} = \frac{i\pi}{E}e^{-\frac{2z^2}{E^2}}\operatorname{erfc}\left[i\sqrt{2}\frac{z}{E}\right]$$

and the error function is defined by

$$\operatorname{erfc}(x) = \int_x^\infty e^{-t^2}dt \tag{17.190}$$

Returning to Eq. (17.187), we see that

$$\begin{aligned}\langle x|G\left(z\right)|x'\rangle &= \frac{1}{z + Ex}\delta\left(x - x'\right) + \frac{1}{z + Ex}\lambda\sqrt{\frac{2}{\pi}}e^{-x^2}f\left(z, x'\right) \\ &= \frac{1}{z + Ex}\delta\left(x - x'\right) + \lambda\sqrt{\frac{2}{\pi}}\frac{e^{-x^2}e^{-x'^2}}{\left(z + Ex\right)\left(z + Ex'\right)}\left(\frac{1}{1 - \lambda\sqrt{\frac{2}{\pi}}F(z)}\right).\end{aligned} \tag{17.191}$$

We now wish to approximate the time behavior of the survival amplitude. The time dependence of the survival amplitude (for one channel) is given by

$$A\left(t\right) = \frac{1}{2\pi i}\int_C \left(\varphi|G\left(z\right)|\varphi\right)e^{-izt}. \tag{17.192}$$

The contour C corresponds to a line running in the complex energy plane from right to left slightly above the real axis. The matrix element $\left(\varphi|G\left(z\right)|\varphi\right)$ is analytic in the upper half plane. We can shift this line continuously and differentially through the real axis into the lower half plane, provided that the contributions of the vertical pieces at $\pm\infty$ vanish. It is clear from Eq. (17.190) and Eq. (17.191) that this is true for the parts of the vertical integrations that lie in the upper half plane. To write the integrand along the new curve below the axis, we must analytically continue $F\left(z\right)$. To do this, we consider (for ξ real)

$$\begin{aligned}F\left(\xi + i\varepsilon\right) - F\left(\xi - i\varepsilon\right) &= \int_{-\infty}^{\infty} dx e^{-2x^2}\left\{\frac{1}{\xi + i\varepsilon + Ex} - \frac{1}{\xi - i\varepsilon + Ex}\right\} \\ &= -2\pi i\int dx e^{-2x^2}\delta\left(\xi + Ex\right) \\ &= -\frac{2\pi i}{E}e^{-\frac{2\xi^2}{E^2}}.\end{aligned} \tag{17.193}$$

This function has an analytic extension in the finite lower half plane, given by

$$F^L\left(z\right) = F\left(z\right) - \frac{2\pi i}{E}e^{-\frac{2z^2}{E^2}}. \tag{17.194}$$

We find, numerically, for a reasonable choice of parameters and a simple assumption for $\varphi(x)$, that the analytic continuation of the function

$$(\varphi|G(z)|\varphi) = \int dx dx' \varphi^*(x) \left(\begin{array}{c} \frac{1}{z+Ex}\delta(x-x') \\ +\lambda\sqrt{\frac{2}{\pi}}\left[\frac{e^{-x^2}e^{-x'^2}}{(z+Ex)(z+Ex')}\right]\left[\frac{1}{1-\lambda\sqrt{\frac{2}{\pi}}F(z)}\right] \end{array} \right) \varphi(x) \,, \tag{17.195}$$

defined by Eq. (17.191) and Eq. (17.194), into the lower half plane, has a pole, inducing an exponential decay term to the amplitude. The function $(\varphi|G(z)|\varphi)$ has a very simple form if we assume that $\varphi(x)$ has the Gaussian form

$$\varphi(x) = \sqrt{\frac{2}{\pi}}e^{-x^2} \,. \tag{17.196}$$

The first term of Eq. (17.195) contains e^{-2x^2}; its integral with the denominator $z + Ex$ is, according to Eq. (17.190), the function $F(z)$. The second term factorizes into two integrals of the same form. It then follows, with this assumption on φ, that

$$\begin{aligned} (\varphi|G(z)|\varphi) &= \sqrt{\frac{2}{\pi}}\, F(z) + \lambda\left(\frac{2}{\pi}\right)\frac{(F(z))^2}{1-\lambda\sqrt{\frac{2}{\pi}}F(z)} \\ &= \frac{\sqrt{\frac{2}{\pi}}F(z)}{1-\lambda\sqrt{\frac{2}{\pi}}F(z)} \,. \end{aligned} \tag{17.197}$$

The analytic continuation of this function into the lower half plane is achieved by the continuation of $F(z)$ to $F^L(z)$. This function has no poles in the finite lower half plane, and hence the pole can only come from the condition

$$g(z) \equiv 1 - \lambda\sqrt{\frac{2}{\pi}}F^L(z) = 0 \tag{17.198}$$

where $F^L(z)$ is the analytic continuation of $F(z)$ defined in Eq. (17.194).

For the value $\lambda/E = 11$, Maple provides us with a unique solution for the position of the pole, $z_0 = -4.446 - .31896 \times 10^{-15}i$, which has, as expected, a very small imaginary part for this reasonably physical choice of parameters. Differentiating Eq. (17.198) implicitly with respect to E, we find that the real part of the pole moves to more negative values as E increases. Since the unperturbed system has mean position of the particle at zero, this shift corresponds to an increase in field-induced polarization with increasing value of the field.

There remains, however, a contribution to the survival amplitude from integration on a line running from $+\infty$ to $-\infty$ on the real part of $z = \xi + i\zeta$, where ζ can be very large and negative. The contribution of this so-called background integral is strongly suppressed by the exponent $\exp(-izt)$ for t large and positive. For small t, however, this suppression is not strong unless $\zeta \to -\infty$. However, in this limit, the integrand is not well defined, since for any large and negative ζ, the contribution of the discontinuity in $F(z)$ strongly suppresses the integrand for ξ small compared with ζ. This suppression is not maintained, however, for the contributions from ξ in the neighborhood of or greater than ζ. Hence the convergence of the contribution on the background is not uniform. We see from the general argument that the derivative of $|A(t)|^2$ vanishes at $t = 0$ (clearly seen in Eq. (17.28)), that the pole contribution cannot represent the result precisely for small t, and therefore the Wigner–Weisskopf treatment, even in this case of unbounded spectrum, cannot result in a pure exponential (semigroup) behavior for the reduced evolution. It is exactly in this respect that the Lax–Phillips treatment provides a result which is closer to the physics of irreversible decay.

We now proceed to apply the techniques of the Lax–Phillips treatment to this model for the Stark effect.

Since the Stark model that we are using has spectrum $(-\infty, \infty)$, we may take for the generator of motion

$$K = H_{\text{Stark}}; \quad K_0 = H_{0\text{Stark}}; \quad V = V_{\text{Stark}} , \tag{17.199}$$

where H_{Stark}, $H_{0\text{Stark}}$ and $V = V_{\text{Stark}}$ are the operators defined above. Since H_0 is proportional to x, we may make use of the canonical commutation relations of the quantum theory to identify the momentum p as proportional to the (foliation) variable of the unperturbed (free) translation representation; i.e. one may take $s = p/E$, implying that $[s, K_0] = i$ (we have taken $\hbar = 1$). We then have

$$e^{-iK_0\tau}|f\rangle_s = |f\rangle_{s-\tau} \tag{17.200}$$

or in differential form,

$$K_0|f\rangle_s = -i\frac{d}{ds}|f\rangle_s. \tag{17.201}$$

The auxiliary Hilbert spaces of the corresponding Lax–Phillips theory are one-dimensional.

The spectrum of K_0, given by $\{-Ex\}$, with $-\infty < x < \infty$, can then be identified with σ of the Lax–Phillips (unperturbed) energy representation. We shall follow this formal identification to develop the Lax Phillips theory of resonances, and return to the original interpretation of x and p to obtain physical information about the resonant state.

The wave operators are defined as

$$\Omega_\pm = \lim_{\tau\to\pm\infty} e^{iK\tau}e^{-iK_0\tau} . \tag{17.202}$$

We shall calculate the matrix elements of the wave operator in the unperturbed energy representation. It will be convenient, moreover, to use directly the measure on the spectrum of x; we therefore use kets of the form $|x\rangle \equiv \sqrt{|E|}\,|\sigma\rangle$.

Following the standard procedure for taking these limits, we find for the representation $\{|x\rangle\}$ that

$$\langle x|\Omega_{\pm}|x'\rangle = \delta\left(x - x'\right) - \lim_{\varepsilon \to 0_+} \left\langle x \left| \frac{1}{H + Ex' \pm i\varepsilon} V \right| x' \right\rangle . \tag{17.203}$$

Since this formula is bilinear in the kets $|x\rangle$ and $|x'\rangle$, we could use equally well the kets $|\sigma\rangle$ and $|\sigma'\rangle$.

The operator multiplying V in Eq. (17.203) is $-G(z)$ for $z = -Ex' \mp i\varepsilon$; the matrix elements of this operator were evaluated in Eq. (17.191). Carrying out the integral for the product $G(z)\,V$ with the help of the definition Eq. (17.190), we find that

$$\langle x|G(z)\,V|x'\rangle = \lambda\sqrt{\frac{2}{\pi}} \frac{e^{-(x^2+x'^2)}}{(z+Ex)\left(1 - \lambda\sqrt{\frac{2}{\pi}}F(z)\right)} . \tag{17.204}$$

The wave operators are then given by

$$\langle x|\Omega_{\pm}|x'\rangle = \delta\left(x - x'\right) + \lambda\sqrt{\frac{2}{\pi}} \frac{e^{-(x^2+x'^2)}}{(E(x-x') \mp i\varepsilon)\left(1 - \lambda\sqrt{\frac{2}{\pi}}F(-Ex' \mp i\varepsilon)\right)} . \tag{17.205}$$

Using a partial fraction decomposition for the product of the GV terms, we easily verify that the operators $\Omega_{\pm}$ are unitary.

We now turn to the construction of the incoming and outgoing translation representations. We define the free outgoing translation as the set of functions with support in s (i.e. p/E in the Stark model) on the positive real axis. By Eq. (17.167), the functions

$$f_+^0(x) = \int_0^\infty e^{ipx} f_+^0(p)\,dp \equiv \langle x \mid f_+^0\rangle \tag{17.206}$$

are in the free outgoing translation representation and are analytic in the upper half x-plane (lower half σ-plane). Since the wave operators intertwine K_0 and K, functions of the (full) outgoing representation are then given by

$$\int_{-\infty}^{\infty} \langle x'|\Omega_+|x\rangle dx f_+^0(x)\,dx = f_+^{\mathrm{out}}\left(x'\right) \in D_+ . \tag{17.207}$$

Given a function of the type $f_+^0(x)$, we can calculate the resulting function $f_+^{\mathrm{out}}(x')$ explicitly by noting that the boundary value of $F(z)$ from below the real axis is

$$F\left(-Ex'\right)_{\mathrm{below}} = \lim_{\varepsilon \to 0_+} \int_{-\infty}^{\infty} \frac{e^{-2x''^2}}{E(x''-x') - i\varepsilon} dx'' = \frac{i\pi}{E} e^{-2x'^2} \operatorname{erfc}\left(-i\sqrt{2}x'\right) . \tag{17.208}$$

For the construction of the incoming translation representation, we use Ω_- and a corresponding set of functions $f_-^0(x)$ with support on the negative half line. The kernel of integration then contains $F\left(-Ex' - i\varepsilon\right)$; this may be obtained from Eq. (17.193):

$$F\left(-Ex'\right)_{\text{above}} = F\left(-Ex'\right)_{\text{below}} - \frac{2\pi i}{E} e^{-2x'^2}. \tag{17.209}$$

We now turn to the calculation of the S-matrix. We see from Eq. (17.176) that

$$\langle x|\mathbf{S}|x'\rangle = \langle x|\Omega +^{-1} \Omega_-|x'\rangle \,. \tag{17.210}$$

We now use the definition Eq. (17.145) for the wave operators. Following the standard methods (Jauch, 1958; Taylor, 1972; Newton, 1976), we find that

$$\langle x|\mathbf{S}|x'\rangle = \delta\left(x - x'\right)\left(1 + \frac{2\pi i}{E} \lim_{\varepsilon\to 0_+} \langle x|T\left(-Ex' + i\varepsilon\right)|x'\rangle\right), \tag{17.211}$$

where $T(z) \equiv V(1 + G(z)V)$. We now compute

$$\lim_{\varepsilon\to 0_+} \langle x|T\left(-Ex' + i\varepsilon\right)|x'\rangle = \lim_{\varepsilon\to 0_+} \int_{-\infty}^{\infty} \lambda\sqrt{\frac{2}{\pi}}\, e^{-x^2+x''^2}$$
$$\left[\delta\left(x'' - x'\right) - \frac{\lambda\sqrt{\frac{2}{\pi}}\, e^{-\left(x'^2+x''^2\right)}}{\left(E\left(x' - x''\right) - i\varepsilon\right)\left(1 - \lambda\sqrt{\frac{2}{\pi}}\, F\left(-Ex' + i\varepsilon\right)\right)}\right] dx''$$
$$= \lim_{\varepsilon\to 0_+} \frac{\lambda\sqrt{\frac{2}{\pi}}\, e^{-\left(x^2+x'^2\right)}}{1 - \lambda\sqrt{\frac{2}{\pi}}\, F\left(-Ex' + i\varepsilon\right)}. \tag{17.212}$$

It then follows that

$$\langle x|\mathbf{S}|x'\rangle = \delta\left(x - x'\right)\left[1 + \frac{2\pi i}{E} \lim_{\varepsilon\to 0_+} \frac{\lambda\sqrt{\frac{2}{\pi}}\, e^{-2x^2}}{1 - \lambda\sqrt{\frac{2}{\pi}}\, F\left(-Ex + i\varepsilon\right)}\right], \tag{17.213}$$

so that we may write

$$\langle x|\mathbf{S}|x'\rangle = \delta\left(x - x'\right) S(x)\,. \tag{17.214}$$

If we write the semigroup evolution, restricted to the subspace $\mathfrak{K}$, as

$$Z(\tau) = e^{-iB\tau}, \tag{17.215}$$

it follows from the contractive semigroup property that the operator B has an eigenvector in the outgoing representation satisfying

$$B|f\rangle_{out} = \mu|f\rangle_{out} \tag{17.216}$$

with μ in the lower half plane, for which the eigenfunctions are of the form (with support in $(0, \infty)$), as in Eq. (17.137),

$$\langle s|f\rangle_{\text{out}} = e^{-i\mu s}\langle 0|n\rangle, \qquad (17.217)$$

where we shall take $\langle 0|n\rangle$ to be a numerical coefficient n (in our Stark model, the auxiliary space is one-dimensional). The eigenfunctions in the outgoing x representation (corresponding to the "energy" variable σ) are then of the form

$$\langle x|f\rangle_{\text{out}} = \frac{in}{x-\mu}, \qquad (17.218)$$

with n a numerical coefficient (Lax and Phillips, 1967). This result provides one of the most important new aspects of the Lax–Phillips theory: the definition of the resonance in terms of a *state* in the quantum Hilbert space.

In the usual framework of standard scattering theory, resonances appear as enhancements of the cross section in scattering cross section or, as often used in applications, large derivatives of the scattering phase shifts, attributed to the presence of poles in the second Rieman sheet of the S-matrix (Jauch, 1958; Taylor, 1972; Newton, 1976). Such a characterization does not, however, provide a vector in a Hilbert space.

The S-matrix, connecting the incoming to outgoing representations, therefore has the form

$$S(x) \sim \frac{r}{x-z_0}, \qquad (17.219)$$

where z_0 is the position of the pole of the diagonal S-matrix in the lower half plane, which we identify with the semigroup exponent μ, and r is the residue. From Eq. (17.213), we see that the pole of the S-matrix corresponds to a zero of the denominator

$$1 - \lambda\sqrt{\frac{2}{\pi}}F(-Ex + i\varepsilon)$$

continued to the lower half plane; i.e. we must find the zero of

$$1 - \lambda\sqrt{\frac{2}{\pi}}F^L(z). \qquad (17.220)$$

This is precisely the pole approximation of the Wigner–Weisskopf theory as in connection with Eq. (17.198). The residue of the pole in Eq. (17.219) is then given by

$$r = \frac{8\pi i}{E^3}\lambda\sqrt{\frac{2}{\pi}}\left(\frac{e^{\frac{-2z_0^2}{E^2}}}{z_0-\lambda}\right). \qquad (17.221)$$

We have studied, in this analysis, a model which provides the possibility of studying in the Wigner–Weisskopf method, the standard technique for such analyses, giving exponential behavior in the pole approximation, but also is accessible in a simple way to the Lax–Phillips framework.

In this example, we have computed the resonant state for a Stark model. The variable x used here corresponds to the "energy" in the Lax–Phillips formal structure but retains its physical meaning, in the result, as position. Note that this is also true in our formulation of the Wigner–Weisskopf model, taking for the unperturbed Hamiltonian the term $-Ex$ in the Hamiltonian Eq. (17.184), since this term is large compared with the term producing the embedded bound state. The position variable occurs in the Hamiltonian, producing a continuous energy spectrum $\{Ex\}$. The interpretation of the poles of the S-matrix, or resolvent, therefore remains, as in the usual formulation of resonance problems, as occurring in the complex energy plane, but the variable x retains its physical meaning as coordinate as well.

In the framework of general Lax–Phillips theory, the resonant state carries the pole as in Eq. (17.218), where the x that appears would be replaced by another symbol, say, σ, associated with unperturbed energy, and the distribution over space of the wave function would reside in the vector n of the auxiliary Hilbert space. In our case, this vector is just a number (one-dimensional), and the space distribution is provided by the equivalence of (unperturbed) energy and the variable x.

The resonance state provided by the Lax–Phillips theory in the "energy" representation actually therefore corresponds to a distribution of x values in the resonant state:

$$|\langle x|f\rangle_{\text{out}}|^2 = \frac{|n|^2}{(x - \text{Re } z_0)^2 + |\text{Im } z_0|^2},$$

a Cauchy distribution with width $|\text{Im } z_0|$. The Cauchy distribution is centered on $\text{Re } z_0$, corresponding to a shift away from the mean value of $x = 0$ in the bound state in the absence of electric field; as we have seen, the pole moves farther to the left with increasing field, so that the center of the wave packet moves to the left.

This example therefore illustrates the approximate exponential decay law in time in the pole approximation of Wigner–Weisskopf theory, and the exact exponential decay law in the Lax–Phillips treatment, with precisely the same exponent.

We have seen in this chapter some of the limitations of the Wigner–Weisskopf method, as well as some of its very useful and robust results. In the Wigner–Weisskopf analysis, the resonance is described by the position of a pole in the complex energy plane but does not have a state in the Hilbert space associated with it. There has been considerable study, as discussed in Section 17.6, of the application of the method of Gel'fand triples, or rigged Hilbert spaces, for the description of resonances which satisfies the property of exact exponential decay (without the "background" corrections to the Wigner–Weisskopf pole approximation) (Bailey and Schieve, 1978; Bohm, 1978, 1980; Parravicini *et al.*, 1980; Bohm and Gadella, 1989; Bohm and Kaldass, 2000). The elements of the Gel'fand triple are not, however, vectors of a Hilbert space (they belong to a Banach space) and have no scalar products. Hence, it is not possible, in general, to compute the expectation value of an observable or to study physical properties, such as localization, of the state.

The Lax–Phillips formulation describes the resonant state as an element of a Hilbert space, and answers to such questions then become accessible.

References

Agam, O., Altshuler, B. L. and Andreev, A. V. (1995). *Phys. Rev. Lett.* **75**, 4389.

Agam, O., Wingreen, N. S., Altshuler, B. L., Ralph, D. C. and Tinkham, M. (1996). *Phys. Rev. Lett.* **78**, 1996.

Aguilar, E. J. and Combes, J. M. (1971). *Comm. Math. Phys.* **22**, 269.

Aharonov, Y. and Vardi, M. (1980). *Phys. Rev.* **21**, 2235.

Antoniou, I., Levitan, J. and Horwitz, L. P. (1993). *J. Phys. A Math Gen.* **26**, 6033.

Bailey, T. and Schieve, W. C. (1978). *Nuovo Cim.* **47A**, 231.

Balslev, E. and Combes J. M. (1971). *Comm. Math. Phys.* **22,** 280.

Baumgartel, H. (1978). *Math. Nachr.* **75**, 133.

Baumgartel, H. (2006). *Rev. Math. Phys.* **18**, 611.

Ben-Ari, T. and Horwitz, L. P. (2004). *Phys. Lett. A* **332**, 168.

Bleistein, N., Handelsman, R., Horwitz, L. P. and Neumann, H. (1977). *Nuovo Cim.* **41A**, 389.

Bohm, A. (1978). *The Rigged Hilbert Space and Quantum Mechanics,* Springer Lecture Notes on Physics **78** (Berlin, Springer).

Bohm, A. (1980). *J. Math. Phys.* **21**, 1040.

Bohm, A. (1981). *J. Math. Phys.* **21**, 2813.

Bohm, A. and Gadella, M. (1989). *Dirac Kets, Gamow Vectors and Gel'fand Triplets,* Lecture Notes in Physics **348** (Berlin, Springer).

Bohm, A. and Kaldass, H. (2000). In *Mathematical Physics and Stochastic Analysis*, pp. 99–113 (River Edge, N. J., World Scientific).

Bohm, A., Gadella, M. and Mainland, G. B. (1989). *Am. J. Phys.* **57**, 1103.

Cahn, E. N. and Goldhaber, G. (1989). *The Experimental Foundations of Particle Physics*, p. 193ff. (Cambridge, Cambridge University Press).

Charpak, G. and Gourdin, M. (1967). CERN Report 67–18, (Geneva, CERN).

Christenson, J. H., Cronin, J. W., Fitch, V. L. and Turlay, R. (1964). *Phys. Rev. Lett.* **13**, 138.

Cohen, E. and Horwitz, L. P. (2001). *Hadron. J.* **24**, 593. Estimates based on *Particle Physics Booklet*; see Hagiwara, K. (2002).

Cornfeld, I. P., Formin, S. V. and Sinai, Ya. G. (1982). *Ergodic Theory* (Berlin, Springer).

Dirac, P. A. M. (1947). *Quantum Mechanics* (London, Oxford University Press).

Eisenberg, E. and Horwitz, L. P. (1997). In *Advances in Chemical Physics,* ed. I. Prigogine and S. Rice (New York, Wiley).

Fanchi, R. (1979). *Phys. Rev. D* **20,** 3108.

Feynman, R. P. (1948). *Rev. Mod. Phys.* **20**, 367.

Feynman, R. P. (1950). *Phys. Rev.* **80**, 440.

Fischer, M. C., Gutiérrez-Medina, B. and Raizen, M. G. (2001). *Phys. Rev. Lett.* **87**, 040402.
Flesia, C. and Piron, C. (1984). *Helv. Phys. Acta* **57**, 697.
Friedrichs, K. O. (1950). *Comm. Pure Appl. Math.* **1**, 361.
Friedrichs, K. O. and Rejto, P. (1962). *Comm. Pure Appl. Math.* **15**, 219.
Gamow, G. (1928). *Z. Phys.* **51**, 204.
Gel'fand, I. M. and Shilov, G. E. (1967). *Generalized Functions* **II** (New York, Academic Press).
Gell-Mann, M. and Ne'eman, Y. (1964). *The Eightfold Way* (New York, W. A. Benjamin).
Giuliani, D., Joos, E., Kiefer, C., Kupsch, J., Stamatescu, I.-O., Zeh, H. D. (1996), *Decoherence and the Appearance of a Classical World in Quantum Theory* (New York, Springer).
Hagiwara, K. *et al.* (2002). *Phys. Rev. D* **66**, 010001.
Höhler, G. (1958). *Z. Phys.* **152**, 546.
Horwitz, L. P. and Marchand, J. P. (1969). *Helv. Phys. Acta* **42,** 1039.
Horwitz, L. P. and Mizrachi, L. (1974). *Nuovo Cim.* **21A**, 625.
Horwitz, L. P. and Piron, C. (1973). *Helv. Phys. Acta* **46**, 316.
Horwitz, L. P. and Piron, C. (1993). *Helv. Phys. Acta* **66**, 694.
Horwitz, L. P. and Sigal, I. M. (1980). *Helv. Phys. Acta* **51**, 685.
Horwitz, L. P., Marchand, J. P. and LaVita, J. (1971). *J. Math. Phys.* **12**, 2537.
Horwitz, L. P., Schieve, W. C. and Piron, C. (1981). *Ann. Phys.* **137**, 306.
Howland, J. S. (1974). *Math. Ann.* **207**, 315.
Howland, J. S. (1979). *Springer Lecture Notes in Physics* **130**, 163.
Itano, N. M., Heinzen, D. J., Bollinger, J. J. and Wineland, D. J. (1990). *Phys. Rev. A* **41**, 2295.
Jauch, J. M. (1958). *Helv. Phys. Acta* **31**, 127, 661.
Joos, E. (1996). In *Decoherence and the Appearance of a Classical World in Quantum Theory*, p. 96ff, ed. D. Giulini, E. Joos, C. Kieffer, J. Kupsch, I. O. Stamatescu and H. D. Zeh (New York, Springer). This book contains useful discussions and references on a wide range of subjects associated with the formulation of decoherence.
Kato, T. (1966). *Perturbation Theory for Linear Operators* (New York, Springer).
Kyprionides, A. (1986). *Phys. Rep.* **155**, 1, and references therein.
Lax, P. D. and Phillips, R. S. (1967). *Scattering Theory* (New York, Academic Press).
Lee, T. D. (1954). *Phys. Rev.* **95**, 1329.
Lee, T. D., Oehme, R. and Yang, C. N. (1957). *Phys. Rev.* **106**, 340.
Martin, B. R. and Shaw, G. (1997). *Particle Physics*, pp. 248ff. (Chichester, Wiley).
Misra, B. and Sudarshan, E. C. G. (1977). *J. Math. Phys.* **18**, 756.
Ne'eman, Y. (1967). *Algebraic Theory of Particle Physics* (New York, W. A. Benjamin).
Newton, R. J. (1976). *Scattering Theory of Particles and Waves* (New York, McGraw-Hill).
Parravicini, G., Gorini, V. and Sudarshan, E. C. G. (1980). *J. Math. Phys.* **21**, 2208.
Piron, C. (1976). *Foundations of Quantum Physics* (Reading, Mass., Benjamin/Cummings).
Reed, M. and Simon, B. (1972). *Methods of Modern Mathematical Physics* **I**, Functional Analysis (New York, Academic Press).
Reed, M. and Simon, B. (1979). *Methods of Modern Mathematical Physics* **III**, Scattering Theory (New York, Academic Press).
Riesz, F. and Sz.-Nagy, B. (1955). *Functional Analysis* (New York, F. Ungar).
Schwinger, S. (1951). *Phys. Rev.* **82**, 664.

Simon, B. (1972). *Comm. Math. Phys.* **27**, 1.
Sz.-Nagy, B. and Foias, C. (1976). *Harmonic Analysis of Operators on Hilbert Space* (Amsterdam, North-Holland).
Strauss, Y. (2003). *Int. J. Theor. Phys.* **42**, 2285.
Strauss, Y. (2005a). *J. Math. Phys.* **46**, 032104.
Strauss, Y. (2005b). *J. Math. Phys.* **46**, 102109.
Strauss, Y. (2005c). In *Group Theoretical Methods in Physics,* Cocoyoc, Mexico, 2–6 August 2004, ed. G. S. Pogosyan, L. R. Vincent and K. B., Wolf, Inst. of Physics Series **185** (Atlanta, Taylor & Francis CRC Press).
Strauss, Y. and Horwitz, L. P. (2000). *Found. Phys.* **30**, 653.
Strauss, Y., Horwitz, L. P. and Eisenberg, E. (2000). *J. Math. Phys.* **41**, 8050.
Strauss, Y., Horwitz L.P. and Volovick, A. (2006). *J. Math. Phys.* **47**, 123505.
Stueckelberg, E. C. G. (1941). *Helv. Phys. Acta* **14**, 322, 588.
Sz-Nagy, B. and Foias C. (1976). *Harmonic Analysis of Operators on Hilbert Space* (Amsterdam, North Holland).
Taylor, J. R. (1972). *Scattering Theory* (New York, Wiley).
von Neumann, J. (1955). *Mathematical Foundations of Quantum Mechanics* (Princeton, Princeton University Press).
Weisskopf, V. F. and Wigner, E. P. (1930). *Z. Phys.* **63**, 54; (1930) **65**, 18.
Wilkinson, S. R., Bharucha, C. F., Fischer, M. C., Madison, K. W., Morrow, P. R., Niu, Q., Sudaram, B. and Raizen, M. G. (1997). *Nature* (London) **387**, 575.
Winstein, B. *et al.* (1997). *Results from the Neutral Kaon Program at Fermilab's Meson Center Beamline 1985–1997,* FERMILAB-Pub-97/087-E, published on behalf of the E731, E773 and E799 Collaboration (Batavia, Ill., Fermi National Accelerator Laboratory).
Wu, T. T. and Yang, C. N. (1964). *Phys. Rev. Lett.* **13**, 380.
Zhang, W. M. and Feng, D. H. (1995). *Phys. Rep.* **252**, 1.

18

Quantum statistical mechanics, extended

18.1 Intrinsic theory of irreversibility

In connection with the previous chapter on Wigner–Weisskopf, quantum irreversibility, Gamov states and the Lax–Phillips theory of decay, we shall examine the recent program in statistical mechanics carried forward by Ilya Prigogine and his colleagues since his early book in 1962 (Prigogine, 1962). This discussion is somewhat out of the focus of the present book, since the work to be discussed has principally been devoted to isolated quantum systems, not those in interaction with "reservoirs." In addition, Prigogine's work is very classical in its content. We shall not attempt the task of reviewing the many changes that have taken place between 1962 and the present. Fine texts describing some of this work are those of Balescu (1963, 1975). A recent critical overview of the "modern view" is in the unpublished thesis of B. C. Bishop of the University of Texas Philosophy Department (Bishop, 1999).

In many ways this is related to the use of Gel'fand triplets (Gel'fand and Vilenkin, 1968), mentioned in Chapter 17.6, to describe irreversible quantum states, i.e. Gamov states. See Chapter 17, the book of Bohm and Gadella (1989) and also the more recent review article by Bohm and colleagues (Bohm *et al.,* 1997). The object in statistical mechanics is to extend the range of the Liouville operator (classical and quantum) such that its eigenvalues are complex and intrinsically irreversible. Statistical mechanics has always been focused on many particle systems and the appearance of a continuum spectrum in the thermodynamic limit, already used in many places in this book. In the classical case Antoniou and Tasaki carried out the adaptation of the Gel'fand triplet approach to the Liouville operator, but this is, apparently, not possible quantum mechanically (Antoniou and Tasaki, 1993). Thus, another route must be taken in what Prigogine has termed, for emphasis, "large Poincaré systems" which have a continuum spectrum (Prigogine, 1997).

We must also say that much of the early development was by the use of perturbation theory and diagrammatic methods, in the use of the master equation of

Chapter 4. A central article in the early work is by Prigogine, George, Henin and Rosenfeld (Prigogine *et al.,* 1973) and must not be overlooked. In a sense, the theoretical effort was formalized here and then clarified by later nonperturbation and more rigorous discussion. A most important physical assumption, called the "$i\varepsilon$ rule" of analytical continuation, was introduced by George at that time. Its importance cannot be underestimated (George, 1971). We will discuss this in the subsequent sections.

18.2 Complex Liouvillian eigenvalue method: introduction

Influenced by the idea of a rigged Hilbert space approach to construct an extension of the Schrödinger wave function to complex eigenvalue spectra, Petrosky and Prigogine (1997) began a similar program with the Liouville operator. They somewhat audaciously proposed a super operator, L, eigenvalue spectra

$$L \mid F_\alpha^\nu \gg = z_\alpha^\nu \mid F_\alpha^\nu \gg, \tag{18.1}$$

and a related adjoint eigenvalue spectra

$$\ll \tilde{F}_\alpha^\nu \mid L = \ll \tilde{F}_\alpha^\nu \mid z_\alpha^\nu, \tag{18.2}$$

assuming bi-orthogonality and completeness. They assumed the super operator L to be diagonalizable and represented by a complex spectral representation as

$$L = \sum_\nu \sum_\alpha \mid F_\alpha^\nu \gg z_\alpha^\nu \ll F_\alpha^\nu \mid . \tag{18.3}$$

Here we will be interested in the quantum version of this approach. (The classical version, possibly simpler and related to classical chaotic dynamics, will be left aside.)

This was examined particularly in the context of scattering theory and the well-known and often-employed Friedrichs model, which has a Hamiltonian of the form already considered in the previous chapter,

$$H = \omega_1 \left|1\right\rangle \left\langle 1\right| + \int_0^\infty d\omega\omega \left|\omega\right\rangle \left\langle\omega\right| + \lambda \int_0^\infty d\omega V_\omega \left(\left|\omega\right\rangle \left\langle 1\right| + \left|1\right\rangle \left\langle\omega\right|\right),$$

representing a single excitation $|1\rangle$ interacting with a continuum $|\omega\rangle$ (Friedrichs, 1948).

To illustrate this, we will follow the somewhat more rigorous formulation of Antoniou (Antoniou *et al.,* 1997). We will reduce the mathematical sophistication of their article, which perhaps is its contribution, but still follow the discussion. The central point is to use a manifest continuum representation rather than the thermodynamic limit of a discrete spectra in Hilbert space, which is more traditional and appears throughout this book. This is in keeping with Prigogine's insistence that the theory is "large Poincaré," i.e. completely non-integrable. This is the main difference from the rigged Hilbert space approach of Bohm and colleagues and from the Lax–Phillips theory.

18.3 Operators and states with diagonal singularity

Van Hove observed that in continuous basis Hilbert spaces

$$A_{\alpha\alpha'} = \left\langle \alpha \left| A \right| \alpha' \right\rangle = A_\alpha \delta\left(\alpha - \alpha'\right) + A_{\alpha\alpha'} \tag{18.4}$$

(Van Hove, 1955, 1962). The matrix elements in the continuum representation $|\alpha\rangle$ have a diagonal singularity for nonvanishing A_α in the first term. Here it will be assumed that the off-diagonal $A_{aa'}$ are compact and correspond to trace class operators usually assumed in statistical mechanics. It will be further assumed that the diagonal part A_α corresponds to a Banach algebra norm with

$$\left\| A^d \right\| \equiv \sup_{\psi \neq 0} \frac{\| A\psi \|}{\| \psi \|}. \tag{18.5}$$

The nondiagonal parts are compact on a Hilbert space. As a consequence, any operator A with a diagonal singularity becomes

$$\begin{aligned} A &= A^d + A^c \\ \text{and } \|A\| &= \left\| A^d \right\| + \left\| A^c \right\| . \end{aligned} \tag{18.6}$$

Antoniou *et al.* prove from these assumptions that the space of A is a Banach space which includes the identity, and further that the algebra is not C^*. A basis is constructed with

$$|\alpha) \equiv |\alpha)^d \equiv |\alpha >< \alpha| , \tag{18.7}$$

and further (note the "round" ket),

$$\left|\alpha\alpha'\right) = \left|\alpha >< \alpha'\right| . \tag{18.8}$$

The representation of A is then

$$A = \int d\alpha A_\alpha^d \mid \alpha) + \int d\alpha d\alpha' A^c_{\alpha\alpha'} \left|\alpha\alpha'\right) . \tag{18.9}$$

The basis $\left|\alpha\alpha'\right)$ is discussed in some detail by Antoniou *et al.* (Please note that we will *not* use the notation of Petrosky and Prigogine.) Now we have

$$\langle\langle A \mid B \rangle\rangle = \mathrm{Tr}\left(A^\dagger B\right).$$

Further, the density operator (state!) may be defined with the scalar product

$$(\rho \mid A) \equiv \langle A \rangle_\rho \tag{18.10}$$

over all A in this Banach space. They are normalized linear functionals having the familiar property

$$\begin{aligned} &\text{(a)} \quad (\rho \mid z_1 A_1 + z_2 A_2) = z_1 (\rho \mid A_1) + z_2 (\rho \mid A_2) \\ &\text{(b)} \quad \left(\rho \mid A^\dagger A\right) \geq 0 \\ &\text{(c)} \quad (\rho \mid I) = 1, \end{aligned} \tag{18.11}$$

since I is included.

The ρ themselves form a subset of the dual space to that of the operators A. Thus,

$$\rho = \rho^d + \rho^c \tag{18.12}$$

for any operator A in the space

$$\begin{aligned} \left(\rho^d \mid A\right) &= \left(\rho^d \mid A^d\right) \\ \left(\rho^c \mid A\right) &= \left(\rho^c \mid A^c\right). \end{aligned} \tag{18.13}$$

Because the off-diagonal states ρ are trace class by assumption,

$$\left(\rho^c \mid A\right) = \mathrm{Tr}\hat{\rho}\hat{A}. \tag{18.14}$$

Antoniou *et al.* prove that in the dual space of A,

$$\|\rho\| = \max\left\{\left(\rho^d \mid I\right), \mathrm{Tr}\hat{\rho}\right\}. \tag{18.15}$$

With this, it may be argued that ρ^d_α are the probabilities of the continuous state $|\alpha\rangle$, and the $\rho'_{\alpha a}$ correspond to correlations, as we had physically expected. Now

$$(\rho \mid A) = \langle A \rangle_\rho = \int d\alpha \rho^d_\alpha A^d_\alpha + \int d\alpha d\alpha' \rho^{c*}_{\alpha a'} A^c_{\alpha a'}. \tag{18.16}$$

By this we identify, in the familiar way,

$$\begin{aligned} (\rho \mid \alpha) &= \rho^{*0}_\alpha = \rho^0_\alpha \\ \left(\rho \mid \alpha a'\right) &= \rho^{c*}_{\alpha a'}. \end{aligned} \tag{18.17}$$

Antoniou *et al.* prove the lemma that $(\alpha|, \left(\alpha a'\right|; |\beta), \left|\beta\beta'\right)$ form a biorthogonal basis with the following properties:

$$\begin{aligned} &1. \quad (\alpha \mid \beta) = \delta(\alpha - \beta) \\ &2. \quad \left(\alpha a' \mid \beta\beta'\right) = \delta(\alpha - \beta)\,\delta\left(\alpha' - \beta'\right) \\ &3. \quad \left(\alpha \mid \beta\beta'\right) = 0 \\ &4. \quad \left(\alpha a' \mid \beta\right) = 0 \end{aligned} \tag{18.18}$$

It must be emphasized that these linear functionals are an extension of the usual quantum theory to states $(\rho|$ with diagonal singularity.

Some further properties must be mentioned. If $\|\rho\| = (\rho \mid I)$, then from the norm condition, Eq. (18.15),

$$\max\left(\left\|\rho^d\right\|, \left\|\rho^c\right\|\right) = \left(\rho^d \mid I\right). \tag{18.19}$$

A pure state of the Hilbert space is the vector ψ where

$$\langle\psi \mid A\psi\rangle = (\rho \mid A), \tag{18.20}$$

if and only if

$$\begin{aligned} \rho_\alpha^d &= \left|\psi_\alpha\right|^2 \\ \rho_{\alpha\alpha'}^c &= \psi_\alpha \psi_{\alpha'}^*, \end{aligned} \tag{18.21}$$

where

$$\psi_\alpha \equiv \langle\alpha \mid \psi\rangle .$$

This expresses the Born rule for calculating quantum probabilities. Further, the representation of operator A by ψ is then, as usual,

$$\langle\psi \mid A\psi\rangle = \int d\alpha \left|\psi_\alpha\right|^2 A_\alpha^d + \int d\alpha d\alpha' \left(\psi_\alpha^* \psi_{\alpha'}\right) A_{\alpha\alpha'}^c .$$

18.4 Super operators and time evolution

Super operators in the form of projection operators and the commutator evolution operator—the Liouville operator—are well known and were used in Chapter 3 for the discussion of time evolution and development of the master equation. However, this was in terms of Hilbert space representations, for instance, the tetradic matrix operator (Zwanzig, 1965). This is extended here by the methods of Section 18.3 to continuous spectra having diagonal singularities.

Define the operation of linear U on ρ with the duality

$$(U\rho \mid A) = (\rho \mid VA) \tag{18.22}$$

in the Banach space for all A, U being the dual of V, $U = V^x$. The diagonal and off-diagonal projection operators are

$$\begin{aligned} (P_d\rho| &= \int d\alpha\, (\rho \mid \alpha)\, (\alpha| = \left(\rho^d\right| \\ (P_c\rho| &= \int d\alpha d\alpha' \left(\rho \mid \alpha\alpha'\right) \left(\alpha\alpha'\right| = \left(\rho^c\right| . \end{aligned} \tag{18.23}$$

The form

$$(P_c\rho \mid A) = \int d\alpha d\alpha' \left(\rho \mid \alpha\alpha'\right)\left(\alpha\alpha' \mid A\right) \tag{18.24}$$

is a tetradic (four-index) multiplication. The Liouville (commutator) operator is for any A

$$(L\rho \mid A) \equiv (\rho \mid [H, A]) \equiv (\rho \mid LA) . \tag{18.25}$$

We consider the super operator eigenvalue problem

$$U f_\nu = z_\nu f_\nu , \tag{18.26}$$

which is for all A

$$(U f_\nu \mid A) = (z_\nu f_\nu \mid A) = z_\nu^* (f_\nu \mid A) . \tag{18.27}$$

The left eigenvector

$$U^+ \bar{f}_\nu = z_\nu \bar{f}_\nu \equiv V \bar{f}_\nu , \tag{18.28}$$

which is for all states ρ

$$\left(\rho \mid V \bar{f}_\nu\right) = \left(\rho \mid z_\nu \bar{f}_\nu\right) . \tag{18.29}$$

Assuming these eigenfunctions of U are biorthogonal, U has the complex spectral decomposition

$$U = \sum_\nu z_\nu \left|\bar{f}_\nu\right) \left(f_\nu\right| . \tag{18.30}$$

For *any* operator V there is a tetradic representation:

$$\begin{aligned}(\rho \mid VA) = &\int d\alpha d\beta \,(\rho \mid \alpha)\,(\alpha \mid V \mid \beta)\,(\beta \mid A) \\ &+ \int d\alpha d\beta d\beta' \,(\rho \mid \alpha)\left(\alpha \mid V \mid \beta\beta'\right)\left(\beta\beta' \mid A\right) \\ &+ \int d\alpha d\alpha' d\beta \left(\rho \mid \alpha\alpha'\right)\left(\alpha\alpha' \mid V \mid \beta\right)(\beta \mid A) \\ &+ \int d\alpha d\alpha' d\beta d\beta' \left(\rho \mid \alpha\alpha'\right)\left(\alpha\alpha' \mid V \mid \beta\beta'\right)\left(\beta\beta' \mid A\right) .\end{aligned} \tag{18.31}$$

With these rules the time evolution may be constructed. Assume the Heisenberg evolution

$$\exp(iLt)\, A \equiv \exp(iHt)\, A \exp(-iHt) ; \tag{18.32}$$

from this we obtain the Heisenberg equation of evolution,

$$\partial_t A = i\,[H, A] = iLA , \tag{18.33}$$

and the Schrödinger representation evolution of the state $(\rho\mid$,

$$(\partial_t \rho \mid A) = (\rho \mid i\,[H, A]) \\ = -i\,(L\rho \mid A)\,. \tag{18.34}$$

This is familiar in form to the evolution in Hilbert space when ρ is trace class.

The assumed spectral decomposition of L is

$$(L\rho| = \sum_{\nu} z_\nu\,(\rho \mid \bar{f}_\nu)\,(f_\nu|\,. \tag{18.35}$$

This is what Petrosky and Prigogine first used. It must be emphasized that in this representation, outlined in detail here, z_ν is complex, similar to but not the same as the Gamov state representation of Arnold Bohm and others.

18.5 Subdynamics and analytic continuation

We introduce a many-body operator Ω similar to the Möller operator of scattering theory mentioned in Chapter 4 and Chapter 17:

$$L\Omega = \Omega\theta \tag{18.36}$$

$$L = \Omega\theta\Omega^{-1}. \tag{18.37}$$

The intertwining relation will be used to construct the spectral decomposition, Eq. (18.35). This was first shown by Petrosky and Prigogine (1991). To do this, we must first introduce creation C^n and destruction D^n super operators, first appearing in the diagrammatic perturbation analysis of the generalized master equation (Prigogine, 1962). We introduce $P_0 = P_d$ and P^1, and P^n where

1. P_0 is the diagonal projector on the states

$$P_0 + P_c = I, \tag{18.38}$$

   P_c being the off-diagonal part, $P_c^2 = P_c$.
2. P^n is a further projection onto states of degree n correlation, $n = 1, 2, \ldots$

   In this, the states are

$$\left(P_d L'^{n'} \rho\right| = 0 \quad n' < n \\ \left(P_d L'^{n'}\right| \neq 0 \quad n' = n, \tag{18.39}$$

where $L'\rho = [V, \rho]$, the Liouvillian of perturbation. We also assume that L_0 is diagonal and hence $P_d = P^0$ in the states of L_0. The minimal power of n, which connects the diagonal to off-diagonal, is the degree of correlation. This is a beginning of a further decomposition, a subdynamics, early used by George (see Prigogine *et al.*, 1973). We have

$$P_c = P^1 + P^2 + \cdots + P^n. \tag{18.40}$$

Now

$$\begin{aligned} P^0 + P^1 + \cdots &= I \\ P^0 P^n &= 0 \\ P^n P^{n'} &= \delta_{nn'} P_n \end{aligned} \tag{18.41}$$

$$\begin{aligned} L_0 P^0 &= P^0 L_0 \\ L_0 P^n &= P^n L_o. \end{aligned} \tag{18.42}$$

We define

$$\theta = \sum_n \left(P^n L P^n + P^n L C^n P^n\right) \equiv \sum_n \theta^n, \tag{18.43}$$

and also, most importantly,

$$\Omega^{-1} = \sum_n \left(P^n + D^n C^n\right)^{-1} \left(P^n + D^n\right), \tag{18.44}$$

where

$$\begin{aligned} C^n &= \left(1 - P^n\right) C^n P^n \\ D^n &= P^n D^n \left(1 - P^n\right). \end{aligned} \tag{18.45}$$

The reader must verify that C^n, D^n obey the operator equations

$$\left[L_0, P^m C^n\right] = \left(P^m C^n - P^n\right) L' \left(P^n + C^n\right) \tag{18.46}$$

$$\left[L_0, D^n P^m\right] = \left(P^n + D^m\right) L' \left(P^m - D^n P^m\right). \tag{18.47}$$

These form the basis of a perturbation analysis. They are equivalent to the resolvent expansion analysis used earlier (Prigogine, 1962; also see Balescu, 1975).

The "subdynamics" is constructed by introducing a transformed projector Π^n:

$$\Pi^n = \Omega^{-1} P^n \Omega. \tag{18.48}$$

It is not Hermitian. Now we may show that

$$\Pi^n = \left(P^n + C^n\right) A^n \left(P^n + D^n\right), \tag{18.49}$$

where

$$A^n \equiv P^n \left(1 + D^n C^n\right)^{-1}. \tag{18.50}$$

Further, $\Pi^n \Pi^m = \Pi^n \delta_{nm}$, and the commutation relation

$$L\Pi^m = \Pi^m L. \tag{18.51}$$

Introducing a transformation of $(\rho \mid A)$ for the arbitrary operator A,

$$\left(\rho' \mid A\right) = (\Pi\rho \mid A), \tag{18.52}$$

we find

$$i\frac{d}{dt}\left(\rho' \mid A\right) = \theta\left(\rho' \mid A\right), \tag{18.53}$$

where we have used Eq. (18.36). This may be further decomposed by the orthogonality of the subspaces:

$$i\frac{d}{dt}\left(P^n\rho' \mid A\right) = \theta^n\left(P^n\rho' \mid A\right); \qquad t \geq 0. \tag{18.54}$$

This is the main result of the subdynamics decomposition of a set of *independent* kinetic Markovian semigroup equations governing the time evolution in the correlation subspaces. This was discussed in detail by Balescu (1975). It represents a considerable development of the master equation methods of Chapter 3.

Now let us comment on the George analytic continuation rule, which is central to the perturbation analysis of the solution to Eq. (8.46) and Eq. (18.47) (George, 1971). The formal solution to the nonlinear equations, Eq. (18.46) and Eq. (18.47), may be written with the time ordering (see Kato, 1966, p. 553; Antoniou and Tasaki, 1993):

$$P^m C^n = i\int_0^{\pm\infty} dt \exp\left(-iL_0t\right)\left(P^m C^n - P^m\right)L'\left(P^n + C^n\right)\exp\left(iL_0t\right)$$

$\lim +\infty$ for $m > n$

$\lim -\infty$ for $m < n$ (18.55)

and

$$D^n P^m = +i\int_0^{\pm\infty} dt \exp\left(-iL_0t\right)\left(P^n + D^n\right)L'\left(P^m - D^n P^m\right)\exp\left(iL_0t\right)$$

$\lim +\infty$ for $n > m$

$\lim -\infty$ for $n < m$. (18.56)

Here, transitions are from n to m in Eq. (18.55) and m to n in Eq. (18.56). Thus, if we choose time running $0 \to \infty$ in Eq. (18.55), the correlation patterns increase in size in the future.

This may be formulated in complex variable space, resulting in the so-called $i\varepsilon$ rule of analytic continuation. We will not pursue this further. See the articles by Antoniou and Tasaki (1993) and by Petrosky and Prigogine (1997).

This time boundary condition in Liouville space should be contrasted with that of Bohm for the wave function in the rigged Hilbert space approach (Bohm *et al.*, 1997; see also Chapter 17). There, it is assumed in interaction with

$$\begin{array}{llll} \text{detection} & \left\langle E \mid \psi^{\text{out}}\right\rangle = \langle -E \mid \psi-\rangle & \varepsilon\mathfrak{L}\cap\mathfrak{H}_{+}^{*}\mid_{R+} & t\geq 0 \\ \text{preparation} & \left\langle E \mid \phi^{\text{in}}\right\rangle = \left\langle +E \mid \phi^{+}\right\rangle & \epsilon\mathfrak{L}\cap\mathfrak{H}_{-}^{*}\mid_{R+} & t\geq 0. \end{array} \tag{18.57}$$

We have a *pair* of rigged Hilbert spaces:

$$\begin{array}{ll} \Phi_{-}\subset\mathfrak{H}\subset\Phi_{-}^{x} & \text{(in states)} \\ \Phi_{+}\subset\mathfrak{H}\subset\Phi_{+}^{x} & \text{(out states).} \end{array} \tag{18.58}$$

Φ_+ is the subspace of the measurement detection, and Φ_- is the subspace of preparation. Analytic continuations are taken consistent with this boundary condition. Time $t = 0$ is taken as the moment state where preparation ends and detection begins, continuing into the future. This separation determines the two regions of Eq. (18.57) and Eq. (18.58). There are two spaces, Φ_- and Φ_+, both of which are the Gel'fand triplets seen in Eq. (18.58). The Φ_-^x and Φ_+^x are further assumed to be Hardy class.

From these states two semigroup continuous evolution operators are constructed:

$$\begin{array}{ll} U_{-}^{x}\Phi_{-}\rightarrow\Phi_{-}^{x} & t\leq 0 \\ U_{+}^{x}\Phi_{+}\rightarrow\Phi_{+}^{x} & t\geq 0. \end{array} \tag{18.59}$$

U_+^x and U_-^x are extensions of $U^\dagger(t)$ to the two spaces Φ_+^x and Φ_-^x. They do not represent evolution from Φ_- *to* Φ_+. Both are semigroups. Here, for instance,

$$U_{+}^{x} = \exp\left(-iE_{R}t\right)\exp\frac{-\gamma t}{2} \text{ for } t\geq 0.$$

In both, the evolution is *toward* the future, decaying in the future. It evolves as a Gamov state decaying into the past, but is interpreted as the preparation growing from $t = -\infty$ to $t = 0$ in the future (see Bohm and Harshman, 1996).

We can see from the $i\varepsilon$ rule, in Eq. (18.55) and Eq. (18.56), that the forward-in-time propagators are used for $m > n$ and the backward-in-time propagators are used for $m < n$ in Eq. (18.55). The direction of the semigroup evolution depends upon the degree of correlation. Both semigroups are intertwined. It cannot be expected that in quantum mechanics the Bohm formulation in rigged Hilbert space will give the same result as the physical extension to a complex eigenvalue decomposition in Liouville space as outlined here. How may we be expected to derive one from the other in quantum mechanics? This is an interesting problem. It would seem that there is no propagation from $-\infty$ to 0 in Eq. (18.55) and Eq. (18.56).

18.6 The Pauli equation revisited

Consider again the $\lambda^2 t$ approximation already mentioned in Chapter 3. The lowest-order contribution, Eq. (18.55), is for P_0 for $m = c > n$:

$$C^0 = -i\lambda \int_0^\infty dt \exp(-iL_0 t)\, P^c L' P_0 \qquad (18.60)$$
$$= \frac{\lambda}{i\varepsilon - L_0} P^c L' P_0 \quad \text{where } \varepsilon > 0.$$

We construct the evolution operator to this order:

$$\theta_0^{(2)} = P_0 L P_0 + P_0 L C^{0(1)} P_0$$
$$= P_0 L P_0 + P_0 L C^{0(1)}.$$

Now $P_0 L P_0 = 0$ by construction. Then

$$\theta_0^{(2)} = \lambda^2 P_0 L' P^c \frac{1}{i\varepsilon - L_0} P^c L' P_0. \qquad (18.61)$$

This is the Pauli operator. The master equation is

$$\frac{d}{dt}\rho^d = -i\theta_0^{(2)} \rho^d. \qquad (18.62)$$

Let us turn again, in this context, to the continuous model of Friedrichs (1948), already met in the previous chapter, as an example of the more general discussion. Here

$$H = H_0 + V \qquad (18.63)$$
$$H_0 = \omega_1 |1\rangle \langle 1| + \int_0^\infty d\omega \omega |\omega\rangle \langle \omega|$$
$$V = \lambda \int_0^\infty d\omega V_\omega \left[|\omega\rangle \langle 1| + |1\rangle \langle \omega|\right].$$

A single level in Hilbert space $|1\rangle$ interacts with the continuum $|\omega\rangle$. The dyadic states previously introduced are defined:

$$|1) \equiv |1\rangle \langle 1| \qquad |\omega) \equiv |\omega\rangle \langle \omega| \qquad (18.64)$$
$$|1\omega) \equiv |1\rangle \langle \omega| \qquad |\omega\omega') \equiv |\omega\rangle \langle \omega'|.$$

The diagonally singular observables (because of the continuum) are written $A = A^d + A^c$ as before:

$$\left|A^d\right) = A_1 \left|1\right) + \int d\omega A_\omega \left|\omega\right) \tag{18.65}$$

$$\left|A^c\right) = \int d\omega' A_{1\omega'} \left|1\omega'\right) + \int d\omega A_{\omega 1} \left|\omega 1\right) + \int d\omega d\omega' A_{\omega'\omega} \left|\omega\omega'\right). \tag{18.66}$$

Correspondingly, we form linear functionals $\left(1\right|, \left(1\omega\right|, \left(\omega 1\right|$, and $\left(\omega\omega'\right|$. A few properties are

$$\begin{aligned} (1 \mid \omega) = (1 \mid \omega\omega') &= (1\omega \mid 1) = (1\omega \mid \omega' 1) \\ (\omega \mid \omega'\omega'') &= (1\omega \mid \omega'\omega'') = 0 \\ (\omega \mid \omega') &= \delta\left(\omega - \omega'\right) \\ (1\omega \mid 1\omega') &= \delta\left(\omega - \omega'\right). \end{aligned} \tag{18.67}$$

With these, we represent the functional $\left(\rho\right| = \left(\rho^d\right| + \left(\rho^c\right|$ where $(\rho \mid A) = \langle A \rangle_\rho$. Now these represent $\left(\rho\right|$:

$$\left(\rho^d\right| = \rho_1 \left(1\right| + \int d\omega \rho_\omega \left(\omega\right| \tag{18.68}$$

$$\left(\rho^c\right| = \int d\omega \rho_\omega \left(\omega 1\right| + \int d\omega' \rho_{1\omega'} \left(1\omega'\right| + \int d\omega d\omega' \rho_{\omega\omega'} \left(\omega\omega'\right|.$$

Here

$$\begin{gathered} \rho_1 = \rho_1^* \qquad \rho_\omega = \rho_\omega^* \qquad \rho_{\omega 1} = \rho_{1\omega}^* \\ \rho_{\omega' w} = \rho_{\omega w'}^* \qquad \rho_1 + \int d\omega \rho_\omega = 1. \end{gathered} \tag{18.69}$$

The relevant super operator projectors are

$$P \equiv \left|1\right)\left(1\right| + \int d\omega \left|\omega\right)\left(\omega\right| \tag{18.70}$$

$$(1 - P) \equiv Q = \int d\omega \left|1\omega\right)\left(1\omega\right| + \int d\omega \left|\omega 1\right)\left(\omega 1\right| + \int d\omega d\omega' \left|\omega\omega'\right)\left(\omega\omega'\right|.$$

The super operator (commutator) $L = L_0 + L_1$ may now easily be written. We have

$$L_0 = \int d\omega \left(\omega_1 - \omega\right) \left|1\omega\right)\left(1\omega\right| + \int d\omega \left(\omega - \omega_1\right) \left|\omega 1\right)\left(\omega 1\right| + \int d\omega d\omega' \left(\omega - \omega'\right) \left|\omega\omega'\right)\left(\omega\omega'\right| \tag{18.71}$$

and the perturbation interaction

$$\begin{aligned}
L' = &\int d\omega V_\omega \left\{ [|\omega 1) - |1\omega)] (1| + \int d\omega V_\omega [|1\omega) - |\omega 1)] (\omega| \right\} \\
&+ \int d\omega' V_{\omega'} \left|\omega'\omega\right) |1\omega) - \int d\omega V_\omega |1) (1\omega| \\
&- \int d\omega' V_{\omega'} \left|\omega\omega'\right) (\omega 1| + \int d\omega V_\omega |1) (\omega 1| \\
&+ \int d\omega d\omega' \left[V_\omega \left|1\omega'\right) - V_{\omega'} |\omega 1)\right] \left(\omega\omega'\right| .
\end{aligned} \tag{18.72}$$

These have the same form as a tetradic multiplication in a discrete Hilbert space representation. The student should show that, for this, the Pauli operator equation gives, taking $A = |1)$,

$$\begin{aligned}
&\frac{d}{dt}(\rho\,|\alpha) = -2\pi\lambda^2 V_m^2 (\rho\,|1) \qquad t \geq 0 \\
&\frac{d}{dt}(\rho\,|\omega) = 2\pi\lambda^2 V_m^2 \delta(\omega - \omega_m)(\rho\,|1) .
\end{aligned} \tag{18.73}$$

The solution is

$$\left(\rho_t \mid 1\right) = \exp\left(-2\pi\lambda^2 V_m^2 t\right) (\rho_0\,|1) . \tag{18.74}$$

The decay of the discrete state is the "golden rule" form, so with $A = \omega$,

$$\left(\rho_t \mid \omega\right) = \left(\rho_0 \mid \omega\right) + \left[\begin{array}{c} 1 - \exp\left(-2\pi\lambda^2 V_m^2 t\right) \\ \times\, \delta\left(\omega - \omega_m\right) \left(\rho_0 \mid \omega_m\right) \end{array} \right], \tag{18.75}$$

which grows with the overlap of $|\omega)$ with $|\omega_m)$. This is, of course, semigroup evolution, as is the operator Pauli equation.

The exact Friedrichs model, to all orders in λ, has been treated (Antoniou *et al.*, 1997). The reader is referred there for the discussion of the complex extension. The result is the same as that of de Haan and Henin (1973).

An exact expression for Θ^0 of Eq. (18.54) is obtained:

$$\Theta_0 = C^0 L P_0 = \left(z_1 - z_1^*\right) |1) \left\{ (1| - \left(z_1^* - z_1\right)^{-1} \int d\omega f(\omega) (\omega| \right\}, \tag{18.76}$$

where

$$f(\omega) = \lambda^2 V_\omega^2 \left[\left(\frac{1}{(\omega - s_1)} \right)_{z_1}^{+} - \left(\frac{1}{(\omega - s)} \right)_{z_1^*}^{-} \right] . \tag{18.77}$$

In Eq. (18.77) the $\pm$ terms arise from the analytical continuation of the form

$$f(z) = \int d\omega \frac{1}{\omega - z} \phi(\omega) \qquad \operatorname{Im} z > 0$$

from above to below (+), and similarly $f(z)$ for $\mathrm{Im}\, z < 0$ from below to above (−). Further, it is assumed that

$$\eta(z) = z - \omega_1 + \int \frac{d\omega V_\omega^2}{\omega - z}$$

has a pole at $\eta_+(z_1) = 0 \quad (\mathrm{Im}\, z_1 < 0)$ and $\eta_-\left(z_1^*\right) = 0$. This is the result of the continuation rules discussed earlier.

We conclude with a reminder to the student that the extension of the Liouville–von Neumann equation, here described briefly, has naturally led to an irreversible set of equations, Eq. (18.54) and the Pauli master equation, which we have met in many forms as a special case and illustration.

References

Antoniou, I. and Prigogine, I. (1992). *Nuovo Cim.* **219,** 93

Antoniou, I. and Tasaki, S. (1993). *Int. J. Quantum Chem.* **46**, 425.

Antoniou, I., Suchanedki, Z., Laura, R. and Tasaki, S. (1997). *Physica A* **241**, 737.

Balescu, R. (1963). *Statistical Mechanics of Charged Particles* (New York, Wiley).

Balescu, R. (1975). *Equilibrium and Non-equilibrium Statistical Mechanics* (New York, Wiley), also published by Elsevier.

Bohm, A. and Gadella, M. (1989). Dirac kets, Gamov vectors and Gel'fand triplets. In *Lecture Notes in Physics* **348** (New York, Springer).

Bohm, A. and Harshman, H. L. (1996). *Irreversibility and Causality,* ed. A. Bohm (Berlin, Springer).

Bohm, A., Maxson, S., Loewe, M. and Gadella, M. (1997). *Physica A* **236**, 485.

de Haan, M. and Henin, F. (1973). *Physica* **67**, 197.

Friedrichs, K. (1948). *Comm. Pure. Appl. Math.* **1**, 361.

Gel'fand, I. M. and Vilenkin, N. Ya. (1968). *Generalized Functions* **4** (New York, Academic Press).

George, C. (1971). *Bull. Cl. Sci. Acad. R. Belge* **56**, 505.

Kato, T. (1966). *Perturbation Theory for Linear Operators* (New York, Springer).

Petrosky, T. and Prigogine, I. (1991). *Physica A* **175**, 146.

Petrosky, T. and Prigogine, I. (1997). *Advances in Chemical Physics* **XCIX**, ed. I. Prigogine and S. A. Rice (New York, Wiley).

Prigogine, I. (1962). *Non-equilibrium Statistical Mechanics* (New York, Wiley).

Prigogine, I. (1997). *The End of Certainty* (New York, Free Press).

Prigogine, I., George, C., Henin, F. and Rosenfeld, L. (1973). *Chemica Scripta* **4,** 5.

Van Hove, L. (1955). *Physica* **21**, 801–923.

Van Hove, L. (1962). *Fundamental Problems in Statistical Mechanics*, ed. E. G. D.Cohen (Amsterdam, North-Holland).

Zwanzig, R. (1965). *Physica* **30**, 1109.

19

Quantum transport with tunneling and reservoir ballistic transport

19.1 Introduction

In all the previous discussions of transport (in Chapters 4, 5, 6, 15 and 16), we have been dealing with the small Knudson number regime, $K = l/d$ (Cercignani, 1969; Kogan, 1969). l is the mean free path between collisions, and d is a system size parameter. Here the major source of irreversibility and the impedance to transport have been internal system collisions. The reservoirs have played a lesser role in this aspect of these discussions. This is also true of the quantum situation.

The reservoirs become more important in the intermediate Knudson regime, and for large l, the collisions in the system become increasingly less important. Classically, the linearized Boltzmann equation (see Cercignani, 1969) has been utilized to discuss this. Much less has been done quantum mechanically from this point of view. The case $K \to \infty$ corresponds to free or ballistic motion. Qualitatively speaking, the Knudson number scales the left-hand noncollision part to the collision term in the Boltzmann kinetic picture. In the large Knudson regime, the character of the boundaries becomes important. In gases near the wall, the thermodynamic constitutive equations do not hold, and in this case there is a formation of the Knudson "layer" and a reduction there in the viscosity. In the pure ballistic regime, there are no local hydrodynamic equations at all.

Recently, with the advent of nanoscience and its technology in condensed matter, the transport in systems of few electrons (molecules) has become an important problem (Datta, 1995, 2005). The discussion of the nanotechnology is not the point here. A recent good reference is the book by Ferry and Goodnick (1997).

R. Landauer was apparently the first to discuss electron ballistic transport in semiconductors. Utilizing a simple model and the ideas of one-dimensional quantum scattering, he took the electrical conductance σ to be

$$\sigma = \frac{e^2}{h}\left(\frac{T}{R}\right), \tag{19.1}$$

where T is the 1–D transmission coefficient of free electrons between two randomizing reservoirs. R is the reflection coefficient (Landauer, 1970). This point of view was extended by Buttiker (1986). A review of this simple scattering point of view is given by Stone and Szafer (1988). They discuss the controversy. In his book, in Chapter 2, Datta (1995) gives extensive discussion of the physical aspects of this and completely ignores the many particle aspects. It is the many particle aspects that we wish to take up here and in the next section, where we will consider the Keldysh Green's function approach to the transport current of electrons through a tunneling junction. The electrons will not be ballistic in the region between the reservoirs. This will give an all-order perturbation theory via the appropriate contour time Dyson equation. An expression for the time-dependent current will be obtained. The purpose is to illustrate the Keldysh perturbation theory as well as to obtain a generalization of Landauer's formula.

19.2 Pauli equation and boundary interaction

The dissipative quantum Pauli equation for a system interacting with reservoirs was derived in Chapter 3, Eq. (3.50) (Peier, 1972):

$$\begin{aligned}\frac{\partial \rho_{snn}(t)}{\partial t} &= -2\pi \sum_{m} \left| H'_{smn} \right|^2 \delta\left(E_n^0 - E_m^0\right) \left[\rho_{snn}(t) - \rho_{smm}(t)\right] \\ &\quad - 2\pi \sum_{m} \sum_{\alpha\beta} \left| H'_{sRn\alpha m\beta} \right|^2 \delta\left(E_n^0 + E_\alpha^0 - E_m^0 - E_\beta^0\right) \\ &\quad \left[\rho_{R\alpha\alpha}(0)\, \rho_{snn}(t) - \rho_{R\beta\beta}(0)\, \rho_{smm}(t)\right]; \\ t &\geq 0.\end{aligned} \tag{19.2}$$

It is the second terms which we will utilize here. Recall that the equation is exact in the singular Van Hove limit, $\lambda \to 0, t \to \infty$; $\lambda^2 t$ is finite. In this case λ characterizes the strength of H'_{sR}. It is a time asymptotic equation for the diagonal elements of the system density matrix $\rho_{snn}(t)$. We will take the system to be a 1–D free non-interacting system of electrons (ballistic). Thus, $|n\rangle \equiv |k\rangle$, $k = (2\pi n)/d$, $n = 0, 1, \ldots$, $H_{snn'} = 0$. The Knudson number is infinite.

Initially, the boundary condition is

$$\rho_0\,(t=0) = \rho_s\,(t=0)\, \rho_R\,(t=0)\,. \tag{19.3}$$

The diagonal elements of the initial reservoir states, $\rho_{R\alpha\alpha}(0)$, are influential at all time. In the ballistic case the total irreversible behavior comes from the system–reservoir interaction, $H'_{sRn\alpha m\beta}$. (Irreversibility and dissipation have been discussed in Chapters 5 and 6.) We further assume that $\left[\rho_R(0)\,, H_R\right] = 0$.

The reservoir is further characterized by being described in a macroscopic thermodynamic limit, $L \to \infty$, $N \to \infty$, $N/L =$ constant. N is here the number of reservoir electron degrees of freedom, and L the size. As first pointed out by Van Hove (1955, 1956, 1957), using perturbation theory, in this limit in Hilbert space a diagonal singularity in $\left(\lim R^0 L_{int} \ldots L_{int}\right)_{\nu\nu}\left(N, N'\right)$ appears. This was discussed in Chapter 16, where we used a direct method of choosing continuum states to deal with this problem. Suffice it to say that here we may use perturbation methods to deal with continuum Hilbert space matrix elements appearing in the thermodynamic limit. This limit eliminates Poincaré recurrences in the system reservoirs, as we shall see.

We will take the two reservoirs as incoherent, being sufficiently spacially separated. Call the two reservoirs l and r, left and right:

$$\begin{aligned} H_R &= H_r + H_l \\ [H_r, H_l] &= 0. \end{aligned} \tag{19.4}$$

Thus, in Eq. (19.2), $\alpha = (l, r)$, and

$$\begin{aligned} H &= H_s^0 + H_l + H_r + H'_{sR} \\ H'_{sR} &= H'_{sl} + H'_{sr}. \end{aligned} \tag{19.5}$$

We might think of the reservoir system interactions to be of the approximate tunneling form (see Datta, 1995):

$$\begin{aligned} H_T &= \sum_{kp\sigma} \left[T_{kp} a^{\dagger}_{k\sigma} a_{p\sigma} + h.c.\right] \\ N_k &= a^{\dagger}_k a_k. \end{aligned}$$

The precise details of H'_{sl} or H'_{sr} do not concern us except that they are short range compared with d and weak ($\lambda \to 0$). Since this is a one-dimentional problem, $H_s^0 = p_k^2/2m$, we will integrate around $k, -k$ since there is present, implicitly, an energy-conserving delta function. Thus, the relevant system diagonal density matrix contributions are integrated around ρ_{kk}, ρ_{-k-k}. We take the left equilibrium reservoir to be Fermi,

$$\rho_{ll}(0) \to f_l(E_l) = \left(\exp\beta\left(E_l + \mu_l\right) + 1\right)^{-1} \tag{19.6a}$$

and the right,

$$\rho_{rr}(0) \to f_r^0(E_r + eV) = \left(\exp\beta\left(E_l + \mu_l + eV\right) + 1\right)^{-1}. \tag{19.6b}$$

The chemical potential is shifted by a voltage parameter. We will not discuss its external measurement but just assert a shift in the chemical potential between the left and right reservoirs.

We may write

$$\rho_R(0) = \rho_l(0)\,\rho_r(0) \qquad (19.7)$$
$$\mathrm{Tr}_l\rho_R(0) = \rho_r(0) \ \text{ etc.,}$$

so

$$\langle r\,|\rho_r|\,r'\rangle = f(E_r)\,\delta_{rr'} = f(E_l + eV)\,\delta_{rr'} \equiv \rho_r(0)\,\delta_{rr'}$$

and

$$\langle n\alpha\,|H'_{sR}|\,n'\alpha'\rangle = \langle nl\,|H'_{sl}|\,n'l'\rangle\,\delta_{rr'} + \langle nr\mid H'_{sr}\mid n'r'\rangle\,\delta_{ll'}.$$

Using these assumptions we have for the left–right equilibrium reservoir–system interaction

$$\begin{aligned}\dot{\rho}_{skk}(t) = {}& -2\pi\sum_{k'l'}\left|H'_{sRklk'l}\right|^2\delta\left(E^0_k - E^0_{k'}\right)\\ & \times\left[\rho_l(0)\,\rho_{skk}(t) - \rho_{l'}(0)\,\rho_{sk'k'}(t)\right]\\ & - 2\pi\sum_{k'r}\left|H'_{sRkrk'r}\right|^2\delta\left(E^0_k - E^0_r\right)\\ & \times\left[\rho_r(0)\,\rho_{skk}(t) - \rho_r(0)\,\rho_{sk'k'}\right].\end{aligned} \qquad (19.8)$$

We have assumed no correlations between the left and right reservoirs; they are independent. Now $H'_{kr,k'r'}$ is invariant under $k, r;\ \ k'r' \to kl;\ \ -k'l'$ and independent of the volume V. The right–left transition rate of l from $k \to k'$ is the same as the right–left of r between k' and $-k$. This is a form of detailed balance.

We assume, further, that the interaction $H'_{sRklk'l'}$ has a resonance or sharp peak at $E_k = E_l$ and $E_k = E_r$. Thus, the dominant contribution of the reservoir is at $f_l(E_k)$ and $f_l(E_k + eV)$. Because the interactions at the two reservoirs are taken to be the same, we have

$$\frac{d\rho_{kk}(t)}{dt} = -2\pi\sum_{k'l}A_{kl;\,k'l}\delta(E_l - E_k)\left[f(E_k)\,\rho_{kk}(t) - \rho_{-k'-k'}f(E_{k'} + eV)\right], \qquad (19.9)$$

where

$$A_{klk'l'} = 2\pi\left|H'_{sRkl,\,k'l'}\right|^2\delta(E_k - E_{k'}).$$

This is the gain–loss Pauli equation for an electron in free state $|k\rangle$. The gain–loss is due to the reservoir's interaction appearing naturally in Eq. (19.9). All dissipative effects are due to this interaction of the macroscopic reservoir pair in thermal equilibrium with differing chemical potential coming from their uncorrelated initial condition, $\rho_R(0) = \rho_r(o)\,\rho_l(0)$.

We remind the reader that the thermodynamic limit is implicit here, since it is necessary to go to the continuum limit ($N \to \infty,\ \ V \to \infty,\ \ N/V =$ constant) to evaluate the delta function.

19.3 Ballistic transport

We identify the overall transition rates as

$$W_{l,r} = 2\pi \sum_{l'k'} A_{kl;-k'l}\delta\left(E_l - E_k\right) f\left(E_l\right) \tag{19.10}$$

$$W_{r,l} = 2\pi \sum_{r'k'} A_{kr;\ -k'r}\delta\left(E_r - E_{k'}\right) f\left(E_r + eV\right).$$

We have again made explicit the energy conservation law between the left and right electron reservoir in this order of perturbation theory. To higher orders, line broadening may appear (see Appelbaum and Brinkman, 1969). We note that $W_{l,r} \neq W_{r,l}$.

A global equilibrium, $\rho_{kk} = \rho_{-k,-k} =$ constant, is achieved only when $eV = 0$ and thus $W_{l,r} = W_{r,l}$. The chemical potentials of the reservoirs are equal, and $\beta_l = \beta_r$.

For the steady state (assumed) flow,

$$\dot{\rho}_{kk} = I_k, \tag{19.11}$$

and the current to the right is $J_k = -eV_k I_k$. The *net* current to the right is

$$J = -eV_k\left(I_k - I_{-k}\right) = \text{constant}. \tag{19.12}$$

The total right flow is $J = \sum_k J_k$, where V_k is the particle velocity in state k.

The entropy behavior was discussed in Chapter 6. It was shown (see Eq. 6.80) for equations of the form of Eq. (19.2) that

$$S_c\left(\frac{P_t p_i}{p_i^*}\right) \geq S_c\left(\frac{p_i\left(t\right)}{p_i^*}\right) \qquad t > 0,$$

where S_c is the conditional entropy. P_t is a Markov operator. $p_i^* = P_t\ p_i$ is the steady solution. $S_c = 0$. Thus the dissipative evolution of the reservoir system without internal system interaction is proved. Such systems as this simple model are dissipative and irreversible. Further, there is a heat flow into the system from the reservoirs:

$$J = \frac{d}{dt}\text{Tr}_R\rho_S\left(t\right) \cdot \log z^{-1} \exp\left(-\beta H_R\right).$$

Because of this, assuming a steady state (not proved), Spohn and Lebowitz (1978) showed that there is a time averaged entropy production, and indicated conditions for the validity of the Onsager symmetry, $L_{kj} = L_{jk}$. See Chapter 6 again for more details. We must say that there is no rigorous proof to date of the existence of such steady states.

Considering Eq. (19.9) and Eq. (19.10), we assume

$$J = \sum_k J_k$$

is constant. Thus, since $A_{kl,-k'l} = A_{-kl,k'l}$, the net steady current becomes

$$\begin{aligned} J &= \sum_k J_k = -e \sum_{kk'} V_k \left(\dot{\rho}_{kk} - \dot{\rho}_{-k'-k'} \right) \\ &= 2\pi \sum_{kl,k'l} V_k A_{kl,k'l} \left(\rho_{kk} - \rho_{-k'-k'} \right) \left[f\left(E_k\right) - f\left(E_{k'} + eV\right) \right]. \end{aligned} \tag{19.13}$$

Now

$$\begin{aligned} N\left(E_k\right) &= \left(\rho_{kk} - \rho_{-k'-k'} \right) f\left(E_k\right) \delta\left(E_l - E_k\right) \\ N\left(E_{k'} + eV\right) &= \left(\rho_{kk} - \rho_{-k'-k'} \right) f\left(E_{k'} + eV\right) \delta\left(E_l - E_{k'} - eV\right) \end{aligned} \tag{19.14}$$

are the net "to the right" and "to the left" distributed particle density in the right and left reservoirs. Thus,

$$J = -2\pi e \sum_{kl,k'l} V_k A_{kl,\,-k'l'} \left[N\left(E_k\right) - N\left(E_{k'} + eV\right) \right]. \tag{19.15}$$

The current depends on the difference in the chemical potentials of the two separated *reservoir boundaries which are Fermi distributions*. It is zero if the potential is zero, $V = 0$.

In the classical limit, the rate would reduce to the particle velocity in state $|k\rangle$. If we expand to lowest order in eV, we may then define a conductance coefficient. We have

$$J = 2\pi \sum_{kl} V_k A_{kl,-k'l} \frac{\partial N\left(E_k + eV\right)}{\partial E_k} |_{V=0}\ V. \tag{19.16}$$

In going to the evaluation of δ functions on energy of the reservoirs, we have, in the thermodynamic limit,

$$\sum \rightarrow \frac{L}{2\pi} \int dl.$$

L is the reservoir length. We may formally perform these integrals and obtain

$$J = \frac{e}{2\pi} L \sum_{kk'} V_k A_{k,-k'} \frac{\partial N\left(E_{k'} + eV\right)}{\partial E_k} |_{V=0}\ V,$$

where

$$V_k A_{k,-k'} = \int dl \; V_k A_{kl,-k'l} \tag{19.17}$$

is the reduced transition rate between states $|k\rangle$ to $|k'\rangle$. Now we introduce density $N = Ln$. We have then, in the limit $N \to \infty$, $L \to \infty$,

$$J = \sigma V, \tag{19.18}$$

where the conductance is

$$\sigma = \frac{e}{2\pi} \sum_{kk'} V_k A_{k,-k'} \frac{\partial n \left(E_{k'} + eV\right)}{\partial E_{k'}} |_{V=0}, \tag{19.19}$$

independent of L. Note that no electric field between the reservoirs has been introduced, just a difference in chemical potentials. Now Eq. (19.17) gives $A_{k-k'}$. The reservoir thermodynamic limit is taken ($N \to \infty$, $L \to \infty$, $N/L = n$). We do not take the thermodynamic limit of the small system in state $|k\rangle$, as has already been emphasized. The discrete sums remain. There is an overall state energy conservation law, so $\delta\left(E_k - E_{k'}\right)$ remains. The transitions are between degenerate states, $k' = \pm k$, with $A_{kk} = 0$. Hence the linear conductance coefficient is

$$\sigma = \frac{e}{h} \sum_{k} V_k A_{k,-k} \frac{\partial n \left(E_k + eV\right)}{\partial k} |_{V=0}, \tag{19.20}$$

where

$$V_k = \frac{1}{\hbar} \left(\frac{\partial E_k}{\partial k} \right).$$

To summarize, resistance is due to the irreversible reservoir–system interaction here to lowest order λ^2. The reservoir is in thermodynamic equilibrium with a Fermi distribution. This illustrates the Knudson regime for few-particle transport. The Landauer notion is quite qualitatively correct. Such a formula may be carried rigorously to higher orders in perturbation. We will discuss the means to accomplish this in the next section.

19.4 Green's function closed-time path theory to transport

We shall now turn to an illustration of the diagrammatic perturbation theory of Keldysh, which was discussed in the previous chapter (Keldysh, 1965; Caroli *et al.*, 1971, 1972; Jauho *et al.*, 1994). At the same time we will consider tunneling transport, which is closely related to the previous section of this chapter. Here we will consider time-dependent theory and strong coupling by means of the Keldysh–Schwinger time path Green's functions.

We will follow closely the discussion of Jauho *et al.* See also the fine book of Ferry and Goodnick (1997). The two reservoirs (called leads) are time dependent, having been turned on at t' $\left(t' \rightarrow -\infty\right)$. The Hamiltonian is

$$H_R = \sum_{k,\alpha=l,r} \varepsilon_{k\alpha}(t)\, c_{k\alpha}^{\dagger} c_{k\alpha}. \tag{19.21}$$

The isolated reservoirs have time-dependent but independent Green's functions,

$$\begin{aligned} g^{<}\left(t,t'\right) &\equiv i\left\langle c_{k\alpha}^{\dagger}\left(t'\right) c_{k\alpha}(t)\right\rangle \\ &= if\left(\varepsilon_{k\alpha}^{0}\right)\exp\left[-i\int_{t'}^{t} d\tau\, \varepsilon_{k\alpha}(\tau)\right], \end{aligned} \tag{19.22}$$

with the equilibrium being established at t'. In the central region now occupied by electrons, $d_m^{\dagger} d_m = N_m^c$, so

$$H_c = \sum_m \varepsilon_m(t)\, d_m^{\dagger} d_m, \tag{19.23}$$

the electrons being in time varying states. The tunneling interaction is taken as

$$H_{Rc} = \sum_{k\alpha=l,r}\left[V_{k\alpha,n}(t)\, c_{k\alpha}^{\dagger} d_n + h.c.\right]. \tag{19.24}$$

We have here a possible simple model of quantum dot tunneling (Ferry and Goodnick, 1997).

The time-dependent electron current from the left lead to the center is

$$J_{\alpha=l} = \frac{-i\,|e|}{\hbar}\left\langle\left[H, N_l\right]\right\rangle.$$

H_l and H_c commute with this, and thus

$$J_l = \frac{+i\,|e|}{\hbar}\sum_{k,n}\left[V_{kl,n}\left\langle c_{kl}^{\dagger} d_n\right\rangle + V_{kl,n}^{*}\left\langle d_n^{\dagger} c_{kl}\right\rangle\right]. \tag{19.25}$$

We define two Green's correlation functions between the reservoir and the center:

$$\begin{aligned} G_{n,k\alpha}^{<}\left(tt'\right) &= i\left\langle c_{k\alpha,n}^{\dagger}\left(t'\right) d_n(t)\right\rangle \\ G_{k\alpha,n}^{<}\left(tt'\right) &= i\left\langle d_n^{\dagger}\left(t'\right) c_{k\alpha}(t)\right\rangle. \end{aligned} \tag{19.26}$$

Thus, the current is

$$J_l(t) = \frac{2e}{\hbar}\,\mathrm{Re}\sum_{kn} V_{kl,n}(t)\, G_{n,kl}^{<}(t,t). \tag{19.27}$$

We need, from diagrammatic analysis, equations of motion for the two-contour time-ordered Keldysh Green's functions. The derivation is given in an appendix of

Jauho's paper (Jauho *et al.*, 1994). Let us look at this analog to the equilibrium Dyson equations.

As proved by Rammer and Smith (1986), the contour-ordered Green's functions utilizing the Keldysh contour perturbation structure diagrams have the same topological structure as the equilibrium $T = 0$ diagrams. $\langle\rangle$ in Eq. (19.26) contains a contour-ordering operator T_c. This Keldysh idea orders operators with later time labels on the contour to the left of operators of an earlier time. With this, one is assured, by analogy with the $T = 0$ equilibrium theory, that a diagrammatic perturbation re-summation may achieve a Dyson equation. However, because the Keldysh Green's functions are matrices, as discussed in the previous chapter (the elements of which are not linearly independent), rules of multiplication are necessary. These rules have been given by Langreth (1976) for some cases and are commonly employed.

Let us briefly describe these rules. There are products in the time contour integrations of the form $C = \int AB$ for which the following prescription holds:

$$C_r\left(t,t'\right) = \int d\tau A_r\left(t,\tau\right) B_r\left(\tau,t'\right) \tag{19.28}$$

$$C^<\left(t,t'\right) = \int d\tau \left[\begin{array}{c} A_r\left(t,\tau\right) B^<\left(\tau,t'\right) \\ +A^<\left(t,\tau\right) B_r\left(\tau,t'\right) \end{array} \right]. \tag{19.29}$$

Similar expressions are used for C_a and $C^>$.

Now we begin with the equations of motion for the $T = 0$ time-ordered Green's functions in the intermediate region:

$$G_{n,k\alpha}\left(t-t'\right) = -i\left\langle T\left\{d_m^\dagger\left(t'\right) d_n\left(t\right)\right\}\right\rangle$$
$$\text{(not } T_c\text{)},$$

which is simply the closed equation

$$-i\frac{\partial}{\partial t'} G_{n,k\alpha}\left(t-t'\right) = \varepsilon_k G_{n,k\alpha}\left(t-t'\right) + \sum_m V_{k\alpha,m}^* G_{nm}\left(t-t'\right). \tag{19.30}$$

Because the reservoirs are non-interacting, the hierarchy is closed at this simple equation. To go to higher orders, we must use coupling to more complicated Green's functions (as discussed in the previous chapter). By defining

$$G_{n,k\alpha} g_{k\alpha}^{-1} = \sum_m G_{nm} V_{k\alpha m}^*,$$

we have the integral equation, by construction:

$$G_{n,k\alpha}\left(t-t'\right) = \sum_m \int d\tau G_{nm}\left(t-\tau\right) \times V_{k\alpha,m}^* g_{k\alpha}\left(t-t'\right). \tag{19.31}$$

Now we generalize this to the Keldysh complex time closed contour, as follows:

$$G_{n,k\alpha}\left(\tau,\tau'\right)=\sum_m\int d\tau_1 G_{nm}\left(\tau,\tau_1\right)\ V^*_{k\alpha,m}\left(\tau_1\right)g_{k\alpha}\left(\tau_1\tau'\right). \tag{19.32}$$

The product form is apparent on the right. We use this for the $<$ function with rule Eq. (19.29) to write

$$\begin{aligned}G^<_{n,kl}\left(tt'\right)=&\sum_m\int d\tau_1 V^*_{kl,m}\left(\tau_1\right)\\&\times\left[G^r_{nm}\left(t\tau_1\right)g^<_{kl}\left(\tau_{1,}t'\right)+G^<_{mn}\left(t,\tau_1\right)g^a_{kl}\left(\tau_1,t'\right)\right].\end{aligned} \tag{19.33}$$

With this we may obtain an expression for the current $J_l(t)$, combining Eq. (19.27) and Eq. (19.33), utilizing the initial values $\left(t'=-\infty\right)$. We have

$$J_l\left(t\right)=-\frac{2\left|e\right|}{\hbar}\int_{-\infty}^t d\tau_1\int\frac{dE}{2\pi}\,\mathrm{Im\,Tr}\left\{\begin{array}{c}\exp\left(-iE\left(\tau_1-t\right)\right)\times\Gamma_l\left(E,\tau_1,t\right)\\\times\left[\mathbf{G}^<\left(t,\tau_1\right)+f_l\left(E\right)\mathbf{G_r}\left(t,\tau_1\right)\right]\end{array}\right\}. \tag{19.34}$$

Here we have taken the continuum limit of the reservoir, as in the earlier discussion,

$$\sum_{kl}\to\int dE\rho_l\left(E\right), \tag{19.35}$$

and have defined

$$\Gamma_{l,mn}\left(E,t',t\right)=2\pi\rho_l\left(E\right)V_{kl,n}\left(t\right)V^*_{kl,m}\left(t'\right)\times\exp\left(\frac{i}{\hbar}\int_{t'}^t d\tau_1 E_{kl}\left(\tau_1\right)\right), \tag{19.36}$$

the level width function. $\mathbf{G}^<$ and $\mathbf{G}_r$ are Keldysh matrices of the central region, the dynamics of which are not yet determined. The second term in Eq. (19.34) is interpreted as the "out" rate, and the first as the "in." These equations are irreversible. This arises from the macroscopic reservoir limit.

For the time-independent steady case, $G^<$ and G_r are functions of τ_1-t, and Γ_l must be assumed time-independent, assuming this with appropriate potential modulation. The integral on $d\tau_1$ may be done immediately. The time-independent current is

$$J_l=\frac{-i\left|e\right|}{\hbar}\int\frac{dE}{2\pi}\mathrm{Tr}\left\{\Gamma_l\left(E\right)\left[G^<\left(E\right)+f_l\left(E\right)G_r\left(E\right)-G_a\left(E\right)\right]\right\}, \tag{19.37}$$

and we obtain a similar result for the right current J_r with $l\to r$. Now, as the steady state is approached in time, which is *not* proved but assumed, we have $J=J_l=-J_r$, and using $2J=J_l-J_r$, we obtain

$$\mathbf{J}=\frac{i\,|e|}{2\hbar}\int dE\,\mathrm{Tr}\left\{\begin{array}{c}[\Gamma_l(E)-\Gamma_r(E)]\,\mathbf{G}^<(E)\\+[f_l(E)\,\Gamma_l(E)-f_r(E)\,\Gamma_r(E)]\\\times[\mathbf{G}_r(E)-\mathbf{G}_a(E)]\end{array}\right\}. \tag{19.38}$$

Eq. (19.38) appears to be the all-order non-equilibrium steady state generalization of the Landauer idea. For applications to time-dependent situations, the student is urged to consult the paper of Jauho (Jauho *et al.,* 1994).

References

Appelbaum, J. A. and Brinkman, W. F. (1969). *Phys. Rev.* **186**, 464.

Buttiker, M. (1986). *Phys. Rev. Lett.* **57**, 1761.

Caroli, C., Conboscott, R., Nozieres, P. and Saint James, D. (1971). *J. Phys. C.*; solid state vol. **4**, 916.

Caroli, C., Conboscott, R., Nozieres, P. and Saint James, D. (1972). *J. Phys. C.*; solid state vol. **5**, 21.

Cercignani, C. (1969). *Mathematical Methods in Kinetic Theory* (New York, Plenum).

Datta, S. (1995). Electronic Transport in Mesoscopic Systems. (Cambridge, Cambridge University Press).

Datta, S. (2005). *Quantum Transport* (Cambridge, Cambridge University Press).

Ferry, D. K. and Goodnick, S. M. (1997). *Transport in Nanostructures* (Cambridge, Cambridge University Press).

Jauho, A. P., Wingreen, N. S. and Meir, Y. (1994). *Phys. Rev. B* **50**, 5528.

Keldysh, L. V. (1965). *Sov. Phys. J. Exp. Theor. Phys.* **20**, 1018.

Kogan, M. N. (1969). *Rarefied Gas Dynamics* (New York, Plenum).

Landauer, R. (1970). *Phil. Mag.* **21**, 863.

Langreth, D. C. (1976). In *Linear and Nonlinear Electron Transport in Solids* **17** (New York, Plenum).

Peier, W. (1972). *Physica* **57**, 229.

Rammer, R. and Smith, H. (1986). *Rev. Mod. Phys.* **59**, 2291.

Spohn, H. and Lebowitz, J. L. (1978). *Adv. Chem. Phys.* **38**, 109, ed. S. A. Rice and I. Prigogine (New York, Wiley).

Stone, A. D. and Szafer, A., (1988). *IBM J. Res. Develop.* **33**, 384.

Van Hove, L. (1955). *Physica* **21,** 517, 901.

Van Hove, L. (1956). *Physica* **22**, 343.

Van Hove, L. (1957). *Physica* **23**, 441.

20

Black hole thermodynamics

20.1 Introduction to black holes

In 1783 John Mitchell wrote, “If the semi-diameter of a sphere of the same density as the sun were to exceed that of the sun in the proportion of five hundred to one, and supposing light to be attracted by the same force in proportion to its vis-inertia with other bodies, all light emitted from such a body would be made to return towards it, by its own gravity” (Mitchell, 1783). Much later, in a prophetic paper, Oppenheimer and Snyder (1939) described the nature of “continued gravitational contraction” of a neutron star. With the nuclear heat gone, the core of the dead star becomes incapable of supporting itself under its own gravitational pull. The final phase is that the high density of the remaining core prevents the escape of the last light. The star disappears from view. Wheeler, later, coined the term *black hole* for such an object in the cosmos (Misner *et al.,* 1973).

What is most remarkable is that today astronomers/astrophysicists have identified, with modern technical skills, numbers (uncountable) of these black holes. There seems no empirical doubt as to their existence. See the incredible visual treat in the volume *The Universe,* edited by Martin Rees (2005). Frolov and Novikov (1998) have given a condensed list of objects, eleven in number, which are binary systems that contain black holes. This comes from X-ray studies of binaries. As pointed out by them, the central arguments for the existence of black holes are: (a) the emission has a compact nature, and (b) the emission makes possible the analysis of the orbital motion, and one obtains the mass of the compact partner. If it is of the order of three solar masses, it is a black hole. See the resultant discussion of Cherepaschuk (1996). The strongest black hole candidates are three in number: GS2023+338, GS2000+25 and XN oph 1997. The first has a period of 6.5 days, a mass of the compact companion is of the order of 10 solar masses, and its luminosity is 6×10^{38} erg/sec.

There is more dramatic evidence for supermassive black holes in galactic centers. In what are called active galactic nuclei, great quantities of energy are emitted from the galactic nuclei in the form of giant jets (luminosity of 10^{47}erg/ sec). Quasars are an example, emitting total energy a hundred times all the other energy in a large galaxy. Estimates of the quasar mass are $1 - 100 \times 10^7$ solar masses, and only a few light-hours in dimension. The Milky Way has an example of a dormant black hole of 3×10^6 solar masses with accretion of 10^{-8} solar masses per year. Also, M31 with 2×10^7 solar masses exists nearby in Andromeda.

All this is quite exciting, but it is not our purpose to review it further, except to say that Einstein's theory of general relativity (Einstein, 1915a, 1915b; Wald, 1984; Rees and Hawking, 1997) gives the prediction of classical black holes. The spherically symmetric solution to Einstein's equation was obtained by Schwarzschild (1916a, 1916b). The solution is

$$d^2s = -\left(1 - \frac{2GM}{c^2r}\right)c^2dt^2 + \left(1 - \frac{2GM}{c^2r}\right)^{-1} d^2r \qquad (20.1)$$
$$+ r^2\left(d\theta^2 + \sin^2 d\theta d\phi^2\right).$$

In this equation, G is the Newtonian gravitational constant, and M the mass of the field source. d^2s is the metric, the solution. t, r, $\theta\phi$ are the Schwarzschild reference frame. In a local Cartesian coordinate system, infinitesimally,

$$\delta x = \left(1 - \frac{2GM}{c^2r}\right)^{-\frac{1}{2}} dr \qquad (20.2)$$
$$\delta y = r d\theta$$
$$\delta z = r \sin\theta d\phi,$$

and the local time

$$d\tau = \sqrt{-g_{00}}dt = \left(1 - \frac{2GM}{c^2r}\right)^{\frac{1}{2}} dt. \qquad (20.3)$$

The free-fall acceleration is

$$a = \sqrt{a^i a^k h_{ik}},$$

where

$$h_{ik} = g_{ik} - \frac{g_{0i}g_{0k}}{g_{00}},$$

and we obtain

$$a = \frac{GM}{r^2\left(1 - \frac{2GM}{c^2r}\right)^{\frac{1}{2}}} \qquad (20.4)$$

along the radius toward the center. The acceleration approaches infinity at $r \equiv r_g = 2[GM/c^2]$, the Schwarzschild radius, in this reference frame. $r_g = 0.9$ cm for the earth and 3 km for the sun. In the Schwartzschild coordinates there are two singularities, $r = r_g$ and $r = 0$. The question is, are they the result of the coordinate choice, or are they physical? We will turn to this shortly. We should, of course, mention that the analysis of the $r > r_g$ solution led to the famous tests of general relativity, the gravitational red shift, precession of the planetary orbits, bending of light, and time delay of radar signals, all of which have been verified. We will not repeat these calculations but refer the reader to the book of Wald (1984).

Our purpose is to find the black holes in the solution, which means we must examine the $r < r_g$ region. We must obtain a description valid *inside* the Schwarzschild sphere. To do this, we will use the Lemaitre reference frame. We choose a reference frame of freely falling particles, with no infinite accelerations, and choose the frame which has zero velocity at spacial infinity. The time coordinate, T, is taken to be a clock fixed to the falling particles. The time of fall from r_1 to r is

$$\Delta T = \frac{2}{3}\left(\frac{r_g}{c}\right)\left[\left(\frac{r_1}{r_g}\right)^{\frac{3}{2}} - \left(\frac{r}{r_g}\right)^{\frac{3}{2}}\right]. \tag{20.5}$$

At $T = 0$ the freely falling ensemble of particles is located at r_1. These are the new radial coordinates of the new frame. The metric may be written

$$ds^2 = -c^2 dT^2 + \frac{dR^2}{B} + B^2 r_g^2 \left(d\theta^2 + \sin^2\theta d^2\phi\right), \tag{20.6}$$

where

$$B = \left[\left(\frac{r_1}{r_g}\right)^{\frac{3}{2}} - \frac{(3cT)}{(2r_g)}\right]^{\frac{2}{3}}, \tag{20.7}$$

and

$$R = \frac{2}{3} r_g \left(\frac{r_1}{r_g}\right)^{\frac{3}{2}} \tag{20.8}$$

is the scaled radial coordinate. The Lemaitre reference frame has eliminated the singularity at $r = r_g$. The frame extends to $r < r_g$, and at $r = r_g$, $B = 1$; $r_g = \frac{3}{2}(R - cT)$.

Comparing motion without the Schwartzschild sphere, we find that particles in the future move to $r = \infty$, whereas in the Lemaitre coordinates they move within the sphere from r_g to the singularity $r = 0$, never outside the sphere. They are invisible outside. This is the Lemaitre description of a black hole. There is some difficulty with this description. However, we will use it for simplicity (see Frolov and Novikov, 1998).

Other coordinates are possible. Wald discusses the Kruskal extension and the geometry of the black and white hole picture obtained (Kruskal, 1960; see Wald, 1984). In the T, X plane there are four regions: I, II, III, IV. The radial null geodesics are 45° lines separating them. For $r > 0$, $X^2 - T^2 > -1$, and $r > r_g$ is region I, corresponding to the original Schwarzschild picture. The singularity $r = 0$ exists both in region II in the future and in region III in the past. A particle (observer) falling into region II (from region I) cannot escape but falls into $r = 0$. Region II is the *black hole*. Region III is delineated by the line $r = r_g, t < -\infty$. A particle within III must, in finite time to the future, leave III, called a *white hole,* and go into region IV, which is a Schwarzschild region also. The Kruskal metric is a spherically symmetric vacuum solution to Einstein's equations. The reader should consult appendix B in the book of Frolov and Novikov (1998) for the proof.

The preceding solution is the vacuum solution, but matter may be included with a pressure of $T_{\alpha\beta}$ in the Einstein equations. The simple model solution is due to Tolman (1934). With this, we can describe the black hole formation due to gravitational collapse. Tolman considered a spherical relativistic dust cloud with zero hydrodynamic pressure. The dust particles move along geodesics. In a co-moving reference frame, with constant R, θ, ϕ, Tolman assumed

$$ds^2 = -c^2 dT^2 + g_{11}(T, R)\, dR^2 + r^2(T, R)\left(d\theta^2 + \sin^2\theta\, d\phi^2\right) \tag{20.9}$$

with

$$\dot{r}^2 = f(R) + \frac{F(R)}{r} \tag{20.10}$$

$$g_{11}(T, R) = \frac{(r')^2}{1 + f(R)}$$

$$\frac{8\pi G\rho}{c^2} = \frac{F'(R)}{r' r^2}.$$

"Prime" indicates R differentiation. $f(R)$ and $F(R)$ are arbitrary and determined by initial conditions at T_0. $R = 0$ is the cloud center with $\dot{r}(0, T) = 0$ with R as the boundary of which $F(0) = 0$ follows. $r(R)$ is monotonic and positive. Thus, $F(R) \geq 0$. The first equation of Eq. (20.10) gives $\ddot{r} = -F/2r^2$. Thus, $\ddot{r}$ is negative, and hence all dust particles with fixed R and $\dot{r} < 0$ reach the true singularity $r = 0$, never leaving the sphere $r = r_g$. This is gravitational collapse of matter into the center of the black hole.

With these introductory remarks, let us turn to the topic of this chapter, the remarkable thermodynamic analogy of the black hole description.

20.2 Equilibrium thermodynamic analogies: the first law

Bekenstein (1972, 1973) first noticed that the area of the event horizon of a black hole, A, has a similarity to thermodynamic entropy, S. The area of a Schwarzschild black hole is

$$A = 4\pi r_g^2. \tag{20.11}$$

A new result, obtained in Austin, was the Kerr solution (Kerr, 1963), which introduced angular momentum J and has an event horizon radius

$$r = r_+ = M + \sqrt{M - a^2}. \tag{20.12}$$

Here we adopt the relativistic units $c = G = 1$, where $a = J/M$. M is the black hole mass (see Frolov and Novikov, 1998, for details of the Kerr solution). The event horizon area in this case is

$$A = \int d\theta d\phi \sqrt{g_{\theta\theta} g_{\phi\phi}} = 4\pi \left(r_+^2 + a^2\right). \tag{20.13}$$

The area may be seen to be a function of the parameters M and J or by inverting and writing

$$M(A, J) = \left[\frac{\pi}{A}\right]^{\frac{1}{2}} \left[\left(\frac{A}{4\pi}\right)^2 + 4J^2\right]^{\frac{1}{2}}. \tag{20.14}$$

An infinitesimal change in A and J leads to an equation for the change of mass dM. We write

$$dM = \frac{k}{8\pi} dA + \Omega^H dJ, \tag{20.15}$$

where

$$k = \frac{4\pi \sqrt{M^2 - \frac{J^2}{M^2}}}{A} \tag{20.16}$$

$$\Omega^H = \frac{4\pi J}{MA}. \tag{20.17}$$

Ω^H is the angular velocity. k is the surface gravity. It is the strength of the gravitational field on a black hole event horizon surface, evaluated by a distant observer. (See Frolov and Novikov, Chapter 6, for a considerable discussion, the details of which we do not need here.) For us it is a constant surface property and a black hole parameter. General derivations of Eq. (20.15) have been given by Bardeen, Carter and Hawking (Bardeen *et al.*, 1973; see also Wald, 1984, 1994). The further introduction of the parameter charge, Q, is possible, utilizing the Kerr–Newman metric.

Comparing Eq. (20.15) with the first law of static thermodynamics, it seems the following association is possible for the black hole, similar to Bardeen *et al.*, for energy E:

$$E \Leftrightarrow Mc^2. \tag{20.18}$$

Dimensionless entropy

$$S \Leftrightarrow \frac{A}{l_{pl}^2}, \tag{20.19}$$

where the Plank length is

$$l_{pl}^2 = \frac{\hbar G}{c^3}$$

and the Hawking temperature is

$$\Theta^H = k_B T^H = \frac{G\hbar}{2\pi c k_B} \cdot k, \tag{20.20}$$

or if $\hbar = c = K_B = G = 1$ (universal units), then

$$\theta^H = \frac{k}{2\pi}. \tag{20.21}$$

Eq. (20.15) then becomes

$$dE = \theta^H dS + \Omega^H dJ. \tag{20.22}$$

This is the analog of the first law of thermodynamics for black holes, governed by infinitesimal changes in the "macroscopic" thermodynamic parameters E, S, J. Before examining A and its analogy to entropy further, let us consider Einstein radiation theory to understand Θ^H. We naively quantize the black hole horizon to be in the "two" -level energy state $|\varepsilon\rangle$, $|g\rangle$. It is taken to be in equilibrium with its surroundings at $T^H = T_{\text{universe}}$.

Let $A_{g\varepsilon}$ be the *spontaneous* emission coefficient for the de-excitation from the excited mass state $|\varepsilon\rangle$. Also, assume induced emission $B_{\varepsilon g}\mu_\nu$ of Bose radiation. μ_ν is the radiation density. By the usual Einstein argument (Louisell, 1973), we assume thermal equilibrium between the black hole and surroundings and write

$$\left(A_{g\varepsilon} + B_{\varepsilon g}\mu_\nu\right)\exp\left(\frac{-E_\varepsilon}{\theta^H}\right) = B_{g\varepsilon}\mu_\nu \exp\left(\frac{-E_g}{\theta^H}\right). \tag{20.23}$$

We argue that $B_{\varepsilon g} = B_{g\varepsilon}$ where the transition rate is $W_{\varepsilon g} = B_{\varepsilon g}\mu_\nu$. By experiment, $W_{\varepsilon g} = W_{g\varepsilon}$.

The *assumed* black hole quantization gives

$$h\nu = \left(M_\varepsilon - M_g\right)c^2, \tag{20.24}$$

and from the equilibrium condition, Eq. (20.23),

$$\mu_{\nu} = \frac{\frac{a}{b}}{\exp\left(\frac{h\nu}{\theta^H}\right) - 1}. \tag{20.25}$$

In Eq. (20.25) a and b are parameters; this is Hawking's famous result (Hawking, 1975a; Parker, 1975).

Hawking's quantum S-matrix field calculation showed that baryon emission of a black hole followed a Plank formula, Eq. (20.25), with the temperature being θ^H. This fundamental result reinforces the interpretation of Eq. (20.20) as truly a macroscopic first law obeyed by the black hole. θ^H is the temperature, related through k, to the surface gravity of the event horizon. A detailed critique of this derivation has been given by Wald (1994).

The process is pair creation. This is possible in the processes which are termed the Hawking model:

(1) Particle 1, energy E, escapes to infinity, and particle $1'$ remains in the black hole.
(2) Particle 2 is captured and does not go to ∞, and $2'$ is created and remains in the hole.
(3) Particle 3, outside, is captured, and $3'$ remains in the hole.
(4) Particles 4 and $4'$ are created inside and remain there.

Thus, particle 1 appears as a spontaneous emission product at ∞. The Einstein argument is used to describe it. Hawking's calculation gives the result for a long time scale. Wald has estimated this as $GM/c^3 = 10^{-5} M/M_0^S$, which is rapid, even on galactic scales.

Bekenstein (Bekenstein and Mukhurov, 1995) has presented the picture of particle 1 passing through a potential barrier near the horizon and there, by interaction at the horizon, achieving the equilibrium state. He and others argue that the horizon area should be quantized in integers. He takes $A = \alpha\hbar n$, α being a pure number and n an integer. He assumes that the degeneracy factor is $g(n) = \exp\frac{\alpha(n-1)}{4}$ being integer, and so with $S = 0$ at $n = 1$, we have $\alpha = 4\ln l$, $l = 2, 3, 4, \ldots, f$. There is a recent article with references to this "atom black hole" approach by Mäkela (2003).

These views of the quantization are phenomenological and are really not part of the long and important history of quantum gravity. See the early reviews of these efforts in the books edited by Isham (Isham *et al.*, 1975, 1981). For more recent work using string theory of black holes, see the reviews of Maldacena (1996); Akhmedo (1997); and Horowitz (1995). A fine recent introduction to string theory with a chapter on black holes is in the book of Becker (Becker *et al.*, 2007). For a brief review, see Chapter 12 in the book by Frolov and Novikov (1998).

To obtain the entropy, the important task is to count the various string excitations. A comparatively simple example is the excitations of the two-dimensional D-branes, assuming that a nonzero area charged black hole may be described by these solitons of a single charge Q. The number of states in flat space-time was found to be $\exp\left(\frac{\pi Q^2}{4}\right)$. This gives the entropy $S = A/4G$ in four and five dimensions, a good answer, agreeing with Hawking and Bekenstein. However, the branes are said to be *extremal,* that is, they are configurations of the highest possible charge, as are the black holes that they are compared with. The extremal branes have the same properties as the black holes. This is interesting but not the complete theory one would desire.

It is beyond the focus of our brief remarks to say more. Certainly, true quantum statistical mechanics depends upon the success of an approach such as string theory. This is the reason to focus on thermodynamics in our comments about black holes. Black holes are possibly one of the most important tests of quantum gravity theories.

20.3 The second law of thermodynamics and black holes

Now let us turn to the classical analog to the second law of thermodynamics obtained by Hawking (1971). We will follow the short argument presented by Wald (1994). To follow this, consider the Raychauduri equation. A congruence of curves is a three-parameter family of curves $x^\mu\left(\lambda; y^i\right)$. y^i is a set of parameters which label the curves. One and only one curve passes through each point. λ is a parameter (proper time!) along each curve. The congruence of timelike curves is a reference frame. There are important properties of these curves. There is a representation

$$\mu_{\alpha;\beta} = \omega_{\alpha\beta} + D_{\alpha\beta} - \omega_\alpha \omega_\beta. \tag{20.26}$$

Here $\mu^\alpha = \frac{dx^\alpha}{d\lambda}\left(\mu^\alpha \mu_\alpha = -1\right)$, $\omega_\alpha = \mu^\beta \mu_{\alpha;\beta}$ is the acceleration, $\omega_{\alpha\beta}$ is the vorticity, and $\nabla_\alpha () = ()_{;\alpha}$, where

$$\omega_{\alpha\beta} = \frac{1}{2}\left(\mu_{\alpha\mu} P^\mu_\beta \quad \mu_{\beta;\mu} P^\mu_\alpha\right). \tag{20.27}$$

The rate of deformation tensor is

$$D_{\alpha\beta} = \frac{1}{2}(\mu_{\alpha;\mu} P^\mu_\beta + \mu_{\beta;\mu} P^\mu_\alpha), \tag{20.28}$$

where

$$P_{a\beta} = g_{\alpha\beta} + \mu_\alpha \mu_\beta$$

is a projection tensor onto the three-dimensional space perpendicular to μ^{α}. The expansion which concerns us here is

$$\theta = \mu^{\alpha}_{;\alpha} = \nabla_{\alpha}\mu^{\alpha}. \tag{20.29}$$

It is a trace, so we have

$$D_{\mu\nu} = \sigma_{\mu\nu} + \frac{1}{3}\theta P_{\mu\nu}. \tag{20.30}$$

The Raychaudhuri equation is

$$\frac{d\theta}{d\lambda} = \omega^{\alpha}_{;\alpha} + 2\left(\omega^2 - \sigma^2\right) - \frac{1}{3}\theta^2 - R_{\alpha\beta}\mu^{\alpha}\mu^{\beta} \tag{20.31}$$

(see Wald, 1984), where

$$\omega^2 = \frac{1}{2}\omega_{\alpha\beta}\omega^{\alpha\beta} \tag{20.32}$$
$$\sigma^2 = \frac{1}{2}\sigma_{\alpha\beta}\sigma^{\alpha\beta}.$$

$\sigma_{\mu\nu}$ is the shear, and $R_{\mu\nu}$ is the Ricci tensor.

For null geodesics, which we are considering here, we generate a null hypersurface, the event horizon. λ is then the affine parameterization of the generators of the event horizon. k^a is the tangent. The expansion is $\theta = \nabla_a k^a$. This is then the local rate of change of the cross section of the area as moved up the geodesic. Thus, $\theta = \frac{1}{A}\left(\frac{dA}{d\lambda}\right)$.

The Raychauduri equation for null geodesics is obtained by Wald (1984, p. 222):

$$\frac{d\theta}{d\lambda} = -\frac{1}{2}\theta^2 - \sigma_{ab}\sigma^{ab} + w_{ab}\omega^{ab} - R_{cd}k^c k^d \tag{20.33}$$

with $\omega_{ab} = 0$. The 1/2 appears because the space of interest is two-dimensional, in the case of null congruences.

The area theorem is immediate. We assume that the stress energy tensor in the Einstein equation satisfies $T_{ab}k^a k^b \geq 0$, and then we have $R_{ab}k^a k^b \geq 0$. Classically, the energy density is nonnegative. We obtain

$$\frac{d\theta}{d\lambda} = -\frac{1}{2}\theta^2. \tag{20.34}$$

Further, it may be proved (see Wald, 1994) that the null geodesics generating the future horizon cannot become infinite on that horizon. From Eq. (20.34), we have

$$\frac{d}{d\lambda}\theta^{-1} \geq \frac{1}{2}.$$

Hence,

$$\theta^{-1}(\lambda) \geq \theta^{-1}(0) + \frac{1}{2}\lambda. \tag{20.35}$$

If $\theta(0) > 0$ (expanding area), then $\theta(\lambda)$ expands for all time in the future. If $\theta(0)$ is negative, then there is a λ_1 such that $\theta^{-1}(\lambda_1) = 0$ or $\theta(\lambda_1) = \infty$, which is not possible by Eq. (20.35). Thus, for all positive time, the area of a black hole must be increasing:

$$\theta = \frac{1}{A}\left(\frac{dA}{d\lambda}\right) \geq 0. \tag{20.36}$$

This was first proved by Hawking (1971).

This result strongly reinforces, classically, the notion that the black hole area A is the *intrinsic* entropy S of the black hole, as has already been suggested by the first law of thermodynamics. The entropy principle indicates an intrinsic dissipation of black hole processes (classically). Moreover, we can associate with this increase the direction of time, λ (time's arrow). This is macroscopic, in contrast to the familiar discussions on a microscopic level (see Chapter 4 of this book). Further, as matter is lost into a black hole, the uncertainty is increased, as seen by the external observer, and thus the area of the event horizon increases. This is consistent with the Shannon information point of view. The information is indelibly lost into the black hole interior. The relationship of the inaccessible information with black hole entropy was first recognized by Bekenstein (1972).

We will close this section by remarking that the entropy of a black hole is enormous. An estimate is

$$S \approx k_B c^3 \hbar^{-1} G^{-1} A \approx 10^{60} \text{erg } K^{-1}$$

for a one-solar-mass black hole.

20.4 Extended entropy principle for black holes

The Hawking area theorem does not hold quantum mechanically, because the $T_{ab}k^a k^b > 0$ condition need not be true. It may be an approximation for a quantum system quasi-classically. We remind the reader of the difficulty of proving a quantum $\mathfrak{H}$ theorem (see Chapter 6). The details as to when the condition $R_{ab}k^a k^b \geq 0$ might be true have not been determined (see Wald 1994). This remains an open question. The radiation surrounding a black hole, plus the black hole itself, might be expected to obey an entropy principle

$$\Delta S = AS^H + \Delta S^{\text{rad}} \geq 0. \tag{20.37}$$

This is called the generalized second law. Bekenstein gave a number of examples which implied the validity of the generalized second law (Bekenstein, 1972, 1973).

Frolov and Page (1993) later gave a limited proof. It is this proof that we shall consider now. Zurek and Thorne (1985) suggested such an approach earlier.

Let the initial density matrix for the black hole and radiation (rather general and unspecified) be

$$\rho_{in} = \rho_{01} = \rho_0 \otimes \rho_1. \tag{20.38}$$

ρ_0 is the density matrix of the radiation (in up modes). ρ_1 is the density matrix of incident radiation from far away and in the past. They are uncorrelated fields (semiclassical).

ρ_{01} interacts with an eternal black hole classical curvature barrier separating the horizon from infinity. The final state after interaction is

$$\rho_f \equiv \rho_{23}, \tag{20.39}$$

where

$$\rho_{23} \neq \rho_2 \otimes \rho_{3.}$$

Here $\rho_2 = \mathrm{Tr}_2\rho_{23}$, and $\rho_3 = \mathrm{Tr}_2\rho_{23}$. ρ_2 is the density matrix of radiation escaping to null infinity. ρ_3 is the radiation completely absorbed by the future horizon. The entropy of these states will be taken as $S = \mathrm{Tr}\rho \ln \rho$. ρ_{01} and ρ_{23} are in the same Hilbert space,

$$\mathfrak{H}_0 \otimes \mathfrak{H}_1 = \mathfrak{H}_2 \otimes \mathfrak{H}_3,$$

and thus $\mathrm{S}^{01} = S^{23}$ are related by a unitary transformation in this space. A fundamental theorem of Araki and Lieb (1970) is utilized. If ρ^{12} is a density matrix on $\mathfrak{H}^1 \otimes \mathfrak{H}^2$, then

$$S^{12} \leq S_1 + S_2. \tag{20.40}$$

From this we may prove, by the relation

$$S_2 + S_3 \geq S_{23} = S_{01} = S_0 + S_1,$$

from Eq. (20.38).

Now, from the first law of black hole thermodynamics, we assume the black hole evolves through the "in to out" process by means of a set of isothermal states such that

$$\Delta S^H = \left(T^H\right)^{-1}(E_3 - E_0). \tag{20.41}$$

Now we define

$$\Delta S = \Delta S^H + \Delta S^{\mathrm{rad}} \tag{20.42}$$

and take

$$\Delta S^{\text{rad}} = S_2 - S_1.$$

From this and the inequality (Eq. 20.40),

$$\Delta S \geq S_0\left(\frac{1}{T}\right) - S_3\left(\frac{1}{T}\right), \tag{20.43}$$

where $S\left(\frac{1}{T}\right) = S - T^{-1}E$ are Massieu functions (see Callen, 1985). Equilibrium maximizes $S\left(\frac{1}{T}\right)$. $S_0\left(\frac{1}{T}\right)$ is the maximum. Thus,

$$\Delta S \geq 0. \tag{20.44}$$

This is the Bekenstein entropy principle for black holes. It is apparent that the properties of black holes enter in the quasistatic temperature T^H of Bekenstein and Hawking. Otherwise, this is a rather simple general thermodynamic argument.

20.5 Acausal evolution: extended irreversible dynamics in black holes

For the purpose of describing radiation and gravitational collapse of a black hole, Hawking introduced a density matrix map,

$$\rho_{2AB} = \sum S_{ABCD}\rho_{1CD}. \tag{20.45}$$

Here ρ_{2AB} is the final density matrix, and ρ_{1CD} the initial one. S_{ABCD} is a generalized (tetradic) scattering matrix between these Hilbert space states. (We have already met tetradic operators in the early chapters of this book. An example was the tetradic Liouville operator L_{abcd}.) Hawking termed S_{ABCD} a superscattering operator. The observed final density matrix is *not* a pure state. In the gravitational collapse, producing an event horizon in the black hole, the interaction region is bounded by an initial and final surface *and* a third "hidden" macroscopic surface, for which only incomplete quantum data are available. Here the rule of equal *a priori* probability is applied and thus introduces the classical probability, making the final state impure and a density matrix. We may write, for pure initial and final states,

$$S_{CDAB} = \frac{1}{2}\left(S_{CA}S^{-1}_{BD} + S^{-1}_{AD}S_{CB}\right).$$

Here S_{CA} is a pure S matrix where $\xi_C = \sum S_{CA}\xi_A$. This relation does not hold for a mixture state black hole. Further, it is assumed that

$$\sum S_{CCAB} = S_{AB} \tag{20.46}$$

$$\sum S_{CDAA} = S_{CD}.$$

The latter may be taken as the result of assuming gravitational *CPT* invariance on the hidden surface. *CPT* invariance implies that black holes must completely evaporate, since they can form spontaneously.

As discussed previously in this chapter, Hawking's calculation showed that particles radiated to infinity from a black hole are in an equilibrium thermodynamic state at the Hawking temperature, and thus described by a mixture density matrix. One of these paired particles disappears into the black hole and cannot be seen by the infinite observer. This is a loss of information to the observer. This information loss was deemed by Hawking to be a special feature of quantum gravity not present in other quantum field theories. He called it the information loss puzzle. Gravity must be quantized consistent with this, an unsolved problem. Here the super operator S-matrix cannot be factorized.

In Chapter 18 we have discussed *extended* statistical mechanics, which introduced super operators and the irreversible time evolution of density matrix states with diagonal singularity. This is a much more complete theory than the early discussion by Hawking. Utilizing the analytic continuation rule, Eq. (18.55) and Eq. (18.56), we may write

$$\langle\alpha\,|C_n|\,\beta\rangle = \frac{\lambda}{\omega_\alpha - \omega_\beta + i\varepsilon_{\alpha\beta}}\langle\alpha\,|(C_n - Q_n)\,L_1\,(P_n + C_n)|\,\beta\rangle \tag{20.47}$$

and

$$\langle\beta\,|D_n|\,\alpha\rangle = \frac{\lambda}{\omega_\beta - \omega_\alpha + i\varepsilon_{\alpha\beta}}\langle\beta\,|(P_n + D_n)\,L_1\,(Q_n - D_n)|\,\alpha\rangle \tag{20.48}$$

with

$$\varepsilon_{\alpha\beta} = \begin{Bmatrix} -\varepsilon \text{ for } d_\alpha > d_\beta \\ +\varepsilon \text{ for } d_\alpha < d_\beta \end{Bmatrix}.$$

$d_{\alpha,}$ d_β measure the degree of correlation. These are operator forms of nonlinear Lippman–Schwinger equations in this theory and play the role of the superscattering operator analogous to that introduced by Hawking. Thus, if we apply the theory of Eq. (20.47) and Eq. (20.48) to quantized gravity, we may expect, from Hawking's argument, that there is a quantum information loss puzzle (Hawking, 1975a, 1981; Wald, 1994).

References

Akhmedov, E. T. (1997). E-print hep-th/9711153.
Araki, H. and Lieb, E. H. (1970). *Commun. Math. Phys.* **18**, 160.

Bardeen, J. M., Carter, B. and Hawking, S. (1973). *Commun. Math. Phys.* **31**, 161.
Becker, K., Becker, M. and Schwartz, J. H. (2007). *String Theory and* M-*Theory* (New York, Cambridge University Press).
Bekenstein, J. D. (1972). *Lett. Nuovo Cim.* **4**, 737.
Bekenstein, J. D. (1973). *Phys. Rev. D* **7**, 949.
Bekenstein, J. D. and Mukhurov, V. F. (1995). *Phys. Lett. B* **360.**
Callen, H. (1985). *Thermodynamics and an Introduction to Thermostatics* (New York, Wiley).
Cherepaschuk, A. M. (1996). *Russ. Phys. Uspekhi* **29**, 759.
Einstein, A. (1915a). *Preuss. Akad. Wiss. Berl.* 778.
Einstein, A. (1915b). *Preuss. Akad. Wiss. Berl.* 844.
Frolov, V. P. and Novikov, I. D. (1998). *Black Hole Physics* (Dordrecht, Kluwer).
Frolov, V. P. and Page, D. N. (1993). *Phys. Rev. Lett.* **71**, 3902.
Hawking, S. (1971). *Phys. Rev. Lett.* **26**, 1344.
Hawking, S. (1975a). *Phys. Rev. D* **14**, 112.
Hawking, S. (1975b). *Commun. Math. Phys.* **43**, 199.
Hawking, S. (1981). In *Quantum Gravity 2*, ed. Isham *et al.* (Oxford, Clarendon), 395.
Horowitz, G. T. (1995). E-print gr-qc/970407.
Isham, C. J., Penrose, R. and Sciama, D. W., eds. (1975). Quantum gravity, Oxford Symposia (Oxford, Clarendon).
Isham, C. J., Penrose, R. and Sciama, D. W., eds. (1981). Quantum gravity, Oxford Symposia (Oxford, Clarendon).
Kerr, R. (1963). *Phys. Rev. Lett.* **11**, 237.
Kruskal, M. D. (1960). *Phys. Rev.* **119**, 1743.
Lamaitre, G. (1933). *Ann. Soc. Brux. A* **53**, 51.
Louisell, W. H. (1973). *Quantum Statistical Properties of Radiation* (London, Wiley).
Mäkela, J. (2003). *Found. Phys.* **32**, 1809.
Maldacena, J. M. (1995). E-print hep-th/9607235.
Misner, C. W., Thorne, K. S. and Wheeler, J. A. *Gravitation*. (San Francisco, Freeman).
Mitchell, J. (1783). *Phil. Trans.* **57**, 234.
Oppenheimer, J. R. and Snyder, H. (1939). *Phys. Rev.* **56**, 455.
Parker, L. (1975). *Phys. Rev. D* **12**, 1519.
Raychadhur, A. (1955). *Phys. Rev.* **98**, 1123.
Rees, M., ed. (2005). *Universe* (London, Dorling Kindersley).
Rees, M. and Hawking, S. (1997). *Before the Beginning* (Cambridge, Mass., Perseus).
Schwarzschild, K. (1916a). *Sitzber. Deut. Akad. Wiss. Berl.* and *K1. Math.-Phys. Tech.* 189.
Schwarzschild, K. (1916b). *Sitzber. Deut. Akad. Wiss. Berl.* and *K1. Math.-Phys. Tech.* 424.
Tolman, R. (1934). *Proc. Natl. Acad. Sci. USA* **20**, 169.
Wald, R. M. (1984). *General Relativity* (Chicago, Univ. of Chicago Press).
Wald, R. M. (1994). *Quantum Field Theory in Curved Spacetime and Black Hole Thermodynamics* (Chicago, University of Chicago Press).
Zurek, W. H. and Thorne, K. S. (1985). *Phys. Rev. Lett.* **54**, 2171.

Appendix 1

Problems

A.1 Comments on the problems

These exercises have been used over a number of years in a one-semester graduate course during the germination of this book. They include homework problems and exam questions. They are approximately equivalent in difficulty and are roughly divided into three topic areas: (1) foundations of quantum statistical mechanics, (2) kinetic dynamics and (3) equilibrium and phase transitions. The outline of the course itself is given in the preface to the book.

Other sets of problems are available. Of those, one must mention the first book of R. Kubo, *Statistical Mechanics* (Kubo, *et al.*, 1965), which has an excellent collection with answers.

In addition to offering the problems written here, we have often called upon the student in this book to "finish a calculation." These challenges, of course, should be used as problems but will not be repeated here.

A.2 "Foundations" problems

1. (a) In the Schrödinger q representation, show that the canonical density matrix $\exp(-\beta H)$ may be written as

$$\left\langle q \left| \exp(-\beta H) \right| q'' \right\rangle = \exp\left[-\beta H \left(\frac{\hbar}{i} \frac{\partial}{\partial q'}, q' \right) \right] \times \delta\left(q' - q'' \right).$$

   (b) Now apply this to a free particle $H = \frac{p^2}{2m}$ obtaining $\left\langle x' \left| \exp\left(\frac{-\beta p^2}{2m}\right) \right| x'' \right\rangle$, showing that it is a Gaussian. Discuss the result.

2. (a) Show for a mixed state with Hermitian operators $\hat{A}, \hat{B}$ that $\Delta \hat{A}\ \Delta \hat{B} \geq \frac{1}{2} \left| \left\langle \left[\hat{A}, \hat{B} \right] \right\rangle \right|$.

   (b) Show that this leads to $\Delta x\ \Delta p \geq \frac{\hbar}{2}$ where $(\Delta x)^2 = \left\langle x^2 \right\rangle - \langle x \rangle^2$.

3. Argue from problem (2.2b) that the uncertainty relationship is consistent with the Wigner function.

4. A system is in an eigenstate of H. Show that the Wigner function is then constant in time.
5. Show that $g = 1$ in the appendix of Chapter 4 leads to the Weyl correspondence rule.
6. (a) Show that the N-body Wigner function may be written for a pure state $\psi\left(x^N, t\right)$

$$w\left(x^N, p^N, t\right) = \left(\frac{1}{\pi\hbar}\right)^{3N} \int dy^N \exp 2i\left(\frac{p^N \cdot y^N}{\hbar^N}\right)\psi^*\left(x_+^N\right)\psi\left(x_-^N\right)dy^N,$$

where $x_\pm = x^N \pm y^N$ and $\psi\left(x^N, t\right)$ obeys

$$i\hbar\frac{\partial\psi\left(x^N, t\right)}{\partial t} = \left[\frac{-\hbar^2}{2m}\sum_{k=1}^{N}\frac{\partial^2}{\partial x_k^2} + V\left(x_N\right)\right]\psi\left(x^N, t\right).$$

(b) From problem (6a) show that $w\left(x^N, p^N, t\right)$ obeys

$$\frac{\partial w\left(x^N p^N, t\right)}{\partial t} = -\sum_k \frac{p_k}{m}\frac{\partial w}{\partial x_k} + \frac{i}{\hbar}\left(\frac{1}{\pi\hbar}\right)^{3N}\int\int dy^N dp^N$$
$$\times \exp\left[\frac{2iy^N \cdot \left(p^N - p'^N\right)}{\hbar}\right]\left[V\left(x_+^N\right) - V\left(x_-^N\right)\right] w\left(x^N, p'^N, t\right).$$

7. From problem (6b) obtain the classical limit $\hbar = 0$, and show $\frac{\partial w}{\partial t} = \{H, w\}$, which is the classical Liouville equation.
8. What are the conditions for the $P\rho$ of the generalized master equation to be constant in time?
9. Derive a time-reversed generalized master equation, that is, an equation evolving to $t = -\infty$ from $t = 0$.
10. On C^2 we have the observables (projections)

$$A = \begin{vmatrix} 1 & 0 \\ 0 & 0 \end{vmatrix}, \quad B = \begin{vmatrix} 0 & 0 \\ 0 & 1 \end{vmatrix}, \quad C = \frac{1}{2}\begin{vmatrix} 1 & 1 \\ 1 & 1 \end{vmatrix}.$$

Examine $A \cap (B \cup C) = (A \cap B) \cup (B \cap C)$, and show that this quantum state is non-Boolean.
11. Construct the density matrix for a quantum particle moving with equal likelihood to the left or right in a box of length L.
12. Obtain the general solution to the Fokker–Planck equation, Eq. (7.34).
13. A mixture state is constructed as

$$\rho = \frac{1}{2}\left|x\right\rangle\left\langle x\right| + \frac{1}{2}\left|y\right\rangle\left\langle y\right|,$$

where

$$\left|x\right\rangle = \sqrt{\alpha}\left|+\right\rangle + \sqrt{1-\alpha}\left|-\right\rangle$$
$$\left|y\right\rangle = \sqrt{\alpha}\left|+\right\rangle - \sqrt{1-\alpha}\left|-1\right\rangle$$

are pure states. Verify both statements, and construct another mixture state from $\left|+\right\rangle, \left|-\right\rangle$.

14. Use the von Neumann equation to show that $\rho(t)$ if pure cannot evolve into a mixture and vice versa.
15. For a beam of spin 1/2, particles $\mathbf{S} = \mathrm{Tr}\rho\sigma$. $\boldsymbol{\sigma}$ are then the Pauli matrices.
    (a) Argue that $\rho = \frac{1}{2}(I + \mathbf{S}\cdot\boldsymbol{\sigma})$.
    (b) Show $\rho = \rho^{\dagger}$, and diagonalize ρ in terms of $|\mathbf{S}|$.
    (c) Argue that $\mathrm{Tr}\rho^2 \leq 1$.
    (d) For an unpolarized beam, obtain ρ. Is it a pure state?
    (e) For complete polarization, show ρ is pure.
16. What is the surface of constant energy for a harmonic oscillator of frequency ν? Find the volume in phase space Γ_0 with energy below E. Quantize this, and find the number of quantum states below E. For large E show that the number of states is $\frac{\Gamma_0}{\hbar}$.
17. Prove that if the entropy $S(x)$ only increases, and if there is a process governing the variable (operator) x is adiabatic

$$H(qp, x) \rightarrow H(q, p, x + \Delta x);\ \frac{dx}{dt} = 0,$$

then $S(x)$ does not change.
18. For a density matrix $\rho_{nm} = \alpha_m^*\alpha_n$ where $\alpha_n = c_n \exp(i\phi_n)$, show that a uniform average over phases ϕ_n gives $\rho_{nm} = c_n^* c_m \delta_{nm}$.
19. Derive by time-dependent perturbation theory (in detail) the Pauli equation (for isolated system).
20. From problem 19, consider the Pauli equation for a beam of two-level atoms entering a uniform magnetic field with interaction $\mu = \mu_b\sigma$, $\hat{H}' = -\boldsymbol{\mu}\cdot\mathbf{B}$ in z direction. Describe the solution. Describe what happens at $\dot{\rho}_{nn} = 0$.
21. Suppose the density operator for a harmonic oscillator is

$$\rho(a, a\dagger) = (1 - \exp(-\lambda)) \exp -\lambda a\dagger a,$$

where $\lambda = \beta\hbar\omega$.
    (a) Show that this maximizes the entropy, subject to the constraint $\mathrm{Tr}\rho = 1$.
    (b) Show also that $\langle H\rangle = \hbar\omega\langle n\rangle$, and as $\hbar \rightarrow 0$, the average energy is $\langle H\rangle = kT$.
    (c) Prove that $\langle n\rangle = \frac{1}{\exp\lambda - 1}$.
22. Let p_s be the probability that a system is in state E_s. The entropy is $S = k\sum_s p_s \ln p_s$. Show by means of Lagrange multipliers that the canonical distribution maximizes S under the conditions $\bar{E} = E$.
23. Examine the energy states of the free particle Schrödinger equation in 3–D for (1) a box of side L with $\psi(0) = \psi(L) = 0$, and (2) periodic boundary conditions.
    (a) What is the spacing of states in the lattice of the two boundary states?
    (b) Obtain the energy density of states $g(E)$ in the two cases.
24. (a) Describe quantum entanglement.
    (b) Give examples of a non-entangled two-atom Q bit and an entangled one. Are they mixtures? Show why or why not.
    (c) Describe the process of teleportation. Give the Bob and Alice example.

25. (a) Write the Pauli equation for $\langle\alpha|\,\rho(t)\,|\alpha\rangle \equiv P(\alpha,t)$ for an isolated system. Explain all terms.
    (b) Outline the derivation of the $\mathfrak{H}$ theorem (entropy principle) from this equation. Discuss the physical results.
    (c) What is the equilibrium solution to the Pauli equation, $\dot{P}(\alpha,t)=0$?

A.3 Kinetic dynamics problems

26. In the KBG approximation to the Boltzmann equation, for the collision term, one takes

$$J(F)=\nu\left(F_L^0-F\right)$$

where

$$\nu=\int d_3vF_L^0g\sigma d\Omega.$$

F_L^0 is the local Maxwellian.
    (a) Derive from this the center of mass hydrodynamic equations in detail, defining also T (also called conservation laws). Now follow the *normal* solution, Chapman–Enskogg (see Huang, 1987). Do in detail each step in your discussion.
    (b) Obtain the lowest order solution and discuss it.
    (c) In the next order obtain in detail formula (5.67) in Huang (Huang, 1987).
    (d) Obtain a formula for viscosity and thermal conductivity, proving their ratio is $\frac{5}{2}C_V$ (the famous result). C_V is the specific heat.
27. (a) Write down the Boltzmann equation.
    (b) Give an intuitive physical derivation.
    (c) Is it reversible? Prove your answer.
28. From the considerations of Eq. (4.46), a Uhlenbeck–Uehling equation for electrons may be obtained. In a quasi-classical approximation,

$$J(f)=\int\left[f^1f_1^1(1+\theta f)(1+\theta f_1)-ff_1\left(1+\theta f^1\right)\left(1+\theta f_1^1\right)\right]\times\mathbf{g}\sigma d\Omega dv_1.$$

Here $\theta=\frac{h^3}{m^3}\times 1$ for bosons, and $\theta=\frac{h^3}{m^3}\times -1$ for fermions. For free photons,

$$\frac{d\mathbf{p}}{h^3}=\frac{4\pi\,(2m)^{\frac{3}{2}}}{h^3}E^{\frac{1}{2}}dE.$$

    (a) Argue why this is a reasonable physical result.
    (b) Show that the steady equilibrium solution is

$$f_0dv=\frac{\frac{dp}{h^3}}{\left[\exp\beta\,(E-\mu)\mp 1\right]}.$$

    (c) Define the $\mathfrak{H}$ function as

$$\mathfrak{H}=\frac{V}{h^3}\int d_3p\left[(f\pm 1)\ln(1\pm f)-f\ln f\right].$$

Assuming f positive, prove in the conventional way that $\frac{d\mathfrak{H}}{dt} \geq 0$.

(d) Show that $\frac{d\mathfrak{H}}{dt} = 0$ implies the equilibrium state f_0.

29. (a) For electron transport in the Krook-Bhatnager-Gross approximation, employ the relaxation time approximation and obtain, in the steady state,

$$-\left(\mathbf{v}\cdot\nabla f + \frac{e\boldsymbol{\varepsilon}}{\hbar}\cdot\frac{\partial f}{\partial \mathbf{k}}\right) = \frac{1}{\tau}\left[f_0 - f\right],$$

where f_0 is the Fermi distribution. Here $f_0\left[\mathbf{E}(k),\ T(\mathbf{x}),\ \mu(\mathbf{x})\right]$ is space dependent.

(b) Solve this equation in perturbation about f_o, assuming the left side is of order f_0. Obtain the following equation for $g(xk) = f(xk) - f_0$:

$$\mathbf{v}\cdot\left(\frac{\partial f_0}{\partial T}\nabla T + \frac{\partial f_0}{\partial \mu}\nabla\mu\right) + e\boldsymbol{\varepsilon}\cdot\mathbf{v}\frac{\partial f_0}{\partial E} = \frac{g(\mathbf{k},\mathbf{x})}{g(E)}.$$

(c) From the solution to problem (29b), obtain the electrical current density in the following approximation:

$$J_e = e\int \frac{d_3k}{4\pi^3}v(k)\, f(k).$$

(d) Obtain the thermal current, defined as

$$J_Q = \int \frac{d_3k}{4\pi^3}(E-\mu)\, v(k)\, f(k).$$

(e) Obtain the Onsager coefficients L_{ij} where

$$J_e = L_{11}\epsilon + L_{12}(-\nabla T)$$
$$J_Q = L_{21}\epsilon + L_{22}(-\nabla T).$$

How is L_{12} related to L_{21}?

30. For the harmonic oscillator $H^0 = \hbar\omega a\dagger a$, take the distribution function in normal ordering to be

$$P\left(\alpha,\alpha^*,t\right) = \mathrm{Tr}\rho(t)\,\delta\left(\alpha^* - a\dagger\right)\delta(\alpha - a),$$

α, α^* being coherent states.

(a) Show that this obeys

$$\frac{\partial P(\alpha\alpha^*,t)}{\partial t} = i\omega\left[\alpha\frac{\partial P}{\partial \alpha} - \alpha^*\frac{\partial P}{\partial \alpha^*}\right].$$

(b) Prove that the general solution is

$$P\left(\alpha,\alpha^*,t\right) = g\left[\alpha\exp(i\omega t),\alpha^*\exp(-i\omega t)\right],$$

where g is an arbitrary function.

31. (a) Write the von Neumann equation in the exact energy representation, $H|\alpha\rangle = E_\alpha|\alpha\rangle$.

(b) Obtain the solution.

(c) Discuss this time evolution.

32. (a) Show that the Vlasov equation is time reversible.
    (b) Define $\mathfrak{H} = \int dxdvF(xv) \ln F(xv)$, where $F(xv)$ is a solution to the Vlasov equation. Is there an entropy principle? Discuss in detail, and compare with the Boltzmann result.

A.4 Equilibrium and phase transition problems

33. (a) Prove for the two quantum ideal gases that the dispersions may be written $(n - \bar{n})^2 = \bar{n}(1 \pm \bar{n})$, where $\bar{n}$ is the average energy level occupation number.
    (b) Also obtain the Boltzmann distribution result. Why do you expect this result?
34. Prove that the magnetic susceptibility obeying *classical* statistical mechanics is zero. Take

$$\mathfrak{H} = \sum_{j=1}^{N} \frac{1}{2m_j} \left\{ P_j + \frac{e_j}{c} A(r_j) \right\}^2 + U(r_1 \ldots r_n).$$

35. Consider an ideal Bose gas composed of particles with internal states as well as translational. Consider only one internal state, ε_1. Determine how the Bose condensation temperature changes as a function of this energy, ε_1.
36. Use the transfer matrix method to solve the 1–D Ising problem. Particularly obtain
    (a) Eq. (14.80) (Huang, 1987), and
    (b) Eq. (14.82) (Huang, 1987).
    (c) Then show in detail that there is no magnetic phase transition in 1–D.
37. For photons of the electromagnetic fields, prove that $\mu = 0$. They are bosons, of course.
38. For fermions (electrons), show that at low temperature, $C_V = \frac{1}{3}\pi^2 k^2 T g(\mu_0)$, where g is the density of states and μ_0 the zero-temperature Fermi energy.
39. The Hamiltonian of an electron in a magnetic field H is $\mathfrak{H} = -\mu_B \boldsymbol{\sigma} \cdot \mathbf{H}$. $\boldsymbol{\sigma}$ is the Pauli matrices. Take H in the z direction. Now calculate the density operator,

$$\hat{\rho} = \frac{\exp(-\beta\mathfrak{H})}{\mathrm{Tr}\hat{\rho}} \qquad \beta = \frac{1}{kT},$$

    as follows:
    (a) Obtain ρ in the diagonal representation of σ_z.
    (b) Obtain ρ in the diagonal representation of σ_x.
    (c) Find $\langle \sigma_z \rangle$, the average of σ_z in both representations.
    (d) Comment on your answer physically.
40. For the one-dimensional nearest neighbor Ising spin model, discuss the mean field approximation as follows.
    (a) Obtain the equation of state for M, the magnetization.
    (b) Prove there is a spontaneous magnetization. Obtain T_C.
    (c) Show that M/N has a critical index 1/2 below the critical point. Obtain the critical index for χ_+ (susceptibility above T_C).
    (d) What are all the mean field critical magnetic indices

41. Assume $\exp\left(\frac{\varepsilon-\mu}{kT}\right) >> 1$ for the Fermi–Dirac and Bose–Einstein distributions.
    (a) What is the meaning of the result?
    (b) Prove in detail that this is valid if $\left(\frac{V}{N}\right)^{\frac{1}{3}} >> \frac{\lambda}{\sqrt{2\pi}}$. λ is the so-called "thermal" de Broglie wavelength.
    (c) Comment on the conditions physically when this is *not* true.
42. Let P_s be the probability that the system is in state E_s. The entropy is $S = k\sum_s P_s \ln P_s$. Show, by means of Lagrange multipliers, that the canonical density matrix arises from maximizing S, subject to $\sum_s P_s = 1$, $\sum_s P_s E_s = E$.
43. (a) Obtain the occupation number of the ground state N_0 of a Bose gas in a three-dimensional harmonic oscillator trap (equal $w_i, i = 1, 2, 3$) as a function of temperature below the critical temperature T_c, having defined T_c.
    (b) Is this a phase transition?
44. (a) Obtain the critical index relations by either Widom or Kadanoff scaling.
    (b) What are the values of the mean field critical indexes? Do they scale?
45. (a) From the quantum microcanonical ensemble and suitable assumptions, derive the equilibrium thermodynamic laws.
    (b) Explain them physically.
46. Consider $H = \mu\mathfrak{H}\sigma_z$ (z-axis is along the magnetic field $\mathfrak{H}_z$). $\boldsymbol{\sigma}_z$ is the z-component Pauli spin operator, $\mathfrak{H}$ is the magnetic field, and μ a constant. Prove, independent of a particular representation for σ_z, that the canonical density matrix gives σ_z as $\langle\sigma_z\rangle = \tanh\beta\mu\mathfrak{H}$.
47. The energy spectrum of a photon is $E(q) = \hbar cq$. $q = |\mathbf{q}|$, $\mathbf{q}$ being the wave vector. Assume no polarization.
    (a) Find the Helmholtz free energy F, integrating in detail.
    (b) Obtain PV, also in detail. Comment on this result physically.
    (c) Obtain the entropy S.
48. (a) Discuss Bose–Einstein condensation for a box of arbitrary dimension. For $D = 3$, obtain the formula for condensation in the ground state for T_C and $T < T_C$.
    (b) Show that there is no condensation for $D = 1, 2$ at finite temperature.

References

Kubo, R., Ichimura, H., Usui, T. and Hashizume, N. (1965). *Statistical Mechanics* (New York, Wiley).

Huang, K. (1987). *Statistical Mechanics,* 2nd edn. (New York, Wiley).

Index

For EU product safety concerns, contact us at Calle de José Abascal, 56–1°, 28003 Madrid, Spain or eugpsr@cambridge.org.

www.ingramcontent.com/pod-product-compliance
Ingram Content Group UK Ltd.
Pitfield, Milton Keynes, MK11 3LW, UK
UKHW052227270726
14060UKWH00004B/650

* 9 7 8 0 5 2 1 8 4 1 4 6 7 *